FARADAY DISCUSSIONS
NO. 111   1998

Molecular Interactions of Biomembranes

The Faraday Division
The Royal Society of Chemistry
London

Organising Committee
Dr L. R. Fisher (*Chairman*)
Professor A. D. Buckingham
Professor J. Finney
Professor J. Holzwarth
Dr M. N. Jones
Professor B. Robinson
Dr M. Sansom

ISBN: 0-85404-869-3
ISSN: 0301-7249

Typeset by Santype International Ltd., Netherhampton Road, Salisbury, Wiltshire and printed and bound in Great Britain by Whitstable Litho Printers Ltd.

A General Discussion

on

Molecular Interactions of Biomembranes

16th, 17th and 18th December, 1998

A General Discussion on Molecular Interactions of Biomembranes was held at the University of Bristol, UK on 16th, 17th and 18th December, 1998.

Contents

1 Introductory Lecture: Energy landscapes of biomolecular adhesion and receptor anchoring at interfaces explored with dynamic force spectroscopy
Evan Evans

17 Collective membrane motions of high and low amplitude, studied by dynamic light scattering and micro-interferometry
Rainer Hirn, Thomas M. Bayerl, Joachim O. Rädler and **Erich Sackmann**

31 Trapping of short-lived intermediates in phospholipid phase transitions: The L_α^* phase
Peter Laggner, Heinz Amenitsch, Manfred Kriechbaum, Georg Pabst and **Michael Rappolt**

41 Sensing isothermal changes in the lateral pressure in model membranes using di-pyrenyl phosphatidylcholine
Richard H. Templer, Saffron J. Castle, A. Rachael Curran, Garry Rumbles and **David R. Klug**

55 Interfacial enzyme activation, non-lamellar phase formation and membrane fusion. Is there a conducting thread?
Félix M. Goñi, Gorka Basáñez, M. Begoña Ruiz-Argüello and **Alicia Alonso**

69 General Discussion

79 Characterization of the physical properties of model biomembranes at the nanometer scale with the atomic force microscope
Yves F. Dufrêne, Thomas Boland, James W. Schneider, William R. Barger and **Gil U. Lee**

95 Modelling and simulation of light-activated membrane proteins: Dynamical transitions in bacteriorhodopsin
Christian Simon, Malika Aalouach and **Jeremy C. Smith**

103 Interactions between poly(2-ethylacrylic acid) and lipid bilayer membranes: Effects of cholesterol and grafted poly(ethylene glycol)
David Needham, Jeff Mills and **Gary Eichenbaum**

111 Membrane electroporation and electromechanical deformation of vesicles and cells
Eberhard Neumann, Sergej Kakorin and **Katja Toensing**

127 Lipid–protein interactions in the membrane: Studies with model peptides
Sanjay Mall, Ram P. Sharma, J. Malcolm East and **Anthony G. Lee**

137 General Discussion

159 Phospholipid chain length alters the equilibrium between pore and channel forms of gramicidin
Toby P. Galbraith and **B. A. Wallace**

165   Protein inclusion in lipid membranes: A theory based on the hypernetted chain integral equation
Patrick Lagüe, Martin J. Zuckermann and **Benoît Roux**

173   Lipid packing stress and polypeptide aggregation: alamethicin channel probed by proton titration of lipid charge
Sergey M. Bezrukov, R. Peter Rand, Igor Vodyanoy and **V. Adrian Parsegian**

185   Structure-based prediction of the conductance properties of ion channels
Oliver S. Smart, Guy M. P. Coates, Mark S. P. Sansom, Glenn M. Alder and **C. Lindsay Bashford**

201   Molecular dynamics simulation of a hydrated diphytanol phosphatidylcholine lipid bilayer containing an alpha-helical bundle of four transmembrane domains of the Influenza A virus M2 protein
Thomas Husslein, Preston B. Moore, Quingfong Zhong, Dennis M. Newns, Pratap C. Pattnaik and **Michael L. Klein**

209   Alamethicin channels in a membrane: molecular dynamics simulations
D. Peter Tieleman, Jason Breed, Herman J. C. Berendsen and **Mark S. P. Sansom**

225   General Discussion

247   Kinetics of the competitive response of receptors immobilised to ion-channels which have been incorporated into a tethered bilayer
Gillian E. Woodhouse, Lionel G. King, Lech Wieczorek and **Bruce A. Cornell**

259   Three-dimensional models of glutamate receptors
Michael J. Sutcliffe, Allister H. Smeeton, Z. Galen Wo and **Robert E. Oswald**

273   Functional immobilization of biomembrane fragments on planar waveguides for the investigation of side-directed ligand binding by surface-confined fluorescence
Michael Pawlak, Ernst Grell, Eginhard Schick, Dario Anselmetti and **Markus Ehrat**

289   Analysis of membrane protein cluster densities and sizes *in situ* by image correlation spectroscopy
Nils O. Petersen, Claire Brown, Anna Kaminski, Jonathan Rocheleau, Mamta Srivastava and **Paul W. Wiseman**

307   Direct measurement of recognition forces between proteins and membrane receptors
Paul F. Luckham and **Kate Smith**

321   Use of a laminar flow chamber to study the rate of bond formation and dissociation between surface-bound adhesion molecules: Effect of applied force and distance between surfaces
Anne Pierres, Anne-Marie Benoliel and **Pierre Bongrand**

331   General Discussion

345   Concluding Remarks
A. Neil Barclay

351   List of Posters

355   List of Participants

359   Index of Contributors

Introductory Lecture

Energy landscapes of biomolecular adhesion and receptor anchoring at interfaces explored with dynamic force spectroscopy

Evan Evans

Departments of Physics and Pathology, University of British Columbia, Vancouver, BC, Canada V6T 1Z1 and Department of Biomedical Engineering, Boston University, Boston, Massachusetts, USA 02215

Received 21st December 1998

Beyond covalent connections within protein and lipid molecules, weak noncovalent interactions between large molecules govern properties of cellular structure and interfacial adhesion in biology. These bonds and structures have limited lifetimes and so will fail under any level of force if pulled on for the right length of time. As such, the strength of interaction is the level of force most likely to disrupt a bond on a particular time scale. For instance, strength is zero on time scales longer than the natural lifetime for spontaneous dissociation. On the other hand, if driven to unbind or change structure on time scales shorter than needed for diffusive relaxation, strength will reach an adiabatic limit set by the maximum gradient in a potential of mean force. Over the enormous span of time scales between spontaneous dissociation and adiabatic detachment, theory predicts that bond breakage under steadily rising force occurs most frequently at a force determined by the rate of loading. Moreover, the continuous plot (spectrum) of strength expressed on a scale of $\log_e$(loading rate) provides a map of the prominent barriers traversed in the energy landscape along the force-driven pathway and reveals the differences in energy between barriers. Illustrated with results from recent laboratory measurements, dynamic strength spectra provide a new view into the inner complexity of receptor–ligand interactions and receptor lipid anchoring.

Introduction

Well-recognized in biology, ligand–receptor interactions are the fundament of nanoscale chemistry in recognition, signalling, activation, regulation, and other processes from outside to inside cells. Thus, following the advent of atomic force microscopy (AFM) a decade ago,[1] it was no surprise that researchers quickly seized the opportunity to test strengths of receptor–ligand bonds. Since then, AFM and other sensitive force probes have been used to pull on a variety of molecules embedded in—or adhesively bonded to—surfaces. Applying these techniques, experimentalists often imagine that probe force establishes a well-defined property of an interaction between molecules. Such expectations originate from the age-old creed of physics, which states that strength is the maximum gradient $-(\partial E/\partial x)_{max}$ of an interaction potential or energy contour $E(x)$ defined along the direction (x) of separation. Hence, it is anticipated that detachment forces for different types of molecular interactions will follow a scale set by the ratio of bond energy to the effective range of the interaction (bond length). This seems consistent with the standard model of biochemistry where the scale for bond strength is the free energy $\Delta G°$ reduction when molecules combine in solution as found from the equilibrium ratio $k_{eq}[\sim \exp(\Delta G°/k_B T)]$ of bound to free constituents. As such, the criterion for a *strong* bond should be simply a binding energy much larger than the thermal energy per molecule $k_B T$. In marked contrast to these two paradigms, we will see that even bonds with binding energies $>40\ k_B T$ can fail under minuscule forces—more than 100-fold lower than the maximum energy gradient implied by energy/distance. Indeed, we will find that measurement of molecular detachment force—no matter how precise the technique or how carefully performed—is not in itself a fundamental property of a molecular interaction. So what is the appropriate framework for describing *strength* of molecular bonds and how can we relate measurements of these forces to nanoscale chemistry?

When we test strength of molecular cohesion or adhesion at surfaces, we determine the maximum level of force that a molecular attachment can support at the instant of failure. Unlike intimate covalent connections within protein and lipid molecules, biomembrane structure and interfacial adhesion bonds involve noncovalent interactions between large macromolecules, which have limited lifetimes and thus will fail under any level of force if pulled on for the right length of time. In other words, when we speak of strength, we should think of the force that is most likely to disrupt an adhesive bond or structural linkage on a particular time scale. At equilibrium for example, bonds dissociate and reform under zero force. Thus, an isolated bond has no strength on time scales longer than its natural lifetime $t_0 = 1/k_{off}^0$ for spontaneous (entropy-driven) dissociation. On the other hand, if detached within the time needed for diffusive relaxation over the range of molecular interaction (*e.g.* $x_\beta \lesssim 1$ nm $\to x_\beta^2/D < 10^{-9}$ s in water), the strength of a bond will reach and even exceed the adiabatic limit $f_\infty \approx |\Delta E|/x_\beta$ set by the maximum gradient in a potential of mean force. This is the situation in molecular dynamics (MD) simulations.[2,3]

From the slow limit set by spontaneous transition (from µs to months) to the ultrafast limit set by diffusive relaxation ($<$ns), strength is governed by thermally activated kinetics under external force and thus depends on how the force is applied over time. Since application of force always requires a finite interval of time, the simplest way to parameterize the history of loading is to treat force as a ramp in time set by a constant loading rate $r_f = \Delta f/\Delta t$. In fact, a ramp of force is what single molecular attachments experience when a force probe and test surface are separated at constant speed (*i.e.* loading rate = probe stiffness × speed). Using this parameterization and some nearly sixty year old physics[4] for Brownian dynamics of chemical reactions in liquids, we have shown that bond dissociation under steadily rising force occurs most frequently at a time determined by the rate of loading.[5] Since loading rate is constant, the time of dissociation specifies the most likely rupture force—strength—which has the same dependence on loading rate. Of particular significance, the continuous plot of strength expressed on a scale of $\log_e$(loading rate) maps the most prominent barriers traversed in the energy landscape to distances along the force-driven pathway and reveals the splitting in energy between barriers. Thus, strength *vs.* $\log_e$(loading rate) establishes the basis for a dynamic force spectroscopy (DFS) to probe the inner world of molecular-scale chemistry. Testing bond strength or structural transitions at different loading rates effectively probes the lifetime of a molecular complex under different levels of force. The experimental challenge is to measure forces over many orders of magnitude in loading rate. This dynamic requirement is dictated by the exponential of the energy difference ΔE_b between the highest and lowest barriers divided by thermal energy, *i.e.* $\exp(\Delta E_b/k_B T)$, which can be enormous!

The strength spectra to be presented here will show that we can now cover six orders of magnitude in loading rate from <0.1 pN s^{-1} to $\sim 10^5$ pN s^{-1} with a rather simple dynamic force probe, which could be extended to $\sim 10^7$ pN s^{-1} with complementary measurements using other probes. But more important than demonstrations of technique, these spectra provide a new level of insight into the complexity of macromolecular interactions and structural linkages. First, results from biotin–(strept)avidin[6] and carbohydrate–L-selectin[7] bond tests will show that a cascade of sharp energy barriers exists in receptor–ligand bonds where each barrier governs strength on a different time scale. We see then that these bonds, cannot be simply idealized by a sole energy barrier, and we cannot rely on the classical intuition about kinetics implicit in the detailed balance $k_{eq} = k_{on}/k_{off}$ where k_{on} and k_{off} are constants. Second, tests of lipid extraction[8] from membranes will show that anchoring of receptors to surface structure plays an important role in adhesion strength and can introduce unexpected transitions in the strengths of receptor–ligand attachments. In other words, we should not assume that a structural linkage of several molecules will fail at a specific weak connection nor that we can uniquely attach *strong* or *weak* labels to bonds in a linkage. Simultaneous kinetics over different energy landscapes in serial molecular linkages can lead to strength or weakness on different time scales. Dynamic crossovers in strength switch the site of failure from one location to another. Taken together, these insights show that mechanical force can tune and switch time scales for kinetics in biomolecular reactions governed by complex energy landscapes, which exposes a potentially new dimension in biochemical regulation and control.

Theory of molecular kinetics under force in liquids

We begin with an abstract of the physics that underlies the kinetics of bond dissociation and structural transitions in a liquid environment. Developed from Einstein's theory of Brownian motion, these well-known concepts take advantage of the huge gap in time scale that separates rapid thermal impulses in liquids ($<10^{-12}$ s) from slow processes in laboratory measurements (*e.g.* from 10^{-4} s to min in the case of force probe tests). Three equivalent formulations describe molecular kinetics in an overdamped liquid environment. The first is a microscopic perspective where molecules behave as particles with instantaneous positions or states $x(t)$ governed by an overdamped Langevin equation of motion,

$$dx/dt = D/k_B T[f - \nabla E + \delta f]$$

The rate of change in state equals the instantaneous force scaled by the mobility of states or inverse of the damping coefficent $\gamma (= k_B T/D)$. The deterministic force ($-\nabla E + f$) includes both the local gradient in molecular interaction potential $E(x)$ and the external force f. An uncorrelated random force δf from thermal impulses modulates the deterministic force and obeys the *fluctuation–dissipation theorem* where the integrated square fluctuation in a window of time can be modeled as a Gaussian distribution $\sim \exp[-\int \delta f^2 dt/(4k_B T)]$ with variance set by temperature and viscous damping.[9] The microscopic physics also defines a stochastic process that has become the foundation of an important computational technique—*Brownian dynamics* or *smart* Monte Carlo (SMC)[10] simulations. In this description, the likelihood $P(x + \Delta x, t + \Delta t | x, t)$ that a state $x(t)$ will evolve to a new state $x + \Delta x$ over a time increment Δt is specified by the product of the equilibrium (long-time) Boltzmann weight for the step and a Gaussian weight for dynamics,

$$P(x + \Delta x, t + \Delta t | x, t) \sim \exp\{-(\Delta E - f \cdot \Delta x)/k_B T\} \exp\{-|\Delta x - (D/k_B T)f\Delta t|^2/(4D \Delta t)\}/(D \Delta t)^{1/2}$$

Finally, on time scales that include many thermal impulses, the overdamped dynamics can be cast in a continuum representation where the density of states $\rho(x, t)$ at location x and time t obeys Smoluchowski transport,[9]

$$d\rho/dt = -\nabla \cdot J$$

where the *flux* of states $J = D[f - \nabla E]\rho/k_B T - \nabla \rho]$ reflects both convection by force and spread by diffusion. Although each description of the ultrafast kinetics brings to light important features, Kramers[4] demonstrated that Smoluchowski transport readily predicts the rate of escape from a deeply bound state when a large number of thermally activated steps are needed to pass a barrier in a dissipative environment.

Escape from a bound state confined by a single barrier

Starting far from equilibrium with all states confined inside the barrier, the kinetics of escape are idealized as a stationary flux of probability density along a preferential path from the deep energy minimum outward past the barrier *via* a saddle point in the energy surface. In real molecular interactions, there can be many such paths and the paths can map out complex trajectories in configuration space. However, application of an external pulling force acts to select the reaction path, which we express by a scalar coordinate x. Assumed to be bounded by steeply rising energy in other directions, the energy landscape $E(x)$ along this coordinate is illustrated schematically in Fig. 1(a). Governed by orientation θ relative to the microscopic reaction coordinate, external force adds a mechanical potential $-fx(\cos \theta)$ that tilts the energy landscape and diminishes the energy barrier E_b at the transition state ($x = x_{ts}$). When the tilted landscape is introduced into the Smoluchowski equation, the stationary solution (J = constant in 1-D or = constant/x^{d-1} in d-

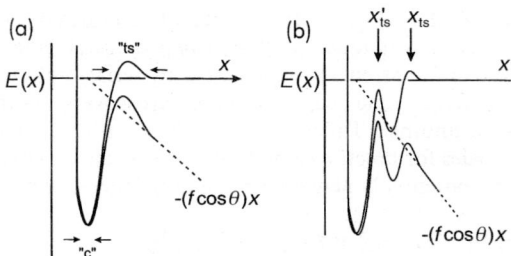

Fig. 1 Conceptual energy landscapes for bound states "c" confined by sharp activation barriers. Oriented at an angle θ to the molecular coordinate x, external force f adds a mechanical potential $-(f\cos\theta)x$ that tilts the landscape and lowers the barrier. For sharp barriers, the energy contours local to barriers—transition states "s"—are highly curved and change little in shape or location under force. (a) A single barrier under force. (b) A cascade of barriers under force. The inner barrier emerges to dominate kinetics when the outer barrier is driven below it by $\geqslant k_B T$.

dimensions) yields a generic expression for rate of escape from bound to unpopulated free states under force,[5]

$$k_{\text{off}} \approx (D/l_c\, l_{\text{ts}})\exp[-E_b(f)/k_B T]$$

The diffusive nature of kinetics in liquids is embodied in the attempt frequency, $D/l_c l_{\text{ts}}$, which is the reciprocal of a characteristic time $t_D = l_c l_{\text{ts}}(\gamma/k_B T)$ set by damping and two length scales. The first length l_c represents confinement in the bound state and defines the entropy gradient ($\partial\rho/\partial x \approx 1/l_c$), which drives escape. In a harmonic approximation, l_c is derived from curvature $\kappa_c = (\partial^2 E/\partial x^2)_c$ of the energy landscape local to the minimum, i.e. $l_c = (2\pi k_B T/\kappa_c)^{1/2}$. The second length l_{ts} is the energy-weighted width of the barrier $l_{\text{ts}} = \int dx \exp[\Delta E(x)_{\text{ts}}/k_B T]$ local to the transition state $x = x_{\text{ts}}$, also determined by curvature $\kappa_{\text{ts}} = (\partial^2 E/\partial x^2)_{\text{ts}}$ of the energy landscape, i.e. $l_{\text{ts}} = (2\pi k_B T/\kappa_{\text{ts}})^{1/2}$. Although force can displace and deform the width of the barrier [i.e. $(\kappa_{\text{ts}}/2\pi k_B T)^{1/2} \approx g(f)$], the major impact of force arises in the thermal likelihood of reaching the top of the energy barrier, $\exp[-E_b(f)/k_B T]$. For a sharp energy barrier, the shape and location of the transition state are insensitive to force but force lowers the barrier in proportion to the thermally averaged projection $x_\beta = <x_{\text{ts}}\cos\theta>$, i.e. $E_b(f) = E_b - fx_\beta$. As such, thermal activation introduces the characteristic scale for force through the ratio of thermal energy to the distance x_β, i.e. $f_\beta = k_B T/x_\beta$, which can be surprisingly small since $k_B T \approx 4.1$ pN nm at room temperature and $x_\beta \approx 0.1$–1 nm. On this scale, the rate of escape increases exponentially with force, $k_{\text{off}} \approx (1/t_0) \exp(f/f_\beta)$, as first postulated by Bell[11] twenty years ago. But in contrast to the resonant frequency of bond excitations described in Bell's model, Kramers showed that the relevant attempt frequency is $1/t_D = (\kappa_c \kappa_{\text{ts}})^{1/2}/2\pi\gamma$ for overdamped transitions in liquids, which is at least 1000-fold slower. With the Arrhenius dependence on initial barrier height and the attempt frequency, Kramers classic result for spontaneous escape in the overdamped limit sets the scale for the transition rate, i.e. $1/t_0 = (1/t_D)\exp(-E_b/k_B T)$.

Escape from a bound state confined by several barriers

Although a naive model of chemical binding, the single-sharp barrier model already captures the profound impact of force on thermally activated kinetics: i.e. exponential amplification of the forward rate for dissociation (and suppression of backward rate for reassociation) characterized by a small force scale $k_B T/x_\beta$ well below the adiabatic limit $> E_b/x_\beta$! However, the energy landscapes of biomolecular bonds are expected to be much more complex because there are many sites of interaction involving large numbers of small molecules. This should produce a *rough* topography of barriers in an energy landscape and many possible pathways for unbinding. If again conceptualized as precipitous (*sharp*) energy maxima along a single pathway, these prominent barriers are predicted to emerge under increasing force and dominate kinetics in succession as demonstrated by the sketch in Fig. 1(b).[5] An inner barrier is exposed when the force exceeds a crossover force $f_\otimes \approx \Delta E_b/\Delta x_\beta$ set by the splitting ΔE_b between barrier heights and separation in projected positions Δx_β. Depending on the difference in barrier energies, the crossovers occur at forces much larger than the local thermal forces given by $k_B T/x_\beta$. Thus, marked by these crossovers, the kinetic rate constant is predicted to rise in a staircase of force-dependent exponentials that amplify the rate of transition less and less with each increase in thermal force scale. The transition rate for escape past a cascade of n sharp barriers is easily predicted with Kramers–Smoluchowski theory,

$$k_{\text{off}}(f) \approx (1/t_0)\exp(f/f_\beta^0)/\{1 + \sum_{i \to n} l_i \exp[f\Delta x_\beta - \Delta E_b]\}$$

which at low force begins with the steepest exponential dominated by the outermost barrier. At larger forces, the rate crosses over to more shallow exponentials. The transition from one exponential regime to the next depends on the ratio of widths $l_i \approx l_{ts}/l_{ts}^0$ $[=(\kappa_{ts}^0/\kappa_{ts})^{1/2}]$ plus differences in location $\Delta x_\beta = x_\beta^0 - x_\beta$ and energy $\Delta E_b = E_b^0 - E_b$ of inner barriers relative to the outermost barrier as defined by E_b^0, l_{ts}^0, and x_β^0. We see then that a major consequence of structured energy landscapes is to make molecular interactions more durable (survive longer) at higher forces.

Theory of force distributions in probe experiments

Even with ultrasensitive probes and high resolution detection, tests of molecular detachment yield a spread in force values. To understand the origin of the intrinsic uncertainty in force, we have to examine the generic process of bond dissociation in laboratory experiments. Typically, a probe decorated with a small amount of ligand—and a substrate studded with specific molecular receptors—are repeatedly touched together through steady precision movement to/from contact. If the surfaces are prepared with a sufficiently low density of reactive sites, point contacts between the probe tip and the test surface will occasionally result in attachments (*e.g.* one attachment for every 5–10 touches). Under controlled touch, infrequent bonding ensures a high probability of forming single molecular bonds ($\sim 95\%$ confidence for 1 attachment out of 10 touches). An attachment is exposed when the force transducer exhibits an extension or deflection Δx_t during surface separation. Identified by rapid recoil at breakage, the rupture force is given by the maximum transducer extension Δx_t, *i.e.* $f = k_f \cdot \Delta x_t$ where k_f is the spring constant of the transducer. Following many measurements, detachment forces are then cumulated into a histogram. The peak in the distribution is the most likely the rupture force, which is labelled bond strength. This approach has been reported many times in the literature over the past decade including studies of bond strength using AFM[12–16] and other techniques.[17,18] The exception to the generic description is that the frequency of attachment in most tests has been one for every touch, which represents many molecular bonds and yields broad force distributions.

Given that only a single molecular attachment forms on contact, the crucial feature of the generic method is that the force experienced by the attachment prior to rupture is not constant but increases in time. This is shown clearly by two traces of attachment force *vs.* time in Fig. 2 taken from our experiments[6] on single receptor–ligand bonds using a biomembrane force probe (BFP).[19] In probe tests like these, the linear rise of force with time is set by the product of separation speed v_t and transducer spring constant k_f, which is called the loading rate $r_f = k_f v_t$. (Note: if soft structures like long polymers link the bond to a stiff probe, the loading history can be nonlinear in time.[20]) Very different levels of force and time frame characterize the two detachment processes in Fig. 2. Comparing these, we see then that bond survival and breakage force depend on the rate of loading in reciprocal ways: *i.e.* high speed loading → short lifetime but large detachment force whereas low speed loading → long lifetime but small detachment force, which is the direct consequence of thermally activated kinetics.

Statistics of transitions under increasing force

To analyse bond breakage under steady loading, we take advantage of the enormous gap in time scale between the ultrafast Brownian diffusion ($t_D \approx 10^{-10} - 10^{-9}$ s) and the time frame of laboratory experiments ($\sim 10^{-4}$ s to min). This means that the slowly increasing force in laboratory experiments is essentially stationary on the scale of the ultrafast kinetics. Thus, dissociation rate merely becomes a function of the instantaneous force and the distribution of rupture times can be described in the limit of large statistics by a first-order (Markov) process with time-dependent rate constants. As force rises above the thermal force scale, *i.e.* $r_f t > k_B T/x_\beta$, the forward transition

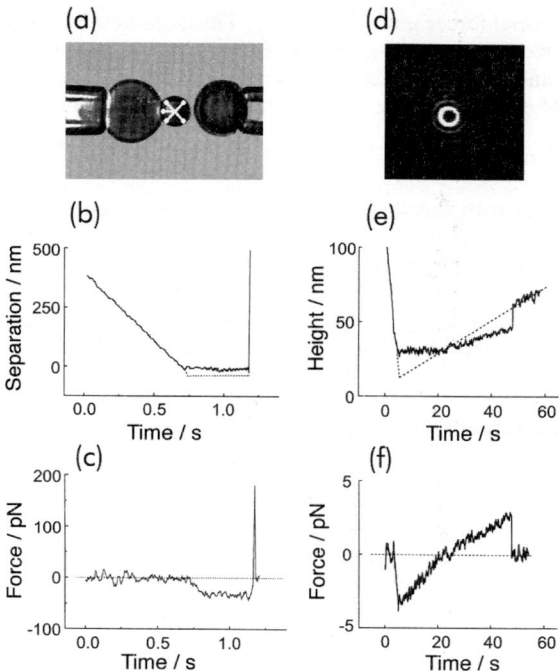

Fig. 2 Testing strength of single molecular attachments with the biomembrane force probe (BFP). The spring component of the BFP is a pressurized membrane capsule.[19] Membrane tension sets the force constant k_f (force/capsule extension), which is controlled by micropipet suction P and radius R_p, $k_f \approx P \times R_p$. Using a red blood cell as the transducer, the BFP stiffness can be selected between 0.1 and 3 pN nm^{-1} to measure forces from 0.5 to 1000 pN. At the BFP tip, a glass microbead of 1–2 μm diameter is glued to the membrane. The probe tip and red cell surfaces are bound covalently with heterobifunctional polyethylene oxide PEG polymers that carry glue components and test ligands.[6] The BFP is operated in two orientations (modes) on the stages of inverted microscopes as illustrated by the following examples of fast and slow bond detachment: (a) first, the BFP (on the left) is kept stationary in the horizontal mode and the microbead test surface (on the right) is translated to/from contact with the BFP tip by precision piezo control. Video image processing is used to track the bead as shown by the simulated cursor; a single high speed ($\sim$1000 frames s^{-1}) scan through the center of the bead is used to track deflection of the transducer (force) on a fast time scale at a resolution of 8–10 nm. Parts (b) and (c) show the BFP tip–substrate separation and force vs. time for rapid bond detachment in the horizontal mode. (b) The test microbead was moved towards the probe tip at a speed of $\sim$500 nm s^{-1}. Stopped for $\sim$0.5 s after sensing contact at a preset impingement force of ~ -30 pN, the test surface was then retracted at speed of $\sim$30 000 nm s^{-1}. (c) Loaded at extremely fast rate, the bond held the tip to the surface for $\sim$0.003 s (spike in force) and broke at $\sim$180 pN as the piezo continued to retract the test surface. The force fluctuations were due to position uncertainties × BFP stiffness. (d) In the vertical mode, reflection interference contrast is used to image the BFP tip as it is translated by piezo control along the optical axis to/from contact with a coverglass test surface. Standard video (30 frames s^{-1}) processing of the circular interference pattern reveals elevation of the tip at a resolution of 2–5 nm. Transducer deflection (force) is obtained from the difference between piezo translation and bead displacement. Parts (e) and (f) show the BFP tip–substrate separation and force vs. time for a slow bond detachment in the vertical mode. (e) The probe was moved towards the coverglass test surface at a speed of $\sim$20 nm s^{-1}. After sensing contact at a preset impingement force of ~ -3 pN, the probe was retracted at slow speed of $\sim$1 nm s^{-1}. (f) Loaded at extremely slow rate, the bond held the tip to the surface for $\sim$24 s and broke at $\sim$3 pN as the piezo continued to retract the probe (dashed trajectory). The fluctuations in tip position were due to thermal excitations of the BFP with mean square displacement $\sim k_B T/k_f$. Stretch of the PEG polymers that linked the bond to the glass surfaces is shown by the slight upward movement of the tip ($\sim$15 nm) under force prior to detachment. Due to polymer compliance, the true loading rate felt by a bond at nominal rates ($k_f v_t$) below 10 pN s^{-1} had to be obtained from the actual force vs. time.

(escape) rate increases extremely rapidly. Moreover, the molecules drift apart faster than diffusion can recombine them from positions beyond the confining barrier so the backward rate for re-association quickly vanishes ($k_{on} \to 0$). Thus, the likelihood $S(t)$ of remaining in the bound state is dominated by the forward process, i.e. $dS(t)/dt \approx -k_{off}(t)S(t)$ or equivalently $S(t) = \exp[-\int_{0 \to t} k_{off}(t')dt']$. The probability density $p(t) = k_{off}(t)S(t)$ for detachment between times t and $t + \Delta t$ describes the distribution of lifetimes. Since instantaneous force is the product of time and loading rate ($f = r_f t$), the probability density $p(f)$ for detachment between forces f and $f + \Delta f$ is given by the distribution of lifetimes $p(t)$,[5]

$$p(f) = (1/r_f)k_{off}(f)\exp[-(1/r_f)\int_{0 \to f} k_{off}(f')df']$$

noting the statistical identity $p(t)dt = p(f)df$. The peak in the distribution of forces defines the force f^* for most frequent transition, which is strength. Analytically, the location of a distribution peak is found from $\partial p(f)/\partial f = 0$ which establishes a transcendental equation that relates the strength f^* to loading rate r_f,

$$[k_{off}]_{f=f*} = r_f[\partial \log_e(k_{off})/\partial f]_{f=f*}$$

Although somewhat forbidding, this expression yields a simple result for strength as a function of loading rate in the case of a single sharp energy barrier,

$$f^* = f_\beta \log_e(r_f/r_f^0)$$

recalling that the rate is modelled by an exponential in force, $k_{off} \approx (1/t_0)\exp(f/f_\beta)$. Governed by a thermal scale for loading rate $r_f^0 = f_\beta/t_0$, the most likely force—strength—simply shifts upward linearly with the logarithm of loading rate multiplied by the thermal force f_β. Similarly, the curvature of the distribution local to the peak, $1/\Delta_f^2 = -[1/p(f)][\partial^2 p(f)/\partial f^2]_{f=f*}$, can be used to estimate a Gaussian width for uncertainty in the force distribution,

$$\Delta_f^2 = 1/\{[\partial \log_e(k_{off})/\partial f]^2 - [\partial^2 \log_e(k_{off})/\partial f^2]\}_{f=f*}$$

For a sharp energy barrier, this again yields a simple result, $\Delta_f = f_\beta$. Hence, even without experimental uncertainty, the distribution of forces is broadened by thermal activation (kinetics)!

Dynamic force spectroscopy

In the context of experiments, the signature of a major sharp barrier is predicted to be a straight line in a plot of most frequent probe force f^* vs. log(loading rate) as illustrated in Fig. 3(a). This linear regime can span orders of magnitude in rate as determined by the ratio of barrier energy E_b

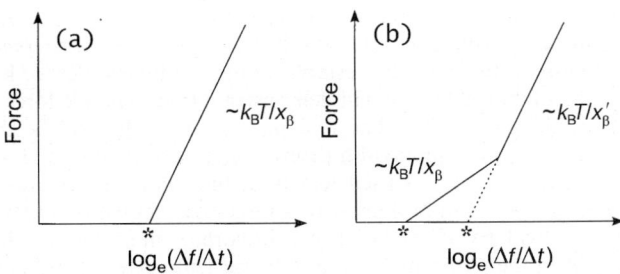

Fig. 3 Dynamic strength spectra defined by most likely bond detachment force f^* vs. $\log_e$(loading rate = r_f/r_f^0), where the loading rate scale $r_f^0 = (f_\beta/t_D)\exp(-E_b/k_BT)$ is set by thermal force $f_\beta = k_BT/x_\beta$, diffusive attempt frequency $1/t_D$, and height E_b of the activation barrier. (a) Linear spectrum predicted for a single sharp energy barrier. The logarithmic intercept at zero force (represented by ∗) is determined by the barrier height and the microscopic diffusion time, $\log_e(r_f^0) = -E_b/k_BT + \log_e(f_\beta/t_D)$. (b) Piece-wise linear spectrum for a cascade of two sharp energy barriers. The abrupt increase in slope from one thermal force scale to the next shows that the outer barrier has been suppressed and that the inner barrier has become the dominant kinetic impedance to detachment [cf. Fig. 1(b)]. The difference between logarithmic intercepts (represented by ∗) is governed by the splitting in barrier energies and the ratio of thermal force scales, $\log_e(\Delta r_f^0) \approx -\Delta E_b/k_BT + \Delta\log_e(f_\beta)$.

to thermal energy $k_B T$. The slope f_β of this line maps the thermally averaged projection of the microscopic transition state to a distance $x_\beta = \langle x_{ts} \cos \theta \rangle$ along the direction of force. Moreover, the logarithmic intercept at zero force reflects the magnitude of barrier energy as given by $\log_e(r_f^0) = -E_b/k_B T + \log_e(f_\beta/t_D)$. Setting the scale for loading rate, the ratio f_β/t_D involves the microscopic attempt frequency $1/t_D$. Assuming that attempt frequency is weakly affected by point mutations, the simple linear-log behavior exposes a unique opportunity to quantitate the resulting chemical modifications in energy and/or location of barriers. Such changes in microscopic properties can be derived from the shift in the logarithmic intercept and/or change in slope, $\Delta E_b/k_B T \approx -\Delta \log_e(r_f^0) + \Delta \log_e(f_\beta)$. Taken together, these features demonstrate that the plot of most frequent probe force *vs.* log(probe loading rate) represents a dynamic spectral image of an activation barrier. [Although unknown, attempt frequency can be estimated from the damping factor indicated by MD simulations. Values for damping factor seem to be typically on the order of $\gamma \approx 10^{-8}$ pN s nm^{-1} (equivalent to Stokes drag on a 1 nm size sphere in water), *e.g.* $\gamma \approx 2 \times 10^{-8}$ pN s nm^{-1} in simulations of biotin–streptavidin separation[2] and $\gamma \approx 5 \times 10^{-8}$ pN s nm^{-1} in simulations of lipid extraction from a bilayer.[3] Since the product of molecular lengths $l_c l_{ts}$ should lie in the range ~ 0.01–0.1 nm^2, the attempt frequency is expected to be in the range $1/t_D \approx 10^9$–10^{10} s^{-1} and the microscopic scale for loading rate in the range $f_\beta/t_D \approx 10^{10}$–$10^{11}$ pN s^{-1}. The effective loading rate in the slowest MD simulations[2,3] is even higher $\geq 10^{12}$ pN s^{-1}.]

As described earlier, the most idealized view of a complex molecular energy landscape is a cascade of sharp activation barriers, which leads to a staircase of exponential increases in the rate constant under force. Using this prescription, the most likely force *vs.* log(loading rate) is predicted to follow a simple spectrum of piece-wise continuous linear regimes with ascending slopes as shown in Fig. 3(b). The abrupt increase in slope from one regime to the next signifies that an outer barrier has been suppressed by force and that an inner barrier has become the dominant kinetic impedance to escape as sketched in Fig. 1(b). These dynamic crossovers occur at somewhat higher forces than the stationary crossovers in rate constant as shown by the analytical approximation,

$$f_\otimes^{dyn} \approx \Delta E_b/\Delta x_\beta + k_B T [\log_e(x'_\beta/x_\beta)]/\Delta x_\beta$$

where $\Delta x_\beta = x'_\beta - x_\beta$ and $\Delta E_b = E'_b - E_b$ represent adjacent prominent barriers. In contrast to the idealized theory, the shape of a strength spectrum could be nonlinear and a challenge to interpret because force can distort physical potentials and molecular structure. Surprisingly, the results from recent probe experiments to be shown next yield linear plots for strength *vs.* log(loading rate) with one or more well-defined regimes, which allows the spectra to be interpreted in terms of sharp activation barriers.

Energy landscapes of receptor–ligand bonds

Not well-appreciated in biology is that energy landscapes of receptor–ligand bonds can be *rugged* terrains with more than one prominent activation barrier. The inner barriers are undetectable in test-tube assays but are important since they establish different time scales for kinetics under force. With two unrelated pairs of molecules, we will demonstrate that dynamic force measurements can be used to reveal these hidden barriers. The first pair of molecules will be the ligand biotin (a vitamin) and the protein receptor streptavidin (from bacteria) or avidin (a closely similar protein from hen egg white).[21] This complex is used widely in biotechnology because it has one of the highest affinity noncovalent bonds in biology with a force-free lifetime on the order of days.[22] The second pair of molecules will be a sialylated (carbohydrate) short peptide ligand† and the L-selectin receptor resident in the outer membrane of blood leukocytes. Although weaker in affinity with a lifetime of ~ 1 s or less, the carbohydrate–L-selectin bond plays a crucial role in the initial capture of leukocytes from blood circulation at sites of injury or infection.[23] In preparation for both experiments, the ligand was covalently anchored to a glass microbead along with a chemical glue for attachment of the bead to the BFP transducer [as noted in Fig. 2(a)]. A similarly pre-

---

† Note: the actual ligand used in the tests was a short peptide chimera of the biological molecule called P-selectin glycoprotein ligand (PSGL1), which was constructed by Genetics Institute and obtained through collaboration with Scott Simon at Baylor College of Medicine. The generic label *carbohydrate* will be used for convenience.

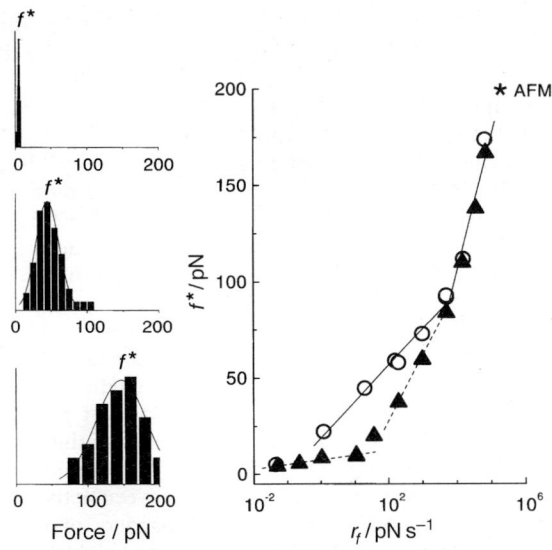

Fig. 4 On the left are examples of force histograms taken from tests of single biotin–streptavidin bonds, which demonstrate the shift in peak location and increase in width with increase in loading rate (top histogram, 0.05 pN s^{-1}, middle histogram, 20 pN s^{-1}, bottom histogram, 60 000 pN s^{-1}). Superposed on the histograms are Gaussian fits used to determine the most frequent rupture force–bond strength. Governed ideally by the thermal force f_β, standard deviations σ_f of the distributions also reflected uncertainties in position Δx and video sampling time Δt_v, i.e. $\sigma_f \approx [f_\beta^2 + (k_f \Delta x)^2 + (r_f \Delta t_v)^2]^{1/2}$. As σ_f increased from ± 1 pN at the slowest rate to ± 60 pN at the fastest rate, the standard error in mean force—the uncertainty in strength—ranged from ± 0.3 to ± 5 pN. On the right are complete dynamic strength spectra for both biotin–streptavidin (open circles) and biotin–avidin (closed triangles) bonds.[6] Defined as thermal energy $k_B T$/distance x_β, the slopes of the linear regimes seen in the spectra map activation barriers at positions along the direction of force. The common high strength regime in the biotin–streptavidin and biotin–avidin spectra place the innermost barrier at $x_\beta \approx 0.12$ nm. Separate intermediate strength regimes place the next barrier at $x_\beta \approx 0.5$ nm for biotin–streptavidin and $x_\beta \approx 0.3$ nm for biotin–avidin (with a slight reduction in slope below 38 pN suggesting that the biotin–avidin barrier extends to ~ 0.5 nm). Only well-defined in the biotin–avidin spectrum, a low strength regime implies a distal barrier at $x_\beta \approx 3$ nm. Also marked ($*_{AFM}$) is the biotin–streptavidin strength measured recently by AFM at $\sim 10^5$ pN s^{-1} using a carbon nanotube as the tip.[14] This and the earlier measurements of biotin–avidin bond strength[13] at loading rates of $\sim 6 \times 10^4$ pN s^{-1} also correlate with the high strength regime shown here.

pared microbead was used as the test surface for probing biotin–(strept)avidin bonds whereas a white blood cell (granulocyte) taken from a small blood sample was used as the test surface for probing carbohydrate–L-selectin bonds.

[Methods: In testing molecular bonds, the density of reactive sites must be reduced significantly as mentioned earlier so that only 1 out of 7–10 touches results in a molecular attachment. Assumed to be governed by Poisson statistics, this ensures that 90–95% of the attachments are single bonds. To obtain strength spectra with the BFP technique, detachment forces are measured over a six order of magnitude range in loading rate from 0.05 pN s^{-1} to 100 000 pN s^{-1}. The loading rate $k_f v_t$ is preselected by setting the transducer force constant k_f in the range 0.1–3 pN nm^{-1} and the piezo retraction speed v_t in the range 1–30 000 nm s^{-1} as described in Fig. 2. From thousands of repeated touches at fixed loading rate, histograms of detachment forces are compiled at many rates and Gaussian fits are used to locate the peak in each histogram. These most probable values of force are then plotted as a function of $\log_e$(loading rate), which yields the dynamic strength spectrum.]

Biotin (strept)avidin bonds

Because of high affinity, the first ligand–receptor pair chosen by researchers for testing with AFM was biotin and streptavidin; which was soon followed by biotin and avidin.[12–14] Deduced from a broad distribution of AFM forces, it was concluded that the strength of a biotin–streptavidin bond lies in a range of ~200–300 pN and somewhat lower for biotin–avidin ~160 pN. However, the examples of force histograms and the strength spectra[6] in Fig. 4 show that biotin–streptavidin (and biotin–avidin) bond strengths fall continuously from ~200 pN to ~pN with each decade increase in time scale for rupture from 10^{-3} to 10^2 s, which clearly demonstrates the thermally activated nature of bond breakage. Moreover, distinct linear regimes with abrupt changes in slope imply sharp barriers, which can be analysed using the idealized theory.[6] First, above 85 pN, there is a common high strength regime for both biotin–streptavidin and biotin–avidin with a slope of $f_\beta \approx 34$ pN. This locates a barrier deep in the binding pocket at $x_\beta \approx 0.12$ nm. Below 85 pN, the $f_\beta \approx 8$ pN slope in the biotin–streptavidin spectrum maps the next activation barrier at $x_\beta \approx 0.5$ nm whereas the steeper slope $f_\beta \approx 13$–14 pN between 38 and 85 pN in the biotin–avidin spectrum indicates that its next barrier maps to $x_\beta \approx 0.3$ nm. Interestingly, a slight curvature and reduction in slope between 38 and 11 pN suggests that the barrier in biotin–avidin extends to ~0.5 nm. Below 11 pN, the biotin–avidin spectrum exhibits a very low strength regime (dashed line) with a slope of $f_\beta \approx 1.4$ pN that maps to $x_\beta \approx 3$ nm. A similar low strength regime is indicated by results from the slowest test of biotin–streptavidin bonds; but it was not possible to perform tests at loading rates below 0.05 pN s^{-1} as needed to verify the existence of this regime. In addition to the map of barrier locations, the logarithmic intercepts found by extrapolation of each linear regime to zero force also yield estimates of the energy differences between activation barriers within each landscape as well as energy differences between related barriers of biotin–avidin and biotin–streptavidin landscapes. However, instead of discussing barrier heights, it is more illuminating to examine how the 1-D map of barrier locations compares with detailed molecular simulations of biotin–(strept)avidin interactions.

In separate MD simulations,[2] biotin was extracted from a binding pocket of streptavidin and avidin by pulling on the outer end with a pseudo-mechanical spring. Consistent with the numerous bonds to small molecules in the binding pocket, simulations yield a fluctuating superposition of many attractions—buffeted by steric collisions—along the unbinding trajectories. This is shown by a profile of instantaneous energy calculated over a *slow* ~500 ps extraction of biotin from avidin [Fig. 5(a), kindly provided by Professor K. Schulten and coworkers, Beckman Institute, University of Illinois]. Even with the enormous and fast changes in energy, simple qualitative features appear in the profile that provide important clues to the thermally averaged free energy landscape relevant on laboratory time scales. In particular, transition states are readily identified by regions with rarified statistics where biotin passes quickly. Taking a simple coarse-grained average over ~20 ps windows [Fig. 5(b)] smooths over the strong rapid fluctuations and exposes locations of activation barriers. First, within an initial displacement of 0.1–0.2 nm, the spring force in the simulations revealed abrupt detachment of biotin from a nest of hydrogen bonds, water bridges, and nonpolar interactions deep in the binding pocket. Next, forces reached maximal

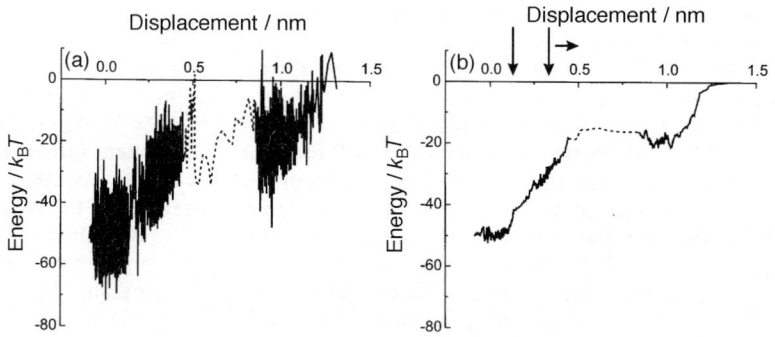

Fig. 5 (a) Profile of instantaneous energy computed for interaction between biotin and avidin over a half-nanosecond extraction from the binding pocket in the simulations of Israilev et al.[2] (kindly provided by Professor K. Schulten and coworkers, University of Illinois). Separating regions of rapid intense fluctuations, locations of rarified statistics coincide with maximal forces in the simulations, which signify the presence of transition states. (b) Coarse-grained average over the fast degrees of freedom which yields an approximate potential of mean force.[5] Arrows mark barrier locations derived from the intermediate and high strength regimes of the spectrum for biotin–avidin in Fig. 4.

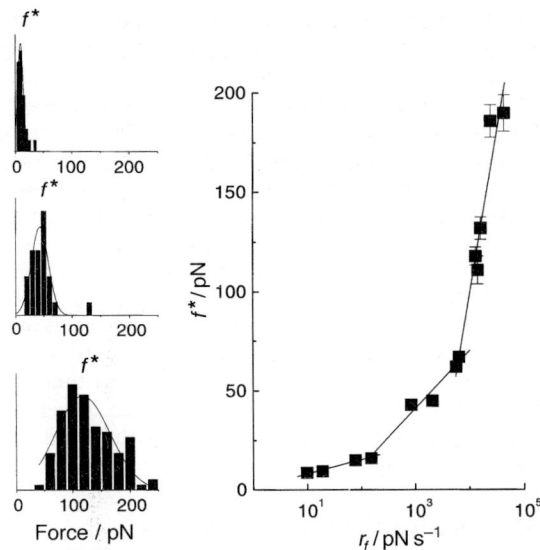

Fig. 6 On the left are examples of force histograms taken from tests of single carbohydrate (sialylated PSGL1 short peptide chimera)–L-selectin bonds, which demonstrate shift in peak location and increase in width with increase in loading rate (top histogram, 10 pN s^{-1}, middle histogram, 850 pN s^{-1}, bottom histogram, 13 000 pN s^{-1}). Superposed on the histograms are Gaussian fits used to determine the most frequent rupture force—bond strength. On the right is the complete dynamic spectrum of strength vs. log(loading rate). Defined as thermal energy $k_B T$/distance x_β, the slope of the high strength regime places the innermost barrier at $x_\beta \approx 0.06$ nm. The intermediate strength regime places the next barrier at $x_\beta \approx 0.3$ nm. The low strength regime implies a barrier further out at $x_\beta \approx 1.2$ nm.

values followed by sudden displacements of biotin at a distance of ∼0.4 nm (and ∼0.5 nm in the biotin–streptavidin simulation). Finally, biotin was observed to still cling to peripheral polar groups at ∼1.4 nm in the avidin simulation. As labelled in Fig. 5(b), the locations of activation barriers derived from the high and intermediate strength regimes of the laboratory spectra in Fig. 4 correlate well with regions of rarified statistics and the qualitative appearance of the energy landscape. The conclusion is that these transition states inferred from the simulations persist on long time scales. However, the outer barrier indicated by the low strength regime is 2–3-fold more distant than the last transition state seen in the MD simulation; this is perhaps due to interactions with the peripheral flexible loops[24–27] which border the channel that leads to the binding pocket. More puzzling, however, extrapolation of the lowest strength regime to zero force implies that bond strength vanishes below a threshold loading rate of ∼0.0006 pN s^{-1} for biotin–avidin. In other words, the spontaneous off rate would be ∼1 per hour. This is 50-fold faster than the rate of ∼1 per 55 hours that we measured for spontaneous dissociation of PEG–biotin from probe tips in free solution and found previously for biotin by others.[22] Hence, some nontrivial effect remains that accounts for the significant increase in rate of dissociation under extremely small forces below 5 pN.

Carbohydrate–L-selectin bonds

In contrast to the high affinity biotin–(strept)avidin bonds, carbohydrate–L-selectin bonds with modest affinity stop white cells at vessel walls in the circulation.[23] Numerous bonds to other surface (*integrin*) receptors then form between the white cell and vessel wall to sustain adhesion and enable subsequent movement into the surrounding tissue. On its initial arrest from the blood flow, the white cell can be subjected to forces of ∼100 pN in a time frame of milliseconds, which implies loading rates of 10^4–10^5 pN s^{-1}. With this functional requirement in mind, we now examine recent tests of carbohydrate–L-selectin bonds under dynamic loading in probe tests.

From the results[7] presented in Fig. 6, we again see a sequence of high, intermediate, and low strength regimes for carbohydrate–L-selectin bonds where strength also falls continuously from

~200 pN to ~pN but over fewer decades in time scale for detachment from 10^{-3} to 1 s. The high strength regime has a very steep slope of $f_\beta \approx 70$ pN that maps an inner barrier to a small distance $x_\beta \approx 0.06$ nm along the direction of force. Although we lack detailed molecular information about L-selectin binding, the small value of x_β seems to imply that the microscopic reaction coordinate deviates significantly from the macroscopic orientation of force. For example, if the ligand was bound to the side of the receptor, pulling parallel to the axis of the receptor along the surface normal could result in a large orientation angle θ relative to the microscopic pathway and weak coupling of force to the energy landscape. Departing from the high strength regime below 70 pN, the intermediate strength regime with a slope of $f_\beta \approx 13$ pN places the next activation barrier at $x_\beta \approx 0.3$ nm. Finally, below ~20 pN, the spectrum exhibits a low strength regime with a slope of $f_\beta \approx 3.4$ pN that sets the outermost barrier at $x_\beta \approx 1.2$ nm. Using the logarithmic intercepts found by extrapolation of each linear regime to zero force, the differences in energy between the inner activation barriers are calculated to be only 2–3 $k_B T$. As for biotin–(strept)avidin bonds, the innermost barrier deep in the binding pocket provides strength on short time scales (<0.03 s), which is sufficient to meet the functional requirements noted earlier. Even though only 4–6 $k_B T$ higher in energy, the outermost activation barrier extends the lifetime of the bond almost 100-fold (to ~1 s) beyond that set by the innermost barrier.

Energy landscapes for anchoring lipids in membranes

Lipids and acylated proteins are anchored in bilayers by hydrophobic interactions. The handbook[28] correlation for free energy of transfer from aggregates (*e.g.* micelles or bilayers) to water is a linear proportionality of ~1 $k_B T$ per aliphatic carbon for lysophosphatidylcholines (PCs) and not quite double ~1.7 $k_B T$ per carbon for diacylPCs, although little evidence exists for diacyl lipids with chain lengths longer than 10–12 carbons. This reinforces the established view that anchoring potential increases with hydrophobic surface area embedded in the bilayer.[29] Partition and solubility provide important static-equilibrium assays but represent energetic measures of strong *vs.* weak anchoring—not strength—which is the force needed to extract a molecule. Based on the hydrophobic energy scale for exposure to water, the energy landscape for hydrophobic anchoring in bilayers should simply rise linearly with displacement along the bilayer normal. Treating the embedded molecule as a cylinder with radius r_m, the surface energy per unit area for creating a water/nonpolar interface and the circumference of the cylinder (*i.e.* energy/length ≈ 30 mJ m^{-2} × $2\pi r_m$ or 7 $k_B T$ nm^{-2} × $2\pi r_m$) suggest naively that the molecular extraction force should be a constant set by molecular size, $f \approx 180$ pN × r_m(nm). Taking a radius ~0.5 nm for a lipid, the anchoring force would be ~100 pN. On the other hand, we will see next that lipids can be extracted from membrane bilayers with forces as small as ~1 pN if performed over seconds! Over the range of anchoring strengths between 0 and 100 pN, the missing ingredient is thermally activated kinetics. By comparison, lipid pull-out forces in MD simulations[3] were >200 pN even under the slowest extraction of ~10^{-8} s and increased with speed apparently due to viscous damping.

Strength of hydrophobic anchoring in fluid membranes was tested by extraction of single receptor lipids from giant bilayer vesicles prepared with two lipid compositions: pure stearoyloleoylphosphatidylcholine (SOPC) (C18:0/1) and a 1:1 mixture of SOPC plus cholesterol (CHOL)—somewhat similar to membranes that encapsulate cells. Doped in the vesicle bilayers at extremely low concentration ($<0.0001\%$), the receptor lipids were a special lipid construct of biotin–PEG–distearoylphosphatidylethanolamine (DSPE) (diC18:0) kindly provided by INEX Pharmaceuticals, Burnaby, B.C., Canada. Plotted in Fig. 7, we see little structure in the spectra for receptor lipid anchoring and much lower forces compared to the spectra in Figs. 5 and 6 for receptor–ligand bonds.[8] Over nearly four orders of magnitude in loading rate, only a single linear strength regime is found for extraction of the receptor lipids from SOPC:CHOL bilayers. The low slope of $f_\beta \approx 2.4$ pN places a barrier at a distance $x_\beta \approx 1.7$ nm along the direction of force. Two linear regimes are found for lipid extraction from pure SOPC bilayers with a modest difference in slopes. The initial slope of $f_\beta \approx 3.4$ pN locates an outer barrier at $x_\beta \approx 1.2$ nm and the second slope of $f_\beta \approx 6.1$ pN implies an inner barrier at $x_\beta \approx 0.7$ nm. Consistent with the simple concept of hydrophobic interaction, the locations of the outermost barriers for both types of bilayers are comparable to (but slightly less than) the hydrophobic half thickness of the bilayer,

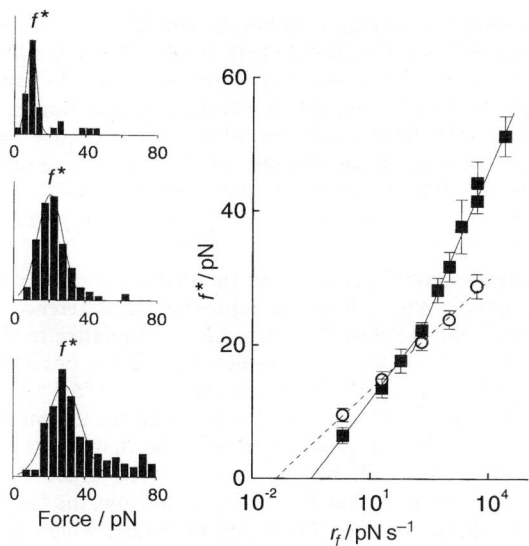

Fig. 7 On the left are examples of force histograms taken from tests of receptor lipid (biotin–PEG–DSPE) extraction from mixed SOPC : CHOL vesicle bilayers, which demonstrate shift in peak location and increase in width with increase in loading rate (top histogram, 2 pN s^{-1}, middle histogram, 200 pN s^{-1}, bottom histogram, 5 000 pN s^{-1}). Superposed on the histograms are Gaussian fits used to determine the most frequent extraction force—anchoring strength. On the right are the complete dynamic spectra of strength *vs.* log(loading rate) for extraction of receptor lipids from SOPC (closed boxes) and mixed SOPC : CHOL bilayers (open circles). Defined as thermal energy $k_B T$/distance x_β, the slopes of the initial linear regimes map activation barriers at $x_\beta \approx 1.2$ nm for extraction from SOPC and $x_\beta \approx 1.7$ nm for extraction from SOPC : CHOL along the direction normal to the bilayer. Not seen in the SOPC : CHOL spectrum, the break in slope for the SOPC spectrum places a weak inner barrier at $x_\beta \approx 0.7$ nm.

which is increased by cholesterol. Addition of cholesterol to SOPC bilayers increases the outer activation barrier by $\sim 2\,k_B T$ as shown by the shift between logarithmic intercepts of the initial regimes for SOPC : CHOL and SOPC. Quite unexpected, the break in slope in the spectrum for SOPC reveals an inner transition state near the middle of the hydrophobic monolayer, which appears to be $\sim 2\,k_B T$ below the outer barrier. Perhaps coincidental, the location of this transition state derived from the thermal force scale correlates with the position of the unsaturated bond in the oleoyl chain of SOPC. Completely speculative, the split in activation barriers could reflect an entropic bottle neck as chains transiently pass the average position of the unsaturated group. Very puzzling, the bilayer residence time of ~ 0.01 s derived from the logarithmic intercept of these spectra at zero force is much shorter than implied by the lack of perceptible dissociation from an isolated vesicle over the the time scale of 1 h. Without an explanation at present, we see again (as for biotin–avidin bonds) that very small forces must strongly affect the shape of the soft outer transition state. In any case, anchoring of lipids and acylated proteins in bilayers will always be weak unless the molecules are extracted very rapidly from the bilayer.

Strong *vs.* weak bonds in serial linkages

For a serial linkage of n identical bonds, the rate of breakage under force is simply increased by a factor n, $k_{\text{off}} \approx (n/t_0)\exp(f/f_\beta)$, if the bond kinetics are uncorrelated. This increases the thermal scale for loading rate, $r_f^0 = n f_\beta/t_0$, and shifts the strength spectrum along the $\log_e$(loading rate) axis by a factor of $\log_e(n)$, which reduces strength at a given loading rate by $-f_\beta \log_e(n)$. In contrast to a simple shift along the log(loading rate) axis, we expect the strength of a multiple linkage of dissimilar bonds to be limited by the weakest bond and naively that strong *vs.* weak should be defined by the energy barriers sustaining the bonds. However, theory shows that this anticipated hierarchy is only correct for some sets of bonds; other sets will exhibit unexpected switching from strong to weak and *vice versa* as loading rate increases. In the determination of strong *vs.* weak at a particular retraction rate, the important parameters are both the spontaneous rates of dissociation set by barrier energies and the thermal force scales that characterize *e*-fold changes in the dissociation rates of bonds under force. Again invoking the simple sharp barrier model, we can easily establish a phase diagram [*cf.* Fig. 8(a)] of the most likely site for breakage in a two-bond linkage. Assuming that both bonds are characterized by the same diffusive time scale t_D for simplicity, the rate of uncorrelated breakage is the combined rates for each bond,

$$k_{\text{off}} \approx (1/t_0)\exp(f/f_\beta)\{1 + \exp[-\Delta E_b/k_B T + f\Delta(1/f_\beta)]\}$$

where $1/t_0$ and f_β specify the spontaneous rate and thermal force scale of the fast bond (smallest barrier energy) as the reference; ΔE_b and $\Delta(1/f_\beta)$ represent the differences in barrier energy and reciprocal thermal force scale for the slow bond (*i.e.* $\Delta E_b > 0$) relative to the fast bond. The combined rate and the predicted strength spectrum predict that the fast bond will remain the expected weak bond so long as the following inequality holds: $\Delta E_b/k_B T > f\Delta(1/f_\beta)$. This will always be the case when the thermal force scale for the fast bond is less than the thermal force scale for the slow bond [*i.e.* $\Delta(1/f_\beta) < 0$ or equivalently $\Delta(x_\beta) > 0$]. On the other hand, if the thermal force scale for the fast bond is larger, then there will be a crossover force where $\Delta E_b/k_B T \leq f\Delta(1/f_\beta)$; this strong $\rightleftharpoons$ weak bond phase boundary is sketched in Fig. 8(a). Now, the fast bond will be the strong bond and the slow bond will be the most likely site of failure, which is not anticipated in the traditional view.

To demonstrate the importance of this concept, imagine that the selectin receptor was linked to a vesicle bilayer by a lipid anchor and then the strengths of carbohydrate ligand bonds to the selectin were probed as in the leukocyte tests. Purely hypothetical, Fig. 8(b) shows that at slow loading rates, the carbohydrate–selectin bond would most likely detach because lipid anchoring is

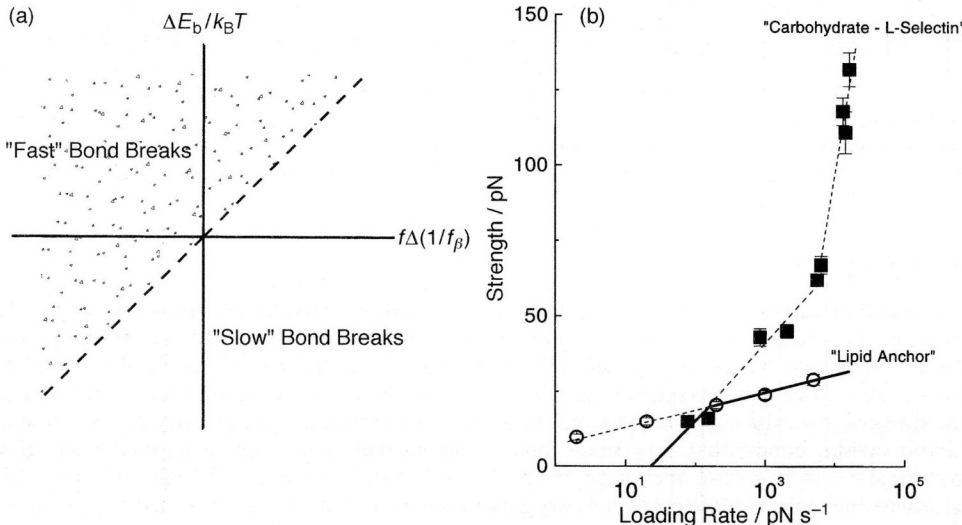

Fig. 8 (a) Phase diagram for definition of strong $\rightleftharpoons$ weak bonds in a serial linkage of two bonds sustained by single sharp energy barriers. The vertical axis is the difference in barrier heights ΔE_b for the two bonds ($\Delta E_b > 0$ characterizes a slow bond relative to a fast bond as defined by spontaneous off rates); the horizontal axis is the product $f\Delta(1/f_\beta)$ of applied force f and the difference $\Delta(1/f_\beta)$ in reciprocal thermal force scales. In the traditional view, strong : weak equates to slow : fast. But for bonds in series, this diagram shows that there can be unexpected switching of these attributes under force. (b) Hypothetical strength of carbohydrate bonds to selectins if linked by lipid anchors to membranes. Simultaneous kinetics over different energy landscapes for the carbohydrate–selectin bond and lipid anchoring predicts a dynamic crossover in site for detachment when pulled on by a probe decorated with the carbohydrate ligand. At slow rates of loading, the lipid anchor is stronger than the carbohydrate–selectin bond, which is the most likely site of detachment. On the other hand, at fast rates of loading, the carbohydrate–selectin bond is stronger than lipid anchoring strength, which then becomes the most likely site of detachment.

stronger. On the other hand, under fast loading, the carbohydrate–selectin bond becomes strong and the lipid anchor weak by comparison. Hence, the lipid-anchored selectin would most likely be pulled out of the membrane by the carbohydrate ligand attached to the probe. In contrast to the image of inner activation barriers in a complex bond, we see that the signature of a strong to weak bond metamorphosis in a serial linkage of bonds is an abrupt reduction in slope from one linear strength regime to the next with increase in loading rate.

Summary

Recent laboratory probe experiments confirm that bond breakage and molecular detachment occur at forces determined by the loading rate. Measured under steadily rising force over an enormous span of loading rates, the spectra of strength *vs.* $\log_e$(loading rate) yield images of the prominent barriers traversed in the energy landscapes along force-driven pathways in unbinding. Simple analysis of the spectra provides a view into the inner complexity of biomolecular interactions and structural cohesion as noted in the following list of highlights.

(1) Examining two unrelated receptor–ligand bonds, we find a similar sequence of linear strength regimes *vs.* log(loading rate). These regimes reveal a cascade of three activation barriers for both receptor–ligand interactions, although quite different in energy scale. The innermost barrier deep in the binding domain is responsible for the high strength perceived on short time scales and the major portion of total activation energy. The more distal barriers lead to weakness on long time scales but significantly extend bond lifetime in the absence of force. The intriguing question is why did nature structure energy landscapes in receptor–ligand bonds and create a sequence of time scales for amplification of kinetics under force? Answering this question is likely to introduce a new perspective of biological chemistry.

(2) No surprise, anchoring strengths of lipids in bilayers are consistent with nearly structureless hydrophobic potentials, although small inner barriers do appear in some cases. Most significant, the small thermal force scale set by acyl chain length results in very weak anchoring strength unless the molecules are extracted extremely rapidly from the bilayer. Still to be confirmed, integral membrane proteins should be much more strongly anchored to membranes since hydrophilic groups at the interfaces will contribute major activation barriers with large thermal force scales.

(3) Dissimilar bonds in a serial linkage can unexpectedly switch from strong to weak and shift the most likely site for failure between bonds as loading rate increases. Such behaviour is not only a major factor in cohesive and adhesive strength but is likely to be important in signalling and regulation of biochemical pathways inside cells.

Acknowledgement

It is important to credit the individuals who carried out the experiments and developed the instrumentation described in this paper since many of the results are yet to be published. Tests of biotin–(strept)avidin bonds were performed by Andrew Leung (University of British Columbia) and Pierre Nassoy (now at l'Institut Curie in Paris). Computer control of the biomembrane force probe assembly and video image processing was developed by Ken Ritchie (now at Nagoya University in Japan). Tests of carbohydrate–L-selectin bonds were also performed by Andrew Leung in collaboration with Scott Simon (from Baylor College of Medicine in Houston) and Dan Hammer (from University of Pennsylvania in Philadelphia). Tests of lipid anchoring in bilayer membranes were performed by Florian Ludwig (University of British Columbia).

The work was supported by grants HL54700 and HL 31579 from the US National Institutes of Health, grant MT7477 from the Medical Research Council of Canada, and the Canadian Institute for Advanced Research Program in Science of Soft Surfaces and Interfaces.

References

1 G. Binnig, C. F. Quate and C. H. Gerber, *Phys. Rev. Lett.*, 1986, **56**, 930; B. Drake, C. B. Prater, A. L. Weisenhorn, S. A. C. Gould, T. R. Albrecht, C. F. Quate, D. S. Cannell, H. G. Hansma and P. K. Hansma, *Science*, 1989, **243**, 1586.
2 H. Grubmuller, B. Heymann and P. Tavan, *Science*, 1996, **271**, 997; S. Izrailev, S. Stepaniants, M. Balsera, Y. Oono and K. Schulten, *Biophys. J.*, 1997, **72**, 1568.

3   S-J. Marrink, O. Berger, P. Tieleman and F. Jahnig, *Biophys. J.*, 1998, **74**, 931.
4   H. A. Kramers, *Physica (Amsterdam)*, 1940, **7**, 284; P. Hanggi, P. Talkner and M. Borkovec, *Rev. Mod. Phys.*, 1990, **62**, 251.
5   E. Evans and K. Ritchie, *Biophys. J.*, 1997, **72**, 1541.
6   R. Merkel, P. Nassoy, A. Leung, K. Ritchie and E. Evans, *Nature (London)*, 1999, **397**, 50.
7   S. Simon, A. Leung, D. Hammer and E. Evans, to be submitted.
8   F. Ludwig and E. Evans, to be submitted.
9   M. Doi and S. F. Edwards, *The Theory of Polymer Dynamics*, Clarendon Press, Oxford, 1986; N. G. van Kampen, *Stochastic Processes in Physics and Chemistry*, North-Holland, Amsterdam, 1981.
10 P. J. Rossky, J. D. Doll and H. L. Friedman, *J. Chem. Phys.*, 1978, **69**, 4628.
11 G. I. Bell, *Science*, 1978, **200**, 618.
12 G. U. Lee, D. A. Kidwell and R. J. Colton, *Langmuir*, 1994, **10**, 354.
13 E-L. Florin, V. T. Moy and H. E. Gaub, *Science*, 1994, **264**, 415.
14 V. T. Moy, E-L. Florin and H. E. Gaub, *Science*, 1994, **264**, 257.
15 P. Hinterdorfer, W. Baumgartner, H. J. Gruber, K. Schilcher and H. Schindler, *Proc. Natl. Acad. Sci. USA*, 1996, **93**, 3477.
16 S. S. Wong, E. Joselevich, A. T. Woolley, C. L. Cheung and C. M. Lieber, *Nature (London)*, 1998, **394**, 52.
17 S. P. Tha, J. Shuster and H. L. Goldsmith, *Biophys. J.*, 1986, **50**, 117.
18 E. Evans, D. Berk and A. Leung, *Biophys. J.*, 1991, **59**, 838.
19 E. Evans, K. Ritchie and R. Merkel, *Biophys. J.*, 1995, **68**, 2580.
20 E. Evans and K. Ritchie, *Biophys. J.*, 1999, in press.
21 N. M. Green, *Adv. Protein Chem.*, 1975, **29**, 85.
22 A. Chilkoti and P. S. Stayton, *J. Am. Chem. Soc.*, 1995, **117**, 10622.
23 R. Alon, D. A. Hammer and T. A. Springer, *Nature (London)*, 1995, **374**, 539; K. D. Puri, S. Chen and T. A. Springer, *Nature (London)*, 1998, **392**, 930.
24 P. C. Weber, D. H. Ohlendorf, J. J. Wendoloski and F. R. Salemme, *Science*, 1989, **243**, 85.
25 O. Livnah, E. A. Bayer, M. Wilchek and J. L. Sussman, *Proc. Natl. Acad. Sci. USA*, 1993, **90**, 5076.
26 S. Freitag, I. Le Trong, L. Klumb, P. S. Stayton and R. E. Stenkamp, *Protein Sci.*, 1997, **6**, 1157.
27 V. Chu, S. Freitag, I. Le Trong, R. E. Stenkamp and P. S. Stayton, *Protein Sci.*, 1998, **7**, 848.
28 D. Marsh, *Handbook of Lipid Bilayers*, CRC Press, Boca Raton, FL, 1990, p. 275–280.
29 C. Tanford, *The Hydrophobic Effect: Formation of Micelles and Biological Membranes*, John Wiley and Sons, New York, NY, 1973.

Paper 8/09884K

Collective membrane motions of high and low amplitude, studied by dynamic light scattering and micro-interferometry

Rainer Hirn,[a] Thomas M. Bayerl,*[a] Joachim O. Rädler[b] and Erich Sackmann[b]

[a] *Universität Würzburg, Physikalisches Institut EP-5, 97074 Würzburg, Germany*
[b] *Technische Universität München, Physik Department E22, 85747 Garching, Germany*

Received 9th October 1998

Undulations of lipid bilayers were experimentally studied for the two limiting cases of high and weak lateral tension using two well established model systems: freely suspended planar lipid bilayers, so-called black lipid membranes (BLM) for high-tension studies and large unilamellar vesicles (LUV) for measurements at weak tension. This variation in tension results in changes of undulation amplitudes from several hundred nm (LUV) down to 1 nm (BLM), thus requiring different physical methods for their detection. We have employed microinterferometric techniques (RICM) for studying the regime of weak tension and dynamic light scattering (DLS) for that of high tension. The dedicated DLS set-up allowed the measurements of undulations over a wide wave vector range of $250 < q/\text{cm}^{-1} < 35\,000 \text{ cm}^{-1}$. This enabled the observation of collective membrane modes in two regimes, the oscillating one at low q and the overdamped regime at high q. The transition between both regimes at the bifurcation point is rather abrupt and depends on the lateral tension of the bilayer, as is demonstrated by comparing the dispersion curves of pure lipid and of lipid–cholestrol BLMs over the same q-range. The DLS measurements allowed a critical test of a hydrodynamic theory of the dispersion behaviour of membrane collective modes under tension. The DLS measurements are compared with RICM results of undulatory excitations of giant vesicles weakly adhering to substrates in the 10^{-6}–2.5×10^{-7} m wavelength regime and at low frequencies (0.1–25 Hz). Experimental evidence for the strong decrease in the relaxation rate by the hydrodynamic coupling of the membrane with the wall is established.

Introduction

Collective motions (undulations) of fluid membranes are crucial for their mutual interaction at the nm lengthscale and thus for the swelling of lipids.[1] The measurement of these motions provides not only a deeper understanding of membrane micromechanic properties down to the molecular scale but also sheds some light on the processes which are essential for the formation of contacts between cells and between cells and solid surfaces.

In the low-frequency regime (up to 100 Hz), time-resolved interferometric microscopy is the method of choice to study collective motions of membranes, of whole cells and of some microorganisms, in great detail. On the other hand, incoherent neutron scattering has been successfully employed for the measurement of collective membrane modes in the THz range.[2] However, there is quite a lack of experimental approaches for their study in length and frequency regimes which are commonly denoted as the mesoscopic range (kHz–MHz). So far, only solid-state NMR techniques have been successful in detecting collective modes in this range,[3] but their application is limited by sensitivity, sample geometry, narrow accessible frequency range and NMR data analysis is hampered by the constraint that one-dimensional solid-state NMR measures temporal correlations only and not spatial correlations.

The availability of a technique which can bridge the wide frequency gap of ca. six orders of magnitude between the well established low-frequency regime and the highest accessible frequency range would be extremely helpful in the testing of a number of theories based on membrane microelasticity, as well as for sorting out the essential collective modes for each time domain.

DLS has been used previously to measure collective motions of BLM and it demonstrated successfully the viscoelastic behaviour of such membranes[4] but their usefulness was limited by the narrow range of accessible undulation modes from 600 to 1 800 cm^{-1}.

Here, we report the first results of DLS measurements on BLMs obtained over a much wider wavevector range from 250 to 35 000 cm^{-1} ($1.8 < \lambda/\mu\text{m} < 250$) and a timescale of four orders of magnitude (from 6×10^{-3} to 6×10^{-7} s) which is sufficient for a critical testing of theoretical predictions. A transition from single to two-exponential relaxation is encountered at high tension, which exhibits much higher frequencies in the vicinity of the bifurcation point. The experiments in this new frequency regime enable one to study the effects of proteins, peptides and polymers on collective membrane dynamics.

Since BLMs exhibit, generally, a high lateral tension, their undulations are dominated by lateral forces, resulting in very low undulation amplitudes. On the other hand, LUV show only negligible or very weak tension and thus a different undulatory behaviour with high undulation amplitudes.[5] The latter is constrained when the LUV comes into close contact with a wall. As a result, a repulsive pressure is established which is gradually reduced with increasing lateral tension. A comparison between tension dominated BLMs and rather tensionless LUV thus allows a critical test of theoretical predictions on membrane micromechanics in the limit of two extremes. We have, additionally, performed time-resolved reflection interference contrast microscopy (RICM) of LUV close to a solid surface. The results show that the relaxation modes are dominated by the hydrodynamic interaction between the LUV and the wall.

Theory of DLS of membranes under high tension

One of the first comprehensive theoretical treatments of the dispersion behaviour resulting from collective motions of thin elastic membranes and the consequences for its dynamic light scattering properties was published by Kramer.[6] The basic assumptions of this theory were: The membrane elastic properties can be described by compression- and shear moduli and a membrane tension. All hydrodynamic equations (i.e. Navier–Stokes) are linear. The fluid symmetrically surrounding the membrane is incompressible. At the membrane/fluid interface the velocities of the two media are the same. Fluid velocity becomes zero at large distances from the membrane. The wavelength of the collective modes is large compared to the membrane thickness and small compared to its diameter.

For a membrane of isotropic molecules separating two compartments of the same liquid, transverse shear was identified as the only DLS sensitive mode. Here, the molecules perform out-of-plane motions only, leading to an effective fluctuation of the membrane area. The dispersion relation of this mode is

$$2m\rho\omega^2 + \gamma q^3(q - m) = 0 \quad (1)$$

where $m = (q^2 - i\rho\omega/\eta)^{1/2}$, $q = 2\pi/\lambda$ is the scattering vector, $\omega = \omega_0 - i\Gamma$ is the complex frequency consisting of the eigenfrequency ω_0 and the damping constant Γ. ρ and η are the density and viscosity of the fluid and $\gamma = \gamma_0 - i\omega\gamma'$ is a complex tension with the real part being the membrane tension and γ' being the surface viscosity.

A plot of f_0 and Γ vs. q of eqn. (1), using tensions typical for a free planar bilayer (BLM), shows the following features (Fig. 1) which were discussed in greater detail by Kramer[6] and later in the work of Earnshaw:[4] For transverse shear there exists an oscillating and an overdamped regime in the mesoscopic range and the transition between them at $q_0 \approx 5000$ cm^{-1} is abrupt (bifurcation point). Above this q value, the single damping constant Γ_0 of the oscillating regime is replaced by two exponential decays, describing two overdamped modes with the slower one characterized by Γ_1 and the faster by Γ_2. No experimental verification of this transition has been provided yet. For

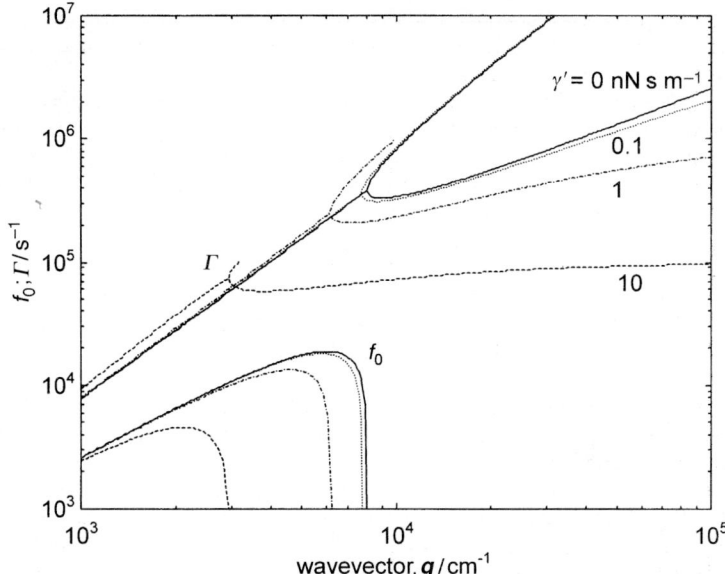

Fig. 1 Theoretical curve according to Kramer's theory [eqn. (1)] with membrane tension $\gamma_0 = 1$ mN m^{-1}, fluid density $\rho = 1$ mg ml^{-1} and fluid viscosity $\eta = 1$ mPs at different surface viscosities γ'.

increasing q-values above the bifurcation point the slower mode Γ_1 approaches the value $\xi = \gamma_0/\gamma'$ asymptotically, while the faster mode Γ_2 merges at the same value of ξ with the bulk mode having the dispersion relation

$$i\omega - \eta q^2/\rho = 0 \qquad (2)$$

Moreover, this bulk mode is insensitive to DLS and, therefore, the faster mode disappears at q-values beyond ξ.

By considering the anisotropy of lipids in a BLM, an additional splay mode coupled to the transverse shear mode was found.[7] The coupling strength between the two modes scales with the total free energy change a single lipid undergoes when its molecular director gets tilted by a certain angle away from that of its neighbour. This leads to a modification of the membrane tension in eqn. (1), now becoming an effective tension $\gamma_{0\text{eff}}$, which includes the curvature energy κ:

$$\gamma_{0\text{eff}} = \gamma_0 + \kappa q^2 \qquad (3)$$

Hence, at higher q the bending modulus κ will dominate undulation behaviour over the membrane tension γ_0 (Fig. 2).

In terms of the effective undulation amplitude u_{eff} of a membrane of area A and driven by a thermal energy kT, this corresponds to the Helfrich[1] equation:

$$\langle u_{\text{eff}}^2(q) \rangle = \frac{kT}{A(\gamma q^2 + \kappa q^4)} \qquad (4)$$

For standard BLM parameters ($\gamma_0 = 1$ mN m^{-1}, $\kappa = 10^{-19}$ J, BLM$_{\text{diameter}} = 3.5$ mm), eqn. (4) gives $u_{\text{eff}} \approx 0.09$ Å at $q = 1000$ cm^{-1} and a maximum amplitude of $u_{\text{eff}} \approx 10$ Å is obtained by summing the contributions over a q range limited by the cut-offs 1/membrane thickness and 1/membrane diameter. The reason for these tiny amplitudes compared to those reported below for LUV is the dominance of tension for BLMs while the flickering systems are tension free and thus κ dominated. For BLMs the effect of κ is negligible up to very high q values of, say, $q = 300\,000$ cm^{-1} ($\lambda = 200$ nm) where the contribution from κ amounts to 1/10 of that of the tension term. Nevertheless, it has been shown experimentally[4] that DLS is sensitive even to these very small amplitudes exhibited by BLMs in the q-range 600 to 1800 cm^{-1}.

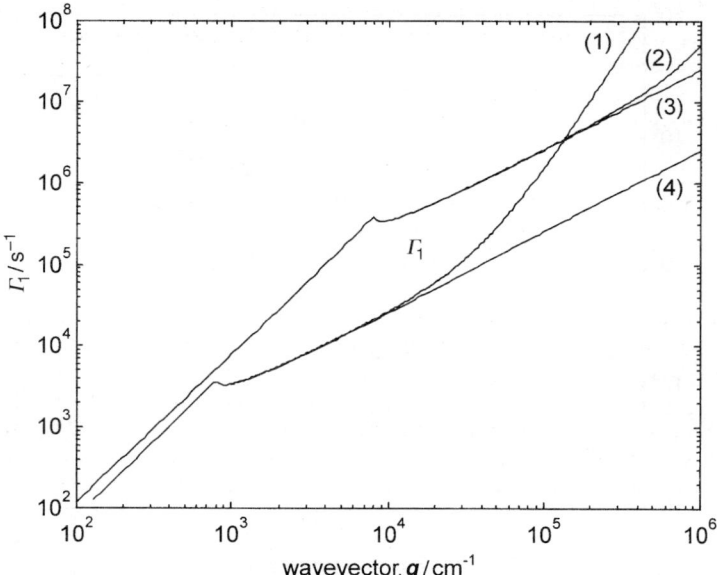

Fig. 2 Comparison of Kramer's [eqn. (1) + (2)] and Fan's dispersion theory [eqn. (1)] for two pairs of γ_0 and κ values. For the sake of clarity, only the slow damping modes Γ_1 are shown. The curves with the increasing slopes at higher q-values (curves 1 and 2) correspond to Fan's theory. The upper two lines were calculated for $\gamma_0 = 1$ mN m^{-1} and $\kappa = 0$ (curve 3) or $\kappa = 10^{-19}$ J (curve 2). The lower two lines are for tension $\gamma_0 = 0.1$ mN m^{-1} and $\kappa = 0$ (curve 4) or $\kappa = 50 \times 10^{-19}$ J (curve 1).

However, to obtain information about the surface viscosity γ' requires DLS measurements over a q-range that extends well into the overdamped regime, since the dependence of Γ_0 and of the eigenfrequency $f_0 = \omega_0/2\pi$ on γ' is almost negligible in the oscillating regime (Fig. 1). Similarly, a representative test of eqn. (1) and (2) is impossible without measurements covering q in both the oscillating and overdamped regimes.

Theoretical predictions for flaccid vesicles near a wall

One of the most striking aspects of bending excitations of flaccid membranes close to a wall is the strong repulsive interaction pressure: $p \sim (k_B T)/\kappa h_0^3$ arising from the entropy loss of a membrane confined to an average membrane–wall distance, h_0. In the presence of membrane tension, however, the steric repulsion is strongly suppressed and decays exponentially $p \sim \exp(-\gamma h_0^2/k_B T)$.

For a weakly adhering liposome, the entropic pressure is balanced by an attractive wall interaction. In this case the external forces experienced by the membrane can be described by an effective harmonic interaction at the equilibrium spacing h_0 from the wall

$$V(h) = 1/2 V''(h - h_0)^2 \qquad (5)$$

where V'' is the second derivative of $V(h)$ at the equilibrium distance h_0. Since we are mainly interested in small-wavelength excitations (q^{-1}), the undulation-induced dynamic surface roughness can be analysed in terms of plane waves

$$h(r, t) = \sum_q h_q(t)\exp\{iqr\} \qquad (6)$$

where q is the undulation wave vector. The amplitudes are determined by the equipartition theorem

$$h_q^2 = \frac{k_B T}{L^2\{\kappa q^4 + \gamma q^2 + V''\}} \qquad (7)$$

where $E(q) = \kappa q^4 + \gamma q^2 + V''$ is the energy of the Fourier mode q and κ and γ are again the membrane bending modulus and tension, respectively. L^2 is the system size over which the Fourier transformation was carried out.

The dynamics of the randomly excited bending modes is characterized in terms of the time correlation function of the Fourier components, which are given by

$$G_q(t) = \langle h_q(t) h_q(0) \rangle = k_q^2 \exp\{-\Gamma(q)t\} \qquad (8)$$

with a relaxation rate $\Gamma(q)$ to be discussed.

In most LUV experiments we deal with membranes at weak tension ($\gamma \approx 10^{-6}$ mN m^{-1}). This situation has been treated theoretically in great detail by Seifert.[8] One has to consider three situations:

(1) At low wave vectors (and thus low frequencies) the local density fluctuations within each monolayer are rapidly equilibrated by lateral diffusion and the bending excitations are damped solely by coupling to the bending-induced hydrodynamic flow in the surrounding fluid.

(2) At increasing q, the equilibration of the local density fluctuations is impeded by the friction between the monolayers which is characterized by a friction coefficient $b = \eta_m/d_m$, where η_m is the 2D membrane viscosity and d_m is the monolayer thickness. The bending modulus is renormalized by the coupling of the bending excitations and the lateral density fluctuations to

$$\kappa^* = \kappa + 2d_m K \qquad (9)$$

where K is the membrane modulus of isotropic compression.

(3) At very high frequencies and q-vectors, the damping is finally determined by the in-plane shear deformation, which is not considered here.

Due to the interplay of bending and density fluctuations, the membrane dynamics (of free and adherent vesicles) is determined by a low-frequency mode and a high-frequency mode. The relaxation of the latter is determined by monolayer friction. The cross-over wave vector between the two regimes is

$$q_{12} = 2\eta K / b\kappa^* \approx 10^7 \text{ m}^{-1} \qquad (10)$$

The number on the right-hand side holds for the data summarized in Fig. 3.

In the presence of a wall the situation is more complex. A third cross-over wave vector has to be considered, since long wavelength excitations are impeded by the interaction of the membrane

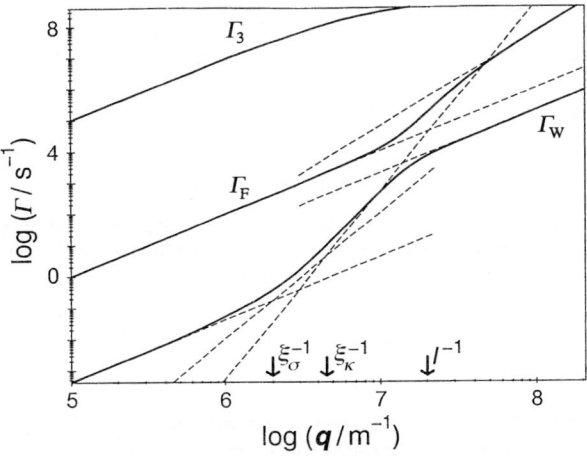

Fig. 3 Theoretical dispersion relation of decay rates of bending excitation of adherent membrane as calculated by Kraus and Seifert.[9] For weak adhesion, corresponding to weak tension, the two lowest frequency modes (out of three) are shown. Γ_W is the decay rate of the slow bending mode with rapid equilibration of local density fluctuations within monolayers. Γ_F is the decay rate of the mode controlled by mutual friction between monolayers. ξ_σ and ξ_κ are defined in eqn. (14) and (15). Note that at $q < \xi_0$ both decay rates scale as $\Gamma \propto q^2$. The curves have been calculated for the following values of the parameters: $h_0 = 1$ nm, $\gamma = 10^{-6}$ J m^{-2}, $V'' = 4 \times 10^6$ J m^{-4}, $b = 5 \times 10^8$ J s m^{-4}, $\eta = 10^{-3}$ J s m^{-3}, $K = 0.1$ J m^{-2}, $\kappa = 0.5 \times 10^{-19}$ J.

with the hard wall. Due to the symmetry break induced by the wall there are two density modes (besides the bending mode):

For negligible tension the undulations are determined by the wall at

$$q \geqslant \xi_k^{-1} = (V''/\kappa)^{1/4} = 2 \times 10^8 \text{ m}^{-1} \quad (11)$$

while for membranes under tension the wall becomes effective at

$$q \geqslant \xi_\gamma^{-1} = (V''/\gamma)^{1/2} = 10^6 \text{ m}^{-1} \quad (12)$$

The transition between the bending and the tension dominated regimes occurs at wave vectors

$$q < \xi = (\gamma/\kappa)^{1/2} = 3 \times 10^6 \text{ m}^{-1} \quad (13)$$

The RICM experiments were performed at a tension $\gamma \approx 10^{-6}$ N m^{-1} and the range of wave vectors studied was $0.6 \times 10^6 < q/\text{m}^{-1} < 4 \times 10^6$. We can thus assume that the tension can still be considered as weak.

Summing up, the above conditions let us conclude that there should be two modes accessible by microinterferometry:

A low-frequency, long-wavelength mode, Γ_W, which is affected by the presence of the wall (being at distance h_0) but which is slow enough to allow for the equilibration of the monolayer density fluctuation and which is not affected by monolayer friction. Its damping constant is

$$\Gamma_W(q) = (V'' h_0^3 q^2 + \gamma h_0^3 q^4)/12\eta \quad (14)$$

This mode dominates the regime of $q \ll (\xi\kappa^{-1}, h_0^{-1}) \approx 10^7$ m^{-1} and is thus observable by the microinterferometric technique. For $h_0^{-1} \gg q \gg \xi^{-1}$, the damping constant is dominated by the bending modulus $\Gamma(q) = \kappa h_0^3 q^6/12\eta$ which is, however, not observable by our technique, since $q_{max} = 5 \times 10^6$ m^{-1}. With the data summarized in Fig. 3 one finds for $q \approx 10^6$ m^{-1} and $h_0 \approx 40$ nm: $\Gamma_W \approx 0.02$ s^{-1}, corresponding to a relaxation time $\tau \approx 50$ s.

The second interesting mode, Γ_F, is controlled by the friction between the monolayers. It decays with a rate

$$\Gamma_F = Kq^2\kappa/2b\kappa^* \quad (15)$$

By inserting the data from Fig. 3 one can estimate a wave vector $q \approx 10^6$ m^{-1}, a decay constant $\Gamma_F \approx 20$–200 s^{-1} or a relaxation time $\tau = 0.05$–0.005 s.

A completely different type of behaviour is expected for permeable membranes, as shown by Prost et al.[10] The undulations can relax by permeation of water through the bilayer. This mechanism is expected to dominate over the hydrodynamic process for wave vectors $q \ll q^* = (\lambda_p \eta)^{1/2} h_0^{-3/2}$, where λ_p is the water permeability of the bilayer. It is of the order $\lambda_p \approx 10^{-6}$ m^2 s kg^{-1} and for $h_0 \approx 40$ nm one obtains $q \ll 3 \times 10^5$ m^{-1}.

The decay rate is

$$\Gamma_p = \lambda_p(\gamma q^2 + \kappa q^4) \quad (16)$$

For $q = 10^6$ m^{-1} one expects, in the case of tension dominated membranes, $(\gamma > 2 \times 10^{-6}$ N m$^{-1})$ $\Gamma_p \approx 1$ s^{-1} and, in the case of bending dominated membranes, $\Gamma_p \approx 0.1$ s^{-1}.

Materials and methods

Substances

Cholesterol and n-decane 99 + % were purchased from Sigma-Aldrich (Steinheim, Germany) and the n-decane was further purified by passing it through an alumina column until all coloured impurities were removed. The phospholipid 1,2-di-elaidoyl-sn-3-glycerophosphocholine (DEPC) was purchased from Avanti Polar Lipids Inc. (Alabaster, AL, USA) and was used without further purification. All light scattering experiments were performed in 20 mM Hepes buffer (Life Technologies Ltd., Daisley, UK) containing 50 mM KCl (Fluka Chemie AG, Buchs, Switzerland) and the water used was from a Milli-Q purification unit (Millipore Corp., Bedford, USA). The buffer was filtered through a sterile filter of 0.1 μm pore size (Millipore Corp., Bedford, USA) and degassed prior to its use in the scattering cell. Microinterferometric measurements used SOPC from Avanti Polar Lipids in Millipore water.

Membrane preparation

The scattering cell was a standard rectangular glass cuvette of 40 × 10 × 10 mm^3 (Hellma GmbH & Co. Mülheim Germany), separated into two compartments by a diagonally inserted Teflon wall (thickness 2 mm)[11] with a circular aperture of 4.5 mm in the centre. Over this hole, a 25 μm thick Teflon foil was spanned using Teflon glue (primer 770 and bonder 406 from Loctite GmbH, München, Germany) which itself featured a hole of 3.5 mm diameter. The latter hole was punched with extreme care to ensure its rim was as smooth as possible (roughness was below the optical resolution limit of a light microscope). This smoothness, together with the use of a very thin foil, ensures the BLM long-time stability (up to 5 days) and prevents positional fluctuations of the BLM as a whole. The scattering cell was placed within a metal cell holder the temperature of which (22 °C for all measurements) was controlled by a water-bath thermostat. Before BLM preparation, the scattering cell was filled with buffer solution to a level of *ca.* 3 mm above the upper edge of the Teflon wall so that both compartments were connected by the water.

BLMs were prepared by sliding a Teflon loop containing the film-forming solution over the hole.[12] Prior to this, the hole was pretreated by spreading a methanol solution of 2 wt.% lipid onto it, followed by methanol removal by drying in air.[8] The term lipid refers to pure DEPC for BLM sample 1 and to a mixture of DEPC with 30 mol% cholesterol for BLM sample 2. The film-forming solution was *n*-decane containing 1 wt.% lipid. BLM formation was confirmed by the formation of a sharp spot of specular reflected incident laser light. A rectangular Teflon pressure cap was then carefully lowered from the top into the scattering cell, down to the water level, thereby eliminating any air bubbles and effectively damping out all surface capillary waves of the water which might couple with the BLM below.

Dynamic light scattering (DLS)

Experimental set-up. The scattering geometry of our set-up is shown in Fig. 4. Note that undulations like the transverse shear mode represent plane waves $u(r, t)$ with

$$u(r, t) = u_0 e^{-i(qr + \omega t)} \quad (17)$$

and can spread in all directions perpendicular to the membrane normal, thus giving rise to both in-plane and out-of-plane scattering of the incident light vector k_i. Thus, in a plane perpendicular to the membrane normal at the inflection point, all possible q-vectors belonging to the same undulation mode are confined within a circle. However, owing to the inclination of the detector

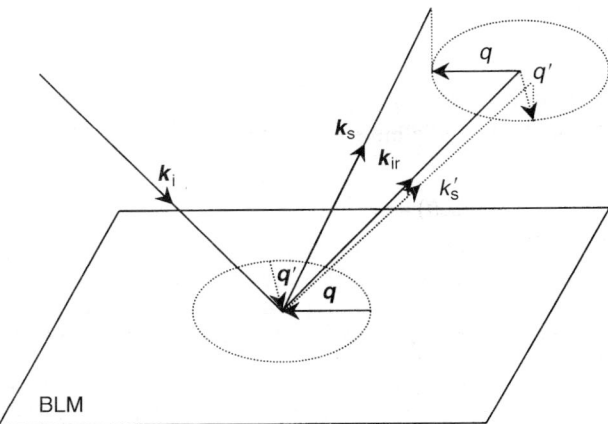

Fig. 4 Schematic depiction of the scattering geometry with the incoming vector k_i and specular reflected vector k_{ir} (scattering angle 45°), the inplane wavevector q, the in-plane scattered vector k_s, (with quasi the same length as k_i, *i.e.* quasi elastic light scattering) the out-of-plane wave vector q' and the out-of-plane scattered vector k'_s.

plane by 45° with respect to the membrane normal (*i.e.* the plane perpendicular to the specular reflected vector k_{ir}), the region of scattered vectors k_s at the detector site can be approximated, to the first order, by an elliptical aperture with an aspect ratio of 1.41 in front of the detector and with k_{ir} going through its centre.

The DLS set-up was mounted on a $3 \times 2 \times 0.2$ m^3 (200 kg weight) laser table (Melles Griot) supported by four air damping modules for mechanical shock protection. A 4 W argon ion laser (Coherent Inc. Santa Clara CA USA) was used, operating at 457.9 nm in TEM$_{00}$ mode with a maximum power of 150 mW. The beam was focused onto the BLM at an incident angle of 0.11° by a lens of $f = 50$ cm to obtain high q-resolution, giving an illuminated BLM spot of 160 µm. Its polarisation was adjusted parallel to the membrane plane to give the highest amount of scattering intensity. At a reflection angle of 45° and at a distance of 30 cm from the BLM the photomultiplier (PM) was arranged at a goniometer arm. Two pinholes between BLM and PM, one right after the BLM ($\varnothing = 700$ µm) and one in front of the PM ($\varnothing = 80$ µm) were used for q-range selection. For measurements at $q > 3400$ cm^{-1}, the PM pinhole was replaced by a vertical slit aperture (2000×200 µm^2) to allow, additionally, the detection of out-of-plane scattering, giving an increase in PM signal by a factor of 80 without any reduction in signal quality compared to the use of two pinholes. The slit aperture represents an approximation of the above-mentioned elliptical shape of the out-of-plane scattering at detector site by simply rotating the PM with respect to the BLM centre, thus replacing the ellipse by its tangent in each point. The error made by this approach compared to the use of different elliptical pinholes for each q-value is 3% at $q = 3400$ cm^{-1}.

The PM signal was preamplified and the heterodyne autocorrelation function calculated in 388 channels using an ALV 3000 correlator (ALV GmbH Langen Germany). The time increments used were in the range 0.1 to 15 µs. Here, the diffuse scattering arising from the molecular roughness of the membrane was used as local oscillator for the heterodyne mode.

Data analysis. For the oscillating regime, the theoretically expected autocorrelation function[2]

$$G'_0(t) = a + b \cos(u'_0 t + \Phi) e^{-\tilde{A}_0 t} + ct \quad (18)$$

was fitted to the experimental data. Here ct is a linear baseline correction and Φ is a phase factor. Corrections of the above autocorrelation function at small q for instrumental effects as suggested in ref. 4 were not performed, since deviations from eqn. (5) were completely negligible for $q > 400$ cm^{-1}.

At the transition point (bifurcation point) between the oscillating and the overdamped regime and within a narrow q-range above it (250–400 cm^{-1}, depending on the membrane used) the data were fitted according to:

$$G_t(t) = a + (b + ct) e^{-\Gamma t} + dt \quad (19)$$

Hence, in this regime, both overdamped modes Γ_1 and Γ_2 were generally detectable. In the overdamped regime, where two damping modes are expected according to eqn. (1), a function of the form

$$G_d(t) = a + b e^{-\tilde{A}_1 t} + c e^{-\tilde{A}_2 t} + dt \quad (20)$$

was fitted to the data, again including a linear baseline correction dt. However, beyond $\Gamma_2 = \gamma_0/\gamma$ the fast overdamped mode becomes undetectable and the remaining slow mode Γ_1 can be fitted by a single exponential. Examples of representative data sets for q-values in the oscillating and the overdamped regime, together with the fitting results according to eqn. (18) and (2), are shown in Fig. 5.

Microinterferometric technique

The microinterferometric technique developed for the analysis of undulatory excitations of weakly adherent vesicles (*i.e.* weak lateral tension) has been described previously by Rädler *et al.*[13] The vesicles are observed by RICM, enabling local measurements of the distance between substrate

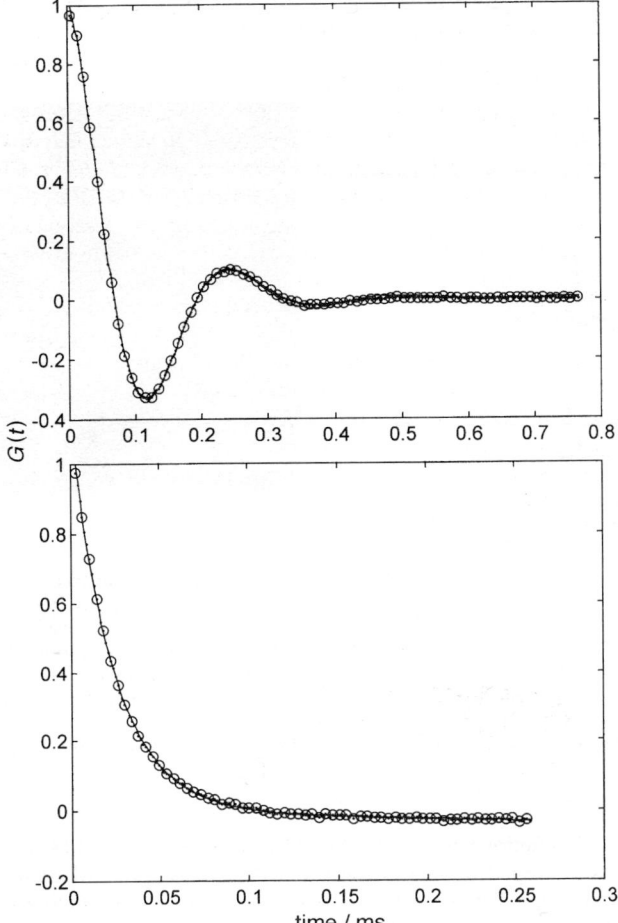

Fig. 5 Examples of typical correlation functions $G(t)$, measured in the damped and in the overdamped regime for a free planar bilayer (BLM). The full lines represent fits to the data according to eqn. (5) and (7). To distinguish the fit from the data more clearly, not all data points are represented by a circle. Note that the overdamped dataset is fitted only with a mono-exponential function.

and vesicle with high precision (± 1 nm). The method is illustrated in Fig. 6(a). The adhering liposome shows interference fringes at the contact rim, where the contour bends away from the surface with a finite contact angle. The flat centre part of the liposome appears dark and exhibits dynamical intensity fluctuations due to height fluctuations. The heights, $h(x, y)$, are obtained from the digitized interference intensities by inverse cosine transform using a Pixelpipeline frame grabber (Perceptics) and NIH image software. A one-dimensional contour line $h_x = \langle h(x, y) \rangle_y$ was evaluated by averaging over the width of a finite stripe as shown in Fig. 6(b). The height profile, h_x, along the stripe with length L was numerically Fourier transformed and the resulting one-dimensional modes correlated in time, $G_{qx}(t) = \langle \tilde{h}_{qx}(t)/\tilde{h}_{qx}(0) \rangle$. In this case the correlation function $G_{qx}(t)$ is given by

$$G_{qx}(t) = \int dq_y \langle h^2_{q_x, q_y} \rangle \exp\{-\Gamma(q_x, q_y)t\} \qquad (21)$$

Note that the correlation function $G_{qx}(t)$ exhibits a superposition of relaxation modes, q_y, which is a result of the data processing due to the integration over the finite width of the evaluated stripe.[9]

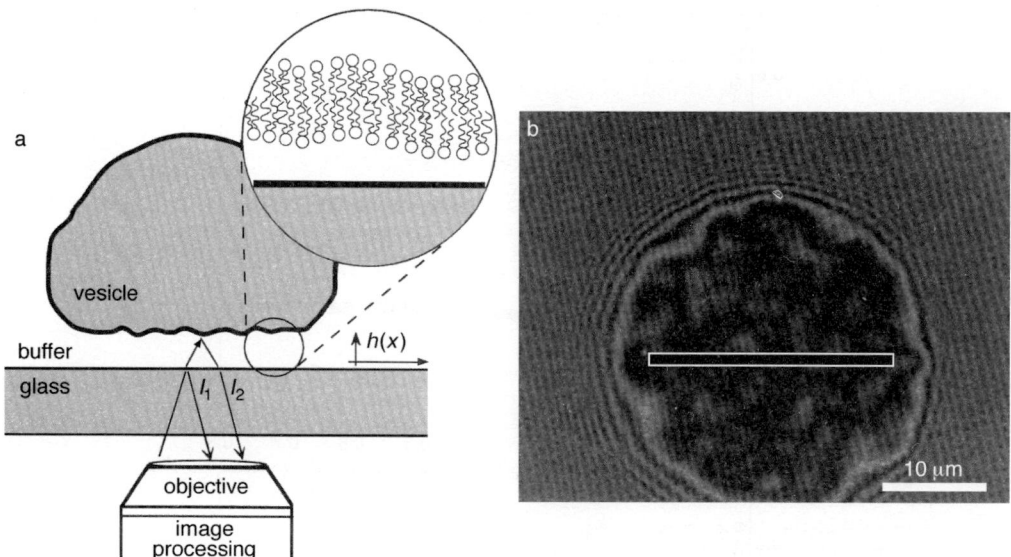

Fig. 6 (a) Analysis of undulations of a flaccid vesicle near a transparent substrate by reflection interference contrast microscopy. The image is generated by interference of light reflected from the substrate and the adhering body, respectively. For contrast enhancement the glass substrate is covered by an MgF_2 film. (b) Snapshot of weakly adherent vesicle. The leopard-like pattern is due to fluctuations of the distance $h(x, y)$ between substrate and vesicle. Bright areas correspond to large and dark areas to small values of $h(x, y)$. The white frame shows the stripe over which a one-dimensional time correlation analysis was carried out.

Results

BLMS measured by DLS

Typical autocorrelation functions measured by DLS at selected q values corresponding to the damped and to the overdamped regime of the transverse-splay mode of BLM (3.5 mm diameter) of DEPC are shown in Fig. 5. The data were fitted according to eqn. (6) and (8), giving the mode frequency $f_0(q)$ and the damping $\Gamma(q)$ as shown in Fig. 7A. For the overdamped data set shown in Fig. 5, a single exponential corresponds to the slow mode Γ_1. However, in the vicinity of the bifurcation point only two-exponential fits [eqn. (7)] gave satisfactory results. This clearly indicates that around the bifurcation point the fast mode Γ_2 is indeed detectable, while it decreases rapidly and merges with the non-detectable bulk mode [eqn. (2)] at higher q. This behaviour strongly suggests a non-negligible effect of the surface viscosity γ being not less than 10^{-7} mN s m^{-1}. The data in Fig 7A are compared with best fits according to hydrodynamic theory [eqn. (1), full lines in Fig. 7A] to obtain the average lateral tension γ_0 and the shear interfacial viscosity γ', acting in the normal direction of the membrane. It is obvious that the agreement between experiment and theory is excellent over almost 3 orders of magnitude of q. The predicted transition from the damped to the overdamped case (*cf.* Fig. 1) is clearly observed experimentally above $q = 3300$ cm^{-1}. Minor deviations of $f_0(q)$ from the theory at lowest measurable q in Fig. 5A can be ascribed to a slight average overall equilibrium deformation of the BLM, giving rise to some diffuse reflection of the incident laser light at the lowest values of q. Moreover, at $q < 400$ cm^{-1} there could be a non-negligible contribution arising from the (Gaussian) instrumental function of the set-up.

From the fits to the data in Fig. 7A, an average value of the lateral tension $\gamma_0 = 0.42$ mN m^{-1} with a maximum deviation of ± 0.03 mN m^{-1} is obtained, while the viscosity γ' was in the range 2×10^{-7} mN s. The rather weak influence of γ' seems justified, considering that the BLM is not expected to exhibit any significant frictional drag between its two constituent monolayers, owing to the presence of retained solvent (decane) between them. The decane retained in the BLM increases the internal volume, thereby effectively reducing the van der Waals interactions between adjacent lipid tails and between the two monolayers. The maximum thickness of this decane layer is *ca.* 2 nm.[11]

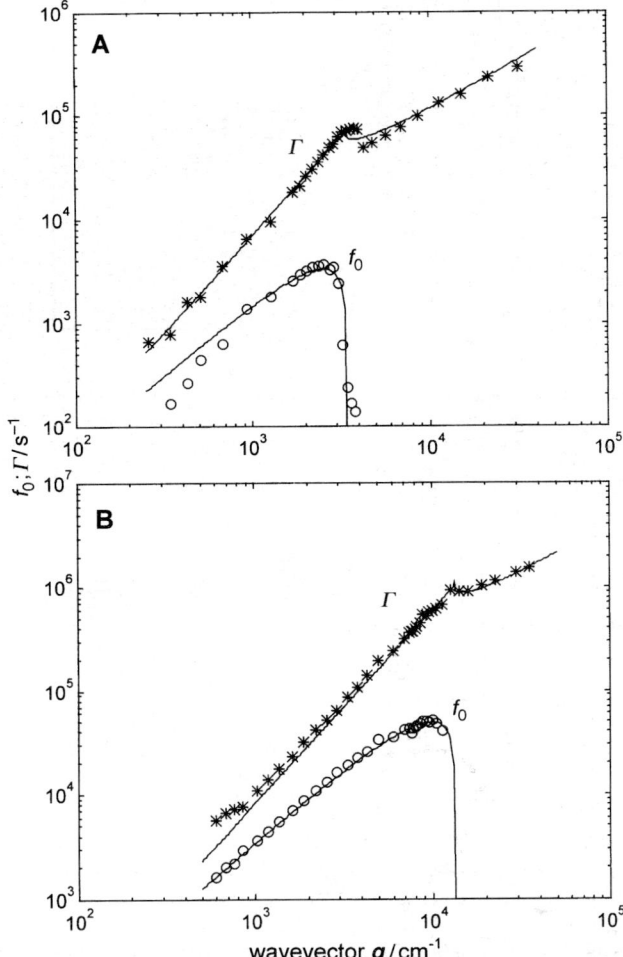

Fig. 7 Mode frequency $f_0 = \omega_0/2\pi$ and damping Γ vs. mode wave vector q of a free planar bilayer (BLM) of DEPC (A) and of DEPC with 30 mol% cholesterol (B), calculated from the autocorrelation functions measured at the corresponding q-values. The full lines represents the theoretical prediction with the average lateral tension γ_0 and the viscosity γ' (see text) as parameters.

In a second experiment we have studied the effect of cholesterol on the collective modes of a BLM as measured by DLS. This steroid is well known from a number of bilayer studies to cause a stiffening of the fluid bilayer and a drastic increase in its molecular order. Fig. 7B shows results for $f_0(q)$ and $\Gamma(q)$ for the case of a DEPC–cholesterol (30 mol%) BLM for a comparable q range as for the pure DEPC BLM from Fig. 7A. The effect of cholesterol manifests itself by a significant increase in the transition of $f_0(q)$ from the damped to the overdamped case and a corresponding change in the damping $\Gamma(q)$. Fits of the data to eqn. (1) now give an average lateral tension of $\gamma_0 = 1.55$ mN m^{-1} with a maximum deviation of ± 0.07 mN m^{-1} while γ' is within the same order of magnitude as for pure DEPC.

LUV measured by RICM

Fig. 8 shows the relaxation measurements of the bending oscillations for a vesicle weakly adhering to the substrate. In a previous analysis by Rädler et al.[13] the equilibrium contour and the static fluctuations of adhering vesicles were evaluated. Under the given conditions it was shown that the equilibrium distance, $h_0 \approx 30$ nm and the membrane tension, $\gamma \approx 10^{-6}$ J m^{-2}. Here, we evaluated

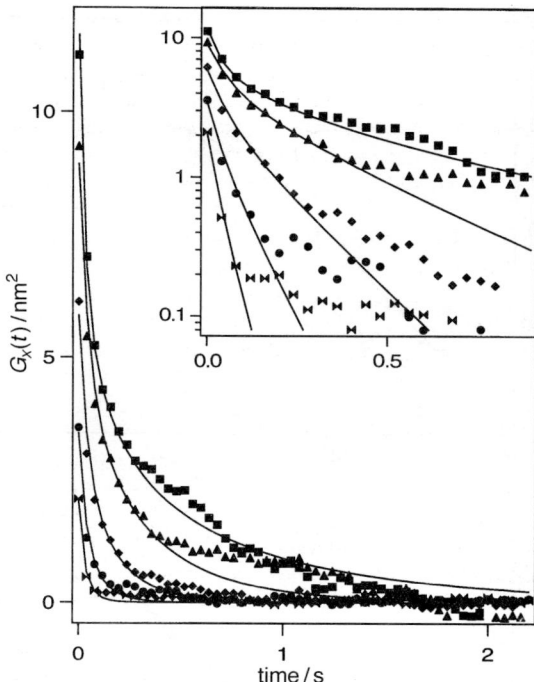

Fig. 8 Time correlation of the Fourier modes h_{qx} of the one-dimensional intensity profile h_x shown in Fig. 6(b). All relaxation curves are fitted using eqn. (14) and eqn. (21) with fixed values $h_0 = 30$ nm, $\gamma = 10^{-6}$ J m^{-2}, and $V'' \approx 5 \times 10^8$ J m^{-4}. The data are in agreement with hydrodynamic modes, Γ_W, which are slowed down by the presence of the wall. The semi-logarithmic presentation of the same data (inset) shows the predicted multi-exponential decay. The first mode (■) corresponds to a wave vector $q_x(1) = 0.67 \times 10^6$ m^{-1}.

the time correlation functions by numerically fitting eqn. (21) to a set of time correlation functions. We tested the different models for relaxation rate, Γ. We found the hydrodynamically damped modes, eqn. (14), fitted best. As shown in Fig. 8, the relaxation of all five modes can be simultaneously fitted with one set of constants $h_0 = 30$ nm, $\gamma = 10^{-6}$ J m^{-2}, $\eta = 10^{-3}$ J s m^{-3} and $V'' \approx 5 \times 10^8$ J m^{-4}. The parameter V'' was left variable to improve the individual fits but the resulting values did not scatter by more than 15%. The fact that the data show indeed a superposition of relaxation modes is demonstrated in the semi-logarithmic plot (inset in Fig. 8), which shows clearly deviations from a straight line at low q. Hence, the data are in good agreement with the slow mode Γ_W described by Kraus and Seifert.[9] The fast, frictional mode, Γ_F, on the other hand, was not observed, even though the theoretical cross-over should have allowed its detection. We have to assume that this mode is too fast to be detected by video microscopy. In fact, even the fourth and fifth mode of the hydrodynamic damping in Fig. 8 are at the signal-to-noise detection limit.

It must be mentioned that the data can be brought to reasonable agreement with the dispersion relation predicted by the permeation model. In this case the parameters $\gamma = 10^{-6}$ J m^{-2} and $\lambda \approx 10^{-6}$ m^2 s kg^{-1} obtained best simultaneous fits. However, the fits did not follow the multi-exponential decay as seen in the semi-logarithmic plot shown in Fig. 8.

Discussion

We have combined two physical techniques and two model systems to study undulations in two limiting cases: high tension (BLM) and weak tension (LUV). The data shown in Fig. 7 provide, for the first time, the viscoelastic dispersion behaviour of a transverse shear mode of a BLM over an exceedingly wide (mesoscopic) q-range. For comparison, previously reported measurements by this method were limited to $q_{max} \approx 1800$ cm^{-1} while, for the present work, we have $q_{max} \approx 35\,000$

cm^{-1}. This allows the experimental observation of the transition from the oscillatory or damped to the overdamped regime of the transverse shear mode and thus offers a critical test of the validity of the hydrodynamic theory suggested previously for such modes.

The general agreement of our DLS data (Fig. 7) with the Kramer theory [eqn. (1)] is quite excellent and thus provides strong support for its validity within the mesoscopic q range covered by our experiments. In particular, we can draw the following conclusions: (1) The dispersion behaviour of the BLMs studied is clearly dominated by the lateral tension, only at highest q values might there be some minor contributions arising from membrane bending rigidity. (2) Accordingly, the transverse shear interfacial viscosity γ' is nearly negligible over the q-range considered, most likely reflecting the fact that $q \gg 1/d$ in the mesoscopic q range (d being the bilayer thickness). (3) In contrast to the theoretically predicted existence of two overdamped modes, our DLS data show in the overdamped regime, with the exception of the vicinity of the bifurcation point, only one mode which is compatible with the slower of the two predicted ones.

Considering the changes in the DEPC-BLM dispersion behaviour upon the addition of cholesterol (cf. Fig. 7B) at an amount (30 mol%) that causes in a fluid bilayer the creation of the so-called liquid-ordered (l_o) phase, we can conclude that this system shows an approximately three-fold higher lateral tension γ_0 while the transverse shear interfacial viscosity γ' remains nearly negligible. The increase in γ_0 is probably caused by the homogeneous distribution of the hydrophobic cholesterol over the bilayer in the l_o-phase. The tails of DEPC molecules adjacent to the stiff steroid body are forced into a state of higher molecular order, thereby reducing their area per molecule projected in the normal direction and thus giving rise to an increase in γ_0. Note that DEPC is a synthetic lipid which can pack rather densely because it exhibits just one *trans*-double bond in each acyl chain. A surprising result is that, in spite of the excellent agreement between experiment and theory, only one of the two theoretically predicted overdamped modes Γ_1 and Γ_2 [eqn. (20)] seems to be detectable by DLS. While Γ_1 describes the slow recovery of the system driven by tension and viscosity, Γ_2 is supposedly driven by the inertia of the system arising, within the approximations of the theory, from the fluid surrounding the BLM. Hence, while Γ_1 is readily detectable in our experiment with the data following the predicted course up to the q_{max} limit, relaxation *via* the Γ_2-process is only seen in the region of the bifurcation point. A rationale for this behaviour is not given here, but there is room for some speculation. One possible reason could be that the occurrence of surface viscosity γ' drives the fast mode Γ_2 into a merger with the bulk mode which, in turn, is not detectable by DLS.

The interferometric measurements are limited to a smaller q-range, but are unique in the analysis of membranes interacting with solid supports. We provided quantitative evidence for hydrodynamic coupling between the undulating membrane and the wall. The hydrodynamic mode was found to be in better agreement than the permeation mode. However, discrimination between the modes is difficult, since in both cases the dominating terms scale as $\Gamma \sim q^2$. It will be necessary for future measurements to extend the q-range of the interferometric microscopy using ultra-fast cameras and two-dimensional image processing. In this case the fast, frictional mode might become visible. On the other hand the fast modes are even more likely to be biased by water permeation.

Conclusions

The dynamic undulations of lipid bilayers reveal a hierachy of relaxation mechanisms if measured over a wide range of wave vectors. These relaxation modes are sensitive to the bilayer tension, the elastic bending modulus as well as the local hydrodynamic environment in case of membranes close to a solid wall. The study of multi-component membranes, where the dominating relaxation parameters can be controlled by incorporation of steroids, transmembrane pores or short membrane binding peptides, will further elucidate the dynamics of membranes.

Acknowledgements

The authors R.H. and T.M.B. are indebted to Professor Lorenz Kramer (Universität Bayreuth) for many helpful discussions.

References

1. W. Helfrich and R-M. Servuss, *Il Nuovo Climento*, 1984, **3D**, 137.
2. W. Pfeiffer, S. König, J. F. Legrand, T. Bayerl and E. Sackmann, *Europhys. Lett.*, 1993, **23**, 457.
3. C. Dolainsky, A. Mops and T. Bayerl, *J. Chem. Phys.*, 1993, **98**, 1712.
4. J. F. Crilly and J. C. Earnshaw, *Biophys. J.*, 1983, **41**, 197.
5. W. Häckl, U. Seifert and E. Sackmann, *J. Phys. France*, 1997, **7**, 1141.
6. L. Kramer, *J. Chem. Phys.*, 1971, **55**, 2097.
7. C. Fan, *J. Colloid. Interface Sci.*, 1973, **44**, 369.
8. U. Seifert, *Adv. Phys.*, 1997, **46**, 13.
9. M. Kraus and U. Seifert, *J. Phys. France*, 1994, **4**, 1117.
10. J. Prost and R. Bruinsma, *Eur. Phys. J. B*, 1998, **1**, 465.
11. J. Dilger and R. Benz, *J. Membrane Biol.*, 1985, **85**, 181.
12. S. White, *Biophys. J.*, 1978, **23**, 337.
13. J. Rädler, T. Feder, H. Strey and E. Sackmann, *Phys. Rev. E*, 1995, **51**, 4526.

Paper 8/07883A

Trapping of short-lived intermediates in phospholipid phase transitions: The L_α^* phase

Peter Laggner, Heinz Amenitsch, Manfred Kriechbaum, Georg Pabst and Michael Rappolt

Institute of Biophysics and X-ray Structure Research, Austrian Academy of Sciences, Steyrergasse 17, A-8010 Graz, Austria

Received 12th August 1998

Time-resolved small-angle X-ray diffraction of liquid-crystalline phospholipid–water systems under temperature or pressure jump conditions has demonstrated the existence of an ordered, intermediate L_α phase, with a sub-second lifetime, designated as the L_α^*-phase. The lamellar repeat spacing is, universally, 0.3 nm smaller than that of the parent phase, irrespective of the lipid composition and of the jump conditions, provided that the jump leads to a net volume expansion of the phase. The presence of salts, most notably LiCl, leads to a prolongation of the lifetime. The results suggest a non-monotonic potential function for the interbilayer water thickness.

I. Introduction

Among the manifold molecular interactions which govern the intra- and intercellular communication in biomembranes, the supramolecular structure and dynamics of lipid constituents deserves particular attention as they provide the structural variability and controlled one- or two-dimensional fluidity essential for membrane function. The wide chemical diversity of membrane lipid constituents, *e.g.* phospho- or glycolipids with varying hydrocarbon chain lengths, saturation or branching, sterols, confers a powerful mechanism for the cell to control the physico-chemical properties, mechanical, electrostatic and chemical, of membranes.[1-4] A crucial point in the discussion of supramolecular lipid aggregates is their cooperativity,[5] *i.e.* the fact that properties and mechanisms are governed by the collective behaviour of cooperative units, rather than by individual molecules. In the extreme, this leads to the concept of generic, physical properties and interactions of bilayer membranes, where the individual molecular structure plays only a subordinate role.[6,7] One of the most interesting features of phospholipid–water systems is their polymorphism, *i.e.* the existence of various ordered, crystalline, gel, or liquid-crystalline phases.[8] In the transitions between these phases, the above-mentioned supramolecular nature becomes particularly apparent.[9] Most transition processes are highly cooperative, such that several hundreds of individual lipid molecules transit simultaneously and discontinuously from one phase to the other.

It has long been an open question how one ordered phase changes into another with different geometry, without disruption of lattice order.[10] This problem is schematically illustrated in Fig. 1 for the simplest case of a one-dimensional lattice, as given, in principle, by a multilamellar liposome structure. It seems that, for highly cooperative transitions between two ordered phases, with a minimisation of structural disruption and disorder, there can only be highly symmetric and localised transition mechanisms, as defined by the martensitic transition type, originally defined for metallurgical phases[11,12] [Fig. 1(b)].

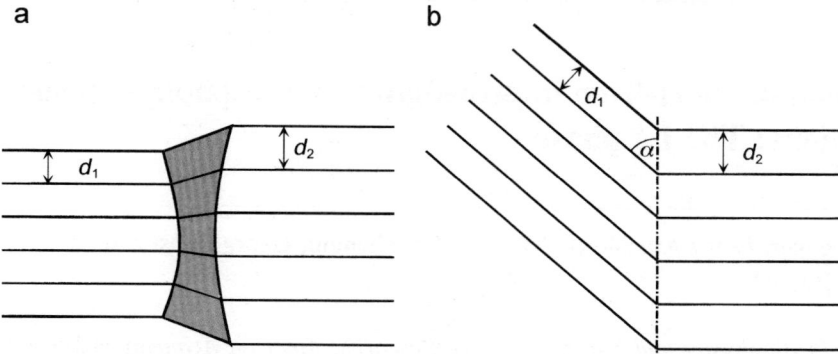

Fig. 1 Scheme of two alternative transition mechanisms in lamellar phases. (a) Transition mechanism with zones of disorder resulting in a loss of coherence. (b) Transition mechanism with minimal loss of order and coherence (martensitic transition).

In addition to the concept of cooperativity, the non-equilibrium nature of transition processes gains relevance in the discussion of supramolecular processes and transitions. Molecular dynamics, diffusivity, flexibility, are normally treated as equilibrium phenomena.[13,14] Close to equilibrium, at minor elevations from the thermodynamic equilibrium potential well, the kinetic and mechanistic behaviour can still be described classically by single-exponential energy and entropy terms. However, under strong jump conditions, at large elevations from the equilibrium potential trough, the system may respond non-linearly, no longer able to be described by single exponential terms.[15,16] The simplest consequence of this notion is that transition processes may follow different pathways, depending on whether they are guided in a slow, isothermal, or in a fast, adiabatic fashion.

Structural intermediates of different lifetime play a significant role in this concept. Such intermediates have been postulated and defined some time ago in studies of membrane fusion by several groups.[17-19] By time-resolved X-ray diffraction experiments on various different lipid phases such intermediates have been demonstrated.[20-22] Although the results have made it possible to propose hypothetical models, it has not been possible, so far, to provide a sound structural analysis of these intermediates to good resolution.

In the present work, this jump–relaxation approach has been focussed on intermediates within one thermodynamic equilibrium phase region, the L_α-phase of phosphatidylcholines. The experiments performed with advanced jump–relaxation techniques and synchrotron X-ray diffraction lead to the notion of a modulated potential *versus* bilayer-distance function, and a universal thickness increment for the interbilayer water space, which closely corresponds to the molecular dimensions of water.

II. Materials and methods

Sample preparation

1-Palmitoyl-2-oleoyl-*sn*-glycero-3-phosphocholine (POPC), 1,2-dipalmitoyl-*sn*-glycero-3-phosphocholine (DPPC), egg-yolk phosphatidylcholine (EYPC) and 1-palmitoyl-2-oleoyl-*sn*-glycero-3-phosphoethanolamine (POPE) were purchased from Avanti Polar Lipids, Birmingham Alabama, and used without further purification. Multilamellar liposomes were prepared by dispersing weighed amounts of dry lipids, typically 20–30 wt.%, in bidistilled water and in solutions of 0.1 and 0.3 M LiCl, respectively. To ensure complete hydration, the lipid dispersions were incubated for *ca.* 1–2 h at least 10 °C above the main transition temperature and, thereafter, vigorously vortexed under a N_2 atmosphere to prevent oxidation. Aqueous dispersions of these lipids displayed narrow, cooperative melting transitions within the limits of published values,[23] thus proving that the lipid purity corresponded to the claimed one of 99%.

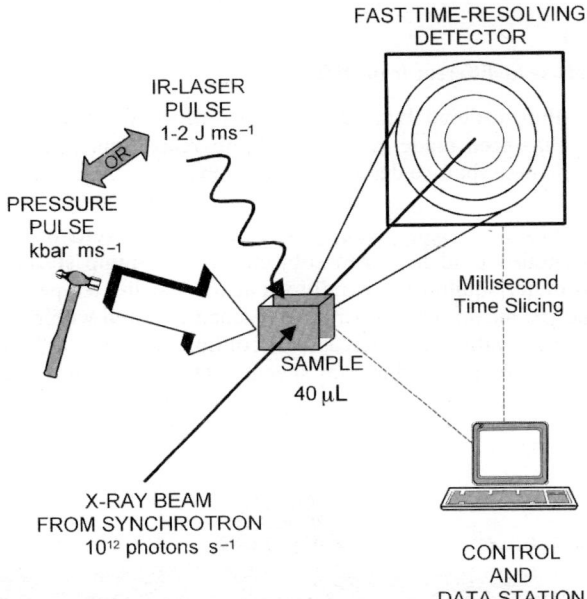

Fig. 2 Schematic view of the set-up for T- and P-jump relaxation experiments. Throughout fast time-resolved X-ray diffraction recordings aqueous suspensions can be studied far from the equilibrium situation. An IR laser with a pulse-characteristic of 1–2 J ms^{-1} provides T-jumps up to 20 °C, and with an in-house-built pressure cell P-jumps up to 3 kbar can be carried out.

Experimental protocol

Fast time-resolved X-ray diffraction experiments on lipid dispersions were carried out at the Austrian small-angle X-ray scattering station at ELETTRA, Trieste.[24-26] During rapid excitation of the lipid–water systems using temperature- (T-) and pressure (P-) -jump techniques the relaxation processes were monitored with a millisecond time resolution (Fig. 2).

For temperature-jump (T-jump) experiments, the lipid dispersions were sealed in a thin-walled 1 mm diameter Mark capillary held in a steel cuvette, which provides good thermal contact to the Peltier heating unit. T-jumps were generated with an erbium-glass laser[27] with an pulse length of 2 ms and a maximum emitted energy of 4 J. Barotropic phase transitions were investigated with a high-pressure X-ray cell[28] using jump amplitudes up to 3 kbar (0.3 GPa) within 10 ms. In particular, P-jump induced phase transitions of the phospholipid DOPE within the temperature region 5–70 °C and pressure region 1–3000 bar were performed.

Data analysis

The raw data of the time-resolved experiments were normalised for the integration time of each time-frame. Each small-angle X-ray diffraction pattern was analysed by fitting the first-order Bragg reflections using a least-squares method based on the Levenberg–Marquardt algorithm. Depending on the number of given phases, the model function was given by one or the sum of two Lorentzians, respectively. The lamellar repeat distances were determined from the corresponding peak positions. The relaxation kinetics of the d spacings are best described by a double-exponential model

$$d(t) = d_0 - A \exp\left(-\frac{t}{\tau_A}\right) - B \exp\left(-\frac{t}{\tau_B}\right) \quad (1)$$

For comparison, single and triple exponential models were also checked, but have been proven by statistical tests (Variance-analysis, F-Test) not to describe the relaxation kinetics as well as the two-component model.

III. Results

The intermediate L_α^*-phase in different transitions

The observation of a thin, ordered lamellar structure, which we denote as the L_α^*-phase, as a transient intermediate, has been made in different jump–relaxation experiments on various lipid classes. In the following, these shall be described separately.

The $L_\alpha \rightarrow L_\alpha^* \rightarrow L_\alpha$ transition (single phase). Fig. 3 shows the results of a typical experiment in which a phosphatidylcholine lipid has been subjected to a T-jump starting from the single L_α phase which it attains under equilibrium conditions at the starting temperature. The jump amplitude was *ca.* 15 °C, and the maximum temperature reached was well within the L_α-phase region. In the jump experiment, the d value first decreases discontinuously by *ca.* 0.3 nm, and then relaxes back within less than 15 s to the equilibrium spacing. The minimum d spacing with a value of

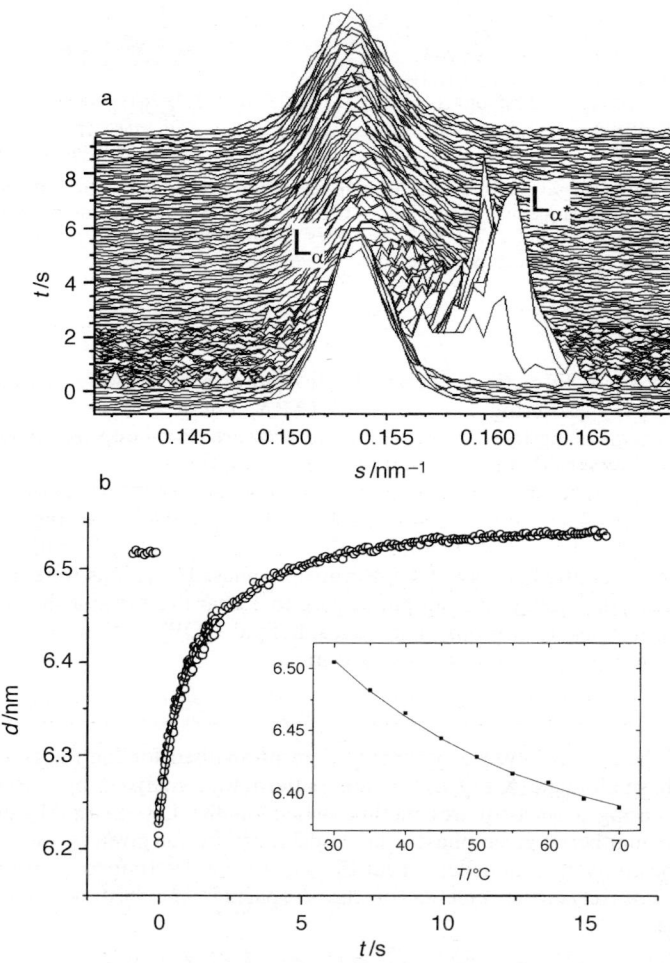

Fig. 3 L_α–L_α^*–L_α transition in POPC (20 wt.%) induced by a 15 °C(2 ms)$^{-1}$T-jump (initial temperature: 30 °C). (a) A series of time-sliced diffraction patterns shows the temporal development of the first-order Bragg peaks (maximum resolution 5 ms). Raw data are given. (b) The evolution of the d spacings in fully hydrated POPC is displayed. Each single diffraction pattern was fitted by a Lorentzian distribution (—). The line gives the best fit to the relaxation model of the L_α^* phase [eqn. (1)]. For comparison, the temperature dependence of the lamellar repeat distance d of POPC under equilibrium conditions is depicted in the insert.

ca. 6.2 nm is clearly thinner than the corresponding lattice parameter under near-equilibrium conditions, which even at a temperature of 70 °C does not drop below 6.37 nm [see insert Fig. 3(b)].

Similar $L_\alpha \to L_\alpha^* \to L_\alpha$ experiments have been performed with other phosphatidylcholine and ethanolamine lipids, and the resulting parameters are listed in Table 1. The result that, with all lipids studied, the non-equilibrium decrease in d value is ca. 0.3 nm, irrespective of the hydrocarbon-chain or head-group composition, emerges as a salient result.

The $L_\alpha \to L_\alpha^* \to L_\alpha$-transition in two coexisting phases. Alkali-metal ions, especially Li$^+$, can induce an equilibrium phase separation of two coexisting L_α-phases.[29] These phases, of which the structural nature and the origin of their coexistence is not yet quite clear, differ in d spacings by ca. 0.6 nm. It was of particular interest, therefore, to examine their behaviour under non-equilibrium jump conditions. Fig. 4 shows that both phases, as demonstrated by the time-course of the lamellar repeat spacings, respond in a parallel fashion to the T-jump, and relax with similar kinetics. As with the single-phase transitions presented above, the discontinuous changes in d spacing amount to ca. 0.3 nm, and the relaxation times are of the order of 8–15 s, i.e. still faster than the thermal equilibration within the sample cell. However, the thermal equilibration does influence the relaxation times, thus the relative behaviour is more relevant than the absolute values in the discussion.

In view of the dramatic effects of Li on the L_α-phase structure, it is necessary to evaluate also the effects of other alkali-metal ions. This has not yet been done comprehensively, due to the excessive requirements on synchrotron beam-time. Some interesting features of salt effects are already visible from Table 2, which summarises the results from all jump–relaxation experiments so far performed in the presence of salts. The general feature of the parameters obtained is the

Table 1 Summary of laser T-jump experiments in the liquid-crystalline phase of various phospholipids

Lipid	T_i/°C	T_f/°C	d_0/nm	Δd/nm	α/degrees
POPC	30	45	6.53	0.31	18
DPPC	70	76.5	6.43	0.32	18
EYPC	25	31.5	6.55	0.26	16
POPE	30	36.5	5.37	0.37	21

All samples were equilibrated in the liquid-crystalline phase at the initial temperature T_i and jumps of 15 and 6.5 °C, respectively, were performed. The d spacing of the L_α phase d_0 reduces directly after the laser-pulse to the value $d_0 - \Delta d$. The theoretical declination angle α between the parent (L_α) and nascent (L_α^*) phase, respectively is described through eqn. (2).

Table 2 The fitting parameters of the relaxation curves of POPC according to eqn. (1)

		0.1 M LiCl		0.3 M LiCl		0.1 M KCl	0.1 M MgCl$_2$
	no salt	phase 1	phase 2	phase 1	phase 2	phase 1	phase 1
A/nm	0.13	0.09	0.10	0.06	0.06	0.06	0.09
τ_A/s	0.5	0.8	0.6	0.3	0.3	1.0	0.5
B/nm	0.18	0.29	0.27	0.27	0.23	0.25	0.23
τ_B/s	3.1	8.3	8.1	15.8	14.1	6.9	6.5
d_0/nm	6.53	6.73	6.07	6.37	5.95	6.37	5.89
Δd/nm	0.31	0.38	0.37	0.33	0.29	0.31	0.32

Increasing LiCl concentration results in longer lifetimes of the intermediate phase L_α^* (see $\tau_A + \tau_B$). The kinetics of the salt-induced phases 1 and 2, respectively, show similar relaxation behaviour. Independent from the salt concentration the d spacings of the liquid crystalline phases always decrease by ca. 0.3 nm directly after the laser pulse ($\Delta d = A + B$). The errors in the parameters are of the order of the last digit given.

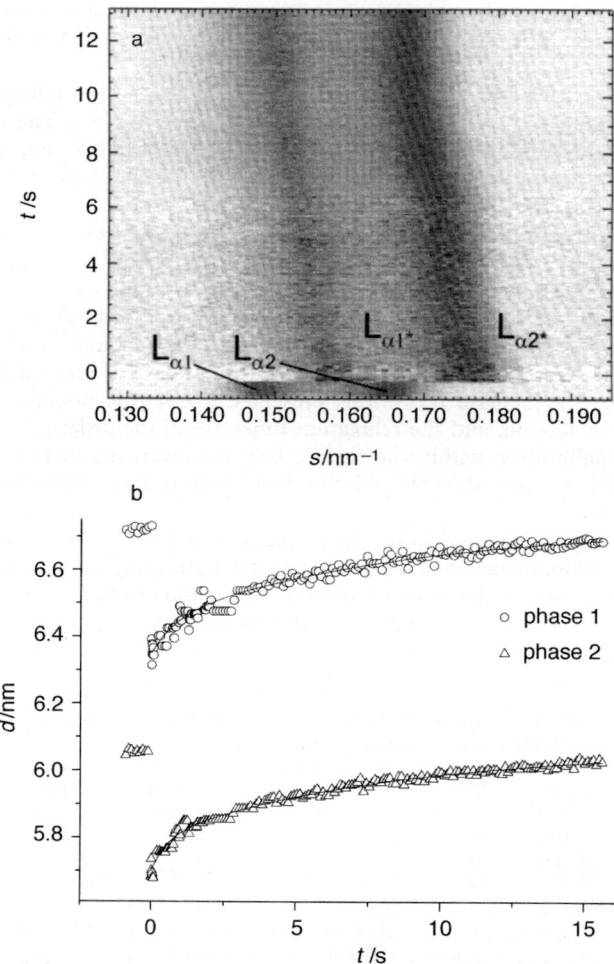

Fig. 4 L_α–L_α^*–L_α transition in POPC (20 wt.% in 0.1 M LiCl) induced by a 15 °C(2 ms)$^{-1}$ T-jump (initial temperature: 30 °C). (a) A series of time-resolved diffraction patterns given in the form of a contour plot demonstrates the kinetics of the salt-induced $L_{\alpha 1}$ and $L_{\alpha 2}$ phases (maximum resolution 5 ms). (b) The d spacings for $L_{\alpha 1}$ and $L_{\alpha 2}$ determined from the fitted Bragg-peak positions (sum of two Lorenztians). Both phases exhibit similar temporal and structural behaviour (see also Table 2).

prolongation of the lifetimes of the intermediates by alkali-metal salts. The prolongation factors reach a value of 5 for the addition of 0.3 M LiCl.

The $L_\alpha \rightarrow L_\alpha^* \rightarrow H_{II}$ transition of phosphatidyl-ethanolamines: pressure jumps. The existence of an intermediate, thin L_α^*-phase has been previously observed by T-jump experiments.[20] This intermediate structure provides the contact conditions of opposing bilayers necessary for fusion and the formation of tubular structures. By P-jump experiments this result has been fully verified in all details [Fig. 5(a)]. The initial, fast step of the transition is the formation of a thin L_α^*-phase which disappears in parallel with the formation of the H_{II}-phase, taking comparatively long times of several seconds to develop fully.

In contrast to the T-jump technique the P-jump technique has the advantage of allowing jumps in either direction. This makes it possible to investigate the reversibility of the transitions and their pathways. Fig. 5(b) shows the pressure-drop experiment from the H_{II} phase first into the L_α-phase and then onwards into the L_β phase. Two observations are particularly noteworthy. First, the transition from the hexagonal phase into the lamellar phase proceeds without intermediates, i.e.

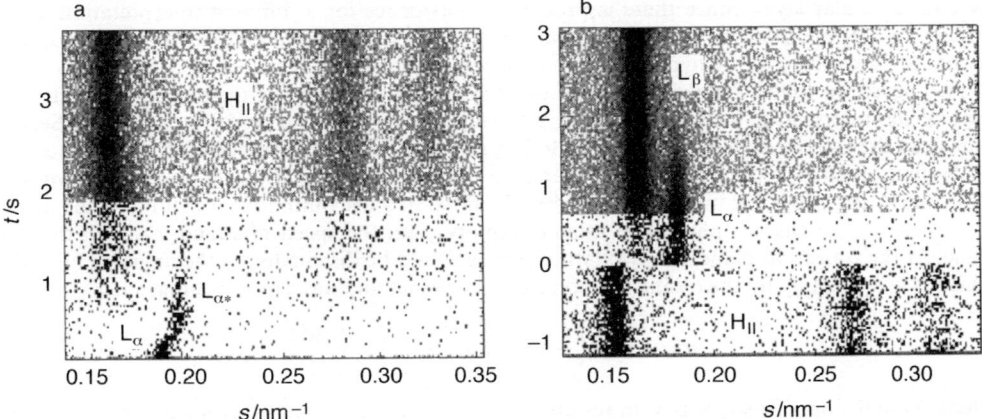

Fig. 5 Time-resolved X-ray diffractograms of fully hydrated DOPE exposed to pressure-jumps recorded with a time resolution of 5, 50 and 500 ms for each frame, respectively. (a) Starting from the lamellar L_α phase at high pressure ($T = 41\,°C$ and $p = 2300$–155 bar) the lattice transforms immediately after a depressurizing jump into the intermediate lamellar L_α^* phase which coexists then with the emerging hexagonal H_{II} phase. (b) Starting in the hexagonal H_{II} phase at low pressure the lattice transforms after a P-jump ($T = 20\,°C$, and $p = 1$–2940 bar) immediately into the lamellar L_α phase. After 500 ms the final L_β phase begins to appear, growing at the expense of the coexisting L_α phase until the phase transformation into L_β is completed. The lattice spacing of all phases remains constant over the entire time interval.

there appears no L_α^*-phase in this direction. Second, the L_α-and L_β-phases coexist over a long period, unlike in the other direction, where the two phases coexist only for the period of the temperature jump *i.e.* 1–2 ms, as described previously.[30] No thin, intermediate L_α^*-phase can be detected in either direction of the $L_\alpha \leftrightarrow L_\beta$ transition.

IV. Discussion

The programme under which the present investigation was performed is primarily aimed at the exploration of methods for prolonging the lifetimes of structural intermediates in phospholipid phase transitions. A benefit from such an achievement could be the better structural description of the intermediates, because longer lifetimes would lead to better precision of diffraction data. Also, there are likely to be biomedical benefits from such results, since the development of agents modulating the dynamics of membrane transformations, such as fusion, is likely to play an important role in many medical applications, *e.g.* liposome-based gene therapy, fertility modulation, or percutaneous drug applications, to name but a few.[31] Strong interest in such intermediates comes also from the field of nanomaterial research,[32] where such structures could serve as templates for new materials, which cannot be obtained under equilibrium conditions.

The main discovery made in this search for trapped intermediates is the demonstration of a discrete, ordered transition state, the L_α^*-phase, which occurs rapidly and cooperatively upon a T- or P-jump from the normal, equilibrium L_α-phase. This phase is always characterised by a *ca.* 0.3 nm lower d spacing than the parent phase, independent of the hydrocarbon chain composition of the phosphatidylcholine species; it also occurs transiently during the P- or T-jump from the L_α into the H_{II} phase. These two facts, the constant Δd, and the composition independence, particularly that of hydrocarbon chain composition, are taken to indicate that the $L_\alpha \to L_\alpha^*$ transition involves primarily a change in interbilayer water thickness. Indirect support for this idea comes from the fact that the same seems to be the case in the thin L_α phases found in the presence of LiCl (also separated by *ca.* 0.3 nm[29]), where the Fourier analysis of the coexisting diffraction patterns of normal and thinner phases results in an invariant bilayer thickness (paper in preparation). It remains uncertain whether the transient L_α^*-phase in the jump experiments is indeed the same as one of the equilibrium structures in the presence of LiCl. However, it is tempting to assume that the two represent the same discrete secondary minimum in the hydration separation of bilayers and that the constant Δd of *ca.* 0.3 nm relates to a change in water thickness by one molecular layer. Since there is presently no evidence for a different interpretation, *e.g.* in terms of a discrete thinning of the hydrocarbon chain thickness, or a change in head-group conformation, we adhere to the hypothesis of a discontinuous hydration change.

Two questions follow immediately from this hypothesis: first, how is the quasi-immediate thinning of the interbilayer water space achieved, while the lattice order is fully preserved? Second, how does the intermediate L_α^* lattice return into the equilibrium L_α structure? A tentative answer to the first question is presented by the martensitic lattice-disclination mechanism, as shown schematically in Fig. 1(b). The transition would be localised in a discontinuous transition plane linking the parent to the nascent phase and moving rapidly, with the speed of sound, through the liposome. At the transition plane, the two lattice planes would be disclinated at the angle α [Fig. 1(b)], which is given simply by the cosine relation between parent and nascent d spacings:

$$\alpha = \arccos\left(\frac{d - \Delta d}{d}\right) = \arccos\left(1 - \frac{\Delta d}{d}\right) \qquad (2)$$

where $\Delta d = 0.3$ nm. This mechanism results in a minimum disruption of lattice order and involves as a diffusion component only the rapid movement of the transition plane, thus providing for maximum transition speed. It should be emphasised that this behaviour is indeed demonstrated by Fig. 3(a), where the Bragg peaks immediately after the jump are very sharp, and no intermediate disordering can be observed.

What happens to the water, where does it disappear to while the bilayer separation decreases? The relative decrease in water layer thickness amounts to *ca.* 15%, assuming a value of 2 nm for the water layer in the fully hydrated L_α structure.[33] A transient increase in water density for the lifetime of the intermediate (of the order of 0.1 s) seems unlikely, considering the typical ps relaxation times of water. An efflux from the liposome structure into the excess water phase through transient defects in the lamellar lattice might be more plausible, but it is again the observation of the very sharp Bragg reflections, right after the jump, which indicates that such defects are not increased and suggests that this is not a likely mechanism. Another possibility would be an increase in bilayer surface area by *ca.* 15%, and a concomitant reduction in water layer thickness, thus conserving the interbilayer water volume. As a consequence the molar phospholipid volume would have to increase by the same proportion to conserve the bilayer thickness, implying an increase in lateral headgroup separation of 7%. Perhaps the simplest way to dispose of the water is through the formation of localized "lentils" or cavities, which would not gravely perturb the multibilayer order (Fig. 6). This would avoid the necessity of increasing the molecular surface area to conserve the bilayer thickness.

While the formation mechanism of the intermediate L_α^* lattice appears to be best described by the discontinuous martensitic mechanism sketched in Fig. 1(b), the return to the equilibrium L_α structure follows a different pathway. As Fig. 3(b) and the decay parameters in Table 2 show, this follows slow (on the experimental timescale) bi-exponential kinetics and, most significantly, passes through a relatively disordered lattice situation, as indicated by the broadening and decrease in intensity of the Bragg peaks, only to increase again with complete re-equilibration. This can be interpreted qualitatively in terms of a model as shown in Fig. 1(a), where zones of disorder link the parent thin lattice with the nascent thicker one. The process could be analysed in terms of a nucleation-and-growth mechanism, but a quantitative evaluation would require more detailed information on the morphology of the liposome particles during the process and is beyond the scope of the present work.

The results of the T-jump experiments in the presence of LiCl (Fig. 4) indicate that the L_α^*-lattice is not a limiting one to smaller thicknesses. There, the initial equilibrium structure is

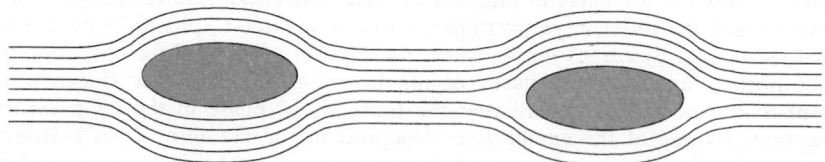

Fig. 6 Schematic view of "lentils" of bulk water in multilamellar liposomes.

already separated into three coexisting phases with discrete d spacings differing by $ca.$ 0.3 nm.[29] The thickest and thinnest ones, respectively, differing in d spacing by $ca.$ 0.6 nm, are the dominant ones. The T-jump experiment shows an essentially identical behaviour for these two coexisting phases. Again, immediately after the jump, a discontinuous shift of both Bragg peaks to smaller (by 0.3 nm) d spacings is followed by a slow, continuous return to the equilibrium situation. We assume that both parent phase structures have the normal molar phospholipid surface areas of approximately 0.63 nm^2.[34] The jump-induced thinning of the water-layer thickness could be again compensated by a transient increase in surface area, or by the formation of "lentils". The parallel return kinetics suggest the same mechanism and thermodynamic driving force for the two phases.

The universality of the 0.3 nm increment in the jump response of hydrated phospholipids is further confirmed by the behaviour of ethanolamine phospholipids in the $L_\alpha \rightarrow H_{II}$ transition in a pressure-release-jump experiment [Fig. 5(a)]. There again, the intermediate L_α^* phase shows a d spacing which is by the same amount thinner as was observed in all other cases so far investigated. An analogous observation has been made previously by a T-jump experiment. Hence, Δd seems to be largely independent of the chemical structure of the phospholipid and also of the physical nature of the jump, as long as it leads to an overall volume expansion. In the pressurising jump through the $H_{II} \rightarrow L_\alpha$ transition, on the other hand, no intermediate was observed [Fig. 5(b)].

Having thus established the model for the L_α^*-phase, it remains to be examined what are the reasons for its formation. Our previous results of an LiCl-induced L_α-phase separation in liquid-crystalline phosphatidylcholines[29] have suggested that there exists a modulation in the interbilayer separation potential, with secondary minima at distances of $ca.$ one molecular water diameter. The present results seem to support this notion, although they are derived from a strongly non-equilibrium approach. This is in contrast to the hitherto generally held belief of a smooth, continuous potential function with only one potential well at the equilibrium separation.[35] Indeed, none of the studies on the hydration dependence of the lamellar repeat spacing has shown any sign of secondary minima.[36-38] The reason might simply lie in the dynamic undulations of the individual bilayers, such that, at thermal equilibrium, all states, differing in energies by a few kT only, are simultaneously occupied, and the Bragg peaks represent the average bilayer separation. The simultaneous energy supplied by the T-jump would lead to a uniform occupation of the secondary potential minimum and, hence, to the observed, sharp Bragg peak of the L_α^* phase.

The interbilayer separation in the L_α phase is $ca.$ 2 nm, corresponding hypothetically to 7 water layers, if one layer is $ca.$ 0.3 nm thick. Despite the highly dynamic state of the water between the bilayers, it is obvious that the restrictions to mobility are much more stringent in the direction normal to the bilayer plane than in the parallel direction. These restrictions, together with a modulated interaction potential between opposing bilayers, would become more pronounced, the closer the bilayers approach, simply from a steric consideration of the discrete water molecule dimensions, despite the flexibility and vertical "bobbing" motions of individual phospholipids.

The fact that the interbilayer water spacing decreases with temperature, even isothermally and not just in the adiabatic jump experiments, shows that the water volume expands thermally in the two dimensions of the bilayer plane. This creates an increased interface area, which is counteracted by the manifold intermolecular phospholipid interactions, the water bridges in the backbone region, and by the hydrophobic effect which, however, decreases with temperature. We want to speculate, therefore, that the 0.3 nm increment is indeed the average thickness of one water layer, in the sense that the interbilayer water space, as it decreases under non-equilibrium conditions, tends to prefer integral multiples of this thickness.

Concerning the underlying intention to explore ways to prolong and eventually trap ordered, intermediate structures, the present experiments with salts hold some promise, although the obtained changes in relaxation times are not as dramatic as one would wish. The most important observation is that the salts used do not change the 0.3 nm increment by which the d spacings vary under jump conditions. The mechanism by which they prolong the lifetimes of the intermediates are still unclear and therefore no predictions in the direction of more potent trapping agents can be made. However, since we assume, as has been elaborated above, that the cause for the discrete intermediate formation is the water structure, it can be speculated that cosmotropic rather than chaotropic agents might be suitable candidates. If one considers the results obtained with

LiCl under equilibrium conditions,[29] with Li$^+$ being at the cosmotropic end of the lyotropic series of monovalent cations, this speculation gains in substance.

Acknowledgements

This work has been supported by the "Elettra-Project" of the Austrian Academy of Sciences. M. Rappolt is the recipient of a long-term grant from the European Commission under the programme "Training and Mobility of Researchers" [Contract no. SMT4-CT97-9024(DG12-CZJU)].

References

1. M. Bloom and O. G. Mouritsen, in *Structure and Dynamics of Membranes*, ed. R. Lipowsky and E. Sackmann, Elsevier, Amsterdam, 1995, p. 65.
2. M. Bloom, E. Evans and O. G. Mouritsen, *Q. Rev. Biophys.*, 1991, **24**, 293.
3. E. Sackmann, in ref. 1, p. 1.
4. P. Kinnunen, *Chem. Phys. Lipids*, 1991, **57**, 375.
5. R. Biltonen, *J. Chem. Thermodyn.*, 1990, **22**, 1.
6. R. Lipowski, in *Festkörperprobleme*, Advances in Solid State Physics, ed. U. Roessler, Vieweg, Braunschweig/Wiesbaden, 1992, vol. 32, p. 19.
7. W. Helfrich, in ref. 1, p. 691.
8. J. M. Seddon and G. Cevc, in *Phospholipids Handbook*, ed. G. Cevc, Marcel Dekker, New York, 1993, p. 403.
9. D. Marsh, *Chem. Phys. Lipids*, 1991, **57**, 109.
10. P. Laggner and M. Kriechbaum, *Chem. Phys. Lipids*, 1991, **57**, 121.
11. J. W. Christian, in *Physical Metallurgy*, ed. R. W. Cahn, North-Holland, Amsterdam, 1970, p. 471.
12. Z. Nishiyama, *Martensitic Transformations*, Material Science Series, ed. M. Fine, M. Meshii and C. Wayman, Academic Press, New York, 1978.
13. A. Blume, in ref. 8, p. 455.
14. P. F. F. Almeida and W. L. C. Vaz, in ref. 1, p. 305.
15. W. Gebhardt and U. Krey, *Phasenübergänge und Kritische Phänomene*, F. Vieweg & Sohn, Braunschweig, 1980.
16. D. Lou, J. Casas-Vasquez and G. Lebon, *Extended Irreversible Thermodynamics*, Springer, Berlin, 1993.
17. P. R. Cullis, M. J. Hope, B. deCruijff, A. Verkleij and C. P. S. Tilcock, in *Phospholipids and Cellular Recognition*, ed. J. F. Kuo, CRC Press, Boca Raton, 1985, p. 1.
18. D. P. Siegel, *Biophys. J.*, 1986, **49**, 1155.
19. D. P. Siegel, *Biophys. J.*, 1986, **49**, 1171.
20. P. Laggner, M. Kriechbaum and G.Rapp , *J. Appl. Crystallogr.*, 1991, **24**, 836.
21. P. Laggner, *J. Phys. IV*, 1993, **3**, 259.
22. M. Rappolt, PhD Thesis, Hamburg, 1995.
23. M. Caffrey, *Lipid Thermotropic Phase Transition Database (LIPIDAT2)*, NIST, 1994.
24. H. Amenitsch, S. Bernstorff, M. Kriechbaum, D. Lombardo, H. Mio, M. Rappolt and P. Laggner, *J. Appl. Crystallogr.*, 1997, **30**, 872.
25. H. Amenitsch, S. Bernstorff and P. Laggner, *Rev. Sci. Instrum.*, 1995, **66**, 1624.
26. S. Bernstorff, H. Amenitsch and P. Laggner, *J. Synchrotron Rad.*, 1998, **5**, 1215.
27. G. Rapp, M. Rappolt and P. Laggner, *Prog. Colloid Polym. Sci.*, 1993, **93**, 25.
28. K. Pressl, M. Kriechbaum , M. Steinhart and P. Laggner, *Rev. Sci. Instrum.*, 1997, **68**, 4588.
29. M. Rappolt, K. Pressl, G. Pabst and P. Laggner, *Biochim. Biophys. Acta*, 1998, **1372**, 389.
30. M. Kriechbaum, G. Rapp, J. Hendrix and P. Laggner, *Rev. Sci. Instrum.*, 1989, **60**, 2541.
31. D. D. Lasic, *Liposomes: From Physics to Applications*, Elsevier, Amsterdam, 1993.
32. S. Mann, *Nature (London)*, 1988, **332**, 119.
33. J. F. Nagle, R. Zhang, S. Tristram-Nagle, W. Sun, H. I. Petrache and R. M. Suter, *Biophys. J.*, 1996, **70**, 1419.
34. S. Tristram-Nagle, H. I. Petrache, and J. F. Nagle, *Biophys. J.* 1998, **75**, 917.
35. J. Israelachvili, in *Intermolecular and Surface Forces*, Academic Press, London, 1992, p. 176.
36. R. P. Rand and V. A. Parsegian, *Biochim. Biophys. Acta*, 1989, **988**, 351.
37. T. J. McIntosh and S. A. Simon, *Biochemistry*, 1993, **32**, 8374.
38. B. Koenig, H. H. Strey and K. Gawrisch, *Biophys. J.*, 1997, **73**, 1954.

Paper 8/06384B

Sensing isothermal changes in the lateral pressure in model membranes using di-pyrenyl phosphatidylcholine

Richard H. Templer,* Saffron J. Castle, A. Rachael Curran, Garry Rumbles and David R. Klug

The Department of Chemistry, Imperial College, London, UK SW7 2AY

Received 17th August 1998

In this work we present data from a homologous series of di-pyrenyl phosphatidylcholine (dipyPC) probes which can sense lateral pressure variations in the chain region of the amphiphilic membrane (lateral pressures are tangential to the interface). The dipyPC has pyrene moieties attached to the ends of equal length acyl chains on a phosphatidylcholine molecule. Ultraviolet stimulation produces both monomer and excimer fluorescence from pyrene. At low dilutions of dipyPC in model membranes the excimer signal is entirely intra-molecular and since it depends on the frequency with which the pyrene moieties are brought into close proximity, the relative intensity of the excimer to monomer signal, η, is a measure of the pressure. We synthesised or purchased dipyPC probes with the pyrene moieties attached to acyl chains having 4, 6, 8 and 10 carbon atoms and then measured η in fully hydrated bilayers composed of dioleoylphosphatidylcholine and dioleoylphosphatidylethanolamine (DOPC and DOPE respectively). Although the resolution of our measurements of lateral pressure as a function of distance into the monolayer was limited, we did observe a dip in the excimer signal in the region of the DOPC/DOPE *cis* double bond. As we isothermally increased the DOPE composition, and hence the desire for interfacial curvature, we observed, as expected, that the net excimer signal increased. However this net increase was apparently brought about by a transfer of pressure from the region around the glycerol backbone to the region near the chain ends, with the lateral pressure dropping above the *cis* double bond but increasing at a greater rate beyond the double bond.

Introduction

In an amphiphilic bilayer the back to back monolayers are restrained to lie flat even though each will in general possess an intrinsic desire for interfacial curvature. The curvature elasticity[1] of the monolayer may be expressed in terms of the bending energy per unit area, g_c,

$$g_c = \tfrac{1}{2}\kappa(c_1 + c_2 - 2c_0)^2 + \kappa_G c_1 c_2 \tag{1}$$

where c_1 and c_2 are the principal curvatures of the region of interest, c_0, is called the spontaneous curvature of the interface, κ is the mean curvature modulus and κ_G is the Gaussian curvature modulus. For the case where the monolayer is held flat the stored curvature elastic energy is simply $2\kappa c_0^2$.

This is precisely the case for cell membranes, so we would anticipate that the amphiphilic fabric of the cell membrane would be storing a curvature elastic energy in proportion to κ and c_0, and it has been suggested that this stored energy may play a role in modulating the behaviour of membrane proteins.[2,3] Evidence that this is the case has been building steadily,[4–8] but the correlation between protein activity and the physical state of the membrane has, for the most part, remained qualitative. In part this has been because our knowledge of κ and c_0 for most amphiphilic lipid systems remains sketchy. However, even if we had a better and more reliable catalogue of values of κ and c_0 we would have to accept that these two parameters are a rather crude attempt to characterise the physical state of the amphiphilic membrane.

The connection between the stored curvature elastic energy and a more complete description of the physical state of the monolayer is, in principle, relatively straightforward. The desire for monolayer curvature arises because of the differential distribution of lateral pressure, Fig. 1. At the polar/apolar interface there is a strong inward pressure as the system attempts to limit the contact between water and hydrocarbon. This is resisted in the chain region where there is a strong outward pressure due to thermally driven collisions between chains. In the headgroup region there may also be positive contributions to the lateral pressure from steric, hydrational and charge effects, but also negative contributions may occur if there is direct hydrogen bonding between headgroups.

Although the integral of the lateral pressures for the monolayer must necessarily come to zero the first moment of the lateral pressure will in general be non-zero. It is this torque tension which represents the monolayer's frustrated desire for interfacial curvature and it is related to κ and c_0 by

$$\int_0^l \pi(z) z \, \mathrm{d}z = 2\kappa c_0 \qquad (2)$$

where l is the monolayer's thickness[9] and we use the convention that c_0 is negative when the monolayer wishes to bend towards the water. Clearly $\pi(z)$ is the most complete description of the average lateral forces exerted by the lipid membrane on proteins inserted into the bilayer. Unfortunately, measurements of κ and c_0 cannot uniquely define $\pi(z)$.

In this work we have therefore attempted to take the alternative approach of trying to measure $\pi(z)$ directly. This is as far as we are aware the first attempt to do this that has been reported in the literature. We have restricted ourselves to probing the pressure in the chain region. We have done this for a number of reasons. Statistical mechanical calculations of the lateral pressure profiles in the chain regions are possible because of the relatively simple nature of the interactions between fluid chains.[10–12] This is not true for the headgroup and interfacial regions. Fluorescent probe molecules tend to be relatively bulky and this militates against their use either in the headgroup region or the interface where it is already known that very small changes can have profound effects on phase behaviour.[3,13] The evidence from the effect of small amounts of hydrophobic additives in bilayers indicates that the effect on the phase behaviour of having the probe in the fluid chain region would be significantly less.[3] Finally, since the chains form an extended region we are most likely to be able to extract some of the details of the shape of the pressure profile here.

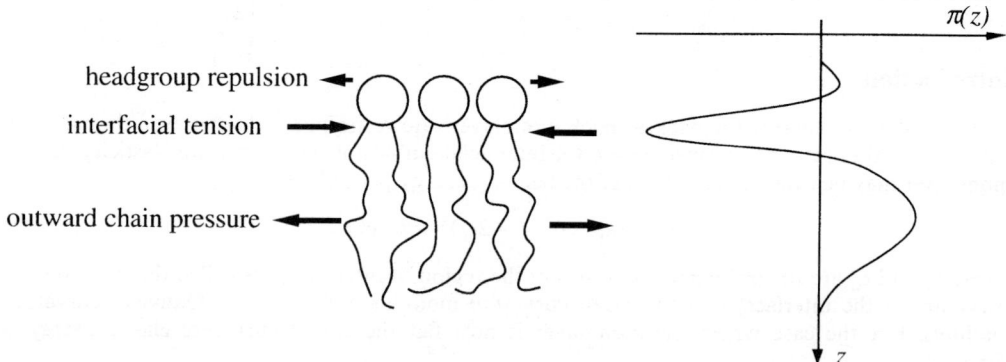

Fig. 1 Distribution of lateral pressure within a flat monolayer. The isotropic lateral pressure, π, can be plotted as a function of depth into the monolayer, z.

Being an extended region of positive lateral pressure, we would expect the magnitude of the pressure and hence the magnitude of variations to be less than in the headgroup and interfacial regions. We may estimate what the magnitude of $\pi(z)$ might be for the relatively well characterised lipid system we have used in this work; dioleoylphosphatidylcholine (DOPC) and dioleoylphosphatidylethanolamine (DOPE). For DOPC and DOPE at 25 °C we have $c_0 = -1/90$ Å^{-1} and $-1/56$ Å^{-1} and $\kappa = 9\ kT$ and $12\ kT$, respectively. The determination of the torque tension is independent of the origin of the integration and hence we choose to place this at the headgroup region so that the torque tension arises predominantly from the pressure in the chain region. We will assume that $\pi(z)$ in the chain region is constant ($= \bar{\pi}_c$) and hence that $\bar{\pi}_c = -4\kappa c_0/l^2$. Since l is approximately 18 Å in the case of DOPC and DOPE the average pressure in the chain region for flat monolayers of DOPC and DOPE will be 5 and 10 MPa respectively. These estimates are of the same order of magnitude as the theoretical calculations of Szleifer and co-workers on saturated chains.[10]

The experiments we present here examine the changes in the lateral pressure profile in the chain region of DOPC/DOPE membranes as a function of the lipid composition. Our probes will therefore reside in a milieu where the pressures are at least one order of magnitude greater than atmospheric, and where we can vary this pressure by a factor of two by changing the relative composition of DOPC and DOPE in excess water. Our report is concerned with the measurement of $\pi(z)$ in the fluid lamellar (L_α) phase, which means that we will only briefly report on our measurements at high DOPE contents, where the inverse hexagonal phase is found.

We chose to use dipyPCs, Fig. 2, to make fluorescence measurements of lateral pressure. In particular we were aiming to make use of intra-molecular pyrene excimer formation as a measure of the lateral pressure. Excimers will occur when a pyrene moiety in its excited state comes into close proximity (≈ 3.5 Å) and in the correct orientation (face to face) to a ground state pyrene and they form an excited dimer or excimer.[14–19] Using a model in which pyrene re-orientation must occur after aggregation Cheng and co-workers have shown that aggregation is the rate limiting step in excimer formation in dipyPCs.[20] Since the frequency with which aggregates form will be proportional to the lateral pressure, their findings imply that the excimer signal in turn will be

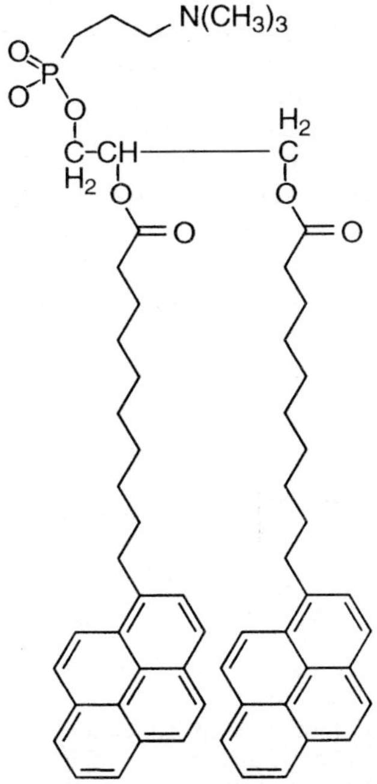

Fig. 2 The 10dipyPC fluorescent probe molecule.

proportional to the lateral pressure. In practice it is more reliable to measure the ratio of the excimer to the monomer fluorescence intensity, η.

Although it has not been stated, there are in fact a number of reports which show evidence that η is a measure of lateral pressure. In the first reports of the synthesis and fluorescence of dipyPC [1,2-bis(1-pyrenyldecanoyl)-L-α-phosphatidylcholine or 10dipyPC] Sunamoto and co-workers showed that low concentrations of 10dipyPC in dipalmitoylphosphatidylcholine and egg lecithin bilayers exhibited increases in η with the addition of cholesterol or increases in temperature.[21,22] Both are known to increase the monolayer torque tension. Cheng and co-workers used 1,2-bis(1-pyrenylmyristoyl)-L-α-phosphatidylcholine (14dipyPC) to examine lipids in both the lamellar and inverse hexagonal phases.[23] Both η and the monomer lifetime were found to be sensitive to the temperature induced lamellar to inverse hexagonal phase transition in DOPE. Butko and Cheng[24] used 14dipyPC to examine mixed dilinoleoylphosphatidylethanolamine/palmitoyloleoylphosphatidylcholine mesophases. Using Birks' description of the kinetics for excimer formation,[14] Fig. 3, they showed that η can be related to the rate constants by

$$\eta = \frac{(k_{fD}/k_{fM})k_{DM}C}{(k_D + k_{MD})} \qquad (3)$$

For dipyPCs at low, fixed concentration, the ratio k_{fD}/k_{fM}, the ratio between the radiative decay parameters of the excimer and monomer and $(k_D + k_{MD})^{-1}$, the fluorescence lifetime of the excimer, were found to be independent of temperature. Hence η was simply proportional to k_{DM}, the rate constant of excimer formation. They used this to determine the activation energy for excimer formation which they found to be lower in the L_α phase. Again this is consistent with the increased lateral pressures in the L_α phase. Sassaroli and co-workers[25] found that η for 10dipyPC decreases with applied hydrostatic pressure in DOPC. They calculated that the activation energy for excimer formation decreased with increasing pressure, in agreement with Butko and Cheng's findings. Cheng and co-workers have also used dipyrenylphosphatidylethanolamine, dipyPE, probe molecules.[20] In these studies they found that η for 4dipyPE was always greater than η for 10dipyPE at any given DOPC/DOPE composition and temperature. This is consistent with the theoretical calculations of the lateral pressure profile, where the lateral pressure is always highest nearest the polar/apolar interface.[10]

Although the data on dipyPCs in bilayers have been the subject of intense analysis, there has been, as far as we are aware, only one mention in the literature of lateral pressures being the source of variation in the photophysical properties of dipyPC. The authors[26] have in fact attempted to determine the variation in c_0 for DOPC/DOPE mixtures in a complex analysis of their spectroscopic data. In doing so they have unwittingly made use of the fact that κ for DOPC and DOPE are so close that almost all of the variation in torque tension is due to the changes in c_0.[27]

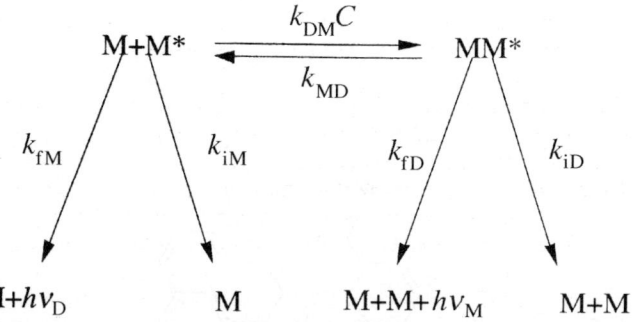

Fig. 3 Birks' kinetic scheme for excimer formation. M is the monomer, M* the excited monomer and MM* the excited dimer. k_{fM} and k_{fD} are the monomer and excimer radiative decay parameters, k_{iM} and k_{iD} are the monomer and excimer non-radiative decay parameters, k_{DM} is the second-order rate constant for excimer formation and C is the pyrene concentration. k_{MD} is the rate constant for the regeneration of excited monomer from the dissociation of the excimer.

With these encouraging data we have set about to record how the lateral pressure profile in the chain region changes as a function of composition in DOPC/DOPE mixtures.

Materials and methods

Materials and synthesis

DOPC and DOPE were purchased from Avanti Polar Lipids (Alabaster, Alabama, USA). They had a stated purity >99% which we confirmed by thin layer chromatography (TLC) before and after use. We found no degradation of these components in our experiments. Vauhkonen and Somerharju[28] have shown that the maximum excimer fluorescence intensity is achieved where the pyrenyl moieties are appended to acyl chains of equal length. DipyPCs having pyrenyl moieties attached to acyl chains with 4 and 10 carbon atoms (4dipyPC and 10dipyPC) were purchased from Molecular Probes Inc. (Eugene, Oregon, USA). They had a stated purity >99% which we confirmed with high pressure liquid chromatography (HPLC).

Since dipyPCs having chain lengths of 6 and 8 (6 and 8dipyPC) were not available commercially we synthesised these ourselves. Their production was broken into two major parts: the synthesis of the pyrenyl fatty acids 6-(pyren-1-yl)hexanoic acid and 8-(pyren-1-yl)octanoic acid; and their subsequent acylation onto glycerophosphocholine. The first part of the synthesis was carried out in our laboratory using a modification of the method described by Sunamoto and co-workers,[21] whilst the final stage of the synthesis was carried out for us by Dr D. M. Phillips of Lipid Products Ltd. (Nutfield, Surrey, UK) using an adaptation of the method used to synthesise 4dipyPC.[29]

Making the pyrenyl fatty acid proceeds *via* a three step synthesis. 0.25 mol $EtO_2C(CH_2)_nCO_2H$ ($n = 6$ or 8) and 0.5 mol thionyl chloride were mixed in 400 ml of low boiling point petroleum ether and left at room temperature until no solid monoethyl adipate remained, about 16 h. The petroleum ether and much of the thionyl chloride was removed on a rotary evaporator using toluene to form an azeotropic mixture with the thionyl chloride. The monoethyl adipoyl chloride or monoethyl suberoyl chloride were then distilled under vacuum at 66–70 °C, 2 mmHg. The yield varied from 70–90%.

The $EtO_2C(CH_2)_nCOCl$ and aluminium chloride (approximately 0.2 and 0.4 mol, respectively) were mixed with 65 ml of dichloromethane and placed in an ice bath. Approximately 0.2 mol of pyrene in 225 ml of dichloromethane was added dropwise to the mixture and stirred for 16 h. Iced water (190 ml) was added slowly with stirring to the orange reaction mixture causing a vigorous exothermic reaction. After 16 h of stirring the dark green, dichloromethane rich layer was removed and dried with sodium sulfate before removal of the dichloromethane on a rotary evaporator. The remaining product was recrystallised from methanol at a yield of 60–80% with respect to pyrene.

Approximately 0.05 mol of 1-(5-carbethoxy-1-oxapentanyl)pyrene or 1-(7-carbethoxy-1-oxaheptanyl)pyrene, 0.18 mol of potassium hydroxide, and 0.17 mol of hydrazine hydrate were mixed with 210 ml of diethylene glycol and refluxed at 162–178 °C for 4 h. After refluxing the excess hydrazine hydrate was removed by distillation at about 100 °C. The remaining mixture was cooled to room temperature and 250 ml water was added dropwise with stirring. The mixture was allowed to stir for 12 h.

We found difficulties in obtaining a pure precipitate of the pyrenyl fatty acids. In the case of the 6-(pyren-1-yl)hexanoic acid we ground the product mixture and stirred with several washings of acetone to dissolve it. The addition of hexane was then used to make the acid precipitate. Mass spectroscopy, infra-red spectroscopy, elemental analysis and TLC indicated that although not pure the sample was predominantly 6-(pyren-1-yl)hexanoic acid. In the case of the 8-(pyren-1-yl)-octanoic acid the potassium salt of the acid had already precipitated out of the reaction mixture. The salt was finely ground and vigorously stirred with a solution of hydrochloric acid and water for two days, resulting in a beige solid which was mostly 8-(pyren-1-yl)octanoic acid. In this case the precipitate was additionally analysed by NMR and though not pure was apparently predominantly composed of 8-(pyren-1-yl)octanoic acid.

It was assumed that the 6-(pyren-1-yl)hexanoic acid was 66% pure, and was treated with 1,1′-carbonyldiimidazole in dry chloroform, for 26 h at 30 °C. After concentrating the solution, it was used to acylate glycerophosphocholine vacuum dried onto a viscose support material, at an assumed molar ratio of 6-(pyren-1-yl)hexanoylimidazole : glycerophosphocholine of 4 : 1. The acylation was carried out, with stirring, at 40 °C for 93 h. The dark orange solution was then chromatographed on Davisil (a silicilic acid) eluting with binary and ternary chloroform, methanol, water solvent systems. Nitrogen bubblers were used in the receivers. The products were monitored and detected by TLC in chloroform : methanol : water 69 : 27 : 4. The disubstituted phosphatidylcholine ran to R_F 0.35, the monosubstituted phosphatidylcholine ran to R_F 0.16. The 6dipyPC was not completely pure by TLC, with five spots seen on the plate, which could not be separated by the column. A small amount of the mono-acylated product was also produced.

With the assumption that the 8-(pyren-1-yl)octanoic acid was 80% pure the same synthesis was used to produce 8dipyPC. The reaction mixture was chromatographed as before showing a major spot at R_F 0.31 due to 8dipyPC, and a small contaminant running just ahead. A small amount of the monosubstituted product was also formed, with a R_F of about 0.11.

To purify these products we used preparative HPLC. A solvent system of chloroform : methanol : water (60 : 40 : 7) was used, with a silica column (250 mm × 9.6 mm). The absorption detector was set to 345 nm, and the flow rate used was 1 cm^3 min^{-1}. Approximately 0.2 mg of crude product was put on to the column. 8dipyPC showed a single peak after treatment (purer than the 4 and 10dipyPC) but 6dipyPC showed a secondary contaminant peak. Fractions of each of the peaks were taken, and examined under UV light for fluorescence. Only the major fraction exhibited fluorescence and fast atom bombardment mass spectroscopy confirmed that this was 6dipyPC.

In subsequent experiments we determined that the contamination was caused by exposure to the UV light during HPLC. We were able to obtain the same effects in all of the dipyPCs, but the 6dipyPC appeared to be peculiarly sensitive to such degradation. Other work with pyrene probes suggests that the use of HPLC grade chloroform stabilised with amylenes causes degradation in the presence of light, but that reagent grade chloroform does not have the same effect.[30]

The 6dipyPC used for all subsequent work was that obtained after purification by HPLC, which we estimate to be better than 90% purity. More importantly perhaps the fluorescence arises entirely from 6dipyPC. All samples were dissolved in reagent grade chloroform, stored at −23 °C in the dark and handled in darkroom lighting conditions to avoid light induced sample degradation.

Sample preparation and instrumentation

Samples were prepared by freeze drying the individual components, *i.e.* DOPC, DOPE and dipyPC. With the assumption that each component was in fact dihydrate the requisite masses of DOPC and DOPE were then weighed under a dry nitrogen atmosphere and co-dissolved in cyclohexane. The dipyPC was also dissolved in cyclohexane at a known concentration and a sufficient quantity added such that the final concentration of dipyPC to total lipid was 0.1 mol%. In agreement with previous work,[20-26,28] we found that at this concentration virtually no intermolecular excimers form.

The resulting mixture was then freeze dried and mixed with an excess of fresh triply distilled, deionised water. We did not use de-oxygenated water since separate experiments indicated that this did not have a detectable effect on the pyrene fluorescence. The mixture was thoroughly homogenised by centrifuging it up and down in a sealed tube. A small amount of sample was then scooped out and placed between two 1 mm thick quartz glass slides separated by a 100 µm thick silicone rubber gasket and sheared until the sample appeared clear and exhibited little if any birefringent textures. The slide was then sealed up along its edges with a high temperature adhesive tape and the fluorescence measurements taken. The silicone rubber gave no fluorescence signal and was not hygroscopic.

These samples were robust, giving identical results after 24 h and being heatable to 80 °C. In our initial experiments we did not shear the sample to surface align the liquid crystalline phase. We found that as a consequence the intensity of the excimer and monomer fluorescence varied widely with time. We assume that this occurred because defect structures created during homogenisation were being slowly annealed and that the lateral pressures in such highly curved regions were quite different from flat ones. By aligning the mesophase the number of defects was lowered to levels where we were unable to detect any variation in the intensity of the excimer and monomer fluorescence over a period of 1 h. Most previous studies of dipyPC fluorescence in amphiphilic bilayers have been done with vesicles. These are convenient since they can be made into clear solutions, but variations in the vesicle diameter, osmotically induced tension[31] and fusion events can all affect the lateral pressure.

For the fluorescence measurements reported here we used a SPEX Fluoromax photon counting fluorimeter running DM3000FL software (Spex Industries Inc., Edison, New Jersey, USA). Excitation was at 345 nm and the emission spectrum was recorded between 370 and 600 nm in 0.5 nm increments integrating at each wavelength for 0.1 s (the excitation and emission bandwidth were both 0.4 nm). All spectra were corrected for variations in detector response with wavelength. The flat samples were held in a temperature controlled cell at an angle of 22° to the incoming excitation beam. This ensured that virtually no scattered excitation light entered the emission monochromator. Sample temperature during these measurements was controlled to within ±0.03 °C by a thermoelectric heater of our own design. A drift in temperature of 0.1 °C resulted in a 2% change in excimer and monomer fluorescence intensities, which were easily detected.

In Fig. 4 we show a typical example of the emission spectra from dipyPC in DOPC/DOPE bilayers. There are two sharp and intense monomeric signals at 377 and 397 nm whilst the broad excimer emission peaks at 477 nm. By converting the spectra to energy scales it is possible to fit the excimer peak to a Gaussian curve and thence determine the ratio of the excimer to monomer signal by integrating the remaining monomer signal and the extracted excimer signal. However we found no advantage in doing this over measuring the intensity of the 377 nm peak and the excimer peak intensity. We did not use the 397 nm peak because our signal deconvolution indicated that there was significant overlap between it and the excimer emission. No matter which technique we used we found that sample to sample reproducibility in the measurement of η was ±5%.

Results

In Fig. 5 we present our measurements of η at 25 °C in DOPC/DOPE membranes in the presence of excess water. Up to 84 mol% DOPE in the binary lipid mixture is in the L_α phase. Over a period of several days at room temperature the system will convert to an inverse bicontinuous

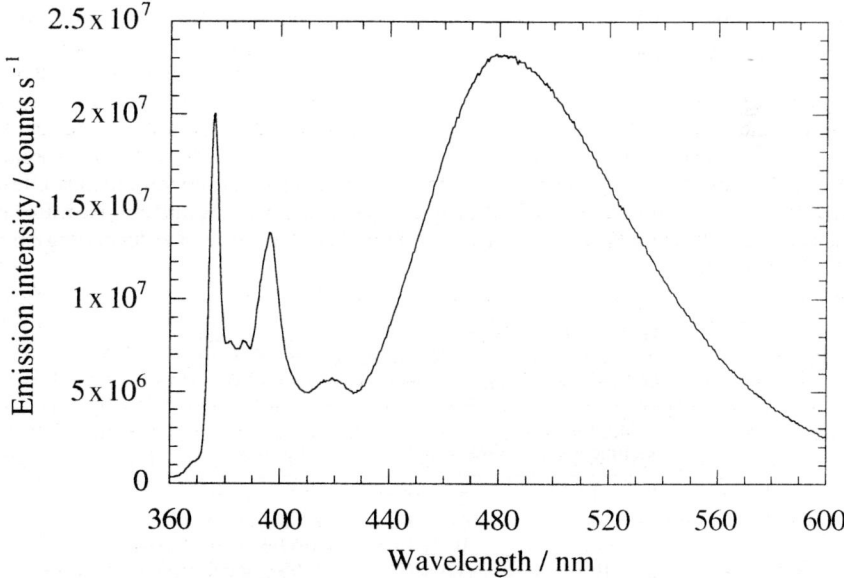

Fig. 4 Fluorescence emission spectrum from 4dipyPC in the L_α phase excited at 345 nm. The sharp peaks in the UV end of the spectrum are due to monomeric pyrene emission, whereas the broad peak in the visible region is from pyrene excimers.

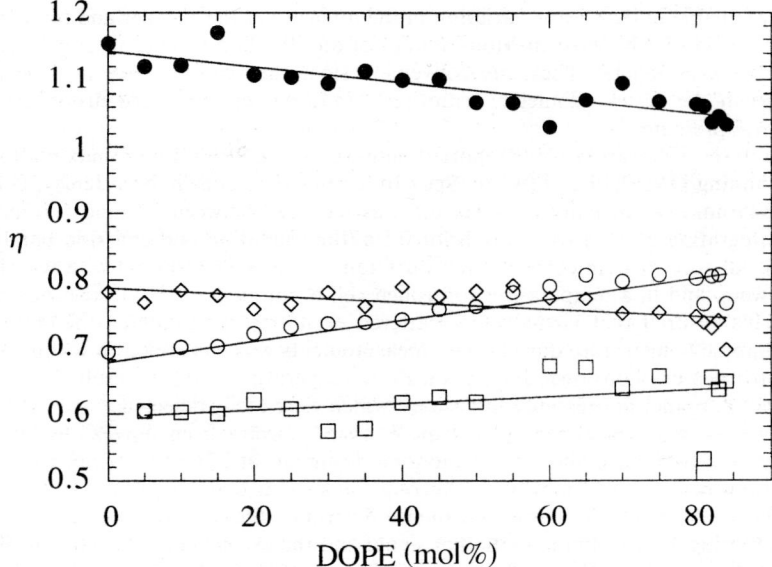

Fig. 5 Variation in η at 25 °C as a function of DOPC/DOPE composition in the L_α phase in excess water. The excimer to monomer ratio has been measured using 4-, 6-, 8- and 10dipyPC (●, ◇, □ and ○, respectively). Linear fits to the data are shown and the coefficients are listed in Table 1.

cubic phase in the region of the phase boundary.[32] Independent checks with polarising microscopy and small angle X-ray diffraction indicated that this had not occurred for our samples in the region of the phase boundary.

We observe both increasing and decreasing signals in the lamellar phase as we increase the amount of DOPE. We have fitted the variation in η to linear functions which are shown in Fig. 5 and reported in Table 1. Cheng and co-workers have also measured η in 4- and 10dipyPC in DOPC/DOPE membranes at 20 and 30 °C.[20] In their measurements they used the intensities at 392 and 475 nm to determine η. These are peculiar values since 392 nm does not correspond to a maxima, but sits on the rising edge of the second most intense monomer peak. Analysing our results at the same wavelengths we obtain good agreement as far as the rate of change of η with respect to composition is concerned, but the absolute values of η disagree by approximately 6%. In fact there is a considerable degree of variation between different groups in the absolute value of η measured. We have found that the ratio is affected by the sample preparation used. For example we were able to achieve as much as a factor of 2 difference in the value of η depending on whether

Table 1 Fitting coefficients for the variation of η as a function of DOPC/DOPE composition at 25 °C in the L_α phase in excess water (see Fig. 5)[a]

dipyPC chainlength	Gradient/10^{-3}	Intercept
4	-1.2 ± 0.2	1.143 ± 0.009
6	-0.55 ± 0.15	0.788 ± 0.008
8	0.5 ± 0.2	0.59 ± 0.01
10	1.3 ± 0.1	0.693 ± 0.008

[a] The data were fitted using the Marquardt–Levenberg algorithm implemented on Kaleidagraph.

we aligned our samples between the glass slides or not. Consistent sample preparation methods are therefore critical to obtaining a self consistent set of data.

Returning to Fig. 5 one might at first glance expect the lateral pressures and hence η to rise at all depths as we increase the monolayer torque tension by adding DOPE. In fact if we sum η at all dipyPC chain lengths we do find this, Fig. 6. What our measurements indicate is that whilst η is dropping for the 4- and 6dipyPC probes it is rising more steeply for 8- and 10dipyPC. Since we know that DOPE increases the torque tension in DOPC/DOPE bilayers, by reducing the average molecular cross-sectional area (72 Å^2 for DOPC (see Fig. 7) and 64.5 Å^2 at 75 mol% DOPE[33]), we should similarly see a drop in η for 4dipyPC if we dehydrate DOPC bilayers, since this also reduces the average molecular cross-section and hence the net lateral pressure. The results of such an experiment are shown in Fig. 7, where we observe a halving in η for a reduction in molecular area from 74 to 68 Å^2. (The differences in the absolute values of η compared to Fig. 5 have arisen because these measurements were made directly on the X-ray capillary samples used to determine molecular dimensions, which are not surface aligned.)

These measurements therefore indicate that in DOPC/DOPE bilayers isothermal increases in torque tension involve not only a net increase in the lateral pressure but also a transfer of pressure towards the chain ends.

In Fig. 8 we have sequentially plotted the profiles of η, calculated from our fit to the data in Fig. 5, at 0, 40 and 80 mol% DOPE. The transfer of pressure to chain ends is clear and our results indicate that this occurs across a minimum in η. If one assumes that the dipyPC and the DOPC/DOPE chains contract by the same proportion upon chain melting then this minimum is in the neighbourhood of the carbon double bond (the pyrene moieties span the C9–C15 positions of DOPC and DOPE when they are in the all-*trans* conformation). Such a dip in the lateral pressure has not been predicted in the case of saturated chains, but recent calculations for unsaturated chains show a dip in the order parameter at the position of the double bond.[34] This lends credence to the idea that the dip in the profile of η is due to a dip in the lateral pressure. We have measured η in the inverse hexagonal phase as a means of providing us with further evidence that we are recording changes in the lateral pressure profile. We would expect that the lateral pressure and hence η would drop at all points between the pivotal surface and the chain ends. Using Chen and Rand's data on the pivotal surface in DOPE[27] and Reiss-Husson and Luzzati's data for the

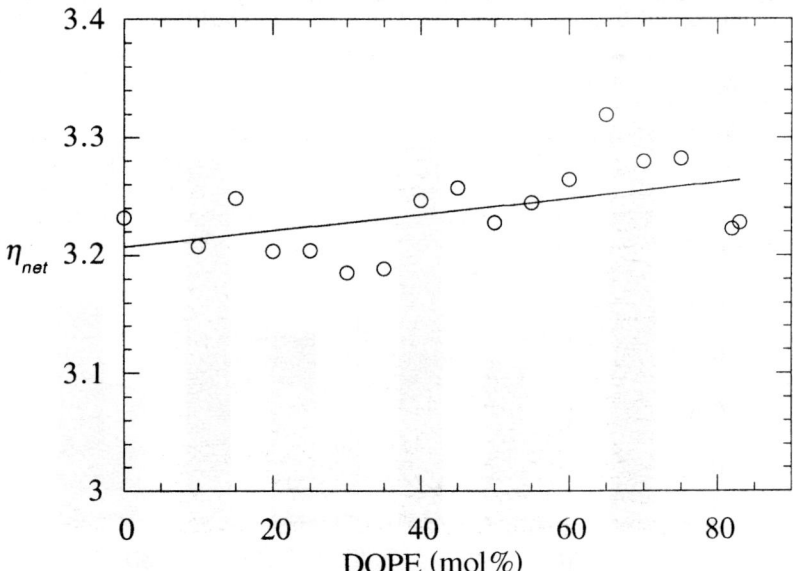

Fig. 6 Variation in the total excimer to monomer signal, η_{net}, at 25 °C as a function of DOPC/DOPE composition in excess water. The total excimer to monomer signal is found by summing the individual values of η from 4-, 6-, 8- and 10dipyPC.

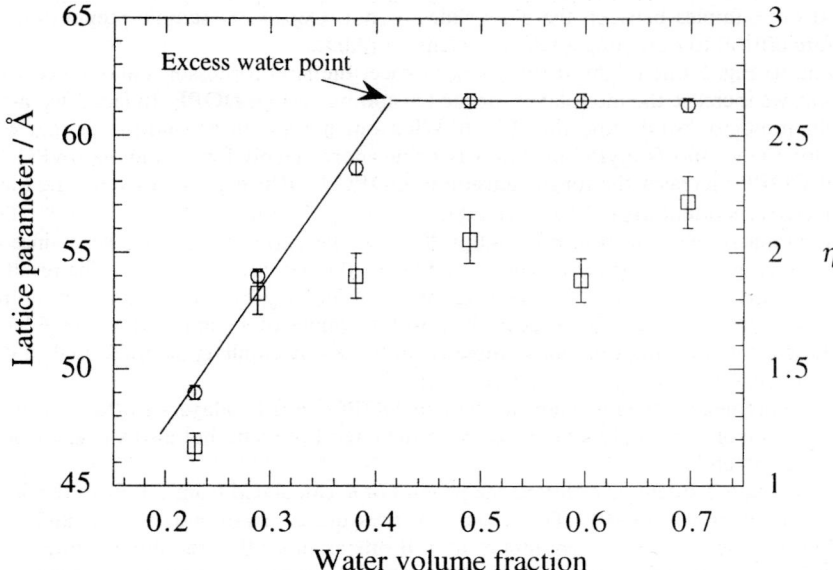

Fig. 7 Changes in unit cell dimensions (○) and η (□), measured with 4dipyPC, in DOPC as a function of water composition. Using Luzzati's approach[36] we have determined the excess water volume fraction, 0.42, and hence the average molecular cross-sectional area in excess water, 72 Å^2, for a molecular volume of DOPC of 1292 Å^3.[27] The samples contained 0.1 mol% 4dipyPC, but this area is in agreement with measurements on pure DOPC by Gruner and co-workers.[13] The same samples were used within two days of being made to measure η. At the lowest water composition we calculate that the molecular area has shrunk to 68 Å^2 and η has declined by a factor of 2.

volume of hydrocarbon moieties in fluid chains,[35] we calculate that the pivotal surface for DOPE is located between C1 and C2 position on the hydrocarbon chain. Hence η should drop for all of our probes upon entering the inverse hexagonal phase and indeed this is what we find, with η being 0.94, 0.62, 0.53 and 0.73 for 4-, 6-, 8-, and 10dipyPC in DOPE at 25 °C. Furthermore, the

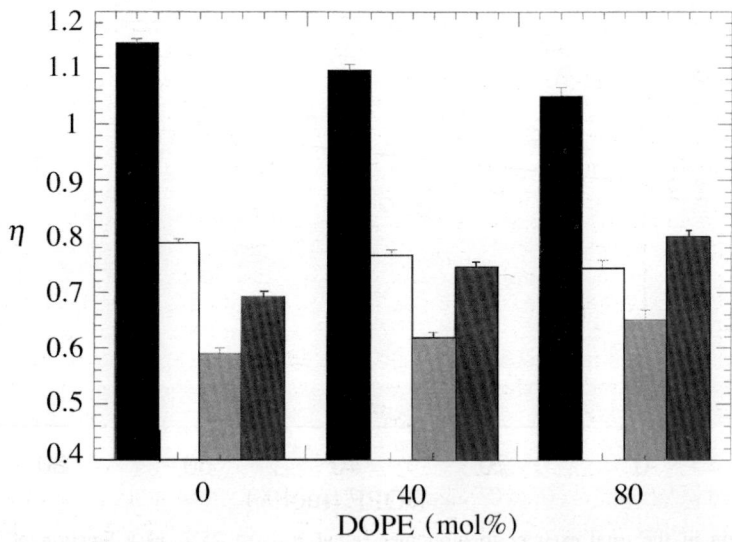

Fig. 8 "Lateral pressure profile" in the excess water L_α phase for three DOPC/DOPE compositions at 25 °C. The excimer to monomer ratio for 4-, 6-, 8- and 10dipyPC are represented by bars of decreasing darkness. The error bars have been determined from the fit to the data in Fig. 5.

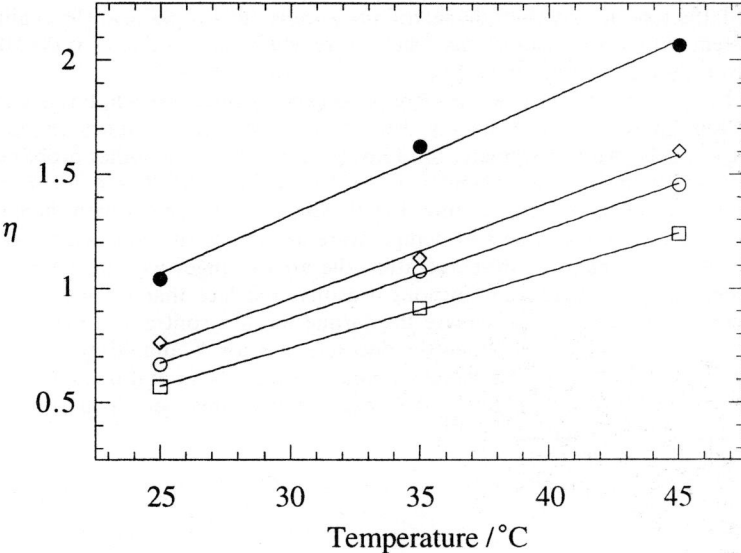

Fig. 9 Response of dipyPC excimer formation to temperature for 80 mol% DOPE. η is plotted for 4-, 6-, 8- and 10dipyPC (●, ◇, □ and ○, respectively). The gradients $d\eta/dT$ for 4-, 6-, 8- and 10dipyPC are 0.051 ± 0.003, 0.040 ± 0.001, 0.035 ± 0.001 and 0.045 ± 0.003, respectively.

calculations of Szleifer and co-workers[10,11] show that the lateral pressure is always highest somewhere around the C3 position on the hydrocarbon chain and drops as one moves to the chain terminus. These observations are compelling evidence that although the profiles of Fig. 8 do not give absolute values of π or z, their qualitative appearance is consistent with the lateral pressure distribution of an oleoyl chain.

Discussion

Although the results we have reported only provide us with a qualitative picture of the lateral pressure profile in part of the chain region, they tell us a number of interesting things.

The first of these has already been stated, and that is that although the net lateral chain pressure increases as we increase the monolayer torque tension, it does so by decreasing the pressure near the head of the chains, but increasing the pressure more rapidly beyond the *cis* double bond. This transfer of lateral pressure away from the region of the pivotal plane increases the torque tension more rapidly than simply increasing the lateral pressure uniformly along the chain.

The increase in monolayer torque tension is brought about by the fact that adding DOPE to DOPC results in a reduction in the average molecular cross-sectional area. Clearly, if the average area per molecule is reduced so is the exposure of hydrocarbon to water and hence the tension (negative pressure) at the polar/apolar interface will fall. Since the pressure in the chain region has increased this must mean that there has been a negative contribution to the lateral pressure in the headgroup region (the total lateral pressure must sum to zero). From this we deduce that phosphatidylethanolamine headgroups impart a negative pressure, that is they exert an attractive force on phosphatidylcholine and -ethanolamine headgroups. We hypothesise that this attractive force is the result of direct hydrogen bond formation between headgroups. The pull between headgroups further increases the desire for the monolayer to bend.

In the case of fully saturated chains Szleifer *et al.*[10] find a single maximum in the lateral pressure. The position of this maximum relative to the length of the molecule moves towards the pivotal interface as the area is reduced. Since it is unlikely that we are collecting any data on the lateral pressure above the C4 position, it is possible that something similar is occurring with the unsaturated oleoyl chains. Furthermore, Szleifer and co-workers' calculations indicate that as the area is reduced, the rate of growth of the lateral pressure is most rapid near the chain ends. This is entirely consistent with our measurements. The feature which does not agree is that there is a drop in lateral pressure; this is not seen in calculations on saturated chains.

We do not believe that the drop we observe is an experimental artefact, for example caused by the relatively bulky pyrene moieties having their motion hindered at shorter chainlengths. Given that η is approximately two times greater for 4dipyPC than any of the other probes it would seem it has little difficulty in forming excimers. Furthermore, we find that η for 4- and 6dipyPC increases rapidly if we raise the temperature, Fig. 9. This is in agreement with theoretical expectations, that the pressure should scale with temperature as long as the cross-sectional area remains constant (for these small changes in temperature the area changes will not be as great as those imparted by changing the lipid composition). Further, the fact that in this case η for 4- and 6dipyPC is seen to increase as we increase the torque tension contradicts the notion that some experimental complication is the cause of the decrease in η for 4- and 6dipyPC when we change the lipid composition. On balance we therefore conclude that the lateral pressure really does drop along some parts of the oleoyl chain as the cross-sectional area per chain is reduced and this remains to be explained theoretically.

Conclusions

This study has revealed a number of qualitative features of the changes in the lateral pressure profile brought about by an isothermal increase in the monolayer torque tension. The results indicate that DOPE drives mixed DOPC/DOPE monolayers into the inverse hexagonal phase by drawing headgroups into closer proximity. This in turn increases the total lateral pressure in the chain region, and this is apparently brought about by the transfer of lateral pressure away from the region close to the pivotal surface to the opposite side of the *cis* double bond. Such subtle re-arrangements of the differential lateral pressure are lost when we use the more familiar curvature elastic parameterisation of the torque tension. It is quite possible that two membranes may have identical torque tensions, but distinctly different lateral pressure profiles. Furthermore, this work implies that the way in which the lateral pressure profile in such membranes changes under identical variations in the environmental conditions is likely to be different. It would be surprising if there were not some membrane proteins which were able to sense such subtle material differences.

The possible oversimplification of correlating membrane protein properties with curvature elastic parameters makes the search for a lateral pressure probe worthwhile. However, any further advances in our measurement and understanding of the micro-mechanics of the monolayer will require that we obtain quantitative measurements of the lateral pressure. To calibrate the probes is experimentally challenging. The rate of change of η with pressure could presumably be found by measurement upon application of an external pressure. However, for this to be meaningful it would be necessary to do this in the L_α phase, since in a solvent excimers can occur with chain conformations which are simply not possible at an interface. The more vexed question is how one might find the value of η at zero lateral pressure. Put another way this would require one to measure η in the limit that the interfacial area per molecule tends to infinity. Of course one cannot do this, and nor could one extrapolate a series of measurements to zero pressure, since the lateral pressure has yet to be established.

It seems to us that a more realistic alternative is to calibrate measurements of η against a model system for which the lateral pressure has been determined from the statistical mechanics. This might also be used to calibrate the region of space probed by the excimer sensor. In the event that we were able to do such a thing there would still remain concerns about the perturbative effect of using a relatively bulky probe moiety such as pyrene on the lateral pressure. In this regard alternative smaller excimer forming moieties such as naphthalene might be better alternatives to pyrene. Finally it should be noted that we estimated a two fold change in the average chain lateral pressure between DOPC and DOPE. At least for the section of the chain that we have examined η changes by no more than a factor of 1.2 suggesting that we are making our measurements in the presence of a sizeable zero pressure excimer signal. It would seem probable that probes such as dipyPC will always suffer from this.

Acknowledgements

We would like to thank Dr D. M. Phillips of Lipid Products Ltd. for his help and advice. This work was supported by an IMPEL Fellowship to S.J.C. and Royal Society University Research Fellowships to R.H.T. and D.R.K.

References

1. W. Helfrich, *Z. Naturforsch.*, 1973, **28c**, 693.
2. S. M. Gruner, *Proc. Natl. Acad. Sci. USA*, 1985, **82**, 3665.
3. J. M. Seddon, *Biochim. Biophys. Acta*, 1990, **1031**, 1.
4. G. Lindblom, A. Wieslander, M. Sjoelund, G. Wikander and A. Wieslander, *Biochemistry*, 1986, **25**, 7502.
5. S. L. Keller, S. M. Bezrukov, S. M. Gruner, M. W. Tate, I. Vodyanoy and V. A. Parsegian, *Biophys. J.*, 1993, **65**, 23.
6. R. M. Epand, *Chem. Phys. Lipids*, 1996, **81**, 101.
7. P. J. Booth, M. L. Riley, S. L. Flitsch, R. H. Templer, A. Farooq, A. R. Curran, N. Chadborn and P. Wright, *Biochemistry*, 1997, **36**, 197.
8. G. S. Attard, R. H. Templer, W. S. Smith, A. N. Hunt and S. Jackowski, *Nature*, 1999, submitted; G. S. Attard, W. S. Smith, R. H. Templer, A. N. Hunt and S. Jackowski, *Biochem. Soc. Trans.*, 1998, **26**, 5230.
9. W. Helfrich, in *Physics of defects*, ed. R. Balian, M. Kléman and J. P. Poirier, North-Holland, Amsterdam, 1981, p. 715.
10. I. Szleifer, A. Benshaul and W. M. Gelbart, *J. Phys. Chem.*, 1990, **94**, 5081.
11. I. Szleifer, D. Kramer, A. Benshaul, W. M. Gelbart and S. A. Safran, *J. Chem. Phys.*, 1990, **92**, 6800.
12. I. Szleifer, D. Kramer, A. Benshaul, D. Roux and W. M. Gelbart, *Phys. Rev. Lett.*, 1988, **60**, 1966.
13. S. M. Gruner, M. W. Tate, G. L. Kirk, P. T. C. So, D. C. Turner, D. T. Keane, C. P. S. Tilcock and P. R. Cullis, *Biochemistry*, 1988, **27**, 2853.
14. J. P. Birks, *Photophysics of aromatic molecules*, Wiley, London, 1970.
15. J. B. Birks and L. G. Christophorou, *Proc. R. Soc. London, Ser. A*, 1963, **274**, 552.
16. J. B. Birks, D. J. Dyson and I. H. Munro, *Proc. R. Soc. London, Ser. A*, 1963, **275**, 575.
17. J. B. Birks and L. G. Christophorou, *Proc. R. Soc. London, Ser. A*, 1964, **277**, 571.
18. J. B. Birks, D. J. Dyson and T. A. King, *Proc. R. Soc. London, Ser. A*, 1964, **277**, 270.
19. J. B. Birks, M. D. Lumb and I. H. Munro, *Proc. R. Soc. London, Ser. A*, 1964, **280**, 289.
20. K. H. Cheng, L. Ruymgaart, L-I. Liu, P. Somerharju and I. P. Sugár, *Biophys. J.*, 1994, **67**, 914.
21. J. Sunamoto, H. Kondo, T. Nomura and H. Okamoto, *J. Am. Chem. Soc.*, 1980, **102**, 1146.
22. J. Sunamoto, T. Nomura and H. Okamoto, *Bull. Chem. Soc. Jpn.*, 1980, **53**, 2768.
23. K. H. Cheng, S-Y. Chen, P. Butko, B. W. Van Der Meer and P. Somerharju, *Biophys. Chem.*, 1991, **39**, 137.
24. P. Butko and K. H. Cheng, *Chem. Phys. Lipids*, 1992, **62**, 39.
25. M. Sassaroli, M. Vaukhonen, P. Somerharju and S. Scarlata, *Biophys. J.*, 1993, **64**, 137.
26. S-Y. Chen, K. H. Cheng and B. W. Van Der Meer, *Biochemistry*, 1992, **31**, 3759.
27. Z. Chen and R. P. Rand, *Biophys. J.*, 1997, **73**, 267.
28. M. Vaukhonen and P. Somerharju, *Chem. Phys. Lipids*, 1990, **52**, 207.
29. H. S. Henderson and P. N. Rauk, *Anal. Biochem.*, 1981, **116**, 553.
30. G. Rumbles, unpublished data.
31. J. Y. A. Lehtonen and P. K. J. Kinnunen, *Biophys. J.*, 1981, **66**, 1981.
32. K. H. Madan, MSc thesis, Imperial College, London University, 1991.
33. R. P. Rand, N. L. Fuller, S. M. Gruner and V. A. Parsegian, *Biochemistry*, 1990, **29**, 76.
34. A. Ben-Shaul, personal communication.
35. F. Reiss-Husson and V. Luzzati, *J. Phys. Chem.*, 1964, **68**, 3504.
36. V. Luzzati and F. Husson, *J. Cell. Biol.*, 1962, **12**, 207.

Paper 8/06472E

Interfacial enzyme activation, non-lamellar phase formation and membrane fusion. Is there a conducting thread?

Félix M. Goñi,* Gorka Basáñez, M. Begoña Ruiz-Argüello and Alicia Alonso

Grupo Biomembranas (Unidad Asociada al C.S.I.C.), Departamento de Bioquímica, Universidad del País Vasco, Aptdo. 644, 48080 Bilbao, Spain.
E-mail: gbpgourf@lg.ehu.es

Received 11th August 1998

Previous studies from this laboratory have shown that the enzymic generation of diacylglycerol in bilayers by phospholipase C may lead to membrane fusion through the formation of transient non-lamellar lipidic intermediates. The present paper intends to explore the correlations existing among the three main processes involved, namely (a) the induction (or inhibition) of lamellar-to-non-lamellar phase transitions in lipid mixtures through the addition of small (<5 mol%) proportions of other lipids, (b) the promotion, by the latter lipids, of fusion in otherwise stable phospholipid vesicles (large unilamellar liposomes) under conditions leading to inverted hexagonal/inverted cubic phase formation in bulk lipid systems, and (c) the modulation, by the same small proportions of lipids, of phospholipase C hydrolysis of phosphatidylcholine in liposome bilayers. It is concluded that phospholipase C may give rise to non-lamellar lipidic structures that in turn permit liposomal fusion to occur, but neither enzyme activity is directly modulated by non-lamellar phase formation, nor will whatever kind of enzyme-induced non-lamellar structure give rise to fusion. Moreover, only under certain kinetic conditions will the enzyme give rise to the organization of non-lamellar structures that are conducive to the fusion event.

1 Introduction

The present view of biomembranes as essentially dynamic structures has been very fruitful in the molecular understanding of many cellular processes. One of the most important membrane-based cellular events is membrane fusion. Membrane fusion is relevant in a wide variety of physiologic (organelle biogenesis, cell secretion, acrosomic reaction) and pathologic (influenza, HIV viral infections) phenomena that take place in all eukaryotic cells.

Although in the cell environment membrane fusion is a carefully regulated process, occurring as a result of the concerted action of numerous proteins, fusion consists essentially of the merging and reorganization of two lipid bilayers, *i.e.* it is mainly a lipid event, although its regulation is protein-dependent. The central role of lipids in a virus-cell membrane event has been put forward recently by Zimmerberg and co-workers.[1]

A number of years ago, our group introduced a model system for the study of membrane fusion, namely the fusion of large unilamellar phospholipid vesicles induced by the catalytic action of phospholipase C.[2,3] Among its merits are that fusion is induced by a *catalytic* agent (in previous models fusogens were added at stoichiometric ratios), that the system is simple enough to allow detailed structural studies of the lipid components, that the fusogen is an enzyme, thus susceptible to regulation, and that phospholipase C has been claimed to be involved in physiological membrane fusion processes.[4,5]

That the catalytic action of phospholipase C, and not the mere presence of the enzyme molecule, is responsible for the fusion event became clear from our early studies.[2] Further data[6,7] have helped to understand the mechanism of this enzyme-induced membrane fusion process. The enzyme appears to play two roles, (a) the rapid, localized and asymmetric production of diacylglycerol (the end-product of the enzyme reaction on phospholipids) acts as *trigger* for the fusion event,[6,8] (b) the generation of diacylglycerol in sufficient amounts permits the formation of *non-lamellar transient structures* that are essential intermediates in the fusion process.[7] Consequently, a strong relationship apparently exists between phospholipase C activity, membrane fusion, and the generation of non-lamellar structures. This is exemplified by the experiments shown in Fig. 1.

The figure combines kinetic (panels A and B) and equilibrium (panel C) data. In Fig. 1A,B, large unilamellar vesicles consisting of egg phosphatidylcholine (PC), egg phosphatidylethanolamine (PE) and cholesterol (Ch) at a 2:1:1 molar ratio are treated with phospholipase C. This is the "control" experiment, as described by Nieva *et al.*[2] Fig. 1A shows the progress of hydrolytic activity with time, and Fig. 1B depicts the parallel (with a few seconds delay) mixing of vesicular aqueous contents, measured with water-soluble fluorescent probes. Fig. 1A,B also describes the effect of small amounts of additives in the control lipid mixture, namely 2 mol% squalene (a linear

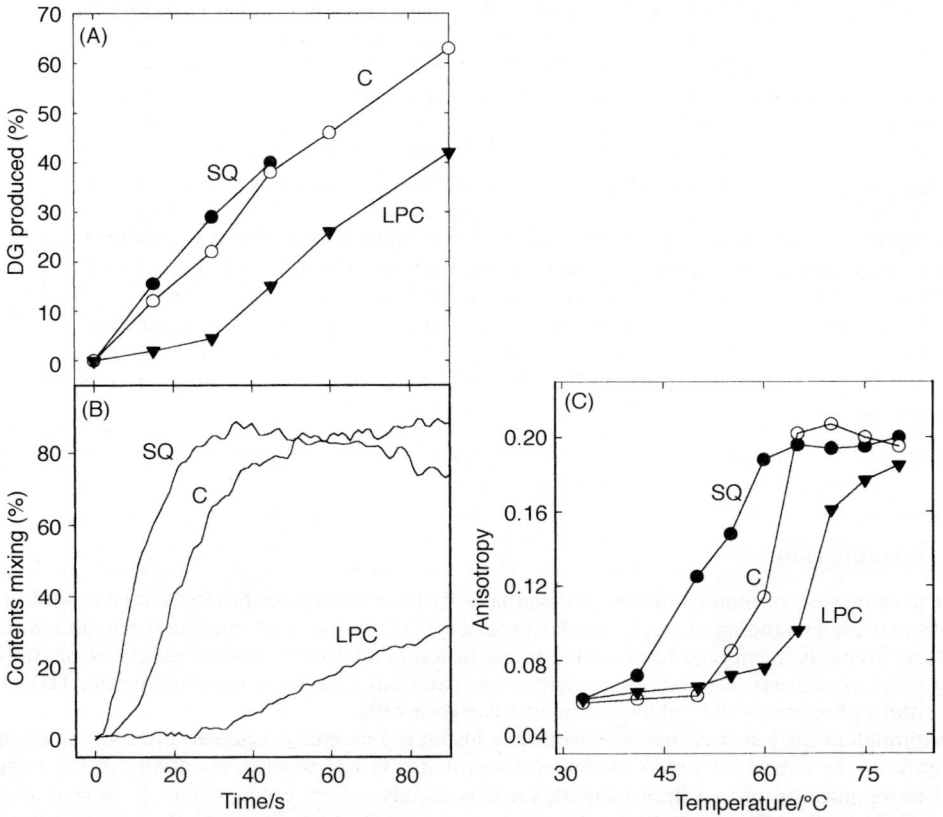

Fig. 1 Phospholipase C-induced liposomal fusion. (A) Enzyme hydrolytic activity, (B) vesicle fusion, measured as contents mixing. Curve C: control (liposomal composition PC : PE : Ch, 2 : 1 : 1 mol ratio); curve SQ: +2 mol% squalene; curve LPC: +5 mol% lysophosphatidylcholine. (C) Lamellar-to-non-lamellar phase transition, detected as an increase in TMA-DPH fluorescence anisotropy. Curve C: control (liposomal composition PC : PE : Ch : diacylglycerol, 47 : 23 : 25 : 5 mol ratio); curve SQ: +2 mol% squalene; curve LPC: +5 mol% lysophosphatidylcholine.

hydrocarbon precursor of cholesterol) or 5 mol% lysophosphatidylcholine (an analogue of PC lacking the $sn-2$ fatty acid). These additives are seen to have a clear effect, stimulatory and inhibitory respectively, on the enzyme phosphohydrolase activity (panel A) and, even more marked, on vesicle fusion (panel B).

Fig. 1C corresponds to a different type of experiment, whose foundations and methodology have been described in detail elsewhere.[9] Essentially the fluorescence polarization of a probe (TMA-DPH) included in a hydrated lipid mixture is measured as a function of temperature. The increase in polarization (anisotropy) under these conditions is related to the transition from a lamellar to a non-lamellar (presumably cubic) phase.[7,9] The lipid composition is intended to mimic the "control" mixture of Fig. 1A,B (PC : PE : Ch, 2 : 1 : 1) once the enzyme has converted 5% of the phospholipids into diacylglycerol. Previous studies[6] have shown that this is the minimum amount of diacylglycerol that allows fusion to occur. Thus the control mixture in Fig. 1C is PC : PE : Ch : diacylglycerol (47 : 23 : 25 : 5, mole ratio). The phase transition occurs at a temperature of $ca.$ 63 °C. When 2 mol% squalene is added to the lipid mixture, the lamellar-to-non-lamellar transition is clearly facilitated, with a decrease in the midpoint transition temperature of ≈ 12 °C. The opposite is found with the enzyme- and fusion-inhibiting lysoPC: addition of 5 mol% to the lipid mixture stabilizes the lamellar phase, increasing the transition temperature by ≈ 6 °C.

The effects of squalene and lysoPC are dose-dependent (data not shown). Similar effects are caused by hexadecane or arachidonic acid, among the positive effectors, and by palmitoylcarnitine or gangliosides, among the inhibitors (data not shown). Thus, a clear correlation appears to exist between phospholipase C activity, vesicle fusion, and the formation of non-lamellar structures in this PC : PE : Ch system. The correctness of this assertion, and its applicability to other lipid mixtures, will be examined in the sections to follow. In this analysis, the three phenomena under study, namely interfacial enzyme activity, formation of non-lamellar structures and fusion of lipid vesicles will be confronted with each other.

2 Enzyme activity and vesicle fusion

This aspect is examined first because the correlation between both phenomena is very clear, both qualitatively and quantitatively, for every enzyme activity-induced fusion system studied to date. Most of the available data correspond to the phospholipase C-induced fusion of PC : PE : Ch (2 : 1 : 1) vesicles described by Nieva *et al.*[2] Fusion occurs as a result of enzyme activity: neither heat-inactivated enzyme, nor enzyme incubated with the specific inhibitor *o*-phenanthroline can induce fusion. The presence of diacylglycerol *per se*, when this lipid is added to the lipid mixture prior to liposome formation, does not allow vesicle aggregation or fusion either: the resulting vesicles are stable for days (see ref. 2 and 6 for experimental details). Moreover, addition of heat-inactivated or *o*-phenanthroline-treated enzyme to vesicles containing up to 20% diacylglycerol does not lead to fusion either (G. Basáñez, unpublished data).

Quantitative variations, either positive or negative, in enzyme activity lead to parallel increases or decreases in the vesicle fusion rates. Some examples are summarized in Table 1. The potencies of the various additives differ considerably. So do their chemical structures. However, in all cases an increase or decrease in phospholipase C hydrolytic rate leads to a corresponding increase or decrease in vesicle fusion rate. The changes in both phenomena are not of the same order of magnitude: small changes, either positive or negative, in enzyme activity are amplified when vesicle fusion is measured for reasons that will be discussed in the next section. All effects described in Table 1 are dose-dependent (data not shown).

Enzyme activity has also been modified by other procedures, such as changes in enzyme concentration, temperature or specific inhibitors. Lowering the temperature from 37 to 25 °C, or decreasing the enzyme concentration below 0.16 U ml^{-1} (the standard concentration in our experiments) diminish considerably the rate of diacylglycerol production and, correspondingly, the rate of vesicle contents mixing (fusion).[8,11]

In spite of this very good correlation between enzyme activity and fusion, special situations exist at very low and very high enzyme activities. The first description of phospholipase C-induced liposomal fusion[2] included an experiment in which fusion was measured, as a function of enzyme concentration (rate of phospholipid hydrolysis). An optimum enzyme concentration was found, for

Table 1 Parallel changes in phospholipase C hydrolytic activity and rate of phospholipase C-induced liposomes fusion (content mixing) as a result of small changes in vesicle lipid composition[a]

Lipid composition	% additive	hydrolysis rate	fusion rate	reference
control	0	100	100	2
+GM3 ganglioside	1	21	3	10
+GM1 ganglioside	1	14	2	10
+GT1b ganglioside	1	4	<1	10
+hexadecane	2	108	215	11
+squalene	2	125	271	11
+PEG-PE	3	80	1	12
+arachidic acid	5	101	122	11
+arachidonic acid	5	119	232	11
+lysoPC	5	88	22	11
+palmitoylcarnitine	5	87	18	11

[a] Data are expressed as percentages with respect to a control lipid mixture consisting of PC : PE : Ch (2 : 1 : 1, mole ratio).

the vesicle concentration used in those measurements, while both above and below certain values fusion was virtually abolished. The lack of vesicle contents mixing at high enzyme concentrations was explained years later,[7] when it was found that the bicontinuous cubic structure that would allow intervesicular mixing of aqueous contents could only be formed within certain limits of diacylglycerol concentration, namely between ≈ 5 and 20 mol% at 37 °C.[6] Beyond a certain enzyme activity, the upper limit of diacylglycerol concentration was reached before any significant fusion could be detected.

The reason why low enzyme activities never lead to vesicle fusion, even after incubation times that allow the formation of appropriate concentrations of diacylglycerol (*i.e.*, between 5 and 20%) is of a kinetic nature. The phenomenon is clearly shown in Fig. 2. In which phospholipase C activity and vesicle fusion are measured in the presence of increasing concentration of the specific enzyme inhibitor *o*-phenanthroline. Both phenomena decrease notoriously in the presence of inhibitor. However, they do not change in parallel: as soon as the enzyme activity decreases below 25% of the native value, fusion is virtually abolished. Our interpretation of this phenomenon is that, as stated in the Introduction, one of the essential roles of the enzyme (perhaps the most

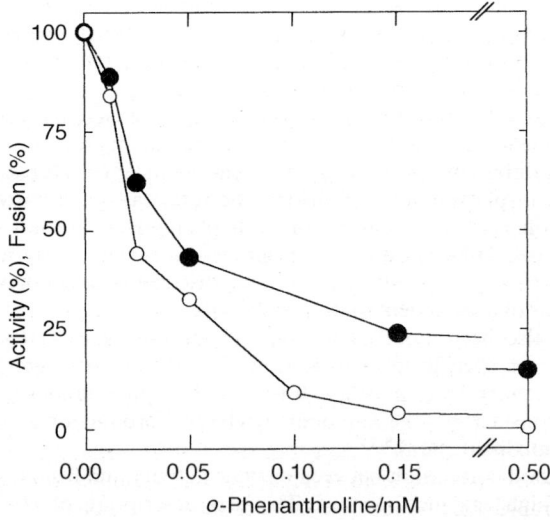

Fig. 2 Percent inhibition by *o*-phenanthroline of phospholipase C activity (●) and of liposomal fusion, measured as contents mixing (○).

essential one) is to act as a trigger for the fusion process. The enzyme triggers fusion by producing a high local diacylglycerol concentration, asymmetrically (the enzyme is present only on one side of the bilayer) and in a short time. The latter point is essential to overcome the spontaneous diffusion of diacylglycerol in the membrane, that will act against the formation of a diacylglycerol-rich patch, in turn a hot point for vesicle aggregation.[13] Low enzyme concentration leads naturally to low rates of diacylglycerol production, that cannot compete with the surface dilution rates of the lipid.[14]

A good relationship between enzyme concentration and fusion rate has been observed by Ruiz-Argüello et al.[8] in a different system, namely the fusion of PC : PE : Ch : sphingomyelin (1 : 1 : 1 : 1 mol ratio) vesicles in the presence of *both* phospholipase C and sphingomyelinase. The fusion rates increase linearly when the enzyme concentrations vary from 0.4 U ml^{-1} each to 1.6 U ml^{-1} each.

In summary, in systems showing enzyme-induced vesicle fusion, the rate of fusion increases with enzyme activity within a certain range of activities, beyond which either kinetic or thermodynamic reasons prevent the formation of the structural intermediates that are required for fusion to occur. In other words, certain phosphohydrolases catalyze vesicle fusion if and when they allow the formation of fusion structural intermediates. The nature of these intermediates and their relationship to the fusion event are discussed in the section that follows.

3 Non-lamellar phase formation and vesicle fusion

The earliest indications of the existence of a "structural intermediate" in phospholipase C-induced liposomal fusion arose from the observations by Nieva et al.[6] of a precise range of diacylglycerol concentrations outside of which no fusion was detected. It was hypothesized that an intermediate of a given lipid composition was involved, at least transiently, in the fusion event. This idea received considerable experimental support from our ^{31}P-NMR and X-ray diffraction studies,[7] and was further reinforced by its accommodation within the so-called "stalk hypothesis" of membrane fusion.[15,16] The stalk is proposed to be a semitoroidal lipidic structure having a negative curvature (the convention is followed that the curvature of a monolayer in the inverted hexagonal H$_{II}$ phase is negative) that would allow the merger of the closest (*cis*) leaflets of apposed membranes.[17]

Moreover, the transient formation of non-bilayer structural intermediates is an unavoidable requirement of membrane fusion. It is also an essential tenet of the stalk hypothesis, since the stalk itself is a non-bilayer structure, in which the monolayers have a negative curvature, such as seen in inverted lipid phases, H$_{II}$ hexagonal, or Q$_{II}$ cubic. "Non-lamellar" has been equated in practice to "reversed hexagonal" in the context of membrane fusion,[16,18] although isotropic ^{31}P NMR signals, which may be compatible with, among others, inverted cubic phases, have also been associated with fusion intermediates.[19-22] Siegel and Epand[23] have recently suggested that TMC (*trans*-monolayer contacts) intermediates play a role in lamellar-to-non-lamellar phase transitions and that they can either rupture to form fusion pores that modulate transitions to Q$_{II}$ inverted cubic phases or assemble into bundles of H$_{II}$ inverted hexagonal phase tubes. Nieva et al.[7] showed a direct correlation between bilayer compositions and temperature giving optimum fusion and those leading to the formation of an "isotropic" component, which was identified with a bicontinuous inverted cubic phase Q^{224} by X-ray diffraction (Fig. 3). Both the stalk and the pore, as predicted by the modified stalk theory, have geometries that can be related to that of the Q^{224} phase. In our previous studies of fusion inhibition by positive-curvature lipids, ganglioside and poly-(ethylene glycol)-modified PE,[10,12] a good correlation has been shown between the inhibitory effects of those lipids and the increased temperatures in the corresponding lamellar-to-non-lamellar transitions. This point has also been explored using a fluorescence polarization technique.[9]

The effects of a variety of single-chain lipids on the lamellar-to-non-lamellar (isotropic, Q^{224}) phase transition of a PC : PE : Ch : diacylglycerol (50 : 25 : 25 : 5, mol ratio) mixture have been studied by fluorescence polarization (Fig. 1C) and ^{31}P NMR.[11] A very good correlation is observed between the modification of phase transition temperature and fusion activity. Squalene and arachidonic acid, which were found to enhance lipid and content mixing, are seen to facilitate the lamellar–isotropic transition, and the opposite occurs with the positive-curvature lipid lysoPC. Arachidic acid was virtually neutral both with respect to fusion and with respect to phase transition. These results are in obvious agreement with the stalk model.

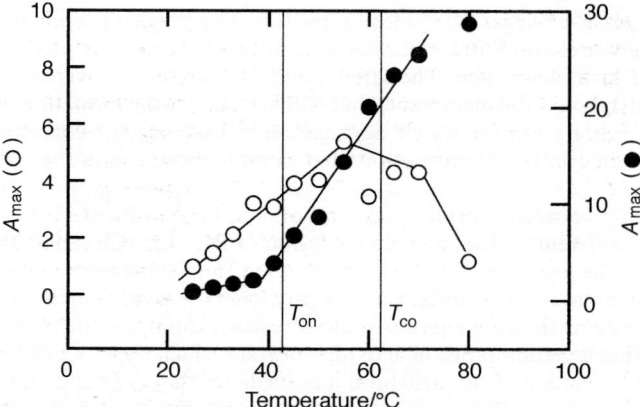

Fig. 3 Relationship between rates of vesicle aggregation, rates of vesicle fusion (content mixing), and lamellar-to-non-lamellar transitions. The maximum rates of phospholipase C-induced vesicle aggregation (●) and fusion (○) are plotted as a function of temperature. Aggregation and lipid mixing change in parallel in this system. The vesicle composition was PC : PE : Ch (2 : 1 : 1 mole ratio). The vertical lines correspond to the onset (T_{on}) and completion (T_{co}) temperatures of the L_α fluid lamellar to the Q^{224} bicontinuous cubic phase transition of a PC : PE : Ch : diacylglycerol (47 : 23 : 25 : 5) mixture. Data are redrawn from Nieva et al.[6,7]

Previously published data on phospholipase C-induced liposomal fusion can be reinterpreted in the light of the predictions of the modified stalk theory. Siegel[16] suggests that when the lipid in the bilayers is very close to the T_h, lamellar-to-hexagonal transition temperature, stalks may form H_{II} phase precursors, and any TMCs that form should have a tendency to radially expand, decreasing the driving force for fusion pore formation. However, the expanded TMC would make a large, comparatively stable lipid connection between opposed bilayers, which would promote extensive lipid mixing. It has been observed that the content mixing rate[18,24–26] often increases with tem-

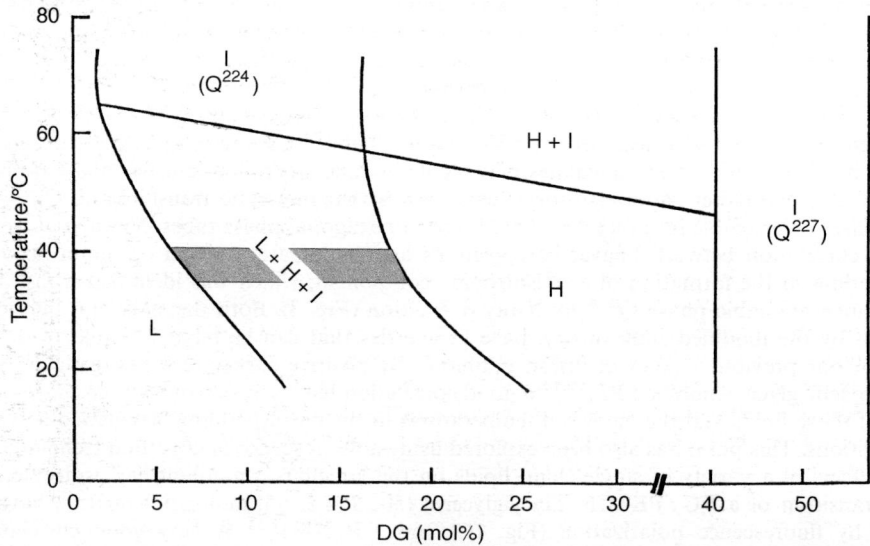

Fig. 4 A pseudo-phase diagram of PC : PE : Ch : DG in excess water, constructed from ^{31}P-NMR data. L, lamellar; H, hexagonal; I, isotropic. In parentheses the nature of the cubic phases, as identified by the X-ray scattering experiments (sample concentration 50% w/w). The shaded area corresponds to the region of temperature and composition at which optimum liposome fusion induced by phospholipase C is observed. Taken from Nieva et al.[7]

perature and goes through a maximum at $T \approx T_h$,[18,24–26] decreasing thereafter, while the lipid mixing rate increases monotonically. Combining our data on phospholipase-induced fusion as a function of temperature[6] with those on the phase behaviour of our lipid mixtures,[7] we can show in Fig. 4 the correlation that is found in our case for fusion rates and transition temperatures.

The figure includes data of vesicle aggregation rates as a function of temperature (under those conditions aggregation and lipid mixing always go in parallel), data of fusion (content mixing) rates as a function of temperature and two vertical lines, marked T_{on} and T_{co}, respectively, corresponding to the onset and completion temperatures of the lamellar–Q^{224} cubic transition of a PC : PE : Ch : diacylglycerol (47 : 23 : 25 : 5) mixture (Fig. 3). This mixture was selected because 5 mol% diacylglycerol is the minimum amount of this lipid that allows fusion to occur.[6] The mixture does not undergo a direct lamellar–cubic transition, but instead, lamellar, hexagonal and cubic phases appear to coexist between T_{on} and T_{co}.[7] Fusion and phase behaviour studies are not strictly comparable, since they consist of kinetic and equilibrium measurements, respectively. Still it can be seen in Fig. 4 that vesicle aggregation increases monotonically with temperature, while content mixing has a maximum in the temperature region corresponding to the lamellar-to-non-lamellar (in our case cubic) transition, in agreement with the above-mentioned predictions and observations. Thus the structural "fusion-intermediate" whose existence was predicted from our kinetic studies[6] corresponds probably to the stalk–TMC–pore.

4 Enzyme activity and non-lamellar phase formation

After having discussed the strict cause–effect relationships existing between enzyme activity and fusion, and between non-lamellar phase formation and fusion, we shall deal with the third side of the triangle, namely the connection between enzyme activity and the formation of non-lamellar phases. One aspect of this problem, *i.e.* whether enzyme activity is responsible for the lamellar-to-non-lamellar transitions, is rather straightforward. In the phospholipase C-induced vesicle fusion system,[3] as well as in the sphingomyelinase-based systems,[8,27] it is obvious that the enzymes are instrumental in modifying the chemical composition of the bilayer, so that new equilibrium conditions settle in, and a phase transition ensues. For example, at 37 °C, the equilibrium phase structure of PC : PE : Ch (50 : 25 : 25, mol ratio) in excess water is lamellar, but when 10% of the phospholipid has been converted into diacylglycerol, and the new composition is PC : PE : Ch : diacylglycerol (43 : 22 : 25 : 10), then the predominant phase structure is non-lamellar ($H_{II} + Q_{II}$).[7] It is thus clear that, in these systems, the non-lamellar phases appear precisely as a result of the enzyme activity.

However, what about the reverse question? Do lamellar-to-non-lamellar transitions somehow modify the activity of interfacial enzymes, such as phospholipase C? The answer to this is less clear and will require the analysis of certain peculiarities of phospholipase C. Phospholipase C, like many other lipases, is a "soluble" (*i.e.*, non-membrane-bound) enzyme, whose substrate is in the solid phase. This is at variance with most enzyme reactions, in which the substrate is in aqueous solution. Also at variance with classical, or soluble, enzyme kinetics, maximum enzyme rates are not the initial enzyme rates. They do not occur as soon as enzyme and substrate are mixed, but rather after a latency period or lag time.[28] The origin of the lag time in the phospholipase C reaction has been examined by us recently.[13,29] The enzyme starts its catalytic work as soon as it meets the substrate, only it does it at a *very* low rate. The low-rate regime continues until a given proportion of diacylglycerol (≈ 10 mol% when the substrate is pure PC)[13] is formed. Then a "burst" of activity occurs, accompanied by vesicle aggregation.[13,28]

Considering that diacylglycerol is notorious for its ability to induce negative curvature in monolayers, thus facilitating lamellar-to-non-lamellar phase transitions,[21,24] it is tempting to assume that a relationship exists between non-lamellar phases and enzyme activation. Indeed this has been proposed by some authors.[30] At first sight, however, our experimental data do not support this hypothesis: ^{31}P-NMR studies of mixtures containing egg PC and egg diacylglycerol (80 : 20 mol ratio), that allow full phospholipase C activity, show spectra compatible with purely lamellar structures in the 20–50 °C temperature range (G. Basáñez, unpublished data). In a more recent series of studies,[28] phospholipase C activity was tested against an extensive series of binary mixtures of egg PC with either PE, Ch or SM. Of these lipids, PE and Ch are known to facilitate non-lamellar phase formation, while SM is a stabilizer of the lamellar phase. It was found that, in

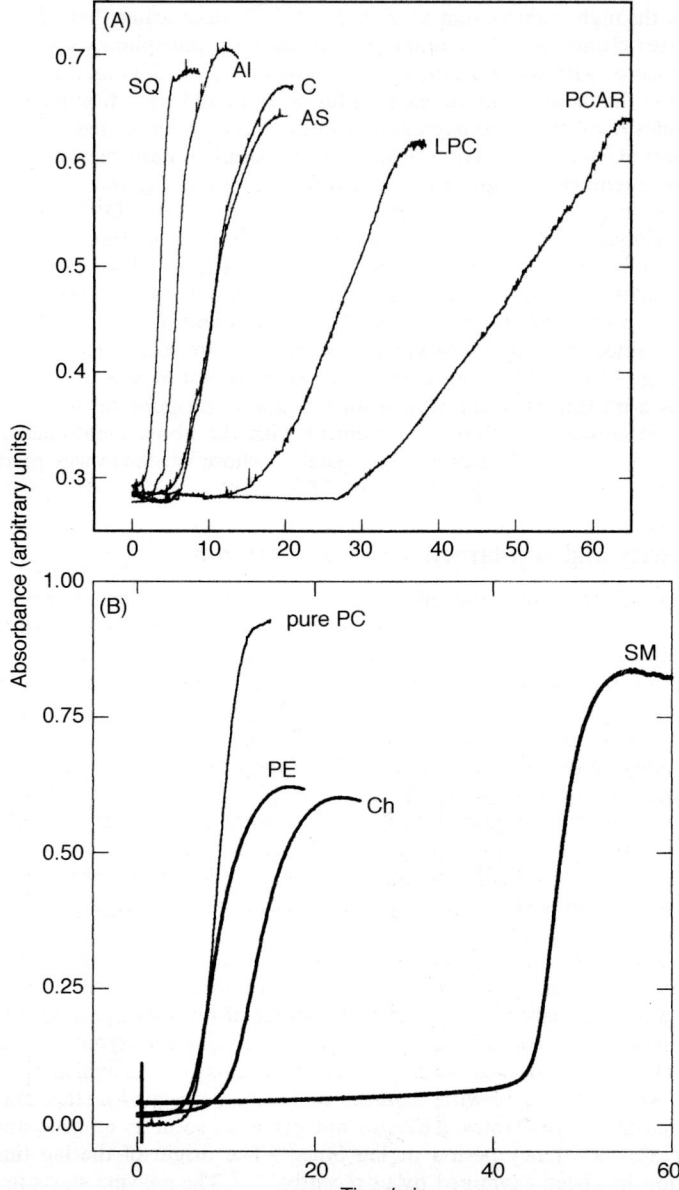

Fig. 5 The effect of small amounts of other, mostly non-substrate, lipids on phospholipase C hydrolysis of egg PC in the form of large unilamellar vesicles. The time course of the reaction is followed as an increase in turbidity (absorbance at 405 nm). (A) Linear, single-chain lipids. C, control (pure egg PC); SQ, +2 mol% squalene; Al, +2 mol% arachidonic acid; AS, +2 mol% arachidic acid; LPC, +2 mol% lysophosphatidylcholine; PCAR, +2 mol% palmitoylcarnitine. (B) More complex lipids. PE, +2 mol% phosphatidylethanolamine; Ch, +2 mol% cholesterol; SM, +2 mol% egg sphingomyelin.

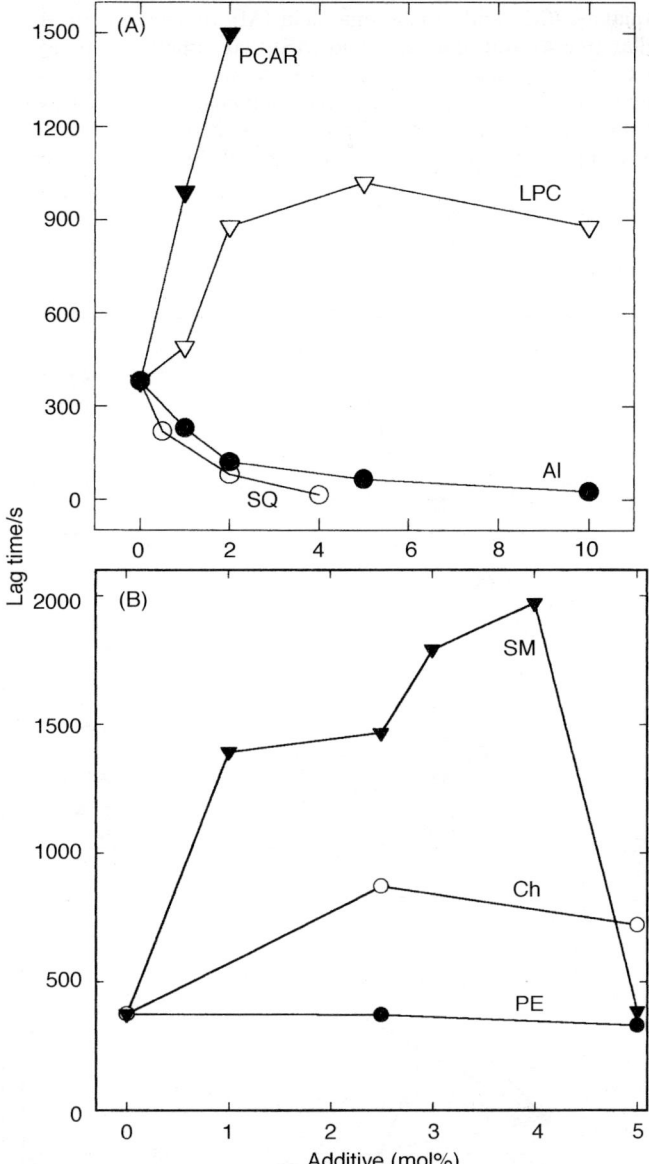

Fig. 6 Lag times of phospholipase C hydrolysis of egg PC to which small amounts of other lipids have been added. Data are taken from experiments of the kind shown in Fig. 5. (A) and (B) are as in Fig. 5.

all cases, the burst of enzyme activity required the formation of the same amounts of diacylglycerol ($\approx$7–10%), while the three systems have a totally different phase behaviour, PC : PE becoming isotropic (probably inverted cubic) with $\approx$10% diacylglycerol while PC : SM mixtures remain purely lamellar even with 20% diacylglycerol and PC : Ch represent an intermediate situation.[28] This is also against an involvement of non-lamellar phase formation in phospholipase C activation.

An additional series of experiments have been carried out, in which the substrate, egg PC, has been doped with small amounts (less than 5 mol%) of lipids that are not enzyme substrates. These are SM, known to preserve the lamellar structure, lysoPC and palmitoylcarnitine (PCAR), that induce a positive curvature in the monolayers, thus opposing the formation of inverted phases, cholesterol (Ch), squalene (SQ) and arachidonic acid (Al), that favour in several ways the transition to non-lamellar phases and arachidic acid (AS) that appears to be neutral in this respect. PE, a substrate for the enzyme *and* a negative-curvature inducer has also been included for comparison. All the mixtures under study are perfectly lamellar at 37 °C, according to ^{31}P-NMR measurements (not shown). The results of phospholipase C activity on these samples are shown in Fig. 5–7. More specifically, Fig. 5 shows turbidity curves of the enzyme activity. Phospholipase C phosphohydrolase activity and vesicle suspension turbidity always change in parallel.[13,28] The effects of single-chain lipids can be seen in Fig. 5A. Squalene and arachidonic acid, lipids with a

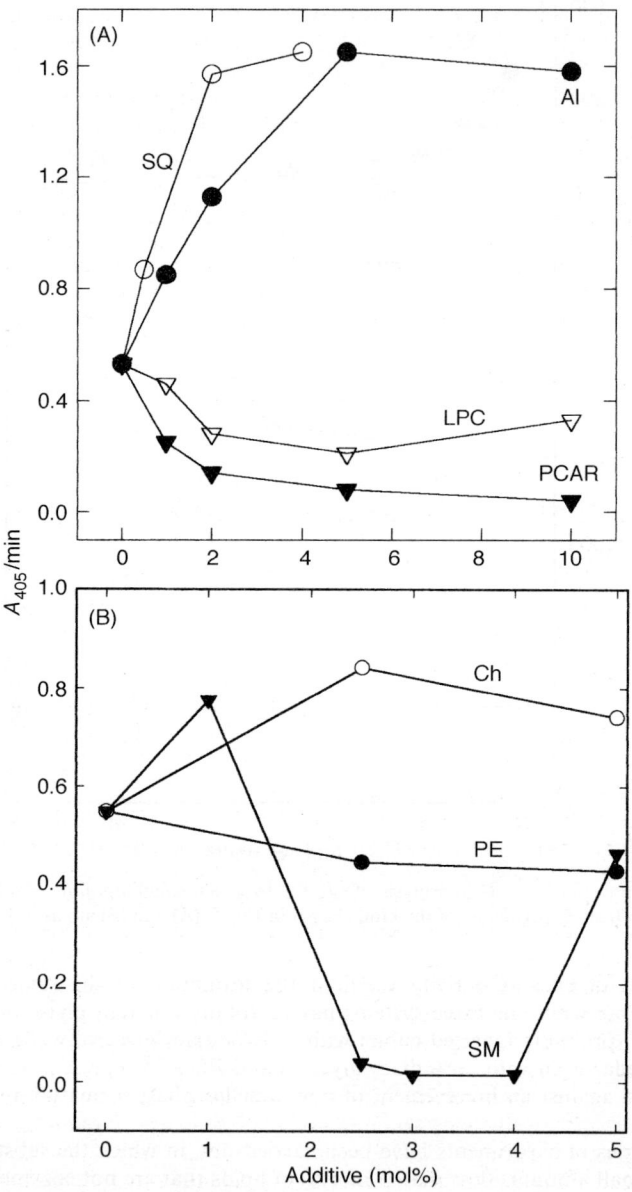

Fig. 7 Rates of phospholipase C hydrolysis of egg PC to which small amounts of other lipids have been added. Data are taken from experiments of the kind shown in Fig. 5. (A) and (B) are as in Fig. 5.

tendency to favour formation of non-lamellar phases, decrease the lag time and increase the maximum rate of phospholipase C. LysoPC and palmitoylcarnitine, stabilizers of the lamellar phase, inhibit the enzyme activity and increase the lag times. Arachidic acid, that has little or no effect on the monolayer curvature, is also inactive on the enzyme. The effects of more complex lipids SM, PE and Ch are shown in Fig. 5B. SM, that stabilizes the lamellar phase in lamellar-to-non-lamellar transitions, increases notoriously the lag time. Interestingly, PE and Ch, that favour the formation of inverted non-lamellar phases, have little effect (PE) or even increase the lag time (Ch). All these effects are dose-dependent (Fig. 6, 7) with the exception of SM, in which case the dose-effect relation is more complex, for reasons that remain as yet obscure. Also unexpected is the effect of cholesterol that increases both the lag times (Fig. 6B) and the enzyme rates (Fig. 7B) of phospholipase C activity. At concentrations above 10 mol%, cholesterol decreases the lag time, in agreement with its role as an enzyme activator.[29] The behaviour at low concentrations, described in Fig. 5–7, may be related to the complexities of the PC–Ch phase diagram, and requires a more detailed investigation.

It should be stressed that all the mixtures in Fig. 5–7 are lamellar at 37 °C, also when 10 mol% diacylglycerol is substituted for PC, to include the effect of phospholipase C activity during the latency period. Again this is not in favour of an association between enzyme activation and non-lamellar phase formation. It could be suggested that the transition from lag to burst would be linked to a particular form of enzyme docking to the membrane, through a non-lamellar lipidic stem. This possibility would be difficult to rule out experimentally, considering that the proportion of lipid involved in the putative stem could be too low to be detected by ^{31}P-NMR. However, because inverted non-lamellar phases are favoured by high temperatures, the hypothetical stem would be more easily formed at these higher temperatures, and as a consequence less diacylglycerol would have to be synthesized during the lag phase. However, phospholipase C assays carried out between 25 and 55 °C show that although the lag times decrease clearly with temperature, the proportion of diacylglycerol generated at the end of the latency period remains constant at ≈ 10 mol%.[13] This observation is clearly against the involvement of non-lamellar phases in the process of interfacial enzyme activation.

Non-lamellar structures could also be related to the post-burst enzyme activity, causing an increase in enzyme rates. This possibility was explored by recording ^{31}P-NMR spectra of aqueous dispersions of ten lipid compositions, namely pure PC, PC : Ch (88 : 12), PC : PE : Ch (1 : 2 : 1), PC : PE : Ch (1 : 1 : 1) and PC : PE : Ch (2 : 1 : 1), plus mixtures derived from these five by substituting part of the phospholipid for diacylglycerol, in the precise proportion that marks the end of the lag period for each mixture. The five mixtures not containing diacylglycerol gave off purely lamellar ^{31}P-NMR signals (not shown). The ^{31}P-NMR spectra of the five diacylglycerol-containing samples are shown in Fig. 8, together with the post-burst enzyme rates observed in each case. The mixtures are ordered from top to bottom, in order of increasing maximal enzyme activities. The phase behaviour does not follow any pattern that can be accommodated to that of the enzyme rates: non-lamellar structures are clearly seen only in a mixture (originally PC : PE : Ch, 1 : 2 : 1) that allows an intermediate enzyme rate. Both the higher and lower enzyme activities occur with the substrate lipids in the lamellar phase. The conclusion of these experiments is that the presence or absence of non-lamellar phases, and the rates of phospholipase C activity, while being both very sensitive to lipid composition, are unrelated phenomena.

What is, then, the explanation for the repeatedly observed phenomenon of the activation of phospholipase C by lipids that favour non-lamellar phase formation, and, conversely, its inhibition by lamellar lipids, as seen in Fig. 1, 5–7, Table 1 and ref. 10–12? Kinnunen[31] has recently suggested a hypothesis, partly related to previous proposals,[32,33] according to which a number of enzymes that interact with membranes as peripheral proteins would be activated by a certain propensity of lipid bilayers to adopt the inverted hexagonal disposition, while remaining in the lamellar phase. Such propensity would be given by the presence of non-bilayer lipids in the membrane, that would induce a frustrated lamellar state. The results in the present paper can certainly be interpreted in the light of this hypothesis. Thus, PLC would join in a large group of enzymes, reviewed in ref. 31, whose activities are enhanced by the presence of non-bilayer lipids in essentially lamellar systems. This hypothesis appears to be physiologically relevant, since it allows the possibility of enzyme regulation in cell membranes without loss of the bilayer structure or its barrier properties.

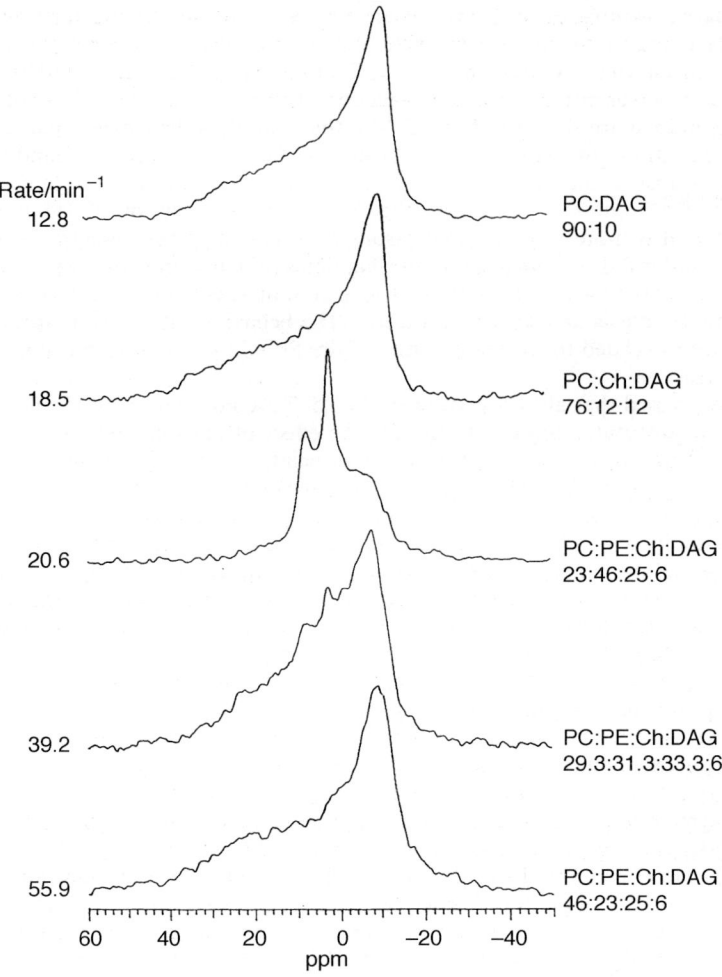

Fig. 8 ^{31}P-NMR spectra of aqueous lipid dispersions representing bilayer compositions at the start of the phospholipase C activity burst. The corresponding post-burst enzyme rates are also indicated for each composition. Line broadening: 80 Hz.

5 Conclusions

(a) When phospholipid-based bilayers are treated with phospholipase C, enzyme-induced generation of diacylglycerol leads ultimately to a phase transition into one or more inverted non-lamellar phases. This occurs under virtually any experimental conditions, at a rate that depends essentially on the enzyme activity.

(b) At least when its substrate is in the form of a low-curvature bilayer, phospholipase C activity can be modulated even by small (<5 mol%) amounts of lipids with the capacity to modulate monolayer curvature. Lipids that favour negative monolayer curvature will stimulate the enzyme activity and decrease the latency times. The opposite effects are caused by lipids that increase the positive curvature of a monolayer. These effects are unrelated to the formation of non-lamellar phases by the lipid.

(c) When the bilayer lipid composition is such that the appropriate non-lamellar phases may be formed, and the rate of generation of diacylglycerol in the bilayer is fast enough to allow the formation of diacylglycerol-rich patches, phospholipase C activity leads to vesicle aggregation and fusion, the latter event being mediated by a non-lamellar lipid structure.

(d) In summary, phospholipase C may give rise to non-lamellar lipidic structures that in turn permit liposomal fusion to occur, but neither enzyme activity is directly modulated by non-lamellar phase formation, nor will whatever kind of enzyme-induced non-lamellar structure give rise to fusion, both kinetic and thermodynamic parameters playing their roles in defining the final outcome of the process.

6 Experimental

Phospholipase C (EC 3.1.4.3) from *Bacillus cereus* was supplied by Boehringer–Mannheim. Egg phosphatidylcholine, egg lysophosphatidylcholine, egg phosphatidylethanolamine and 1,2-diacylglycerol obtained by phospholipase C hydrolysis of egg PC were grade 1 from Lipid Products (South Nutfield, UK). Egg sphingomyelin was purchased from Avanti Polar Lipids Inc. (Alabaster, AL). Cholesterol, free fatty acids, hexadecane, squalene and palmitoylcarnitine were from Sigma (St. Louis, MO). 8-Aminonaphthalene-1,3,6-sulfonate (ANTS) and *p*-xylenebis(pyridinium bromide) (DPX) were purchased from Molecular Probes (Eugene, OR).

Lipid dispersions were prepared by rehydrating lipid films dried from organic solvents under high vacuum. Large unilamellar vesicles (LUV) were prepared by the extrusion method[34] using Nuclepore filters of 0.1 μm pore diameter, at room temperature.

Enzyme assays were carried out in 100 mM NaCl, 10 mM $CaCl_2$, 10 mM Hepes, pH 7.0. Assays were performed at 37 °C, and with continuous stirring. Lipid concentration was 0.3 mM, and enzyme was used at 0.16 U ml^{-1} unless otherwise stated. Phospholipase C was assayed by measuring phosphorus contents[35] in the aqueous phase of an extraction mixture (chloroform : methanol, 2 : 1) after addition of aliquots from the reaction mixture at different times. Enzyme activity was also monitored through changes in the vesicle suspension turbidity (absorbance at 500 nm) in a Cary 3 Bio Varian UV-vis spectrophotometer. Mixing of aqueous vesicle contents was estimated using the ANTS/DPX fluorescent probe system.[25]

^{31}P-NMR spectra were recorded in a KM300 Varian spectrometer operating at 121.4 MHz for ^{31}P. Spectral parameters were 45° pulses (10 μs), 3 s pulse interval, 16 kHz sweep width, and full proton decoupling. Two thousand free induction decays were routinely accumulated from each sample; the spectra were plotted with a line broadening of 80 Hz. Samples ($\approx$ 200 mM in lipid) were equilibrated at 37 °C for 10 min before data acquisition. Phase transitions were also detected through changes of TMA-DPH fluorescence polarization.[9]

Acknowledgements

We are grateful to Ms Sara López for her help with the NMR experiments. G.B. and M.B.R.A. were predoctoral students supported by the Basque Government. This work was supported by grants from DGICYT (No. PB96-0171) and The Basque Government (No. PI96/46).

References

1 L. V. Chernomordik, V. A. Frolov, E. Leikina, P. Bronk and J. Zimmerberg, *J. Cell Biol.*, 1998, **140**, 1369.
2 J. L. Nieva, F. M. Goñi and A. Alonso, *Biochemistry*, 1989, **28**, 7364.
3 F. M. Goñi, J. L. Nieva, G. Basáñez, G. D. Fidelio and A. Alonso, *Biochem. Soc. Trans.*, 1994, **22**, 839.
4 E. R. S. Roldán and R. A. P. Harris, *Biochem. J.*, 1989, **259**, 397.
5 B. Spungin, I. Margalit and H. Breitbart, *J. Cell Sci.*, 1995, **108**, 2525.
6 J. L. Nieva, F. M. Goñi and A. Alonso, *Biochemistry*, 1993, **32**, 1054.
7 J. L. Nieva, A. Alonso, G. Basáñez, F. M. Goñi, A. Gulik, R. Vargas and V. Luzzati, *FEBS Lett.*, 1995, **368**, 143.
8 M. B. Ruiz-Argüello, F. M. Goñi and A. Alonso, *J. Biol. Chem.*, 1998, **273**, 22977.
9 G. Basáñez, J. L. Nieva, E. Rivas, A. Alonso and F. M. Goñi, *Biophys. J.*, 1996, **70**, 2299.
10 G. Basáñez, G. D. Fidelio, F. M. Goñi, B. Maggio and A. Alonso, *Biochemistry*, 1996, **35**, 7506.
11 G. Basáñez, F. M. Goñi and A. Alonso, *Biochemistry*, 1998, **37**, 3901.
12 G. Basáñez, F. M. Goñi and A. Alonso, *FEBS Lett.*, 1997, **411**, 281.
13 G. Basáñez, J. L. Nieva, F. M. Goñi and A. Alonso, *Biochemistry*, 1996, **35**, 15183.
14 G. M. Carman, R. A. Deems and E. A. Dennis, *J. Biol. Chem.*, 1995, **270**, 18711.
15 M. M. Kozlov and V. S. Markin, *Biofizika*, 1983, **28**, 255.
16 D. P. Siegel, *Biophys. J.*, 1993, **65**, 2124.

17 L. V. Chernomordik, *Chem. Phys. Lipids*, 1996, **81**, 203.
18 D. P. Siegel, J. Banschbach and P. L. Yeagle, *Biochemistry*, 1989, **28**, 5010.
19 L. C. M. van Gorkom, S. Q. Nie and R. M. Epand, *Biochemistry*, 1992, **31**, 671.
20 P. L. Yeagle, F. T. Smith, J. E. Young and T. D. Flanagan, *Biochemistry*, 1994, **33**, 1820.
21 H. Ellens, D. P. Siegel, D. Alford, P. L. Yeagle, L. Boni, L. J. Lis, P. J. Quinn and J. Bentz, *Biochemistry*, 1989, **28**, 3692.
22 V. Luzzati, *Curr. Opin. Struct. Biol.*, 1997, **7**, 661.
23 D. P. Siegel and R. M. Epand, *Biophys. J.*, 1997, **73**, 3089.
24 S. Leikin, M. M. Kozlov, N. L. Fuller and R. P. Rand, *Biophys. J.*, 1996, **71**, 2623.
25 H. Ellens, J. Bentz and F. C. Szoka, *Biochemistry*, 1986, **25**, 285.
26 H. Ellens, J. Bentz and F. C. Szoka, *Biochemistry*, 1986, **25**, 4141.
27 G. Basáñez, M. B. Ruiz-Argüello, A. Alonso, F. M. Goñi, G. Larlsson and K. Edwards, *Biophys. J.*, 1997, **72**, 2630.
28 A. D. Bangham and R. M. C. Dawson, *Biochem. J.*, 1959, **72**, 486.
29 M. B. Ruiz-Argüello, F. M. Goñi and A. Alonso, *Biochemistry*, 1998, **37**, 11621.
30 S. W. Hui, T. P. Steward, L. T. Boni and P. L. Yeagle, *Science*, 1981, **212**, 921.
31 P. K. J. Kinnunen, *Chem. Phys. Lipids*, 1996, **81**, 151.
32 A. Sen, T. V. Isac and S. W. Hui, *Biochemistry*, 1991, **30**, 4516.
33 H. De Boeck and R. Zidovetzki, *Biochemistry*, 1989, **28**, 7439.
34 L. D. Mayer, M. H. Hope and P. R. Cullis, *Biochim. Biophys. Acta*, 1986, **858**, 161.
35 C. S. F. Böttcher, C. M. Van Gent and C. Fries, *Anal. Chim. Acta*, 1961, **1061**, 297.

Paper 8/06352D

General Discussion

Dr Sansom opened the discussion of Prof. Bayerl's paper: Please could you say how much protein (Streptavidin) was added *i.e.* what fraction of the membrane surface was coated by protein? Have you been able to perform experiments in which the surface was incompletely covered with protein?

Prof. Bayerl responded: For streptavidin, we worked under excess protein conditions, *i.e.* we can safely assume that all biotin groups exposed by the black lipid membrane (BLM) surface are coupled to a streptavidin molecule. We did not perform experiments at low streptavidin resulting in incomplete coverage of the BLM.

Prof. Holzwarth asked: What do you expect as the time resolution of the dynamic light scattering method you applied?
Is there any angle dependence in your light scattering signal? (0, 30, 60, 90°)?

Prof. Bayerl responded: The best time resolution which we can achieve now is 0.1 μs. An improved setup which is presently under construction will allow us to push this limit one order of magnitude. In answer to your second question, no angular dependence has been observed yet since we are restricted to measurements around the specular angle.

Prof. Neumann asked: Are those proteins which increase the transversal shear undulations by themselves conformationally flexible such that they enforce their fluctuations to the membrane?

Prof. Bayerl responded: It is generally possible that protein internal degrees of motional freedom couple to the BLM motion. However, to be detectable by dynamic light scattering (DLS) a coherent internal motion of all proteins would be required, which appears rather unlikely.

Dr Pawlak asked: When you immobilize biomembranes (*e.g.* as part of a vesicle) on a solid support, how is the frequency spectrum of undulations modified upon the surface contact, spontaneous contact or receptor mediated contact?
Will the undulations disappear? If not, how far can undulations compete energetically with *e.g.* low affinity attractive forces of receptor-mediated surface contacts?

Prof. Bayerl responded: We did some preliminary experiments with bilayers on a solid support (mostly planar silica) using solid state NMR and infrared techniques. We found that the undulations are rendered undetectable in the mesoscopic range of wavevectors due to the presence of the support. This is understandable considering the fact that the average distance between bilayer and solid surface is in the range of 15 Å (neutron reflection results). Hence, there are just a few (bound) water molecules separating bilayer and support establishing a high surface viscosity. Only for highest frequencies (upper GHz range) we observed by coherent quasi-elastic neutron scattering that low amplitude undulations still persist for fully hydrated multilamellar bilayer stacks on a planar silicon support.

Dr Amblard said: What the proteins exactly do, and how they alter the mechanical properties of the membrane is not clear. Nevertheless, it seem that some effects could be due to the properties of the contact between individual proteins and the lipid heads "below" them, while others could involve lateral interactions between the protein molecules. These two situations could be resolved by studying how the mechanical effect varies with the surface concentration of adsorbed protein. Did you study that?

Prof. Bayerl responded: We have not yet undertaken a systematic study of this effect. However, the results we obtained for the so called S-layer proteins (two-dimensionally crystallized protein layers) from bacteria (*Bacillus coagulans*) seem to indicate that at protein concentrations at the bilayer surface which are insufficient for two-dimensional crystallization, protein–protein interaction is not a dominant feature for the change of the undulation pattern. It is conceivable that this might change at higher protein concentrations, but we have no experimental proof yet. This problem is currently being studied in our laboratory and we should have the data soon.

Dr Jones said: You describe the adsorption of flaccid vesicles to a glass surface that shows that some spreading occurs, and bilayer undulations in the region of contact (Fig. 6). How large were these vesicles and how frequent was this event? Also is there a lower limit to the vesicles size to observe this phenomenon and can you observe vesicle disruption by your technique?

Prof. Bayerl responded: Typical vesicle diameters used in this study are in the range of several tens of micrometers. The lower size limit for observation by this technique is given by the lateral resolution of the microscope, thus vesicles below 2 µm are hardly accessible since insufficient numbers of fringes will be resolvable. Nevertheless, disruption of smaller vesicles at the solid surface is detectable by the interferometric technique since the formation of a supported bilayer at the disruption/fusion site can be clearly observed.

Dr P. N. Edwards commented: The changes in undulation behaviour on binding proteins to membranes in tight-packed arrays is likely to, in part, reflect protein–protein interactions. Pegylation, *in situ*, should reduce and might eliminate that influence, thus allowing protein–lipid effects to dominate any residual changes.

Prof. Bayerl responded: Protein–protein interaction contributes to the measured vertical shear motions only if the coupling between adjacent proteins indirection normal to the BLM is strong, like in a two-dimensionally crystallized protein layer. We have not yet encountered such situations for the two proteins studied.

Prof. Holzwarth said: It is well known that cholesterol clusters in bilayer membranes depend on cholesterol concentration as we demonstrated.[1] Can you deduce any information about the size of clusters induced by cholesterol in lipid membranes and do you see any change of lipid mobility in the surrounding of cholesterol clusters which might be reflected in the elastic modulus of the membrane?

1   A. Genz, J. F. Holzwarth and T. Y. Tsong, *Biophys. J.*, 1986, **50**, 1043.

Prof. Bayerl responded: We assume that the DLS method is not sensitive to the clustering of cholesterol. However, we know from our in-plane neutron scattering work on DPPC–cholesterol mixtures in the liquid ordered phase (40 mol% cholesterol) that if clusters exist in this phase they must be exceedingly small (less than 1 nm diameter). So I cannot comment on the effects the cluster size may have on the elastic modulus of the membrane.

Dr Bezrukov asked: From your measurements you deduce the values of lateral tension for lipid and lipid–cholesterol membranes; how well do they compare to the values obtained by other methods? Also, you employ a variation of the Montal–Mueller technique for lipid bilayer formation, but use decane as the lipid solvent. Decane is rather heavy and not particularly volatile. For this reason it is most probably incorporated into bilayer structures in your experiment and influences their mechanical properties significantly. Can you comment on that?

Prof. Bayerl responded: Note that the tension values deduced from our DLS measurements is the tension due to the vertical shear motion and thus a dynamical tension. It is not useful to compare this dynamical value with the static tension values measured by other (static) techniques. Decane is certainly a major contributor to the virtually zero viscosity value obtained by the DLS data analysis in terms of Kramer's theory. We did some preliminary experiments where decane was substituted by squalane, which is supposed to leave the BLM completely after sufficient equilibration. For this case we observed a significant decrease of the tension, *i.e.* decane seems to pose a hindrance to vertical shear motion.

Prof. Roux said: Your results indicate that the continuum model of Kramer[1] provides a very good description of the dynamical fluctuations of pure membranes, and of the influence of proteins associated with the membrane on those fluctuations. Similar continuum models are currently used to describe the microscopic details of the membrane in the neighbourhood of a protein (*e.g.* mattress model, hydrophobic mismatch, *etc.*). Could you provide an upper bound in wave number, $q_{max} = 2\pi/\lambda_{min}$, up to which you trust the validity of a continuum description of membranes?

1 L. Kramer, *J. Chem. Phys.*, 1971, **55**, 2097.

Prof. Bayerl responded: According to the extension of Kramer's theory by Fan[1] we should expect measurable deviations due to bending energy effects at wavenumbers beyond 10^5 cm^{-1}. This is well above the maximum of 3×10^4 cm^{-1} which are currently accessible by our DLS technique.

1 C. Fan, *J. Colloid Interface Sci.*, 1979, **44**, 363.

Dr L. Fisher commented: In *Faraday Discussion* No. 81, Parker, Haydon and I reported interferometric measurements of the interaction between a pair of freely suspended glycerol monooleate bilayers. One result of these measurements was the establishment of a highly reproducible critical separation (30 ± 5 nm), below which the aqueous film between the bilayers collapsed and the bilayers fused in a manner described in detail in the original paper.

Models available at the time suggested that the collapse was induced by mutual reinforcement of independent bilayer fluctuations, inducing a catastrophic increase in the amplitude of these fluctuations. Do your dynamic light scattering studies cast any further light on the possible mechanisms of the process?

1 N. S. Parker, D. A. Haydon and L. R. Fisher, *Faraday Discuss. Chem. Soc.*, 1986, **81**, 249.

Prof. Bayerl responded: We have not yet studied the approach between two membranes or a membrane and a solid surface by this technique. We have to establish a device for precise interferometric distance measurements in the DLS setup first. This is presently under construction. Therefore I cannot suggest any new aspects of this behaviour at the moment.

Dr Amblard opened the discussion of Prof. Laggner's paper: In your paper, your suggest that the quasi-immediate thinning of the interbilayer water space is achieved by a martensitic lattice disclination mechanism. It is clear from Fig 1b that this type of transition involves very large deformations. How is that possible given the viscosity of the environment?

Prof. Laggner responded: Our suggestion of a martensitic transition is based upon an indirect argument: the absence of any significant lattice perturbations immediately after the jump, as seen from the sharp Bragg peaks, leads to this notion. We have no direct information about the ensuing deformations and movements of the liposome. This could be done, and will be done, by time-resolved video microscopy.

Prof. Holzwarth said: You investigated the influence of Li$^+$ ions on the lifetime observed in your system. Did you ever use Cs$^+$ ions, which occupy just the space of a water dipole? I would expect a big difference between the two alkali ions because their ionic radii are different and more importantly the lifetime of their coordinated water differs by six orders of magnitude as Eigen and co-workers demonstrated in the 1960s[1] and as we were able to show for electron transfer reactions.[2]

A temperature jump in aqueous systems is always accompanied by a pressure jump, if you don't work around 4 °C. Can you distinguish between the pressure and the temperature influence on the relaxation time observed in your system?

1 H. Diebler, M. Eigen, G. Ilgenfritz, G. Maaß and R. Winkler, *Pure Appl. Chem.*, 1969, **20**, 93.
2 H. Bruhn, S. Nigam and J. F. Holzwarth, *Faraday Discuss. Chem. Soc.*, 1982, **74**, 129.

Prof. Laggner replied: So far, we have only undertaken static diffraction experiments with CsCl. They show no qualitative difference to what is observed with LiCl, but a different concentration dependence. The L_α-phase separation does occur also with Cs, and I assume, that transient L_α^*-phase will also be observable. Our intention with the present study was to prolong the lifetime of the intermediates, and therefore the shorter lifetime of Cs–water coordination would work in the other direction, I believe. The dynamic experiments with Cs will be done, however.

The sample cell also contains an air volume, so that the volume expansion is practically unhindered. I only expect a shock wave which runs through the sample capillary of about 5 cm length in some 10 µs, and thus by about two orders of magnitude faster than the time resolution of the diffraction experiment.

Dr Dijkstra asked: In the model presented (LiCl-nuclei with lamellar layers, with more outer layer undulation freedom and thus larger d-spacing) wouldn't one expect a distribution of d-spacings between the two extremes, *i.e.* outer and inner most layers, rather than two discrete peaks?

The observed smaller d-spacing of the transient L_α^*-state is a striking reminder of the different d-spacings observed in fully hydrated lamellar phosphatidylcholines (PCs) depending on whether the PC is hydrated by excess liquid water or by 100% vapour pressure. The hypothesized "lentils" of water in the L_α^*-state could evaporate in 100% vapour pressure situations, resulting in stabilized trapping of the L_α^*-state and correlating to the lower d-spacings observed in 100% vapour pressure vs. liquid water. Do you have any arguments/data that refute or augment this hypothesis?

Prof. Laggner replied: Yes, one would indeed expect that, but the two discrete and sharp peaks show that reality, in this case, does not follow our expectations. To answer your second question, this similarity to the 'vapour pressure paradox' is very much along our lines of thinking. Recent experiments with different liquid concentrations and on the coagulation of liposomes have shown that this suppresses undulation in those liposomes that form the inner part of the coacervate, leading to a smaller d-value and sharper peaks, and that the larger d-values of the coexisting L_α-phase come from the outer, unrestricted surface.

Prof. Peterson asked: How certain are you about the structural identification of the transient phase? In particular, have you performed any measurements on ordered lipid systems (*e.g.* cast lipid films) to verify the expected angular relationship between the original and transient phase reflections? Secondly, can you exclude the transient phase being one of the known L_β-phases?

Prof. Laggner responded: We have not yet done any experiments on cast lipid films, but this is planned for the near future. Now, we can only be positive that the transient phase is lamellar, rather than non-lamellar, from the observation of up to three integral orders of Bragg peaks and the absence of any additional, non-integral orders.

Prof. Evans opened the discussion of Dr Templer's paper: Could the low pressure sensed by the short dipyrenyl PC probe reflect preferential partitioning in regions of positive (splay) curvature and coupling to undulations?

Dr Templer responded: Indeed we cannot rule out the possibility that the short di-pyrenyl PC probe may partition preferentially towards regions of curvature, and this would indeed reduce the value of η (the ratio of the excimer to the monomer fluorescence intensity). We see precisely such a reduction when we make measurements in the inverse hexagonal phase (results not presented in our paper). This is a fundamental problem with regard to probes such as these, namely that the bulk of the probe moieties may have a sizeable effect on the very conditions one is hoping to measure.

Prof. Holzwarth asked: What do you understand by a low pyrene ratio and have you tested whether there is any disturbance of the system in the environment of the pyrene molecules?

Dr Templer responded: We are interpreting a decrease in the ratio of pyrene to excimer fluorescence ratio in terms of a concomitant decrease in the lateral pressure in the region of the probe.

We are of course worried about the disturbance the probe may cause on its surroundings; something which can be said of almost all molecular probes used in soft condensed systems. However, we have not been able to devise a way of testing if there is a significant disturbance.

Prof Roux commented: Assuming that the decay rate of the excited monomer and dimer are the same, what you really detect is the population fraction in associated dimer vs. the separated monomers. Exactly how this population is correlated with the local pressure, at the microscopic level, is not so clear. Have you performed similar measurements in bulk solvent, in which the hydrostatic pressure can be set externally? In other words, could you comment on the calibration of such measurements?

Dr Templer responded: Indeed we believe that η is detecting the population of probe molecules that are in an excimeric state vs. the separate state. We have not performed measurements on dipyPCs in bulk solvent to calibrate the measurement of η with respect to applied pressure. Doing such measurements is not directly related to the lateral pressure, since a measurement in bulk solvent will tell us about the response to a three-dimensional pressure, whereas the lateral pressure at an interface is a two-dimensional pressure under significant geometrical constraints.

We believe that it may be possible to make such a calibration on vesicle dispersions. This will tell us how η varies with pressure, but we will still not know what the value at zero lateral pressure is. We have made a first attempt, without success, to determine the value at zero lateral pressure by measuring η on a Langmuir film as we increase the molecular area.

Prof. Almgren commented: It is an interesting idea to try monitor the lateral pressure variation in bilayer membranes in this way. However it is far from clear that the excimer to monomer emission intensity ratio is proportional to the pressure, as assumed. Modifying eqn. (3) in your paper somewhat (mainly stressing that it is a proportionality and that the intramolecular excimer formation is governed by a first order rate constant, k_{DM}, the mean encounter frequency of the two pyrene moieties) we have

$$\eta \propto \frac{k_{fD} k_{DM}}{k_{fM}(k_D + k_{MD})}$$

In this equation all the rate constants could be pressure dependent. From studies of the pressure dependence of excimer formation of pyrene molecules in different solvents[1] it was concluded that the main pressure dependence was in the rate constants k_{DM} and k_{MD}. Quenching by oxygen can occur in the present case, which would effect k_D and also make this rate constant pressure dependent.

A test of the method using varying pressures appears important. This could maybe be achieved perhaps by using liposomes in a simple pressure cell.

1   M. Okamoto and M. Sasaki, *J. Phys. Chem.*, 1991, **95**, 6548.

Dr Templer responded: First may we thank Prof. Almgren for noticing the typographical error in eqn. (3) which has been rectified in the final version of our paper. It is indeed correct that if oxygen quenching is occurring in our system then our measurement of η will not be directly proportional to the lateral pressure. We have undertaken tests to see if our samples were "contaminated" with oxygen. Freeze dried samples of both DOPC and DOPE with 4dipyPC at 0.1 mol% were prepared in a nitrogen glove box. Water was added that had either been bubbled with argon for 15 min or left exposed to air and the excimer to monomer ratio then measured. To within the precision of our measurements we were unable to discern any difference in the samples, with η being 1.07 ± 0.03 in DOPC and 0.91 ± 0.03 in DOPE.

We agree entirely that a pressure calibration is the next important step to demonstrate whether or not this is really a probe of lateral pressure.

Prof. Bohne asked: Could the smaller values for the excimer-to-monomer ratio at the position close to the double bond be due to an effect of the structural environment on the radiative rate constants for the monomer and/or excimer? Secondly, are the spectra for the monomer and excimer emission the same for the pyrene at different positions in the lipid? Finally, I was surprised by your comment that the pyrene fluorescence is not sensitive to the presence of oxygen. Could you expand on whether the oxygen effect was determined for all probes and if the lack of an effect is on the absolute monomer and excimer emission intensities, or on the excimer-to-monomer ratios.

Dr Templer responded: We are aware that the polarizability of the local environment affects the relative intensities of the monomer peaks. We have so far been unable to detect such changes in relative intensities as a function of chain length. Similarly we do not observe significant differences in relative monomer peak heights with temperature or composition. Furthermore we do not see variations in the peak position within the 0.5 nm resolution of our measurements of the first monomer peak at 376.5 nm.

Yes, it is perhaps surprising that we are not able to detect any effects of oxygen. We only tested for the effect on 4dipyPC and it is possible that the effect is only to be seen at the longer chain lengths. Our measurements were made with respect to η not the absolute values of the fluorescence intensity.

Prof. Laggner said: Intuitively it would seem to me that the lateral pressure and its profile along the lipid molecules is related to the flexibility/mobility gradients, which has been extensively studied. How do your present results fit into this picture?

Cholesterol is perhaps the most widely studied moderator of flexibility/mobility in lipid bilayers, have you done experiments with cholesterol and what were the results?

Dr Templer responded: I am not clear what a mobility gradient is, but certainly the flexibility of the monolayer is related to the first moment of the lateral pressure. The first moment of the lateral pressure of the monolayer is proportional to κH_0 where κ is the bending modulus and H_0 is the spontaneous curvature. Our measurements indicate that the total lateral pressure in the chain region appears to be rising as we add DOPE to DOPC. This is entirely consistent with our knowledge of the product κH_0 which is increasing as we add DOPE to DOPC.

We have not done experiments with cholesterol yet.

Dr P. N. Edwards said: Excimer formation involves a reduction in the lipid-accessible surface area, particularly π-surface area, of pyrene molecules. Thus any differential interaction energy between saturated and olefinic lipid part-structures will alter the monomer : excimer ratio independent of pressure effects. The interaction energy will be significantly greater in the region of the *cis*-double bond and this will reduce excimer formation. Solution partial molar volumes of pyrene (or a more soluble derivative) in cyclohexane and cyclohexene could be used to demonstrate such effects.

Dr Templer responded: This is a subtle idea that I have to admit we had not thought of. As I understand, the statement you are making is that the dip we observe in our profile of η with respect to chain position may be an artefact of the probe's interaction with the oleic double bond. The idea would be that pyrene is more readily solvated by carbon double bonds so that there is hindrance to pyrene forming excimers as one approaches regions where the double bond is present. I have to accept that this is indeed a possibility and we will go on to test the idea.

Prof. Lee asked: Why is the order parameter low at the double bond in dioleoyl phosphatidylcholine? Isn't it a purely geometric effect due to the orientation of the C=C with respect to the long axis of the fatty acyl chain?

Is the excimer ratio higher for pyrene at the 4 position than at the 6 position *etc.*, not due to a difference in pressure in the membrane but rather due to tethering of the fatty acyl chains to the glycerol backbone, keeping the tops of the chains closer together than the bottoms of the chain?

Dr Templer responded: Yes, the low order parameter calculated by Fattal and Ben-Shaul[1] at the double bond is a geometric effect, but one might anticipate that the orientation of the bond relative to the chain segments above and below would lead to an enhancement of the lateral pressure below the double bond.

I am not sure that I understand the second question correctly. The excimer formation requires that the monomer is first excited and that before the monomer decays, that it comes into proximity with its (unexcited) companion. At this point it can form the excited dimer. Certainly the number of allowable chain conformations which give rise to an excimer are quite different between the top and bottom of the chain. It is these differences which give rise to the variation in the lateral pressure as we probe at different depths in the monolayer.

1   D. R. Fattal and A. Ben-Shaul, *Biophys. J.*, 1994, **67**, 983.

Dr Sansom asked: Can you distinguish between an effect of percentage DOPE on the lateral pressure along the bilayer normal as opposed to an effect on the degree of conformational freedom of the chains of dipyPC, unrelated to lateral pressure gradients but more related to changes in order parameter profile, which would also be expected to increase the probability of excimer formation?

Dr Templer responded: Once again I think the answer to this question is that the lateral pressure in a monolayer is a combination of several effects. First, the pressure is two-dimensional and in the plane of the monolayer because we have an interface. Second, colliding CH_2 groups are not disembodied particles since they are constrained by the configuration of all the CH_2 groups above them. This latter statement makes it clear that the lateral pressure is therefore related to the order parameter profile. Formally the chain lateral pressure is defined as the negative change in the free energy of the chains with respect to changes in the cross sectional area per chain. To calculate the lateral pressure one allows the lateral volume at some interval along a chain to increase whilst the lateral volumes at all other points and for all other chains is held constant and the change in the number of allowed conformations is calculated.

Prof. Bohne commented: The excimer-to-monomer ratio will depend on the monomer lifetime in the absence of excimer formation. Have you done time-resolved experiments for the monosubstituted lipids? In addition, time-resolved experiments when excimers are formed could be helpful to interpret if the decrease for the excimer-to-monomer ratio for 6dipyPC is due to a change in lateral pressure or to a change in the excimer and/or monomer photophysical behaviour.

Dr Templer responded: We have not yet done time resolved experiments. They are on our 'to do' list, and I could not agree more with the comments.

Prof. Bayerl said: In comparing order parameter profiles obtained by your approach with those of NMR, one has to consider the intrinsic timescales of the methods. What is the characteristic timescale of your method?

Dr Templer responded: The excited pyrene monomer has a lifetime in cyclohexane of 400 ns. We have not of course measured it in our system, but nevertheless one would anticipate a lifetime in the hundreds of nanoseconds region.

Prof. Svetina asked: Is there any reason to expect that the lateral pressure in the middle of the phospholipid membrane approaches zero as the sketch in Fig. 1 in your paper suggests?

Dr Templer responded: A sketch is of course only that, a sketch. Nevertheless, the statistical mechanical calculations made on bilayers by Szleifer and co-workers[1] do all show the lateral pressure to be zero at the middle of the bilayer. But a calculation is of course only a calculation.

1   I. Szleifer, A. Benshaud and W. M. Gelbart, *J. Phys. Chem.*, 1990, **94**, 5081.

Prof. Holzwarth opened the discussion of Dr Goñi's paper: Did you find a correlation between the activity of the enzyme and the induction of membrane fusion? Both processes should be temperature dependent and they are also dependent on the fluidity of the membrane.

It was demonstrated by several groups (Tsong *et al.*[1] and by Holzwarth *et al.*[2,3]), that lipid bilayers are especially sensitive to cross membrane processes in the middle of the main phase transition, because there are different states of order coexisting and lipid vesicles show a facetted surface instead of a spherical surface which is observed above the main phase transition. Our own experiments clearly showed that lipid vesicles can be handled for weeks, if they are kept several degrees above the main phase transition temperature, but the same preparations become very unstable if kept around the main phase transition temperature.

Your system is very complex and I wonder if you might not have some internal phase separation of the different components, which could induce fusion processes. Especially, cholesterol is known to cluster in membranes.

1   A. Genz, J. F. Holzwarth and T. Y. Tsong, *Biophys. J.*, 1986, **50**, 1043.
2   R. Groll, A. Böttcher, J. Jäger and J. F. Holzwarth, *Biophys. Chem.*, 1996, **58**, 53.
3   J. F. Holzwarth, *The Enzyme Catalysis Process*, ed. A. Cooper, J. L. Houben and L. C. Chien, Plenum, New York, 1989, pp. 383–412.

Dr Goñi responded: The relationship between enzyme activity and induction of membrane fusion is described in Fig. 3 in our paper. Both processes are seen to be temperature dependent. Under our conditions, the lipids (egg PC, egg PE and cholesterol) are always in the fluid state, *i.e.* well above the main gel–fluid transition. The transition that is referred to in Fig. 3 occurs from the fluid lamellar to the inverted cubic phase.

The complexity of the system is reflected in the rather complex phase diagram in Fig. 4. The transient formation of microdomains enriched in certain lipids cannot be excluded. In fact, the geometry of the fusion intermediate or stalk would favour a certain enrichment in 'conically-shaped' lipids, such as PE or cholesterol.

Prof. Svetina commented: The facilitation of vesicle fusion by the activity of phospholipase could also occur owing to the redistribution of diacylglycerol in between the two membrane leaflets, and the consequent release of the membrane elastic energy in the fusion process.[1] Do your observations support such a possibility?

1   S. Svetina, A. Iglič and B. Žekš, *Ann. N.Y. Acad. Sci.*, 1994, **710**, 179.

Dr Goñi responded: Redistribution of diacylglycerol between the two membrane leaflets may be very important in the latter stages of the fusion event, *i.e.* transition between stages C and D in your model. However, for the initial steps of vesicle aggregation and (putative) stalk formation, we propose that asymmetric, localized spots of diacylglycerol are essential.

Dr P. N. Edwards said: The flip-flop kinetics in your systems are important in the understanding of concentration gradients of diacylglycerol produced in membranes by the action of phospholipase C. What is known about such rate constants?

Dr Goñi responded: No precise data on the transmembrane or flip-flop rate of diacylglycerol in our system are available. However, Hamilton *et al.*[1] have measured, using NMR methods, a rate constant of *ca.* 62 s^{-1} ($t^{1/2}$ *ca.* 11 ms) at 38 °C for DAG transbilayer movement in a phosphatidylcholine bilayer.

1   J. A. Hamilton, D. T. Fujito and C. F. Hammer, *Biochemistry*, 1991, **30**, 2894.

Dr Templer commented: Although the evidence you have presented indicates that the activity of phospholipase is not affected by the presence of non-lamellar phases, doesn't your data indicate that it is affected by the curvature elastic stresses which are built up in the lipid bilayer? In other words the presence of a non-lamellar phase indicates that at least some of the curvature elastic stress has been reduced, and this might reduce the activity of phospholipase relative to a lamellar phase where the curvature elastic stress is still present.

Dr Goñi responded: Elastic stresses may well constitute one, perhaps the main, factor of what has been called the 'frustrated lamellar state' (ref. 31 of our paper). It is also true that, once the enzyme has been acting for some time, its rate decreases. However, the obvious effect of inhibition by substrate accumulation may, under those circumstances, be more important than the release of curvature elastic stress that will accompany the lamellar-to-nonlamellar phase transition.

Prof. Laggner said: A suggestion on the correlation between fusion propensity and phospholipase-C activity. The first step in fusion is the drainage of water from the space between opposing monolayers. I think polyethylene glycol could facilitate this local dehydration, which is also a prerequisite for phospholipase-C action.

Dr Goñi responded: The correlation between fusion propensity and phospholipase-C activity is discussed at various stages in the paper (see in particular Fig. 1–3 and Table 1). Removing water from the intermembrane space is certainly an important step in fusion. We have studied in detail the process of liposome fusion induced by polyethylene glycol that is partly driven by the ability of the polymer to remove free water from the aqueous suspension.[1] Interestingly, when polyethylene glycol is covalently bound to the lipid bilayer (ref. 12 in the paper) water cannot be removed from the interbilayer interstice, and fusion is inhibited. We have not tested as yet the combined actions of polyethylene glycol and phospholipase-C.

1   A. Alonso, R. Saez and F. M. Goñi, *FEBS Lett.*, 1982, **30**, 137.

Prof. Almgren said: In connection with the discussion of the effects of the bilayer lipid structure on the activity of the enzyme, it would be of interest to know if the curvature of the bilayer, for example a transformation to a reversed cubic membrane, affects the enzyme activity (and not only the presence of lipids with a tendency to induce such a transformation).

I would also like to draw attention to cubosomes[1] and other dispersed particle[2] of reversed phases as possible model systems in this field.

1   K. J. Larsson, *J. Phys. Chem.*, 1989, **93**, 7304; J. Gustafsson, H. Ljusberg-Wahren, M. Almgren and K. Larsson, *Langmuir*, 1997, **13**, 6964.
2   J. Gustafsson, T. Nylander, M. Almgren and H. Ljusberg-Wahren, *J. Colloid Interface Sci.*, 1999, **211**, 326.

Dr Goñi responded: Data such as those shown in Fig. 8 of this paper argue against a direct role of phase transitions in the regulation of enzyme activity. However, a more detailed exploration of this point would be welcome, and, in that context, cubosomes and related systems may be very useful.

Prof. Robinson asked: Can you provide some further information concerning experimental protocols? How, for example, is the enzyme introduced into the system?

The data in Table 1 shows that ganglioside additives have a very dramatic effect on the hydrolysis and fusion rates. Can you offer an explanation for these observations?

Dr Goñi responded: Further experimental details may be found in ref. 2, 6 and 7 of the paper. The enzyme is added to the preformed large unilamellar vesicles in suspension, at the start of the reaction.

The effect of gangliosides on enzyme activity has been explored in detail by Daniele et al.[1] Gangliosides do not affect the adsorption process of the enzyme. Instead, their modulatory effect occurs at the level of the interface, perhaps modifying the electrostatic field configuration across the interface, through their polar head group dipoles.

Further to their inhibitory properties on the enzyme hydrolytic activity, gangliosides exhibit a direct inhibitory effect on fusion itself, due to their ability to stabilize the lamellar phase, thus opposing the formation of any nonlamellar fusion intermediates.

1   J. J. Daniele, B. Maggio, I. D. Bianco, F. M. Goñi, A. Alonso and G. D. Fidelio, *Eur. J. Biochem.*, 1996, **239**, 105.

Miss Bucak communicated: What is the role of cholesterol in your membrane? Do you use it to make the membrane more rigid or because you want the membrane to be more biological as you carry out an enzymatic reaction?

Dr Goñi responded: Cholesterol is essential for fusion to occur under our conditions (see. ref. 2 in our paper). Our hypothesis is that cholesterol facilitates the formation of inverted nonbilayer structures (hexagonal, cubic), thus contributing to the architecture of the nonbilayer fusion intermediate or 'stalk' (ref. 7 and 15).

Characterization of the physical properties of model biomembranes at the nanometer scale with the atomic force microscope

Yves F. Dufrêne,[a,b,†] Thomas Boland,[a] James W. Schneider,[a] William R. Barger[a] and Gil U. Lee*[a]

[a] *Chemistry Division, Code 6177, Naval Research Laboratory, Washington, DC 20375-5352, USA. E-mail: glee@stm.2.nrl.navy.mil*

[b] *Center of Marine Biotechnology, University of Maryland, Columbus Center, 701 E. Pratt Street, Baltimore, MD 21202, USA*

Received 1st October 1998

Interaction forces and topography of mixed phospholipid–glycolipid bilayers were investigated by atomic force microscopy (AFM) in aqueous conditions with probes functionalized with self-assembled monolayers terminating in hydroxy groups. Short-range repulsive forces were measured between the hydroxy-terminated probe and the surface of the two-dimensional (2-D) solid-like domains of distearoyl-phosphatidylethanolamine (DSPE) and digalactosyldiglyceride (DGDG). The form and range of the short-range repulsive force indicated that repulsive hydration/steric forces dominate the interaction at separation distances of 0.3–1.0 nm after which the probe makes mechanical contact with the bilayers. At loads <5 nN the bilayer was elastically deformed by the probe, while at higher loads plastic deformation of the bilayer was observed. Surprisingly, a short-range repulsive force was not observed at the surface of the 2-D liquid-like dioleoylphosphatidylethanolamine (DOPE) film, despite the identical head groups of DOPE and DSPE. This provides direct evidence for the influence of the structure and mechanical properties of lipid bilayers on their interaction forces, an effect which may be of major importance in the control of biological processes such as cell adhesion and membrane fusion. The step height measured between lipid domains in the AFM topographic images was larger than could be accounted for by the thickness and mechanical properties of the molecules. A direct correlation was observed between the repulsive force range over the lipid domains and the topographic contrast, which provides direct insight into the fundamental mechanisms of AFM imaging in aqueous solutions. This study demonstrates that chemically modified AFM probes can be used in combination with patterned lipid bilayers as a novel and powerful approach to characterize the nanometer scale chemical and physical properties of heterogeneous biosurfaces such as cell membranes.

† Current address: Unité de Chimie des Interfaces, Université Catholique de Louvain, Place Croix du Sud 2/18, B-1348 Louvain-la-Neuve, Belgium.

Introduction

Many cellular phenomena, including cell adhesion and membrane fusion, involve specific molecular interactions and non-specific interactions between bilayer membranes. While the function of (glyco)proteins, such as integrins,[1] selectins[2] and lectins,[3] in these interactions has long been recognized, there is now growing evidence supporting the involvement of lipids. Gangliosides are charged glycosphingolipids found in most animal membranes which are thought to have important functions in cell adhesion, molecular recognition, oncogenesis and differentiation.[4-6] On the other hand, phosphatidylcholine, phosphatidylethanolamine and galactolipids, the most common lipids in animal and plant membranes, prevent membranes and vesicles from coming into contact and reacting, adhering, or fusing, due to strong repulsion between their highly hydrated head groups.[7-10]

Interaction forces between opposing supported lipid bilayers have been directly measured using the surface forces apparatus (SFA). For uncharged galactolipid bilayers in NaCl solutions, two types of forces have been identified, i.e. long-range attractive van der Waals forces and short-range repulsive hydration/steric forces.[8] A doubling of the range of the repulsive force was found when comparing monogalactosyldiglyceride with digalactosyldiglyceride. For phosphatidylethanolamine and phosphatidylcholine bilayers in the presence of millimolar levels of Ca^{2+} and Mg^{2+}, a third force was detected in agreement with theory, i.e. long-range electrostatic double-layer.[9] A second approach, the osmotic stress (OS) method, has been widely used to probe short-range hydration forces between lipid bilayers in multilamellar systems.[11-14] SFA and OS studies have provided valuable insight into membrane–membrane interactions, but they are not suited for probing local properties of complex biosurfaces, i.e. surfaces made of multiple component mixtures and exhibiting lateral organization at the nanometer scale.

Since its invention in 1986, atomic force microscopy (AFM, Fig. 1) has been widely used to image the surface of materials down to molecular resolution.[15] AFM has been applied to many biological systems due to its high resolution and ability to operate in aqueous environments. Subnanometer scale resolution has been achieved on proteins,[16-18] DNA–protein complexes[19,20] and bilayers,[21-24] but the resolution of cell surfaces has been limited to 10–30 nm[18,19,25,26] due to the complex interplay between topographic, mechanical and frictional forces. One solution to this problem is to reconstruct more rigid models for lipid bilayer membranes using the Langmuir–Blodgett (LB) deposition technique.[27,28] Such an approach has enabled several research groups to image supported lipid membranes at high resolution under physiological conditions.[22,24,29]

Besides imaging, the very high force sensitivity of the AFM has been used to measure forces with nanometer scale lateral resolution. Quantitative measurements have been made of van der Waals and electrostatic forces,[30-32] hydration forces,[31,33,34] steric repulsive forces,[35,36] specific

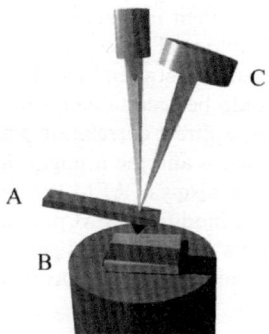

Fig. 1 Schematic of an atomic force microscope. A molecular resolution image of a surface is constructed by rastering a microfabricated probe over a surface under constant, low loads. The critical elements of the instrument are the microfabricated cantilever-probe (A), piezoelectric electromechanical actuator (B), optical-lever displacement detector (C) and digital feed-back loop (not shown). The dimensions of the cantilever can be modified to select a spring constant, and cantilevers have been microfabricated with spring constants of 100–0.0001 N m^{-1}. Microfabrication gives the force transducer a high temporal resolution due to the minute mass of the cantilever.

intermolecular interactions[37,38] and molecular mechanics.[39] Of particular interest is the mapping of interaction forces with nanometer scale resolution, by recording force–distance curve maps in the (x,y) plane.[40–42] A key issue that must be addressed when measuring forces with the AFM is the probe surface chemistry. The most commonly used probes for surface imaging are made of silicon oxynitride. These probes have a complex surface chemistry made of ionizable silanol and amine groups but also of contaminants.[43] Reproducible measurements of forces attributable to interactions between specific functional groups on a surface require rigorous control of the surface chemistry of the probe. Sharp probes bearing single chemical moieties, such as alcohol or aliphatic groups, can be produced using self-assembly of functionalized alkanethiols onto Au-coated probes. This method has allowed mapping of the spatial arrangement and interaction forces on surfaces patterned with different chemical groups.[44,45]

In this paper, we describe the nanometer scale topography, mechanical properties and surface forces, of lipid bilayers that model biological membranes in aqueous solution. To measure unambiguously short-range forces in the absence of electrostatic double-layer contribution, we used lipid bilayers with a net zero surface charge and probes modified by self-assembly of alkanethiol monolayers terminating in hydroxy groups. Bilayers made of phosphatidylethanolamines and galactosyldiglycerides were selected for several reasons: (i) these lipids are the most common in animal and plant membranes, respectively; (ii) controlled deposition of these bilayers on mica using the LB technique has been described previously;[9] (iii) data on surface forces between such bilayers are available from other techniques for comparison.[46] These results represent the first systematic analysis of short-range interaction forces of lipid bilayers using chemically defined AFM probes, and set the stage for the characterization of the physical properties of complex supported bilayers and possibly even living cells with molecular resolution.

Materials and methods

Preparation of thiol monolayers

Cleaved [0001] muscovite mica and AFM probes were functionalized with self-assembling alkanethiols using conditions optimized for the formation of organized monolayers.[47,48] To this end, mica surfaces and probes were coated with a 2 nm thick Cr layer followed by 15 nm of Au using electron beam thermal evaporation at high vacuum (CVC Products, Inc., Rochester, NY, USA). The gold coated surfaces were immediately immersed for 18 h in 0.2 mM ethanolic solutions of 16-hydroxyhexadecane-1-thiol (C_{16}-OH) and hexadecane-1-thiol (C_{15}-CH_3)[49] and then rinsed with ethanol to remove excess alkanethiol. The chemical composition of the C_{16}-OH coating was determined by X-ray photoelectron spectroscopy (XPS), using Au/Cr-coated silicon as a standard (Table 1). Sulfur was detected in significant concentrations at the C_{16}-OH surface, and a dramatic increase in the carbon surface concentration was observed, correlating with a decrease in the gold surface concentration. These results are consistent with previous XPS studies of alkanethiol monolayers.[47]

Table 1 Surface chemical composition [atomic fraction (%)] determined by XPS for Au/Cr-coated silicon and Au/Cr-coated silicon after self-assembly of C_{16}-OH[a]

Sample	Au $4f_{7/2}$	C 1s	O 1s	Cr $2p_{3/2}$	Si 2s	S 2p
Au/Cr-silicon	57	29	11	2	<0.5	<0.5
Au/Cr-silicon + C_{16}-OH	37	52	8	1	<0.5	3

[a] n-Type ⟨100⟩ silicon wafers (Wafernet, San Jose, CA, USA), cleaved into 1×1 cm^2 chips, were submitted to the same coating treatments as the cantilevers and directly analyzed by XPS (220iXL, Fisons, East Grinstead, UK). The pressure during analysis was $\sim 10^{-7}$ Pa. Spectra were collected with monochromatic Al-Kα X-rays with a nominal 150×800 μm spot size. The pass energy was 20 eV and the take-off angle was 90°. The atomic concentrations were calculated using the standard sensitivity factors with a program supplied with the instrument. Results are expressed as atomic fractions (%) with respect to the sum of all elements except hydrogen.

Preparation of supported bilayers

The lipids used in the study, dioleoylphosphatidylethanolamine (DOPE), distearoylphosphatidylethanolamine (DSPE) and digalactosyldiglyceride (DGDG) were obtained from Sigma Chemical Co. (St. Louis, MO, USA) and Matreya, Inc (Pleasant Gap, PA, USA) and their purity was confirmed with electrospray mass spectroscopy. For DGDG, the hydrocarbon chains were saturated and mostly made of 18 carbons. All lipids were dissolved at 0.5 mM in chloroform–methanol (4 : 1). Monolayers were spread at 25 °C at the air/water (triply distilled) interface of a KSV5000 Langmuir–Blodgett system (KSV Instruments, Helsinki, Finland), compressed at a rate of 1 mN m^{-1} min^{-1}, and transferred at a constant surface pressure of 25 mN m^{-1}. Mixed monolayers (1 : 1 molar ratio) of DSPE/DOPE and DGDG/DOPE were transferred onto mica for imaging in air or onto DSPE-coated mica substrates for imaging in water as described previously.[50] Water was replaced by 0.15 M NaCl solutions before AFM measurements. Pure lipid monolayers were deposited onto silicon wafers with a 250 Å thick oxide layer (Transition Technology Interactions, Sunnyvale, CA, USA) for ellipsometry.

Atomic force microscopy

Measurements were made at room temperature (20–25 °C) using an optical lever AFM equipped with a liquid cell (Nanoscope III, Digital Instruments, Santa Barbara, CA, USA). Thermal oxide sharpened microfabricated Si_3N_4 probe-cantilevers (Park Scientific Instruments, Mountain View, CA, USA) with a nominal tip radius of 25 nm were used. Topographic images were taken in the constant-deflection mode with an applied force kept as low as possible, *i.e.* < 1 nN, and scan rates of ~4 Hz. The sensitivity of the AFM detector was estimated using the slope of the retraction force curve, in the region where the probe and sample are in contact. Adhesion maps were obtained by reconstructing 64 × 64 force–distance curves per image and displaying the pull-off force measured for each force curve. Images were then resampled to 512 × 512 pixels. For quantitative interpretation of the force–distance curves, the non-linear response (displacement/applied voltage) of the piezoceramic actuator of the scanner was corrected using a capacitance-based displacement calibrator (Capacitec, Boston, MA). The cantilever spring constants, ranging from 0.01 to 0.6 N m^{-1}, were determined using micromachined reference cantilevers of precisely controlled spring constants following the procedure of Tortonese and Kirk.[51]

Ellipsometry

Ellipsometric measurements of monolayer properties were made in air at 44 wavelengths between 400 and 800 nm at angles of incidence of 65, 70, 75 and 85° (M-44 spectroscopic ellipsometry, J. A. Woollam Co, Inc.). The rotating analyzer ellipsometer uses a broad band light source, with a monochromator built into the solid state detector. The calcite Glan Taylor polarizer is mounted in a high precision rotation stage with accurate positioning below 0.01° and a maximum beam deviation of 0°1′. The angle of incidence, phase (Δ) and amplitude (Ψ) were known with an uncertainty of less than 0.005, 0.02 and 0.01°, respectively.

The optical properties of the films were fit with a three-layer model.[52] The four unknown physical parameters are the film thickness, film coverage and the real and imaginary parts of the permittivity of the film. For each wavelength, the ellipsometer yields the two reflection parameters Δ and Ψ, thus measurements at two angles of incidence should in principle be sufficient to solve for all unknowns. In our case, Ψ is largely independent of the layer thickness and experiments at additional angles were necessary. The optical constants of the initial bare gold or silicon surface were used as substrate references, and the optical constants of the organic layer were fit with a two parameter Cauchy model with no absorptive component ($k = 0$). Film thickness and optical constants of the film were then evaluated in an iterative procedure to satisfy variation with wavelength and angle of incidence.

Results

Knowledge of the physical and chemical properties of lipid bilayer membranes at high spatial resolution is a key step towards a better understanding of complex biological surfaces and cellular events such as cell adhesion and membrane fusion. Functionalized AFM probes have been used to investigate the nanometer scale surface properties of mixed lipid bilayers under aqueous conditions. To evaluate the relative contributions of short-range surface forces and mechanics in the measured forces, we first made force measurements between model hydrophilic and hydrophobic surfaces in water, *i.e.*, between C_{16}-OH functionalized AFM probes and either C_{15}-CH_3 or C_{16}-OH surfaces (Fig. 2A). The C_{16}-OH functionalized probes were then used to study the nanometer scale surface properties of mixed phospholipid/glycolipid bilayers prepared as follows: either DSPE or DGDG (saturated hydrocarbon chains) were mixed with DOPE (unsaturated hydrocarbon chains), and the mixed monolayers were transferred as a second layer onto DSPE-coated surface, yielding bilayers with hydrophilic phosphatidylethanolamine and galactoside head groups exposed to the aqueous phase (Fig. 2B).

Alkanethiol monolayers

The alkanethiol films were characterized with ellipsometry and contact angle measurements. The ellipsometric thickness of the C_{15}-CH_3 and C_{16}-OH films was 2.1 ± 0.1 nm (refractive index, $n = 3$) and 2.4 ± 0.1 nm ($n = 3$), respectively. The thickness of the C_{15}-CH_3 film is in quantitative agreement with earlier ellipsometric studies and confirms the alkanethiol forms a closed packed monolayer tilted 30° to the surface normal.[47] The hydroxy terminated C_{16}-OH film is 0.3 nm thicker than the methyl terminated C_{15}-CH_3 monolayer, which we attribute to the formation of a water monolayer on the C_{16}-OH film under ambient conditions. The water contact angle on the C_{16}-OH and C_{15}-CH_3 monolayers was $\sim 5°$ and $110 \pm 2°$, respectively, which is consistent with close packed monolayers.

The forces measured between the C_{16}-OH and C_{15}-CH_3 surfaces and C_{16}-OH probe in water are plotted as a function of surface separation in Fig. 3. The form of the force curve was not affected by the maximum force applied to the surface, for forces up to 20 nN. Only repulsive forces were observed as the C_{16}-OH surface was approached to and retracted from the C_{16}-OH probe, Fig. 3A. The lack of significant adhesive force suggests a strong repulsive force counterbalances the attractive interactions. In contrast to the C_{16}-OH/C_{16}-OH measurements, significant hysteresis was observed in the retracting force curve between the C_{15}-CH_3 surface and C_{16}-OH probe, Fig. 3B. 'Pull-off' forces of ~ 7.5 nN magnitude were observed, which are indicative of a strong attractive force when the probe is in contact with the surface.

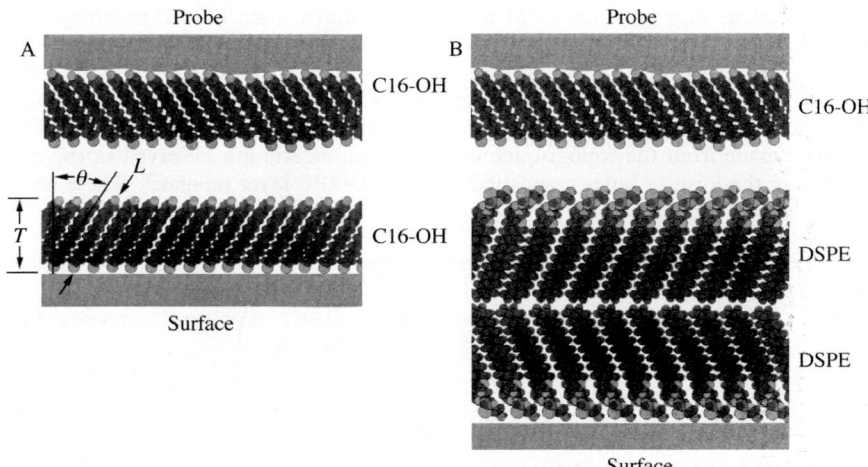

Fig. 2 Schematic diagrams of forces measurements made between (A) a C_{16}-OH modified probe and surface, and (B) a C_{16}-OH modified probe and a homogeneous DSPE/DSPE lipid bilayer. The space filling models of C_{16}-OH and DSPE were constructed using an energy minimization molecular dynamics algorithm (Chem3D, CambridgeSoft, Cambridge, MA, USA). The theoretical length (*L*) of the C_{16}-OH, DSPE and DGDG molecules are 2.4, 3.3 and 3.7 nm, respectively. The tilts (θ) of the DSPE and C_{16}-OH in this image were set at 25 and 30°, respectively, to reflect the measured thickness (*T*) of the monolayers.

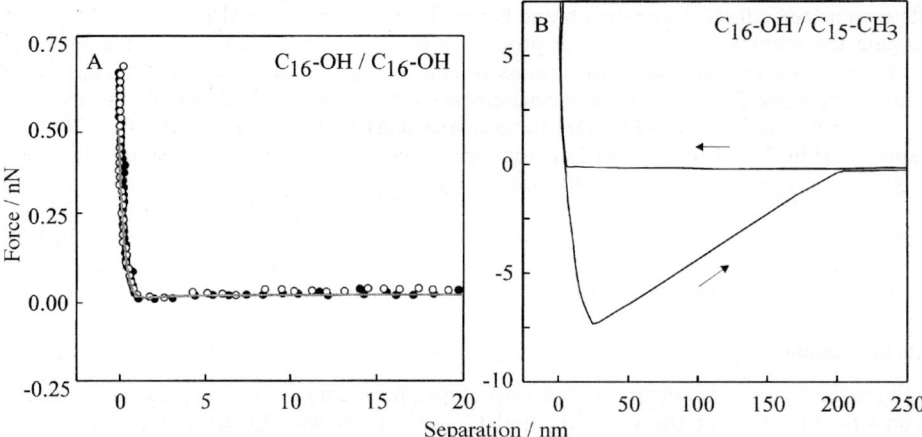

Fig. 3 Force measurements between a C_{16}-OH coated AFM probe and (A) C_{16}-OH and (B) C_{15}-CH$_3$ surfaces in water. The force and separation resolution of the two measurements were varied by an order of magnitude in order to capture the hydration force region of the C_{16}-OH surface and the adhesive force region of the C_{15}-CH$_3$ surface. Data in (A) are shown with circular symbols (1 symbol/5 data points) and the theoretical fit is shown with a solid line.

Supported lipid bilayers

Fig. 4 shows the surface pressure vs. area (π-A) isotherms of DSPE, DOPE, DSPE/DOPE (1:1), DGDG and DGDG/DOPE (1:1) monolayers at the air/water interface. While DSPE and DGDG monolayers show a two-dimensional 'solid-like' behavior, the DOPE monolayers show a 2-D 'liquid-like' behavior due to the unsaturated hydrocarbon chains of the molecule. The area per molecule, surface compression modulus and theoretical space filling (SF) thickness of pure lipid films were determined from the surface pressure diagrams at the deposition pressure of 25 mN m^{-1} and are summarized in Table 2.

Topographic images made of mixed DSPE/DOPE and DGDG/DOPE monolayers on DSPE-coated mica are presented in Fig. 5A and 6A, respectively, with topographic cross-sections. These images are consistent with previous AFM images of saturated lipid/DOPE monolayers[50] in that all the bilayers show phase-separation in the form of elevated microscopic domains surrounded by a continuous matrix. In light of the surface fractions covered by the domains in the different images as well as the molecular lengths calculated from SF models, the higher level domains in the images are ascribed to DSPE and DGDG and the lower level domains to DOPE. Two observations can be made from the topographic images. First, defects are observed in the continuous DOPE phase, in the form of holes associated with the DOPE layer images.[50] These holes appar-

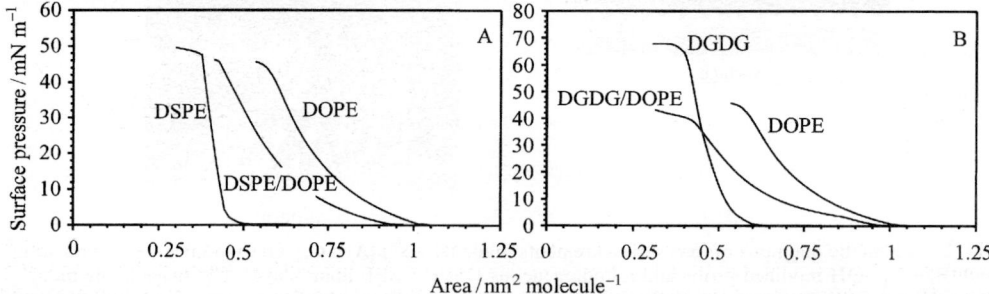

Fig. 4. Surface pressure–area (π–A) isotherms for pure and mixed (1:1) lipid monolayers at the air/water interface at 25 °C: (A) DSPE, DOPE and DSPE/DOPE mixture; (B) DGDG, DOPE and DGDG/DOPE mixture.

Table 2 Physical properties of the LB deposited lipid monolayer and bilayer films

Lipid	Area molecule^{-1}/ nm$^{2\,a}$	Surface compression modulus/ mN m$^{-1\,a}$	SF thickness/ nm^b	Ellipsometry		AFM step height	
				Monolayer thickness/ nm	Cauchy parameters A, B^c	Monolayer/ nm^d	Bilayer/ nm^e
DOPE	0.67	79 ± 1	1.9	1.9 ± 0.1	1.865, −0.0866	—	—
DSPE	0.41	183 ± 1	3.0	3.3 ± 0.1	1.456, −0.00856	1.8 ± 0.2	3.3 ± 0.3
DGDG	0.47	193 ± 1	2.7	2.9 ± 0.1	1.6045, −0.0416	2.1 ± 0.1	4.9 ± 0.4

^{a} At the deposition pressure of 25 mN m^{-1} of the films. ^{b} The theoretical space filling (SF) thicknesses of the monolayers (T) were determined from the area per molecule (a) of the lipids using a constant specific volume (v), i.e. $T = v/a$. A specific volume of 1.03 cm^3 g^{-1} was chosen; the molecular weights of the DGDG, DOPE and DSPE lipids are 778, 744 and 748, respectively. ^{c} The wavelength (λ) dependence of the refractive index (n) of the lipid films was fitted with a Cauchy form $n = A + B/\lambda$. ^{d} Mixed DSPE/DOPE and DGDG/DOPE monolayers supported on mica were made by raising freshly cleaved mica vertically through the air/water interface and imaged in air at forces of ∼1 nN. The step heights between DSPE and DGDG domains and the DOPE matrix were measured from cross-sections in the topographic images. Mean value and standard deviation of height differences of three independent experiments. ^{e} Step heights measured between the DSPE and DGDG domains and the DOPE matrix, for bilayers made of mixed DSPE/DOPE and DGDG/DOPE monolayers deposited onto DSPE-coated mica and imaged in 0.15 M NaCl solutions.

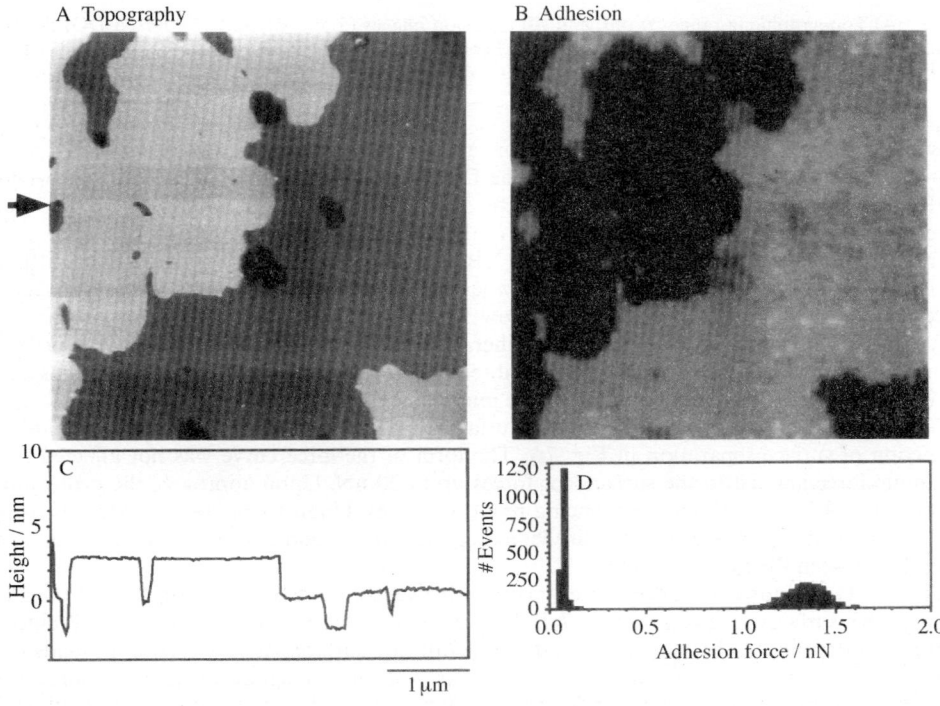

Fig. 5 (A) Topographic (z-range 20 nm) and (B) adhesion images (5 μm × 5 μm) of a mixed DSPE/DOPE monolayer onto DSPE-coated mica in a 0.15 M NaCl solution. Lighter levels in the images correspond to higher height and adhesion. The topographic image was obtained at an applied force of <1 nN, and the maximum force applied to the sample for adhesion force mapping was ∼1 nN. A cross-section (C) taken along a line indicated by the arrow is shown beneath the topographic image, while the histogram of the magnitude of the adhesion force (D) (4096 events per image) is shown beneath the adhesion map.

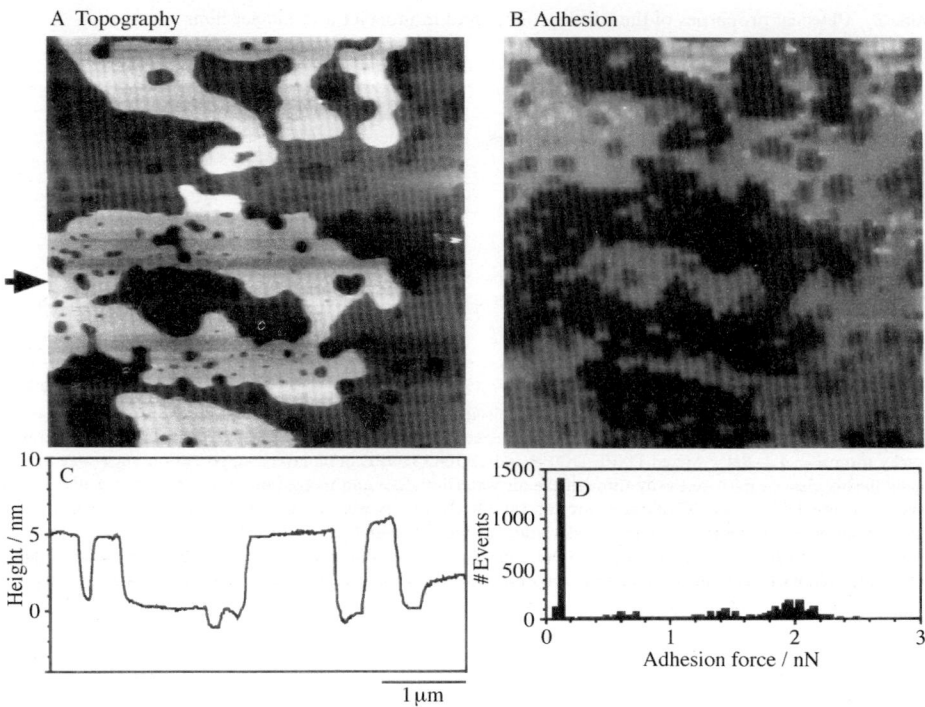

Fig. 6 (A) Topographic (z-range 20 nm) and (B) adhesion images (5 μm × 5 μm) of a mixed DGDG/DOPE monolayer onto DSPE-coated mica in 0.15 M NaCl solution. A cross-section and adhesion map are shown in Fig. (C) and (D), respectively.

ently form during or shortly after deposition and appear to be stable for over 24 h periods. Second, the step heights measured between the DSPE and DGDG domains and the surrounding DOPE matrix in the topographic images are 3.3 ± 0.3 and 4.9 ± 0.4 nm, respectively, clearly larger than predicted from the SF thicknesses of the bilayers.

The adhesion force maps shown in Fig. 5B and 6B exhibit high contrast, the DSPE and DGDG domains presenting systematically low levels of adhesion compared to DOPE. Indeed, the corresponding adhesion force histograms show two discrete levels of adhesion, one centered at about 0.1 nN for DSPE and DGDG, the other centered between 0.5 and 2 nN for DOPE (the magnitude of the adhesive forces can not be directly compared between the different films due to the variations in the radius of curvature of the microfabricated probe).

The force measured between the DOPE surface and a hydroxy-terminated probe is plotted as a function of surface separation in Fig. 7A. The form of the force curve was not affected by the maximum force applied to the surface, for forces up to 20 nN. Upon approach, the probe jumps-to-contact at 4.5 nm, without experiencing repulsive forces. Upon retraction, a hysteresis or pull-off force is observed, of ∼5 nN magnitude, indicating contact and a strong attractive interaction adhesion between the probe and film.

A set of typical force–distance curves recorded between a modified AFM probe and the DSPE and DGDG surfaces is given in Fig. 7 B–E. The curves clearly differ from those obtained on DOPE and depend on the maximum load applied on the surface. At low maximum loads (<1.5 nN), the approach and retraction curves are similar. The lack of significant jump-to-contact suggests that a strong repulsive force balances the attractive interactions between the lipid surface and the probe. Upon retraction no significant adhesion pull-off forces are observed. These two observations suggest that at low loads, molecular contact is not established between the probe and surface of the DSPE and DGDG domains.

In contrast, when the maximum applied force is increased the probe experiences a steep short-range repulsion upon approach, until it jumps into contact when the force is greater than 5 nN.

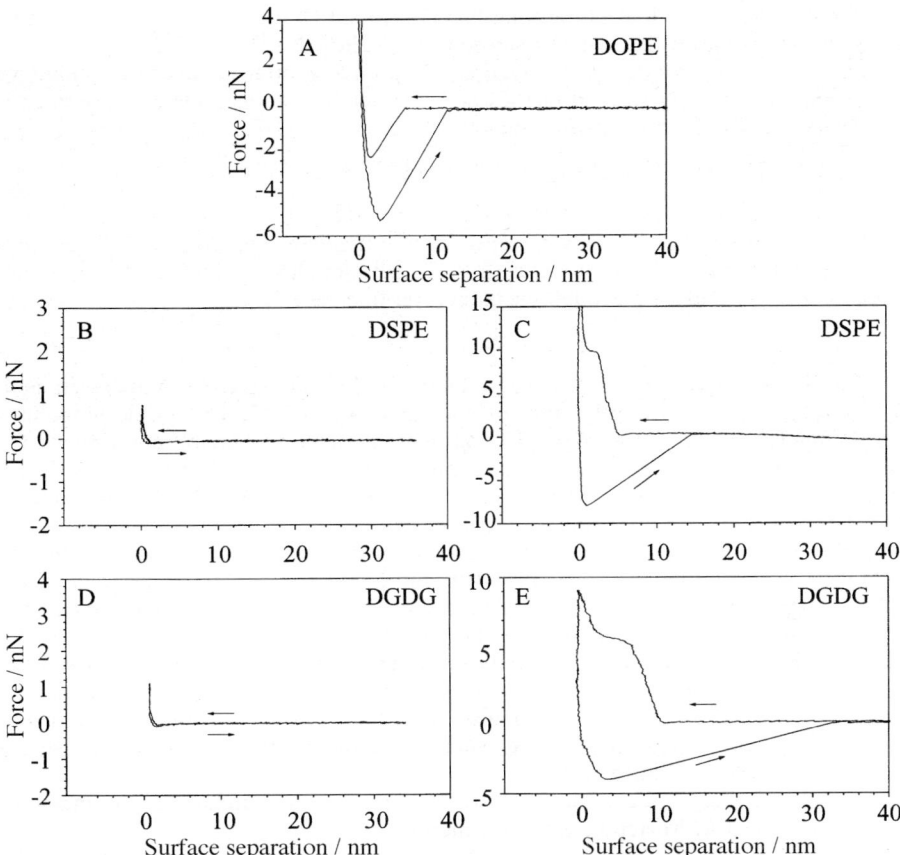

Fig. 7 Typical force–distance curves recorded over the (A) DOPE, (B,C) DSPE, (D,E) DGDG surfaces in 0.15 M NaCl solutions; (B,D): Low maximum applied loads (<1.5 nN), (C,E) high maximum applied forces (>1.5 nN).

The occurrence and form of the jump-to-contact were found to be probe dependent, which is consistent with other measurements of short-range forces with the AFM.[53] Upon retraction, adhesion forces of ~8 and 4 nN magnitude are found for DSPE and DGDG, respectively, suggesting that the lipid and alkanethiol head groups are in direct contact. The repulsive portion of the approach curves of the DSPE and DGDG can be fitted to an exponential form at low loads with decay lengths of 0.5 ± 0.2 nm and 0.8 ± 0.2 nm, respectively. At high loads, the data points no longer fit through an exponential function but through a straight line.

Discussion

Image contrast mechanisms of supported lipid monolayer

We have shown that the topographic contrast of mixed fatty acid and lipid monolayers in air is determined by three factors, *i.e.* length of the molecule (L), tilt of the molecule (θ) and the mechanical properties of the film.[50,54] The step height (ΔZ) measured between two lipid phases is

$$\Delta Z_M = \Delta T + (\delta_M - \delta_{M'}) \qquad (1)$$

where ΔT is the difference between the thickness of the two films, hereafter referred to as relative thickness, and δ is the deformation of film under the load of the AFM probe (M and M' designate the thinner and thicker film, respectively).

The thickness of the pure monolayers has been independently determined using a simple SF model and ellipsometry, Table 2. Two trends are evident in the data. First, both techniques produce similar thicknesses, second the solid-like DSPE and DGDG films are of similar thickness and are significantly thicker that the more loosely packed liquid-like DOPE film.

Using contact mechanics the elastic deformation produced by the probe under a known load can be determined if the physical properties of the system are known. Given an applied load P and probe radius R, the Hertzian deformation is

$$\delta = (9P^2/16RE^{*2})^{1/3} \quad (2)$$

where $E^* = [(1 - v_1^2)/E_1 + (1 - v_2^2)/E_2]$ and E_1, v_1 and E_2, v_2 are Young's modulus and Poisson's ratio of the probe and sample, respectively.[55] If volume compressibility is neglected, the isothermal thickness compression modulus (E_T) of a LB monolayer is related to the surface compression modulus (E_a, Table 2) through the monolayer thickness[56,57]

$$E_T = E_a/T \quad (3)$$

Using ellipsometric thicknesses the E_T of the DOPE, DPSE and DGDG monolayers is 4.3, 6.1 and 7.1×10^7 N m^{-2}, respectively, and for soft materials $v = 0.5$. The elastic modulus of the alkanethiol film is about 2×10^{10} N m^{-2} [58] and thus will make a negligible contribution to E^*. If one assumes the mechanical properties of the supported monolayer are dominated by the LB film, the calculated deformation produced by a 50 nm radius probe at 0.5 nN of load in the DOPE, DSPE and DGDG films is 1.0, 0.8 and 0.7 nm, respectively.

Using the ellipsometric thickness of the films and the Hertzian film deformations, the theoretical AFM step height of the mixed DSPE/DOPE and DGDG/DOPE monolayers in air (ΔZ_M) is 1.6 and 1.3 nm, respectively. The DSPE/DOPE monolayer ΔZ_M and measured step height (Table 2) is in reasonable agreement, but ΔZ_M of the DGDG/DOPE film is 0.8 nm smaller than the measured step height. We attribute the significant difference between the theoretical and the measured DGDG/DOPE step height to a water film between the lipid head groups and mica surface. This nanometer scale water film is produced by the repulsive hydration–steric interactions between the lipid head groups and surface, which make the water film's thickness highly dependent on the form of the lipid head group.

Fig. 8A schematically presents how the water film's thickness influences the measured step height. The theoretical AFM step height of the monolayers is

$$\Delta Z_M = \Delta T + (\delta_M - \delta_{M'}) + (\delta_{L'} - \delta_L) \quad (4)$$

where δ_L and $\delta_{L'}$ is the thickness of the water film associated with the thin and thick lipid film, respectively. Interpretation of the force–distance curves (see discussion below) indicates that the range of the hydration–steric interaction (δ_F) of the DGDG, DSPE and DOPE phase is ≈ 1.2, 0.3 and 0 nm, respectively. If we assume that the water film thickness is equal to the range of the repulsive hydration–steric force, ΔZ_M for the DSPE/DOPE and DGDG/DOPE monolayers is 1.9 and 2.5 nm, respectively. The theoretical step heights are in good agreement with their respective measured step heights (Table 2). This discussion indicates that the measured step height of mixed lipid monolayers is determined from molecular length, molecular tilt, mechanical properties and the nature of the lipid head group–surface interaction. The contribution of all these factors to the AFM imaging mechanism makes step height a sensitive probe of the physical state of a bilayer.

Image contrast mechanisms of supported lipid bilayers

The step heights measured between the DSPE and DGDG domains and the surrounding DOPE for the bilayers in water are much larger than the step heights measured for the corresponding monolayers in air (Table 2). The increase in the bilayer step height is unexpected, as no new factors have been introduced into the image contrast mechanism. However, comparison of topographic images of the bilayers (Fig. 5A and 6A) and monolayers[50] indicates that deposition of the mixed monolayer in water changes its physical state. Specifically, the formation of holes in the DOPE phase indicates that DOPE desorbs from the mixed monolayer during or shortly after deposition.

A mechanism emerges to explain the image contrast of the mixed bilayers from this observation (Fig. 8B). At the low imaging force used in this study (<1 nN), topographic imaging of the DOPE region is conducted with the probe indenting into the film. We expect the deformation of

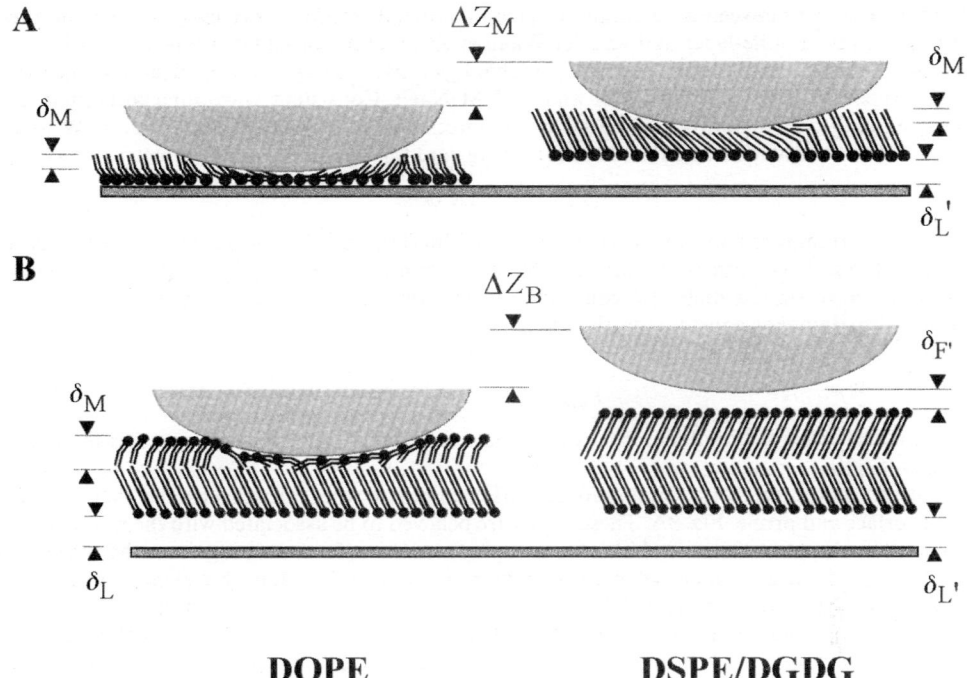

Fig. 8 Schematic diagram of the probe position on the lipid monolayer in air (A) and lipid bilayer in water (B). The unsaturated and saturated lipid–probe interaction is presented on the left and right side of the diagrams, respectively. δ_M is the deformation of the lipid monolayers, δ_L is the thickness of the water layer, ΔT is the relative thickness of the films, δ_F is the distance of the probe from the surface due to repulsive forces and ΔZ is the step height measured by the AFM.

the DOPE phase by the probe to be significantly larger in water than it was in air, as reorganization will produce a significant decrease in elastic modulus of this phase (Fig. 4). Reorganization of the DOPE phase does not appear to change the DSPE and DGDG internal surface pressure as the surface area occupied by these phases does not increase significantly from air to water images.[50] Imaging of the DSPE and DGDG regions appears to take place in a short-range repulsive force mode with the probe some distance off the surface of the lipid film. Noncontact imaging has also been reported using repulsive long-range double layer interactions.[32,42,59] The theoretical step height of the bilayer resulting from this model is

$$\Delta Z_B = \Delta T + \delta_M + \delta_{F'} \qquad (5)$$

where ΔT is the relative thickness of the monolayer. Using the ellipsometric thickness of the films and assuming the probe penetrates through the entire DOPE film ($\delta_M \approx 1.9$ nm), ΔZ_B for the mixed DSPE/DOPE and DGDG/DOPE bilayers is 3.6 and 4.1 nm, respectively. The theoretical and measured step heights are in reasonable agreement considering the assumptions that have been made in estimating δ_M and δ_F. Accordingly, the image contrast mechanism of lipid bilayers in water is controlled by the same factors that have been identified for monolayers in air, but the physical properties of the mixed lipid film is significantly changed by deposition onto a DSPE monolayer on the down stroke.

Surface force measurements between alkanethiol monolayers

Both imaging and interaction force measurements on the mixed lipid bilayers were made with a C_{16}-OH functionalized probe. To gain an understanding of surface forces between alkanethiol surfaces in water, we made force measurements between C_{16}-OH functionalized probes and model hydrophilic (C_{16}-OH) and hydrophobic (C_{15}-CH$_3$) surfaces.

Forces measured between amphiphilic surfaces in aqueous solutions typically result from long-range electrostatic double-layer and van der Waals (vdW) interactions, and short-range hydration/steric interactions. The role of electrostatic double-layer interactions was tested by making force measurements on each surface in water and 0.15 M NaCl. Consistent with our expectation that the surfaces were uncharged, no difference in the forces could be detected. As a rule, vdW forces between surfaces of identical chemistry are attractive, and for a sphere on flat geometry

$$F_{vdW} = -AR/6r^2 \quad (6)$$

where A is the Hamaker constant, R is the radius of curvature of the sphere and r is the distance between surfaces.[60] As both probe and sample surfaces are made of hydrocarbon based monolayers, which have similar dielectric constants ($\varepsilon \sim 2$), we anticipate that the vdW forces will be attractive with Hamaker constants in the range of $4-7 \times 10^{-21}$ J.[61,62] Hydration/steric forces are short-ranged and have the form

$$F_{hydration} = F_o\, e^{-(r/\lambda)} \quad (7)$$

where F_o is the preexponential factor, r is the separation between the surfaces and λ is the hydration decay length.

Repulsive hydration/steric interaction clearly dominate the force measured between the C_{16}-OH surface and probe, Fig. 3A. These forces are believed to be associated with the reorientation of water molecules at surfaces[12–14,63] and entropic fluctuation forces arising from the overlap of thermally excited surface modes of molecular vibrations, including long-wavelength undulation forces and molecular scale protrusion forces.[10,64,65] A reasonable fit of the total force curve (solid line in Fig. 3A) is obtained for decay lengths of 0.3 nm, a preexponential factor of 2 nN, a vdW radius of 100 nm and Hamaker constant of 6×10^{-21} J, if the vdW origin is set 0.6 nm behind the point of zero separation (PZS) of the measured forces (the AFM measures relative displacement and PZS is defined as the position of the probe at maximum force). A vdW radius larger than the specified AFM probe was used for the vdW force calculation, which is justified due to the pyramidal shape of the probe. Locating the vdW plane behind the PZS suggests that a water layer is bound to the surfaces at the PZS, and a distance of 0.6 nm corresponds to two monolayers of water. These results are similar to the SFA and OS measurements between lipid bilayers in which repulsive hydration forces have an exponential form with 0.3 nm decay length.[8,9,11–14]

Significant hysteresis is observed in the retracting force curve between the C_{15}-CH_3 surface and C_{16}-OH probe, Fig. 3B. The strong adhesion between the C_{15}-CH_3 surface and C_{16}-OH probe appears to result from a decrease in the magnitude of the hydration force between these surfaces and/or an increase in the vdW forces. These results are similar to force measurements made between identical hydrophobic surfaces in that strong adhesion is observed between the surface but differ in that the long-range attractive hydrophobic effect is not observed.[66]

Forces between lipid bilayers and alkanethiol-functionalized probes: interplay of hydration/steric forces and mechanical effects

As in the case of alkanethiol monolayers, the interaction forces between the C_{16}-OH probe and lipid bilayers is expected to result from a balance between repulsive hydration/steric and attractive vdW interactions. At low loads the surface forces on the solid-like DSPE and DGDG films appear to be dominated by repulsive hydration/steric interactions, Fig. 7B and D. This seems consistent with measurements of surface forces between symmetric lipid films with the SFA and OS, but on closer examination there are several important differences. First, exponential fits of the repulsive portion of the DSPE and DGDG forces curve result in decay lengths of 0.5 ± 0.2 and 0.8 ± 0.2 nm, respectively, which are 2–5 times larger than the decay lengths measured with the SFA and OS.[8,9,11–14] Second, the probe jumps to contact over a distance of 5–8 nm at loads >5 nN, Fig. 7C and E. If the jump-to-contact were associated with an attractive vdW force this would place the observed repulsive force beyond the range typically associated with hydration/steric forces.

The AFM and SFA instruments are based on the same physical principles and use similar geometries, *i.e.*, the equivalent of a sphere on a flat, thus they should generate similar results. On closer consideration, however, there is a significant difference in the pressure ranges over which the AFM and SFA typically make measurements. The theoretical Hertzian pressure generated by a ~50 nm radius sphere at 1 nN of load is 2×10^7 Pa. SFA measurements are typically made at pressures less than 10^5 Pa due to compliance of an epoxy used in the mechanical loop. Direct comparison to OS measurements is complicated by differences in geometry and the isotropic nature of the osmotic force.

The high pressures in the AFM suggests that the mechanical properties of the bilayer may contribute to the measured forces. The role of bilayer mechanics is supported by fitting the repulsive portion of the DSPE and DGDG force curve with models that combined hydration/steric forces with mechanical deformation. A reasonable fit of the repulsive portion of the DSPE force curve, Fig. 9A, is achieved with the combined model

$$\delta = \lambda \ln(P/F_o) + (9/16RE^{*2})^{1/3} P^{2/3} \qquad (8)$$

where $\lambda = 0.35$ nm, $F_o = 0.8$ nN, $R = 50$ nm and $E^* = 1.7 \times 10^8$ N m^{-2}, if mechanical contact (Hertian plane) takes place 0.3 nm after the hydration/steric force can first be detected (hydration/steric plane). The repulsive portion of the DGDG force curves, Fig. 9B, is also reasonably well fitted with a combined model for $\lambda = 0.3$ nm, $F_o = 0.15$ nN, $R = 50$ nm and $E^* = 2.1 \times 10^8$ N m^{-2}, if the Hertzian plane is placed 1.2 nm inside the hydration/steric plane. The distance separating the hydration from the vdW origin reflects the range of the hydration/steric force. The difference in the range over the DSPE and DGDG films is directly correlated with differences in the measured step heights, thus confirming the imaging mechanism discussed above.

The different behaviors observed for DSPE and DGDG surfaces may be understood in terms of differences in hydration and steric properties. The fact that the mechanical component of the force is applied to DSPE after only 0.3 nm may be explained by the relatively low degree of hydration of the phosphatidylethanolamine (PE) head group.[12,46] This has been attributed to the fact that PEs form a compact lattice that is stabilized by strong hydrogen bonding between the ammonium and phosphate groups.[67] In contrast, the fact that the mechanical model only becomes relevant at a separation of ~1 nm for DGDG is consistent with the mobility of the digalactoside head group and the large number of hydration water molecules associated with head group.[67]

As the load is increased mechanical deformation of the bilayers by the probe plays an increasingly important role in the measured forces of the bilayer surfaces. Several observations suggest that the jump-to-contact results from a transition in the mechanical behavior of the bilayer from elastic to plastic deformation, representing the penetration of the probe through the bilayer to the

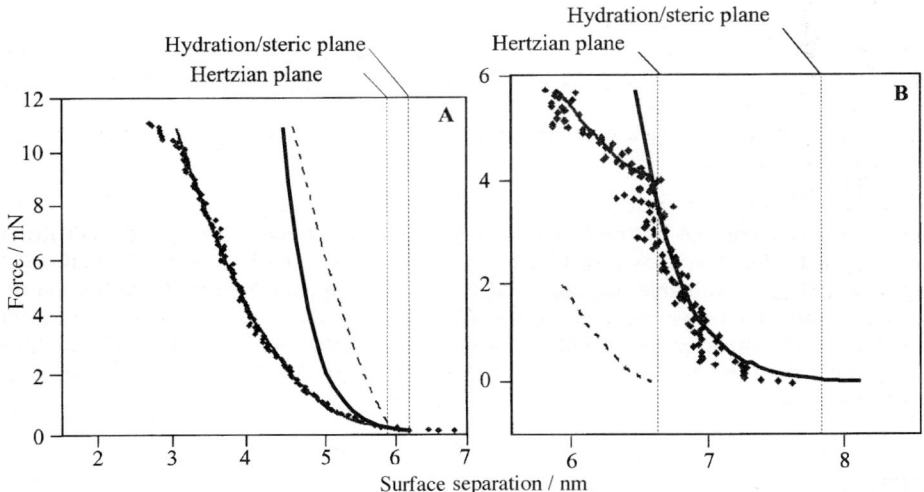

Fig. 9 Approaching force curve recorded over (A) DSPE and (B) DGDG domains with theoretical fits: ♦ raw data, ---- Hertzian contribution, —— hydration/steric forces and —— combined theoretical force curve. The relative separation axis has been set with its origin at the point at which the probe jumps-to-contact with the surface.

mica surface. First, the distance over which the jump to contact takes place is in reasonable agreement with the bilayer thickness, i.e., $T_{DSPE/DSPE} = 6.6$ nm and $T_{DSPE/DGDG} = 6.2$ nm. Second, the strong adhesive forces between the probe and surface after jump-to-contact are consistent with the forces measured between the hydrophilic and hydrophobic surfaces (Fig. 3B). Third, nanoindentation studies of thick LB films suggest the hardness of the film to be on the order of 10^8 N m^{-2}.[68] If we assume the yield pressure of the films is equivalent to hardness, the transition from elastic to plastic deformation should take place at loads >5 nN. A similar conclusion was reached by Ducker and Clarke[69] for self-assembled zwitterionic surfactant monolayers in water. Accordingly, these results lead us to conclude that on DSPE and DGDG the probe experiences repulsive forces that are a combination of hydration/steric forces and mechanics.

Summary

Although the surface properties and interaction forces of lipid bilayers have been investigated for many years, direct information at high lateral resolution has been inaccessible up to now. This study demonstrates that chemically functionalized AFM probes can be used to characterize the surface properties (surface morphology, surface forces and mechanics) of heterogeneous lipid bilayers on the nanometer scale.

Short-range repulsive forces are detected between the hydroxy-terminated probe and the 2-D solid-like DSPE and DGDG domains. The ranges and exponential decay lengths of the forces are larger than those found so far between amphiphilic bilayers using the OS and SFA techniques, which may be attributed to the fact that mechanical properties of surfaces are sampled at the high loads that can be achieved with the AFM. This highlights the important role that mechanics may play in biomolecular interactions. The direct correlation between the range of the repulsive force over the two solid lipid domains and the step heights in the topographic images validates the quantitative analysis of the measured forces and sheds new light on the fundamental mechanisms of AFM imaging of lipid bilayers in aqueous solutions. Surprisingly, no repulsive force is found over the 2-D liquid-like DOPE region, which is related to the low packing density of this phase. Taken together, these results provide direct evidence for the influence of the structure of lipid bilayers on their interaction forces, a behavior that may be of prime importance in the control of cellular interactions.

The methodology developed here has promising applications for the nanometer scale mapping of the properties (mechanical properties and molecular interactions) of biosurfaces and the study of surface forces in complex, asymmetric systems. Cell surface components such as gangliosides, integrins, and lectins can be incorporated into a patterned bilayer and characterized with AFM probes functionalized with relevant (bio)chemical moieties. This would greatly improve our understanding of the molecular mechanisms of cellular processes such as cell adhesion, membrane fusion, molecular recognition, and intercellular communication.

Acknowledgements

This research was supported by the Office of Naval Research (ONR) and by a NATO Research Fellowship (Y.F.D.). We also thank M. Fletcher, David L. Allara and R.J. Colton and all members of NRL Code 6177 for valuable discussion, M.L. Stevens and J.-B. Green for calibration of the AFM, M.D. Porter for providing some of the alkane thiols, M. Tortonese for supplying micromachined reference cantilevers, N.H. Turner for performing XPS analysis, K.Lee for programming associated with the adhesion images and J. Callahan and M. Shahgholi for electrospray mass spectroscopy analysis.

References

1 R. O. Hynes, *Cell*, 1992, **69**, 11.
2 A. Varki, *Proc. Natl. Acad. Sci. USA*, 1994, **91**, 7390.
3 N. Sharon and H. Lis, *Science*, 1989, **246**, 227.
4 W. Curatolo, *Biochim. Biophys. Acta.*, 1987, **906**, 137.

5   N. Sharon and H. Lis, *Sci. Am.*, 1993, January, 74.
6   S-I. Hakomori, *Prog. Brain Res.*, 1994, **101**, 241.
7   R. G. Horn, *Biochim. Biophys. Acta*, 1984, **778**, 224.
8   J. Marra, *J. Colloid Interface Sci.*, 1985, **107**, 446.
9   J. Marra and J. Israelachvili, *Biochemistry*, 1985, **24**, 4608.
10 J. Israelachvili and H. Wennerström, *Nature (London)*, 1996, **379**, 219.
11 D. M. LeNeveu, R. P. Rand and V. A. Parsegian, *Nature (London)*, 1976, **259**, 601.
12 R. P. Rand and V. A. Parsegian, *Biochim. Biophys. Acta*, 1989, **988**, 351.
13 S. Leikin, V. A. Parsegian, D. C. Rau and R. P. Rand, *Annu. Rev. Phys. Chem.*, 1993, **44**, 369.
14 T. J. McIntosh and S. A. Simon, *Annu. Rev. Biophys. Biomol. Struct.*, 1994, **23**, 27.
15 G. Binnig, C. F. Quate and Ch. Gerber, *Phys. Rev. Lett.*, 1986, **56**, 930.
16 A. L. Weisenhorn, B. Drake, C. B. Prater, S. A. C. Gould, P. K. Hansma, F. Ohnesorge, M. Egger, S.-P. Heyn and H. E. Gaub, *Biophys. J.*, 1990, **58**, 1251.
17 H. G. Hansma, M. Bezanilla, F. Zenhausern, M. Adrian, R. L. Sinsheimer, *Nucleic Acids Res.*, 1993, **21**, 505.
18 M. Fritz, M. Radmacher and H. E. Gaub, *Biophys J.*, 1994, **66**, 1328.
19 M. Radmacher, R. W. Tillmann, M. Fritz and H. E. Gaub, *Science*, 1992, **257**, 1900.
20 H. G. Hansma and J. H. Hoh, *Annu. Rev. Biophys. Biomol. Struct.*, 1994, **23**, 115.
21 H-J. Butt, K. H. Downing and P. K. Hansma, *Biophys. J.*, 1990, **58**, 1473.
22 J. A. N. Zasadzinski, C. A. Helm, M. L. Longo, A. L. Weisenhorn, S. A. C. Gould and P. K. Hansma, *Biophys. J.*, 1991, **59**, 755.
23 J. H. Hoh, G. E. Sosinksky, J-P. Revel and P. K. Hansma, *Biophys. J.*, 1993, **65**, 149.
24 S. W. Hui, R. Viswanathan, J. A. Zasadzinski and J. N. Israelachvili, *Biophys. J.*, 1995, **68**, 171.
25 H-J. Butt, E. K. Wolff, S. A. C. Gould, B. Dixon-Northern, C. M. Peterson and P. K. Hansma, *J. Struct. Biol.*, 1990, **105**, 54.
26 J. H. Hoh and C-A. Schoenenberger, *J. Cell Sci.*, 1994, **107**, 1105.
27 L. K. Tamm and H. M. McConnell, *Biophys. J.*, 1985, **47**, 105.
28 E. Sackmann, *Science*, 1996, **271**, 43.
29 J. Mou, J. Yang and Z. Shao, *J. Mol. Biol.* 1995, **248**, 507.
30 W. A. Ducker, T. J. Senden and R. M. Pashley, *Nature (London)*, 1991, **353**, 239.
31 H-J. Butt, *Biophys. J.*, 1991, **60**, 1438.
32 T. J. Senden, C. J. Drummond and P. Kékicheff, *Langmuir*, 1994, **10**, 358.
33 S. J. O'Shea, M. E. Welland and T. Rayment, *Appl. Phys. Lett.*, 1992, **60**, 2356.
34   J. P. Cleveland, T. E. Schäffer and P. K. Hansma, *Phys. Rev. B*, 1995, **52**, R8692.
35 S. Biggs, *Langmuir*, 1995, **11**, 156.
36 R. M. Overney, D. P. Leta, C. F. Pietroski, M. H. Rafailovich, Y. Liu, J. Quinn, J. Sokolov, A. Eisenberg and G. Overney, *Phys. Rev. Lett.*, 1996, **76**, 1272.
37 G. U. Lee, D. A. Kidwell and R. J. Colton, *Langmuir*, 1994, **10**, 354.
38 E-L. Florin, V. T. Moy and H. E. Gaub, *Science*, 1994, **264**, 415.
39 G. U. Lee, L. A. Chrisey and R. J. Colton, *Science*, 1994, **266**, 771.
40 D. R. Baselt and J. D. Baldeschwieler, *J. Appl. Phys.*, 1994, **76**, 33.
41 M. Ludwig, W. Dettmann and H. E. Gaub, *Biophys. J.*, 1997, **72**, 445.
42 C. Rotsch and M. Radmacher, *Langmuir*, 1997, **13**, 2825.
43 G. U. Lee, L. A. Chrisey, C. E. O'Ferrall, D. E. Pilloff, N. H. Turner and R. J. Colton, *Isr. J. Chem.*, 1996, **36**, 81.
44 C. D. Frisbie, L. F. Rozsnyai, A. Noy, M. S. Wrighton and C. M. Lieber, *Science*, 1994, **265**, 2071.
45 J-B. D. Green, M. T. McDermott and M. D. Porter, *J. Phys. Chem.*, 1995, **99**, 10960.
46 J. Marra, *J. Colloid Interface Sci.* 1986, **109**, 11.
47 R. G. Nuzzo and D. L. Allara, *J. Am. Chem. Soc.*, 1983, **105**, 4481.
48 C. D. Bain, E. B. Troughton, Y-T. Tao, J. Evall, G. M. Whitesides and R. G. Nuzzo, *J. Am. Chem. Soc.*, 1989, **111**, 331.
49 R. G. Nuzzo, L. H. Dubois and D. L. Allara, *J. Am. Chem. Soc.*, 1990, **112**, 558.
50 Y. F. Dufrêne, W. R. Barger, J-B. D. Green and G. U. Lee, *Langmuir*, 1997, **13**, 4779.
51 M. Tortonese and M. Kirk, *SPIE*, 1997, **3009**, 53.
52 H. G. Tomkins, *A User's Guide to Ellipsometry*, Academic Press, Inc., San Diego, 1993.
53 S. J. O'Shea, M. E. Welland and T. Rayment, *Appl. Phys. Lett.*, 1992, **61**, 2240.
54 D. D. Koleske, W. R. Barger, G. U. Lee and R. J. Colton, *Mat. Res. Soc. Symp. Proc.*, 1997, **464**, 377.
55 K. L. Johnson, *Contact Mechanics*, Cambridge University Press, Cambridge, 1985.
56 E. A Evans and R. M. Hochmuth, in *Current Topics in Membranes and Transport*, ed. A. Kleinzeller and F. Bronner, Academic Press, New York, 1978, pp. 1–10.
57 E. A. Evans and P. Kwok, *Biochemistry*, 1982, **21**, 4874.
58 K. J. Tupper, and D. W. Brenner, *Langmuir*, 1994, **10**, 2335.
59 S. Manne, J. P. Cleveland, H. E. Gaub, G. D. Stucky and P. K. Hansma, *Langmuir*, 1994, **10**, 4409.
60 S. Nir, *Prog. Surf. Sci.*, 1976, **8**, 1.
61 D. Gingell and V. A. Parsegian, *J. Theor. Biol.*, 1972, **36**, 41.

62 D. B. Hough and L. R. White, *Adv. Colloid Interface Sci.*, 1980, **14**, 3.
63 S. Marcelja and N. Radic, *Chem. Phys. Lett.*, 1976, **42**, 129.
64 J. N. Israelachvili and H. Wennerström, *Langmuir*, 1990, **6**, 873.
65 J. N. Israelachvili and H. Wennerström, *J. Phys. Chem.*, 1992, **96**, 520.
66 J. N. Israelachvili and R. M. Pashley, *Nature (London)*, 1982, **300**, 341.
67 J. M. Boggs, *Biochim. Biophys. Acta*, 1987, **906**, 353.
68 T. P. Weihs, Z. Nawaz, S. P. Jarvis and J. B. Pethica, *Appl. Phys. Lett.*, 1991, **59**, 3536.
69 W. A. Ducker and D. R. Clarke, *Colloids Surf. A*, 1994, **93**, 275.

Paper 8/07637G

Modelling and simulation of light-activated membrane proteins: Dynamical transitions in bacteriorhodopsin

Christian Simon,[a] Malika Aalouach[a] and Jeremy C. Smith*[a,b]

[a] *Laboratoire de Simulation Moleculaire, Section de Biophysique des Protéines et des Membranes, DBCM, CEA-Saclay, 91191 Gif-sur-Yvette CEDEX, France*
[b] *Lehrstuhl für Biocomputing, IWR, Universität Heidelberg, Im Neuenheimer Feld 368, D-69120 Heidelberg, Germany*

Received 2nd September 1998

Many of the functions of membranes are carried out by proteins associated with them. A knowledge of atomic-detail membrane protein structures and dynamics is required for a full understanding of these functions. We briefly discuss recent progress in this field using modelling and simulation. One of the best characterised membrane proteins, bacteriorhodopsin, undergoes dynamical transitions with temperature. Here we present preliminary results of molecular dynamics simulation of this protein as a function of temperature, indicating the presence of dynamical transitions at approximately the temperatures seen experimentally.

Introduction

Many of the functions of biological membranes are performed by proteins bound to them. Among the roles of these proteins are the reception/transmission of messages and/or the transport of materials. However, due to difficulty in their crystallization only a small number of atomic-detail three-dimensional structures exist for membrane-spanning proteins. Among the few known structures are those of the light-transducing proteins, the photosynthetic reaction centre and bacteriorhodopsin.[1-4] In the present paper we briefly review some recent progress in the modelling and simulation of light-activated membrane proteins before presenting new results on dynamical transitions in bacteriorhodopsin as a function of temperature.

The paucity of crystallographic structures has led to a bottleneck in structural membrane protein research and adds impetus to the development of computer modelling techniques for determining their structures. One of these techniques is homology modelling: it can be possible to determine an unknown protein structure by using a known X-ray structure as a three-dimensional template, if there is sequence homology between the two. The higher the sequence homology the higher the probability of obtaining a reliable model structure.

An example of homology modelling at high sequence identity is the recent work on the photosynthetic reaction centre protein from the bacterium *Rhodobacter capsulatus*.[5] This protein has been the subject of a considerable amount of molecular biological and spectroscopic work aimed at improving our understanding of the primary steps of photosynthesis. A structural model was derived by combining information from the experimental structure of the highly homologous (54% sequence identity) reaction centre from *Rhodopseudomonas viridis*[1] with molecular mechanics and simulated annealing calculations. In the *Rb. capsulatus* model the orientations of the bacteriochlorophyll monomer and bacteriopheophytin cofactors on the pathway inactive in electron transfer differ significantly from those in the reaction centre of *Rps. viridis*. The orientational difference was found to be in agreement with linear dichroism measurements.[6] Moreover, the pattern of cofactor hydrogen-bonding to the protein was found to be in agreement with optical spectroscopic experiment.[7] The *Rb. capsulatus* model was used to provide an explanation as to why a partially symmetrized mutant *Rb. capsulatus*, which has been of particular interest for experiments on primary excited states in photosynthesis, lacks an electron acceptor bacteriopheophytin (BPh_L).[8–10] Conformational energy calculations on the partially symmetrised mutant and several BPh_L-binding revertants also provided an explanation for the relative BPh_L-binding properties of the proteins, in terms of interactions involving two residues in the binding pocket, these being a tryptophan and a methionine.[10]

Modelling at lower sequence homology, although less reliable, can be useful for suggesting experiments as part of an iterative procedure to obtain structural information on a membrane protein of particular interest. An example of this is the recent modelling of the photosystem II reaction centre core in plants for which a model was constructed by exploiting homology existing with the bacterial reaction centre proteins.[11]

In the rare cases where high-resolution experimental structures do exist modelling and simulation can be undertaken so as to refine structural detail and to understand physically how structure leads to function. A good example of such a system is bacteriorhodopsin (bR), a membrane protein that functions as a light-driven proton pump in the purple membrane of the bacterium *Halobacterium halobium*.[12] The light-absorbing chromophore in bR is a retinal molecule that is covalently bonded *via* its Schiff base to the ε-amino group of Lys 216.[13] The characteristic purple colour of bacteriorhodopsin is due to absorption by the chromophore. The absorption is red-shifted with respect to that of related model compounds in solution, an effect that has been proposed to originate from interactions between the retinal and its polar environment in the protein.[14]

The retinal interactions may include hydrogen bonds with the Schiff base. Structures for bR at high resolution have been obtained.[2,3] These revealed a channel through the protein that includes the Schiff base. Site-directed mutagenesis experiments suggest that the channel contains the pathway for proton transfer through bR.[15–18] A considerable amount of data exist that suggest that the proton transfer channel is at least partially hydrated. Low resolution neutron diffraction using contrast variation has indicated that about four water molecules are present in the neighborhood of the Schiff base although their positions in the direction perpendicular to the membrane plane could not be accurately determined.[19] There is, however, considerable other evidence that water molecules are directly associated with the Schiff base. A resonance Raman study suggests that a negatively charged counterion located near the Schiff base group is stabilized by water molecules.[20] Solid state ^{13}C and ^{15}N NMR experiments led to a model being proposed in which a water molecule is directly hydrogen-bonded to the Schiff base.[21] Other solid state ^1H and ^{15}N NMR experiments suggest that there is a direct exchange of the Schiff base NH hydrogen with bulk water.[22] A recent resonance Raman study of the Schiff base hydrogen–deuterium exchange also led to the conclusion that a water molecule is directly hydrogen bonded to the Schiff base NH proton.[23] Finally, the recent crystallographic structure of Pebay-Peyroula *et al.* has directly identified some water molecules associated with the Schiff base.[3]

Clearly, a detailed understanding of Schiff base hydrogen bonding in the various stages of the photocycle will be required for a complete description of bR function. Computational chemistry has an important role to play in resolving such questions, by identifying and quantifying hydrogen-bonding geometries and energies of pertinent model systems. For example, quantum chemistry and molecular mechanics techniques have been combined to determine the geometries and energetics of retinal–water interactions.[24,25] *Ab initio* molecular orbital calculations were used to determine potential surfaces for water–Schiff base hydrogen bonding and to characterize the energetics of rotation of the C—C single bond distal and adjacent to the Schiff base NH group. The *ab initio* results were combined with semiempirical quantum chemistry calculations to produce a data set used for the parameterization of a molecular mechanics energy function for retinal. Using the resulting molecular mechanics force field the hydrated retinal and associated bR protein environment were energy minimized and the resulting geometries examined. Two distinct sites were found in which water molecules can make hydrogen-bonding interactions: one near the NH group of the Schiff base in a polar hydrophilic region directed towards the extracellular side, and the other near a retinal CH group in a relatively hydrophobic region directed towards the cytoplasmic side.

To enable further investigations of internal hydration in bR and other systems a statistical mechanical formulation was derived that can be employed using molecular dynamics (MD) simulation to calculate the free energy of transfer of a small molecule from one environment to a specific site in another using molecular dynamics simulation.[26] The method was used to calculate the free energy of transfer of water molecules from the bulk to individual sites in the proton transfer channel of bR. The channel contains a region lined primarily by nonpolar side-chains. The results obtained indicate that the transfer of water molecules from bulk water to this apparently hydrophobic region is thermodynamically favorable. The presence of two water molecules in direct hydrogen-bonding association with the Schiff base was also found to be thermodynamically allowed.

Once a complete structural model of bR is obtained theoretical investigations into the photocycles of this protein can be envisaged. One interesting aspect of this in bR is the phenomenon of dark-adaptation, in which retinal is found to exist in both *all-trans* and (13,15) *syn* conformations, in approximately equal proportions. A theoretical investigation into dark-adaptation has been initiated. Initial free energy molecular dynamics calculations on a model of the isolated retinal suggested that the *all-trans* form is strongly favoured *in vacuo*.[27] Calculations of factors influencing the conformational free energy difference in the protein are now in progress.

To fully understand bR function, structural and thermodynamic examination must necessarily be complemented by dynamical investigations. Several analyses using molecular dynamics have been reported.[28–33] In this respect it is of considerable interest that dynamical transitions have been found in bR as a function of temperature, and have been correlated with function.[34,35] In what follows we present preliminary results on the transitions investigated with molecular dynamics. The atomic position mean-square displacements are computed from a number of simulations at temperatures between 20 K and 300 K. Dynamical transitions are observed in the simulations at ~ 150 K and ~ 240 K.

Methods

Molecular dynamics

The model system consists of 3544 atoms of bR of which 1806 are hydrogens. Four internal water molecules were included, placed according to crystallographic data[3] and each within 1 Å of the positions derived in ref. 27. The model system was subjected to molecular dynamics simulation using version 25 of the CHARMM program[36] with the potential function described in ref. 37, 26 and 27. The function includes bonded interactions (bond stretches, bond angle bendings and dihedral and improper torsions) and nonbonded pairwise interactions represented by 12–6 Lennard-Jones and Coulombic electrostatic potentials, which were cut-off at 12 Å. The Coulomb term was smoothed by multiplying by a switching cubic function between 8 and 12 Å.

The bR model molecule was simulated without an explicit environment. To approximately mimic the effect of the environment the relative permittivity was set to 1 and the α carbon atoms of the residues most surface exposed were harmonically restrained:[28] the force constant used for this has a small value of 0.2 kcal mol^{-1} Å^{-2}. This value was chosen so as to prevent gross deviation from the experimental structure while allowing internal flexibility. The system was energy minimized using 500 steps of Steepest Descent minimization followed by 2500 steps of Adopted Basis Newton–Raphson.[36] The final RMS gradient was 0.15 kcal mol^{-1} Å^{-1}. The energy-minimized structure was used as a starting point for the MD simulations.

The equations of motion were solved using the Verlet algorithm. The SHAKE algorithm was applied to fix the lengths of the bonds involving hydrogen atoms. A 2 fs time step was used for integration of the equations of motion. MD simulations were performed at thirty temperatures, from 10 to 300 K at 10 K intervals as follows. The minimized structure was heated to 10 K over 1 ps, equilibrated over 5 ps then 10 ps of production was performed in microcanonical ensemble. The final frame of the production was then used for 1 ps heating to 20 K and the procedure

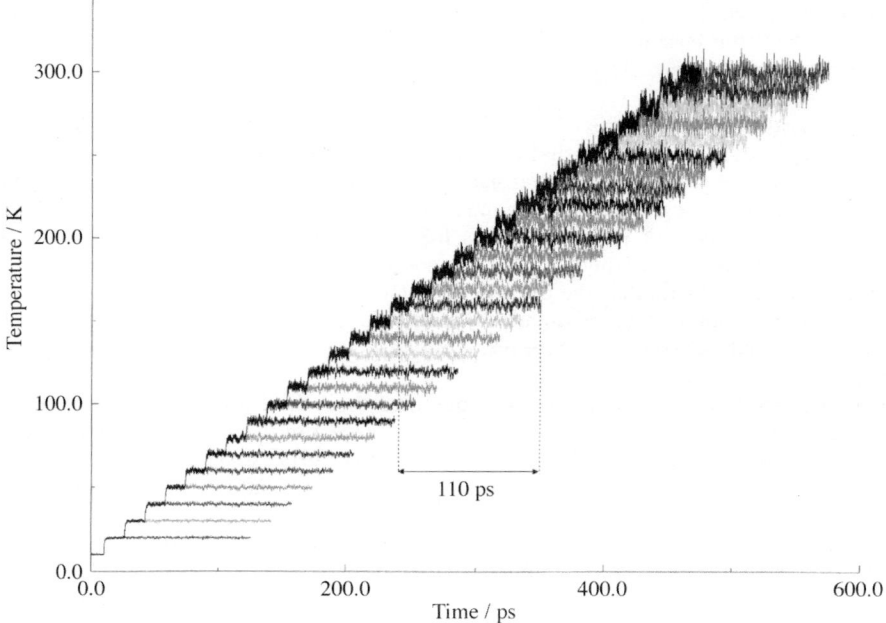

Fig. 1 Temperature–time series calculated from the MD simulations.

repeated. Subsequently the production runs were each extended by 100 ps, yielding thirty 110 ps-long production trajectories at each temperature. The configurations were dumped to disk every 100 fs (50 steps) *i.e.*, 1100 conformations per trajectory.

Experimental connection

Dynamical transitions of bR have been observed by incoherent neutron scattering experiments. The measurable quantity in incoherent neutron scattering, the dynamical structure factor, is the time Fourier transform of the intermediate scattering function $I(q, t)$, where q is the scattering wavevector and t is the time. $I(q, t)$ is the sum of the $I(q, t)$ of the individual atoms.[38] In the case of scattering by a single atom in a harmonic potential $I(q, t)$ is Gaussian in q:[39,40]

$$I(q, t) = \exp[-q^2\gamma(t)] \quad (1)$$

where $\gamma(t) = \frac{1}{6}\langle d^2(t)\rangle$ and $\langle d^2(t)\rangle$ is the mean-square displacement, defined as:

$$\langle d^2(t)\rangle = \langle [R(t) - R(0)]^2\rangle \quad (2)$$

where $R(t)$ is the atomic position vector, and $\langle \cdots \rangle$ indicates an ensemble average. The infinite time limit of $\langle d^2(t)\rangle$ is:

$$\langle u^2\rangle = \lim_{t\to\infty}\langle d^2(t)\rangle \quad (3)$$

Assuming $\lim_{t\to\infty}\langle R(t)R(0)\rangle = 0$, then

$$\langle u^2\rangle = 2\langle \delta R^2\rangle \quad (4)$$

where $\delta R(t) = R(t) - \langle R\rangle$, the displacement from the mean at instant t. The quantities $\langle d^2(t)\rangle$ and $\langle \delta R^2(t)\rangle$ can be extracted from molecular dynamics trajectories by replacing the ensemble average with a time average. However, the infinite time $\langle u^2\rangle$ can be obtained only from simulations long enough to sample all the motions involved. Correspondingly, the experimental $\langle u^2\rangle$ can be obtained only when the instrumental energy resolution is sufficiently good to resolve all the contributing motions. The neutron cross-section of a protein is dominated by the hydrogen atoms.

Results

Stability of the simulation

Fig. 1 shows time series of the temperature for each simulation. No significant drift is seen at any temperature. The total energy was also found to be stable. The root mean square positional deviation (RMSD) between each bR conformation of the trajectory at each temperature and the initial, energy-minimized structure is plotted in Fig. 2. The RMSD increases with temperature, but remains approximately constant with time for most temperatures.

Existence of dynamical transitions

In Fig. 3, the simulation-derived hydrogen-atom $\langle u^2 \rangle$ is plotted against temperature. The increase in $\langle u^2 \rangle$ between 20 and 300 K is approximately 0.6 Å^2, in accord with the neutron results obtained for dry purple membrane (PM).[41] An inflection is seen at ~150 K, a temperature at which a dynamical transition in bR has been experimentally reported.[35,41]

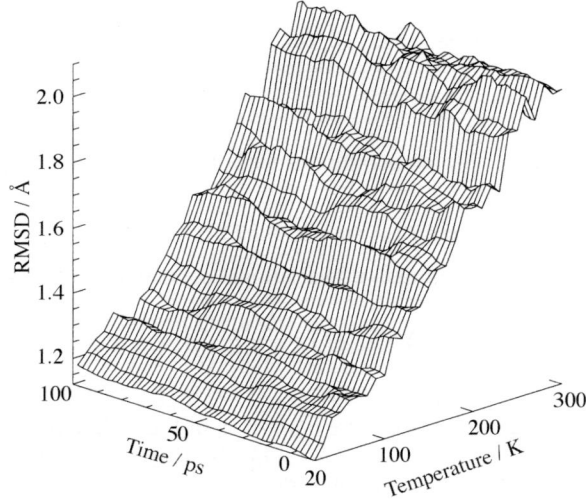

Fig. 2 Time series of the RMSD from the initial energy minimized structure, at each temperature averaged over all the bR atoms.

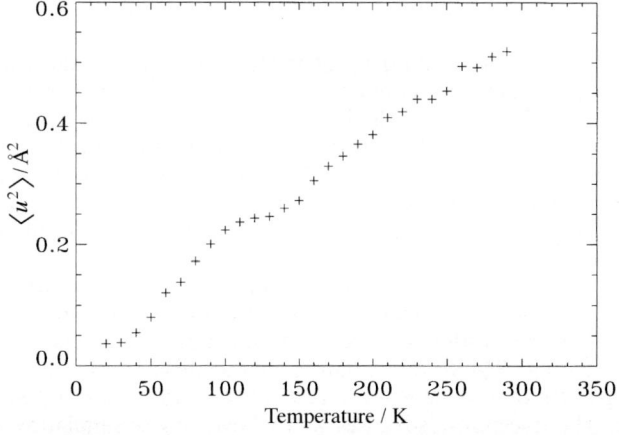

Fig. 3 Simulation-derived mean square displacement averaged over the hydrogen atoms, vs. temperature.

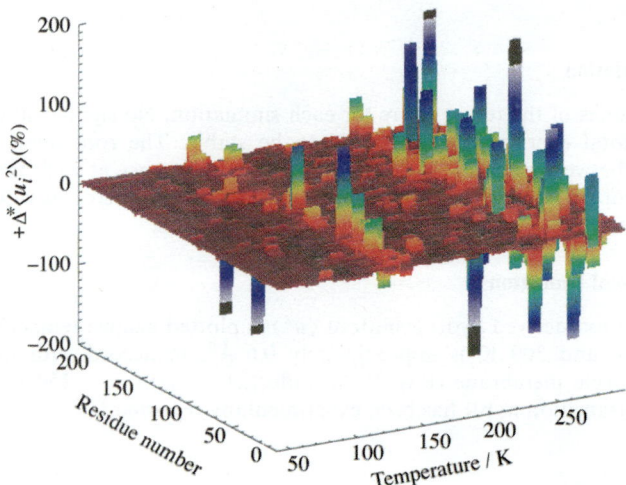

Fig. 4 Temperature dependance of the normalized variational contribution of the individual residues to the mean square displacement. The residues displacements were averaged over the hydrogen atoms.

We introduce the variation with temperature of $\langle u_i^2 \rangle$, the mean-square displacement of residue i:

$$\Delta \langle u_i^2 \rangle (T) = \langle u_i^2 \rangle (T + \delta T) - \langle u_i^2 \rangle (T) \qquad (5)$$

where δT is 10 K in this case. The normalized variational contribution of residue i is:

$$\Delta^* \langle u_i^2 \rangle (T) = \frac{\Delta \langle u_i^2 \rangle (T)}{\Delta \langle u^2 \rangle (T)} \qquad (6)$$

where $\Delta \langle u^2 \rangle = \sum_i \Delta \langle u_i^2 \rangle$.

$\Delta^* \langle u_i^2 \rangle$ is plotted against T and the residue number, i, in Fig. 4. Below 150 K all residues have approximately equal $\Delta^* \langle u_i^2 \rangle$, close to zero. At ~140 K a dynamical transition is revealed by variation of $\Delta^* \langle u_i^2 \rangle$ for some residues. A second transition occurs at ~240 K and above involving larger variation of $\Delta^* \langle u_i^2 \rangle$ than the ~150 K transition, and concerning a larger number of residues. A dynamical transition at ~240 K has also been reported in neutron scattering work, and has been correlated to the activation of bR function.[34]

Conclusions

The modelling and simulation of membrane protein structures and dynamics is still in its infancy, but will be of growing importance as more and more sequences of membrane proteins are determined. As structural research progresses the investigation of associated dynamical properties can also be expected to gain importance. The presence of a dynamical transition with temperature in water-soluble proteins has been recognized for some while. The recent neutron results on bacteriorhodopsin have demonstrated the presence of transitions also in a membrane protein.[35,41] The present preliminary results indicate that transitions may also be apparent in molecular dynamics simulation.

More research will be required to test further the present findings. In particular, simulations of bR with an explicit membrane environment *i.e.* with the protein in trimer form with lipid and water surroundings rather than with harmonic constraints can be envisaged, although they will be computationally demanding. Significant variation in the dynamical transition properties with environmental changes has been documented. Tests of the temperature hysteresis would also be interesting to make. The question also arises as to how long a simulation would have to be performed at any given temperature to obtain converged mean-square displacements. Finally, an examination using MD of the effect of the application of the Gaussian approximation [Eqn. (1)] to the intermediate scattering function would be of interest.

Two transitions are seen in Fig. 4, at ~ 150 and ~ 240 K. These are close to the temperatures at which transitions were seen experimentally. The higher-temperature transition is at about the water-melting temperature. But the present simulation was performed in the absence of water, indicating that water is not required for it. However, the ~ 240 K transition is not apparent in the $\langle u^2 \rangle$ data in Fig. 3, and may not therefore correspond to that seen experimentally. Work on this and other related questions is in progress.

Acknowledgements

We thank E. Pebay-Peroula for providing the crystallographic structure and D. Mihailescu, J. Baudry and B. Costescu for useful discussions and preliminary calculations.

References

1. J. Deisenhofer, O. Epp, K. Miki, R. Huber and H. Michel, *Nature (London)*, 1985, **318**, 618.
2. R. Henderson, J. M. Baldwin, T. A. Ceska. F. Zemlin, E. Beckmann and K. H. Downing, *J. Mol. Biol.*, 1990, **213**, 899.
3. E. Pebay-Peyrouola, G. Rummel, J. P. Rosenbusch and E. M. Landau, *Science*, 1997, **277**, 1676.
4. Y. Kimura, D. G. Vassylyev, A. Miyazawa, A. Kidera, M. Matsushima, K. Mitsuoka, K. Murata, T. Hirai and Y. Fujiyoshi, *Nature (London)*, 1997, **389**, 206.
5. N. Foloppe, M. Ferrand, J. Breton and J. C. Smith, *Proteins: Structure, Function Genetics*, 1995, **22**(3), 226.
6. J. Breton, E. J. Bylina and D. C. Youvan, *Biochemistry*, 1989, **28**, 6423.
7. T. Mattioli, personal communication.
8. M. H. Vos, F. Rappaport, J. C. Lambry, J. Breton and J. L. Martin, *Nature (London)*, 1993, **363**, 320.
9. S. J. Robles, J. Breton and D. C. Youvan, *Science*, 1990, **248**, 1402.
10. N. Foloppe, M. Ferrand and J. C. Smith, *Chem. Phys. Lett.*, 1995, **242**, 238.
11. B. Svensson, C. Etchebest, P. Tuffery, P. van Kan, J. C. Smith and S. Styring, *Biochemistry*, 1996, **35**, 14486.
12. D. Oesterhelt and W. Stoeckenius, *Nature (London), New Biol.*, 1971, **233**, 149.
13. K. J. Rothschild, P. V. Argade, T. N. Earnest, K-S. Huang, E. London, M-J. Liao, H. Bayley, H. G. Khorana and J. Herzfeld, *J. Biol. Chem.*, 1982, **257**, 8592.
14. R. A. Mathies, S. W. Lin, J. B. Ames and W. T. Pollard, *Annu. Rev. Biophys. Biophys. Chem.*, 1991, **20**, 491.
15. T. Mogi, L. J. Stern, N. R. Hackett and H. G. Khorana,.*Proc. Natl. Acad. Sci. USA*, 1987, **85**, 5595.
16. T. Mogi, L. J. Stern, T. Marti, B. H. Chao and H. G. Khorana, *Proc. Natl. Acad. Sci. USA*, 1988, **84**, 5595.
17. L. J. Stern and H. G. Khorana, *J. Biol. Chem.* 1989, **264**, 14202.
18. T. Marti, H. Otto, T. Mogi, S. J. Rösselet, M. P. Heyn and H. G. Khorana, *J. Biol. Chem.*, 1991, **266**, 6919.
19. G. Papadopoulos, N. Dencher, G. Zaccaï and G. Büldt, *J. Mol. Biol.* 1990, **214**, 15.
20. P. Hildebrandt and M. Stockburger, *Biochemistry*, 1984, **23**, 5539.
21. H. J. M. De Groot, S. O. Smith, J. Courtin, E. van der Berg, C. Winkel, J. Lugtenburg, R. G. Griffin and J. Herzfeld, *Biochemistry*, 1990, **29**, 6873.
22. G. S. Harbison, J. E. Roberts, J. Herzfeld and R. G. Griffin, *J. Am. Chem. Soc.*, 1988, **110**, 7221.
23. H. Deng, L. Huang, R. Callender and T. Ebrey, *Biophys. J.*, 1994, **66**, 1129.
24. M. Nina, B. Roux and J. C. Smith, in *Structures and Functions of Retinal Proteins*, ed. J. L. Rigaud, Colloque INSERM/John Libbey Eurotext Ltd., 1992, vol. 221, pp. 17–20.
25. M. Nina, J. C. Smith and B. Roux, *J. Mol. Struc. (THEOCHEM)*, 1993, **286**, 231.
26. B. Roux, M. Nina, R. Pomes and J. C. Smith, *Biophys. J.*, 1996, 670.
27. J. Baudry, S. Crouzy, B. Roux and J. C. Smith, *J. Chem. Inf. Comp. Sci.*, 1997, **37**(6) 1018.
28. M. Ferrand, G. Zaccaï, M. Nina, J. C. Smith, C. Etchebest and B. Roux, *FEBS Lett.*, 1993, **327**, 256.
29. W. Humphrey, I. Logunov, K. Schulten and M. Sheves, *Biochemistry*, 1994, **33**, 3668.
30. W. Humphrey, E. Bamberg and K. Schulten, *Biophys. J.*, 1997, **72**(3), 1347.
31. I. Logunov and K. Schulten, *J. Am. Chem. Soc.*, 1996, **118**(40), 9727.
32. D. Xu, C. Martin and K. Schulten, *Biophys. J.*, 1996, **70**(1), 453.
33. D. Xu, M. Sheves and K. Schulten, *Biophys. J.*, 1995, **69**(6), 2745
34. M. Ferrand, A. J. Dianoux, W. Petry and G. Zaccai, *Proc. Natl. Acad. Sci. USA*, 1993, **90**, 9668.
35. V. Réat, H. Patzelt, M. Ferrand, C. Pfister, D. Oesterhelt and G. Zaccai, *Proc. Natl. Acad. Sci. USA*, 1998, **95**, 4970.
36. B. R. Brooks, R. E. Bruccoleri, B. D. Olafson, D. J. States, S. Swaminathan and M. Karplus. *J. Comput. Chem.*, 1983, **4**, 187.
37. A. D. MacKerell, Jr., D. Bashford, M. Bellot, R. L. Dunbrack, Jr., J. D. Evanseck, M. J. Field, S. Fischer, J. Gao, H. Guo, S. Ha, D. Joseph-McCarthy, L. Kuchnir, K. Kuczera, F. T. K. Lau, C. Mattos, S.

Michnick, T. Ngo, D. T. Nguyen, B. Prodhom, W. E. Reiher, III, B. Roux, M. Schlenkrich, J. C. Smith, R. Stote, J. Straub, M. Watanabe, J. Wiórkiewicz-Kuczera, D. Yin and M. Karplus, *J. Phys. Chem. B*, 1998, **102**, 3586.
38 M. Bée, *Quasi Elastic Neutron Scattering: Principles and Applications in Solid State Chemistry, Biology and Materials Science*, Adam Hilger, Bristol, 1998, UK.
39 A. Rahman, K. Singwi and A. Sjölander, *Phys. Rev.*, 1962, **126**(3), 986.
40 A. Rahman, *Phys. Rev.*, 1963, **140**(4), 1334.
41 V. Réat, G. Zaccai, M. Ferrand and C. Pfister, in *Biological Macromolecular Dynamics. Proceedings of a Workshop on Inelastic and Quasielastic Neutron Scattering in Biology*, ed. S. Cusack, H Büttner, M. Ferrand, P. Langan and P. Timmins, Adenine Pres, New York, 1996, p. 117.

Paper 8/06840B

Interactions between poly(2-ethylacrylic acid) and lipid bilayer membranes: Effects of cholesterol and grafted poly(ethylene glycol)

David Needham,* Jeff Mills and Gary Eichenbaum

Department of Mechanical Engineering and Materials Science, Duke University, Durham, NC 27708-0300, USA

Received 9th November 1998

The exchange of the protonatable polymer, poly(2-ethylacrylic acid) (PEAA), has been studied with vesicle membranes containing cholesterol from 0 to 60 mol% or PEG2000-lipid (5 mol%). The release of an entrapped dye from 100 nm extruded liposomes was used as an assay for membrane perturbation by the polymer as a function of pH. The inclusion of cholesterol was found to reduce the pH at which the polymer caused release of the dye from the lipid vesicles, and the degree of polymer protonation (*i.e.*, degree of hydrophobicity) correlated well with the increase in elastic expansion modulus of the vesicle bilayer. The results are discussed in terms of a balance between polymer solubility and membrane expansion. With respect to the PEG barrier, the presence of 5 mol% PEG2000, which represents full surface coverage, did not prevent PEAA from inducing contents release, demonstrating that highly hydrated polymeric layers are not effective barriers for other water soluble polymers, and may point to some association between the two polymers.

In pure lipid bilayer systems the aqueous solubility for membrane components is so low (of the order of 10^{-10} M) that bilayer mass can essentially be considered constant, at least during the time of an experiment. In contrast, in the presence of other membrane compatible materials like lysolecithin and bile salts, which can have a significant solubility in the aqueous phase as monomers and micelles, the composition and properties of the bilayer can be changed in seconds and, at concentrations in excess of the surfactant critical micelle concentration (c.m.c.), can lyse the membranes causing them to fail.[1–4] Another important class of molecules that shows partitioning into bilayers are pH-sensitive polymers such as poly(ethylacrylic acid) (PEAA).[5] Here, protonation of the carboxyl groups of the polymer reduces the solubility of the molecule in aqueous media, making it more hydrophobic. Consequently, this promotes solution of the polymer in the hydrophobic interior of lipid bilayers.[5] This class of molecule shows a well defined cooperative conformational transition from an expanded coil to a collapsed coil that is dependent on pH. The existence of this transition (when compared to the continuous condensation of simple polyelectrolytes such as polyacrylic acid) is a direct result of additional attractive interactions between hydrophobic moieties such as the ethyl group in PEAA. For polymers dissolved in aqueous solution, the midpoint of this transition is at pH 6.2.[6] Interestingly however, in the presence of Lα-dipalmitoylphosphatidylcholine (DPPC) vesicles, the midpoint of the transition is shifted up to pH 6.5.[6] This shift indicates some adhesive interaction between the polymer and the membrane at pHs above its natural critical pH for the transition, and a role of the bilayer in inducing this transition.

Up to now, most of the work on this polymer has been carried out with DPPC bilayers. The work presented here was motivated by a desire to examine the way in which the properties of the bilayer and its interface may influence the polymer–bilayer interaction. We therefore chose to study bilayers containing cholesterol and bilayers containing a grafted layer of a different water soluble polymer, poly(ethylene glycol) (PEG). The inclusion of cholesterol in lipid bilayer membranes causes a condensation of the interface and a dramatic decrease in elastic area compressibility of the bilayer as well as its permeability to water.[7–9] The presence of PEG grafted to the bilayer interface, *via* the incorporation of PEG–lipids into the bilayer, can inhibit the close approach of globular macromolecules and micelles.[10–12] It is of interest then to determine if these two different kinds of barriers can hinder the approach and interaction of a water-soluble polymer that can be triggered to undergo a conformational transition due to a lowering of solution pH making it less soluble in aqueous media and more soluble in lipid bilayers, which it can then permeabilize.

In the present paper then, we report the measurement of the release of a fluorescent dye entrapped within lipid vesicles that is induced by the exposure of the vesicles to PEAA solutions as a function of solution pH. These measurements are carried out for lipid vesicle bilayers containing either cholesterol (from 0 to 60 mol%) or PEG-lipid (5 mol%). Based on previous mechanical property data,[8] the expectation was that cholesterol-rich membranes would possibly inhibit the polymer–bilayer interaction. For the PEG-lipid membranes, interest centered on whether the presence of a PEG covered surface (at the mushroom-brush transition) would inhibit approach and intercalation of the PEAA into the membrane and therefore prevent release of internal contents.

Experimental

Materials

Stearoyloleoylphosphatidylcholine (SOPC) and cholesterol (from Avanti polar lipids) in chloroform (10 mg mL^{-1}) were mixed to give cholesterol at 0, 20, 30, 40, 50 and 60 mol%. SOPC and distearoylphosphatidylcholine (DSPE)-PEG2000 (10 mg mL^{-1}) were mixed to give DSPE-PEG2000 at 5 mol%. 20 mM Phosphate buffered saline (PBS) [0.8% w/w saline (0.137 M)] was prepared using sodium phosphate monobasic, sodium phosphate dibasic, and sodium chloride from Mallinckrodt. Carboxyfluoroscence (6-CF) was obtained from Aldrich, Inc.

Liposome preparation

The prepared SOPC and cholesterol solutions in chloroform were dried using a Rotavapor onto a 100 mL round-bottomed flask.[13] Lipid vesicles were prepared by hydrating the dried lipid film with a solution of 50 mM 6-carboxyfluoroscene (6-CF) (high enough concentration for fluorescence to be self-quenched), in 20 mM PBS.[13] The hydrated lipid mixture was extruded through two 100 nm Poretics polycarbonate membrane filters using a Lipex extruder to give a homogeneous sample of unilamellar lipid vesicles of 100 nm diameter,[14] containing the entrapped carboxyfluoroscene dye in a quenched form. Lipid vesicles were then filtered using a Sephadex G-50 column (Sigma) to remove unencapsulated 6-CF.[13]

Contents release measurements

20 mM PBS was prepared at pHs ranging from 5.5 to 7.5. PEAA was dissolved in 20 mM PBS at 33.3 mM (1 mg mL^{-1}) over the same pH range. Both sets of solutions were osmotically matched, using glucose, to the 6-CF solution used to hydrate and form the liposomes, to ensure there were no osmotic pressure gradients across the lipid membranes. 10 µL of liposome solutions were suspended in 1 mL of buffer containing PEAA (time zero). Fluorescence (λ_{ex} = 436 nm, λ_{em} = 520 nm) intensity was measured every 5 s for 15 min, at 30 min and 1 h. (Aminco-Bowman Series 2 Luminescence Spectrometer or a Shimadzu RF-1501 Spectrofluorophotometer were used to measure fluorescence.) As 6-CF leaked out of the liposomes and was diluted by the suspending buffer, self-quenching was eliminated resulting in an increased fluorescence signal. After 1 h, 20 µL of Triton X-100 was added to completely solubilize the membranes and a final fluorescence reading was made. This final reading was used as a maximum, or 100% release reference, and all

other readings were divided by this value to obtain a measure of the "percent released". This procedure was similar to that used by Goñi *et al.* to study the effects of surfactants on release from liposomes.[15] In separate experiments, the fluorescence of 6-CF in Triton X-100 was found to be unaffected by the presence of surfactant but was dependent on pH. The pH-dependent intensity was controlled by making control measurements at the same pH.

Results and discussion

Cholesterol-containing membranes

The effect of cholesterol on the pH dependence of PEAA-induced release of contents from liposomes was investigated by measuring release of 6-CF (quenched inside the liposome), as indicated above, for liposomes of various cholesterol compositions. First, consider the release from liposomes composed of pure SOPC.

Fig. 1 shows a plot of percent release of 6-CF *vs.* time in min for 100 nm liposomes composed of pure SOPC in a solution of 1 mg mL^{-1} PEAA. Each trace is the cumulated release of carboxyfluoroscene fluorescence as it leaked out of the liposome sample over a period of 1 h. Each curve represents the pH of the bulk media in the range of 5.5–7.5. As shown on the plot, the curve at pH 6.7 was the lowest pH at which the liposomes were stable in the polymer suspension. In polymer solutions more basic than or equal to pH 6.7, baseline release was observed, rising from 10% to only 30% release over the time course measured. Similar values are noted for control solutions which did not contain PEAA. In contrast, polymer solutions more acidic than pH 6.7 exhibited a burst release reaching 100% of entrapped 6-CF within a few minutes.

By comparison, Fig. 2 shows a plot of percent release of 6-CF as a function of time for 100 nm SOPC/cholesterol liposomes containing 40 mol% cholesterol. Here, release was not observed at pH 6.7, but in solutions more acidic than pH 6.4. Thus, the pH at which onset of release occurred was shifted from near pH 6.7 (for 0 mol% cholesterol) to near pH 6.4 (for 40 mol% cholesterol).

Then, by taking the steady state release values (60 min) at each pH for SOPC systems with cholesterol composition 0, 20, 30, 40, 50 and 60 mol%, as shown in Fig. 3, a plot of steady state release as a function of pH for each composition was constructed. This figure shows that all compositions achieved 100% release if the pH was acidic enough. Importantly, it also shows that the pH at onset of release was monotonically shifted to more acidic pH values with increasing mol% cholesterol, from pH 6.7 for a pure lipid system, to pH 6.0 for a cholesterol-saturated (*i.e.*, ~60 mol%) system.

Thus, the uptake of protonated polymer by the bilayer membrane is increasingly inhibited as the bilayer is condensed with cholesterol. What these results then suggest is that there is indeed a

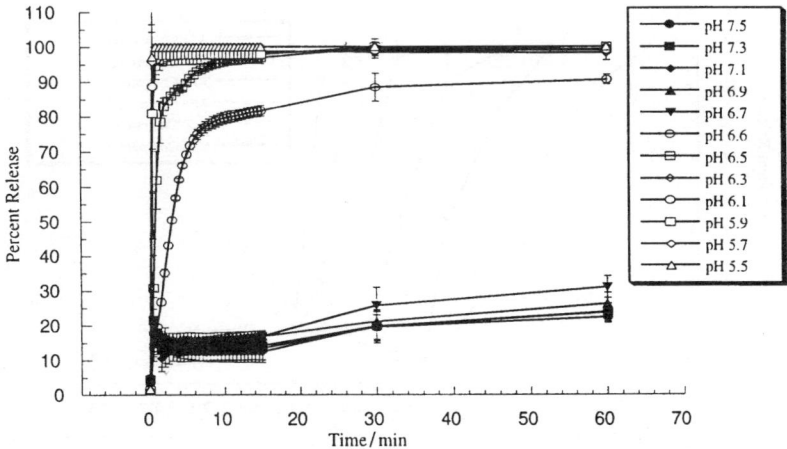

Fig. 1 Plot of percent release of 6-CF *vs.* time in min for a system of pure SOPC, 100 nm liposomes, suspended in solutions of 1 mg ml^{-1} PEAA at pHs varying from 5.5 to 7.5. Error bars on each data point represent the standard deviation from the mean for 2 measurements.

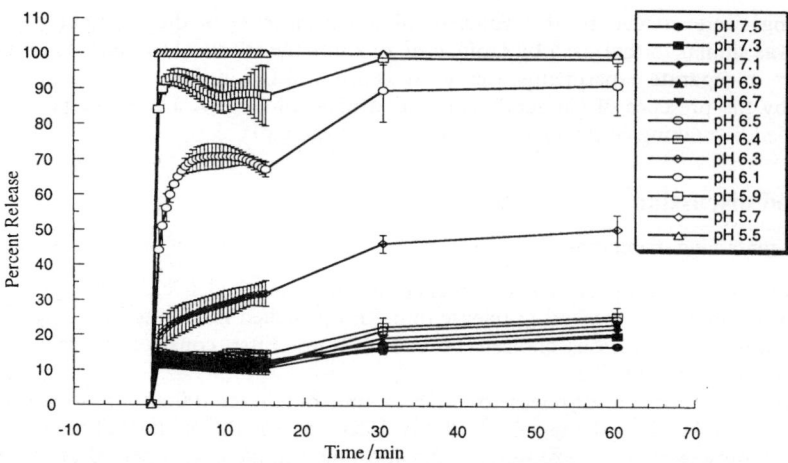

Fig. 2 Plot of percent release of 6-CF *vs.* time in min for a system of 100 nm SOPC liposomes with 40 mol% cholesterol incorporated in the membrane, suspended in solutions of 1 mg ml^{-1} PEAA at pHs varying from 5.5 to 7.5. Error bars on each data point represent the standard deviation from the mean for 2 measurements. The intensity drop noted in some of the curves during the first 15 min can be attributed to photobleaching of the 6-CF, but this artifact did not affect the steady state measurements.

correlation between the compressibility of the lipid bilayer (determined by the presence of cholesterol) and the interaction of the polymer with the bilayer which is controlled by pH, *i.e.*, the degree of protonation of the polymer, which in turn determines its degree of hydrophobicity.

In order to develop this correlation further, what is needed is a measure of the percent protonation of the polymer as a function of pH. Tirrell and colleagues have characterized the pH-dependent solution behavior of the pure polymer.[5,6] Titration of the polymer indicates that it has a pK_a of approximately 5.4.[16] Graphically, then, as shown in Fig. 4 as the pH is lowered in the range 7.4–3.4, the percent of polymer that becomes protonated (and therefore hydrophobic) increases from close to zero to 100%. Critical pHs for some important features of polymer behavior and polymer–bilayer interaction, from the previous literature and from the experiments reported here, are summarized by this plot.

At pH 6.2, as observed by Borden *et al.* using pyrene fluorescence, the free polymer in aqueous solution undergoes a sharp conformational transition.[6] Thus ∼15% of the polymer needs to be

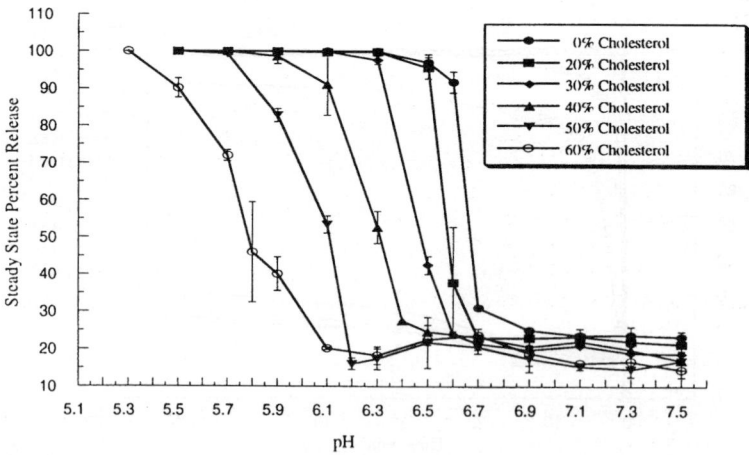

Fig. 3 Plot of steady state release of 6-CF (60 min after liposome addition to the polymer solution) *vs.* pH for 100 nm liposomes composed of SOPC and containing 0, 20, 30, 40, 50 and 60 mol% cholesterol. Error bars on each data point represent the standard deviation from the mean for 2 measurements.

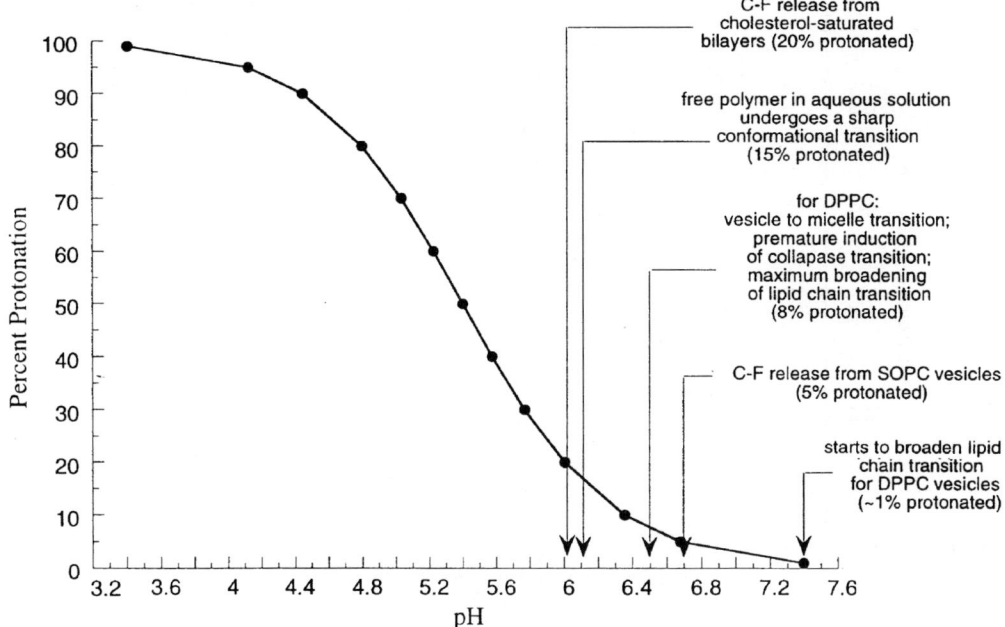

Fig. 4 Plot of percent protonation of the polymer *vs.* pH calculated from the Henderson–Hasselbach equation using the data from Tirrell *et al.*,[16] which measured the pK_a as 5.4. Shown are critical pHs for several features of polymer and polymer–lipid behavior.

protonated in order for the polymer to undergo the collapse transition. At pH 7.4, the polymer was found by Seki and Tirrell,[17] to cause a broadening of the main acyl chain gel to liquid crystalline transition temperature for DPPC vesicles. Thus, at relatively high pH, where the polymer is only marginally (1%) protonated, it is intimately associated with the bilayer. This association is sufficient to cause perturbation of the gel/liquid crystalline transition, but not enough to cause the polymer collapse transition, indicating that at this degree of protonation the interaction with the bilayer is insufficient to "catalyze" the collapse. Also, it seems that this expanded coil polymer cannot traverse to the inner layer of the bilayer in order to cause bilayer disruption. For DPPC vesicles, Tirrell and colleagues showed that at pH 6.5, three events were correlated: PEAA induced a sharp vesicle to micelle transition signifying massive reorganization of the lipid;[18] pyrene fluorescence increased sharply signifying the premature induction of the expanded coil to collapsed coil transition due to binding of polymer to the bilayer;[6] and the DPPC lipid acyl chain transition showed maximum broadening.[6] Taken together, these results implicate the collapse transition as a key event that leads to membrane disruption.

In our experiments, at pH 6.7, the polymer is ~5% protonated when it causes release of dye from the SOPC vesicles (a slightly higher pH than that measured for reorganization of DPPC vesicles,[5] and, by inference and comparison with the DPPC data of Thomas and Tirrell,[5] this signifies the premature induction of the collapse transition by SOPC bilayers. At pH 6.1, where the polymer is 20% protonated it causes release of dye from cholesterol-saturated vesicles. At this pH, the polymer is actually collapsed in solution which represents the maximum potential for the polymer to disrupt the bilayer.

Using K_A, the data from Fig. 4 and previous measurements of the elastic area expansion modulus for SOPC/cholesterol bilayers,[8] Fig. 5 shows that the fractional increase in K_A is in fact found to be directly proportional to the increase in percent protonation required to cause release of dye from the vesicles.

In evaluating this data in terms of energetic considerations, there is an energy state for the polymer in aqueous solution that is pH dependent, and there is an energy state for the polymer in the bilayer which may be changed by the presence of cholesterol and which determines the release

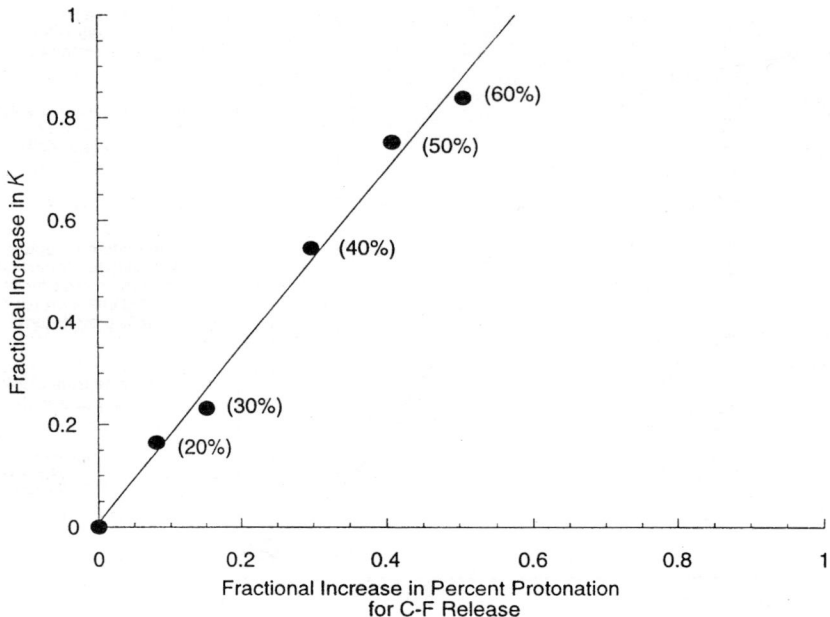

Fig. 5 "Fractional change plot" showing the fractional increase in percent protonation of the polymer (over and above that which occurs at pH 6.7) at the pH where dye is released, *vs.* the fractional change in the elastic area expansion modulus (over and above that for the pure SOPC bilayer) for bilayers containing increasing amounts of cholesterol (in parentheses next to each data point).

of CF. This latter energy determines the ability of the polymer to create defects and perturb the structure of the bilayer. Assuming that cholesterol does not decrease the pH at the bilayer interface relative to the bulk solution, its role is therefore limited to its ability to increase the cohesiveness of the membrane. Then, a greater amount of work is required to perturb the structure of the membrane and induce leakage of contents which has to be compensated by increasing the protonation of the polymer. The assay used here was simply the release of the marker dye (*i.e.*, the end point of the interaction) and so does not provide any information on the amount of polymer

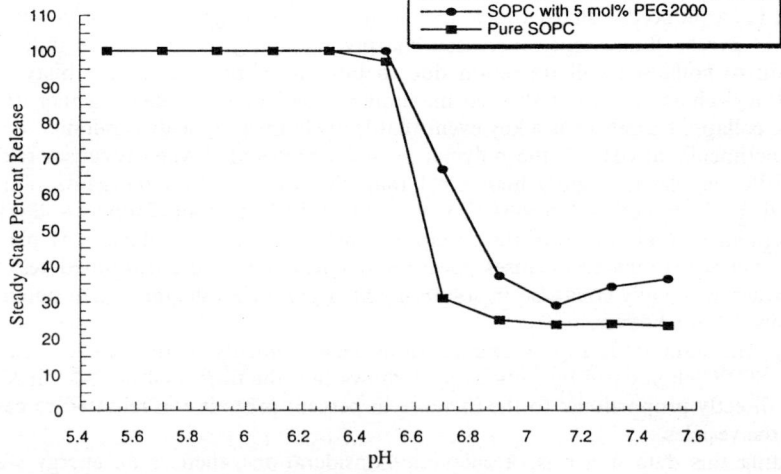

Fig. 6 Plot of steady state release of 6-CF *vs.* pH for 100 nm liposomes composed of pure SOPC with 5 mol% PEG2000, as well as 100 nm liposomes composed of only SOPC.

bound. Thus, at this stage in the work it is not possible to distinguish between the effect of cholesterol on polymer partitioning and the formation of defects that result in CF release. Critical information that needs to be obtained includes the concentration dependence of the contents release and a lipid to polymer ratio for the failure event.

Grafted PEG

Data for CF release from the PEG-grafted liposomes was also obtained as described above. A compilation of the results is shown in Fig. 6, as steady state release *vs.* pH.

The curve for the PEG-grafted liposomes was nearly identical to the plot for the pure SOPC, non-PEGylated vesicle case, with release occurring near pH 6.8 and reaching 100% by pH 6.6. Thus, despite the presence of a complete layer of PEG on the bilayer surface, [enough to significantly retard the approach of globular molecules (avidin) and micelles (MOPC)], PEAA can still approach close enough to the bilayer interface to bind and cause membrane disruption. This indicates an ability of the PEAA polymer to reptate through the PEG polymer layer. As mentioned in the Introduction, this same PEG layer cannot prevent MOPC monomer from adsorbing into or desorbing from such PEG-grafted bilayers, suggesting that monomer can in fact diffuse through the PEG layer.[10] With a cross-sectional area of 40 $Å^2$, the polymer has a similar size to MOPC of 34 $Å^2$. A binding interaction between PEG and PEAA is also not ruled out.

Concluding remarks

The results of this study then suggest that membrane compressibility dominates the interaction between lipid bilayers and PEAA such that highly expandable membranes are more likely to take up hydrophobic polymer molecules than less expandable membranes with more condensed interfaces. We plan to obtain additional data on the binding and uptake of polymer by lipid bilayers, including the use of the micropipet manipulation technique to measure any area change that may accompany polymer insertion into the membrane at pHs slightly higher than those required to initiate membrane disruption, to observe CF release directly for a single vesicle, and to measure any changes in the membrane compressibility and tensile strength for polymer-adsorbed, but stable, bilayers. For the PEG-lipid containing membranes, no additional barrier effect was observed, leading to the conclusion that either these aqueous-soluble polymers can in fact entangle, such that the PEAA can come into intimate contact with the lipid interface, or, an attractive interaction occurs between PEG and PEAA, especially for the protonated form of PEAA.

This work was supported by grant GM 40162 from the NIH and by the National Science Foundation, Engineering Research Center Grant: CDR-8622201. The authors would also like to acknowledge useful discussions with Drs. Evan Evans, Tom McIntosh, Sid Simon and Doncho Zhelev, and Mr. Patrick Kiser.

References

1. E. Evans, W. Rawicz and A. F. Hoffman, *Bile Acids in Gasteroenterology Basic and Clinical Advances*, ed. A. F. Hoffman, G. Paumgartner and A. Stiehl, Kluwer Academic Publishers, Dordrecht, Boston, London, 1994, pp. 59–68.
2. D. Needham and D. V. Zhelev, *Ann. Biomed. Egr.*, 1995, **23**, 287.
3. D. V. Zhelev, *Biophys. J.*, 1996, **71**, 257.
4. D. V. Zhelev, *Biophys. J.*, 1998, **75**, 321.
5. J. L. Thomas and D. A. Tirrell, *Acc. Chem. Res.*, 1992, **25**, 336.
6. K. A. Borden, K. M. Eum, K. H. Langley and D. A. Tirrell, *Macromolecules*, 1987, **20**, 454.
7. M. Bloom, E. Evans and O. G. Mouritsen, *Q. Rev. Biophys.*, 1991, **24**, 293.
8. D. Needham and R. S. Nunn, *Biophys. J.*, 1990, **58**, 997.
9. D. Needham and D. V. Zhelev, *Vesicles*, ed. Rosoff, Marcel Dekker, New York and Basel, 1996.
10. D. Needham, T. J. McIntosh and D. V. Zhelev, *Liposomes, Rational Design.*, ed. A. Janoff, Marcel Dekker New York, 1998, in press.
11. D. Needham, N. Stoiceva and D. V. Zhelev, *Biophysical*, 1997, **73**, 2615.
12. D. Noppl-Simson and D. Needham, *Biophys. J.*, 1996, **70**, 1391.
13. R. R. C. New, *Liposomes: A Practical Approach.*, ed. R. R. C. New, Oxford University Press, New York, 1990, ch. 2.

14 M. J. Hope, R. Nayay, L. D. Mayer and C. P. R. Tilcock, *Liposome Technology: Liposome Preparation and Related Techniques.*, ed. G. Gregoriadis, CRC Press, Boca Raton, Fl., 1993, ch. 8.
15 F. M. Goñi, M. A. Urbaneja and A. Alonso, *Liposome Technology: Entrapment of Drugs and Other Materials*, ed. G. Gregoriadis, CRC Press, Boca Raton Fl., 1993, ch. 15.
16 K. H. Langley, U. K. O. Shröder and D. A. Tirrell, personal communication.
17 K. Seki and D. A. Tirrell, *Macromolecules*, 1984, **17**, 1692.
18 K. A. Borden, K. M. Eum, K. H. Langley, J. S. Tan, D. A. Tirrell and C. L. Voycheck, *Macromolecules*, 1988, **21**, 2649.

Paper 8/08717B

Membrane electroporation and electromechanical deformation of vesicles and cells

Eberhard Neumann, Sergej Kakorin and Katja Toensing

Physical and Biophysical Chemistry, Faculty of Chemistry, University of Bielefeld, P.O. Box 100 131, D-33501 Bielefeld, Germany

Received 17th August 1998

Analysis of the reduced turbidity ($\Delta T^-/T_0$) and absorbance ($\Delta A^-/A_0$) relaxations of unilamellar lipid vesicles, doped with the diphenylhexatrienyl — phosphatidylcholine (β-DPH pPC) lipids in high-voltage rectangular electrical field pulses, demonstrates that the major part of the turbidity and absorbance dichroism is caused by vesicle elongation under electric Maxwell stress. The kinetics of this electrochemomechanical shape deformation (time constants $0.1 \leqslant \tau/\mu s \leqslant 3$) is determined both by the entrance of water and ions into the bulk membrane phase to form local electropores, and by the faster processes of membrane stretching and smoothing of thermal undulations. Moreover, the absorbance dichroism indicates local displacements of the chromophore relative to the membrane normal in the field. The slightly slower relaxations of the chemical turbidity ($\Delta T^+/T_0$) and absorbance ($\Delta A^+/A_0$) modes are both associated with the entrance of solvent into the interface membrane/medium, caused by the alignment of the dipolar lipid head groups in one of the leaflets at the pole caps of the vesicle bilayer. In addition, ($\Delta T^+/T_0$) indicates changes in vesicle shape and volume. The results for lipid vesicles provide guidelines for the analysis of electroporative deformations of biological cells.

1 Introduction

The method of membrane electroporation is widely applied in cell biology and medicine to introduce effector substances and genes into biological cells and tissue.[1] The mechanisms of electric pore formation in lipid bilayer membranes and the resulting transport facilitation for macromolecules are not yet well understood. Basic elements of the electric field effects have already been derived from electro-optic and conductometric data for model systems such as unilamellar lipid vesicles, doped with optical membrane probes.[2,3] Here, the lipid probe 2-[3-(diphenylhexatrienyl propanoyl]-1-hexadecanoyl-sn-glycero-3-phosphocholine (β-DPH pPC) has been used, where one of the hydrocarbon chains of the PC is replaced by the DPH residue. DPH — doped bilayer membranes exhibit characteristic absorbance changes in polarized light when subjected to electric fields.[4] Previously, the negative absorbance dichroism of DPH-doped lipid vesicles was analysed in terms of structural rearrangements of lipids and of DPH molecules in the wall of electropores, assuming the DPH site to be the origin of pore formation.[5]

On the other hand, the electro-optic data with β-DPH pPC can only be rationalized quantitatively if the major part of the absorbance dichroism is attributed to global vesicle deformation under the electric Maxwell stress.[2,6]

Here, we develop the theoretical and analytical framework to differentiate the contributions of the field-induced membrane structural changes and of the vesicle shape deformation to the turbidity and absorbance dichroisms. It appears that the two characteristic turbidity ΔT^- and absorbance ΔA^- modes, associated with orientational processes, are rate-limited by the electric pore formation, whereas the chemical turbidity ΔT^+ and absorbance ΔA^+ modes reveal further details such as the entrance of water and ions into the interface regions of the lipid head groups.[7,8] Additionally, ΔA^- reflects local chromophore displacements in the membrane.

2 Materials and methods

2.1 Vesicle suspensions

Large unilamellar phospholipid vesicles and vesicles doped with the optical lipid probe β-DPH pPC, $M_r = 782$, were prepared by the extrusion method, as described by Toensing et al.[9] The vesicle mean diameters, determined by dynamic light scattering measurements (data not presented), are $\Phi = 100 \pm 36$ nm after extrusion through the 100 nm filter; in line with the size distribution of extruded vesicles determined by Mayer et al.[10] During vesicle preparation and the following electro-optic relaxation measurements, care was taken to protect the DPH samples from photolysis. The final total lipid concentration used for the optical and electro-optical measurements was $[L_T] = 1$ mM, corresponding to a vesicle density of ca. 7.4×10^{15} L^{-1} for vesicles of radius $a = 50$ nm. Under these conditions, the average distance between the anionic surfaces of single vesicles is about ca. 0.53 μm, qualifying the suspension as dilute, with practically no vesicle–vesicle contacts. Actually, at the maximum field $E = 8$ MV m^{-1} and at the effective permittivity of vesicle $\varepsilon_{eff} \approx \varepsilon_L = 2.5$, where ε_L is the permittivity of the lipid membrane, the minimum characteristic time t_{app} of approach of vesicles due to induced dipole forces, taking vesicle–vesicle hydrodynamic interactions into account, is $t_{app} \approx 610$ μs.[11,12] Therefore, at the pulse duration of $t_E = 10$ μs and at field strengths $E \leqslant 8$ MV m^{-1}, we may safely neglect vesicle–vesicle interactions.

2.2 Electro-optical relaxation spectrometry

Rectangular pulses of field strength up to 8 MV m^{-1} and of duration up to $t_E = 10$ μs are applied by cable discharge to the sample cell equipped with parallel planar graphite electrodes, thermostatted at $T = 293.0 \pm 0.1$ K (20 °C). The field-induced changes in the transmittance of plane-polarized light were measured at the wavelength $\lambda = 365$ nm (Hg-line; highest accuracy).

The light intensity change ΔI^σ, caused by the electric pulse and measured at the polarization angle σ relative to the direction of the applied external field vector $\boldsymbol{E}$, is related to the optical density change by $\Delta \Gamma^\sigma = \Gamma^\sigma(E) - \Gamma_0^\sigma = -\log(1 + \Delta I^\sigma/I^\sigma)$, where $\Delta I^\sigma = I^\sigma(E) - I^\sigma$ is the light intensity change from I^σ (at $E = 0$) to $I^\sigma(E)$ in the presence of E, $\Gamma^\sigma(E)$ and Γ_0^σ are the optical densities at E and at $E = 0$, respectively. Generally, $\Gamma = A + lT$, comprising both absorbance (A) and turbidity (T) along the light path length l. The absorbance A^σ of the reporter lipids β-DPH pPC in the bilayer of the vesicles is given by the difference $\Delta \Gamma^\sigma$ between $\Gamma^\sigma(V,D)$ of the doped vesicles and $\Gamma^\sigma(V)$ of the vesicles without the reporter lipid, but at the same total lipid concentration and vesicle size: $A^\sigma = \Gamma^\sigma(V,D) - \Gamma^\sigma(V)$. The field-induced optical change may be decomposed into a deformational/orientational part $\Delta\Gamma^\sigma_{OR}$ and a structural/chemical part $\Delta\Gamma^\sigma_{CH}$ according to

$$\Delta \Gamma^\sigma = \Delta \Gamma^\sigma_{OR} + \Delta \Gamma^\sigma_{CH} \tag{1}$$

The field-induced changes $\Delta\Gamma^\parallel$ and $\Delta\Gamma^\perp$ at the two light polarization modes $\sigma = 0°$ ($\parallel$, parallel to the external field vector $\boldsymbol{E}$) and $\sigma = 90°$ ($\perp$, perpendicular to $\boldsymbol{E}$) are given by $\Delta\Gamma^\parallel = \Gamma^\parallel - \Gamma_0$ and $\Delta\Gamma^\perp = \Gamma^\perp - \Gamma_0$, respectively. As outlined for the absorbance dichroism[7] both the consumptive dichroism (ΔA) and the conservative scattering dichroism (ΔT) are originally defined for optical density changes of purely deformational/orientational origin.[13,14] In the notation used here, the optical density dichroism is classically defined by

$$\Delta\Gamma = \Delta\Gamma^\parallel_{OR} - \Delta\Gamma^\perp_{OR} \tag{2}$$

On the other hand, the difference $\Delta\Gamma^-$, either ΔA^- or ΔT^-, of the actually measured changes $\Delta\Gamma^\parallel$ and $\Delta\Gamma^\perp$, is given by

$$\Delta\Gamma^- = \Delta\Gamma^\parallel - \Delta\Gamma^\perp = \Delta\Gamma + (\Delta\Gamma^\parallel_{CH} - \Delta\Gamma^\perp_{CH}) \tag{3}$$

In the case of small chemical contributions or if $\Delta\Gamma^{\|}_{CH} \approx \Delta\Gamma^{\perp}_{CH}$, we may approximate the difference mode $\Delta\Gamma^-$ by the dichroism $\Delta\Gamma$, i.e.:[7]

$$\Delta\Gamma^- = \Delta\Gamma \quad (4)$$

The equations $\Delta A^+ = \Delta A_{CH}$ and $\Delta T^+ = \Delta T_{CH}$ are straightforward; analogous to the expression for ΔA_{CH} [7,8] we obtain: $\Delta\Gamma_{CH} = (\Delta\Gamma^{\|} + 2\Delta\Gamma^{\perp})/3$. Therefore, $\Delta\Gamma_{CH}$ generally refers to changes in the scattering cross-section or the immediate environment of the absorbing chromophore due to entrance of water and small ions. Note that, if the scattering contribution is negligibly small, we have $\Delta\Gamma = \Delta A$, and outside the absorbance bands we use $\Delta\Gamma = l \Delta T$. The absorbance dichroism is a quantitative measure of rotational displacements comprising both global shape deformations as well as local chromophore shifts in the lipid bilayer. The turbidity term $\Delta T_{CH}/T_0$ reflects changes in the refractive index of the membrane due to entrance of water and ions, as well as changes in vesicle shape and volume.

3 Results

The β-DPH pPC membrane probes greatly increase the absorbance of light in the doped vesicles in the light wavelength range $300 < \lambda/\text{nm} < 400$ (Fig. 1). The optical density and the absorbance both increase linearly with the concentration of β-DPH pPC, suggesting that at zero external field there are no optically detectable interactions between the "reporter lipids" in the vesicle membrane (Fig. 2). In the electric field pulse, the very rapid (ca. 1 μs) increases in the absorbance terms $\Delta A^-/A_0$ and $\Delta A^+/A_0$ (Fig. 3) and the turbidity modes $\Delta T^-/T_0$ and $\Delta T^+/T_0$ (Fig. 4) are followed by similarly rapid field-off relaxations. Outside the absorption band the electro-optic relaxations of labelled and non-labelled vesicles are practically identical (Fig. 5). These data show that the membrane probes incorporated in the lipid molecules change neither membrane electroporatability nor mechanical properties such as bending rigidity or membrane spontaneous curvature. At larger β-DPH pPC concentration there is a weak dependence of the chemical and dichroitic absorbance modes on the probe content in the membrane, especially at higher field strengths (Fig. 6). However, no changes are observed in the turbidity terms $\Delta T^-/T_0$ and $\Delta T^+/T_0$ outside the absorbance band with increasing concentration of β-DPH pPC in the vesicle membrane (Fig. 7). At the field strengths $2 \leqslant E/\text{MV m}^{-1} \leqslant 8$, the chemical turbidity and

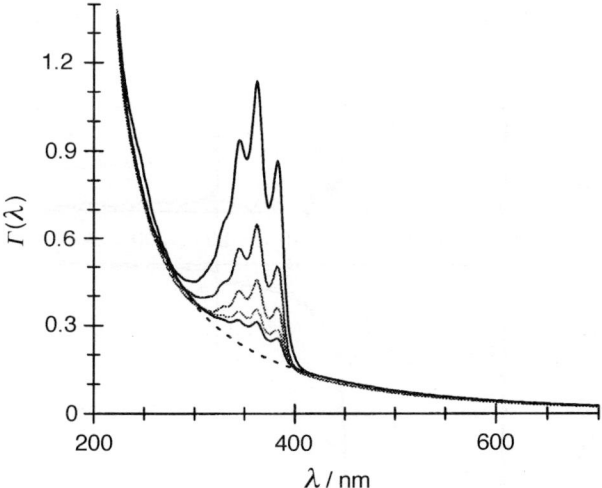

Fig. 1 Optical density of a suspension of unilamellar vesicles composed of L-α-phosphatidyl-L-serine (PS) and 1-palmitoyl-2-oleoyl-sn-glycero-3-phosphocholine (POPC) in the molar ratio PS : POPC of 1 : 2 (---) and the same suspension doped with the optical probe β-DPH pPC, $M_r = 782$ at the concentrations [β-DPH pPC]/μM = 2.5, 3.3, 5, 10, 20 (from bottom to top) as a function of the wavelength λ. Vesicle radius $a = 50$ nm, vesicle density $\rho_v = 7.4 \times 10^{15}$ L^{-1}; total lipid concentration [L$_T$] = 1.0 mM; 0.66 mM HEPES-Na (pH 7.4), 0.13 mM CaCl$_2$, $T = 293$ K (20 °C).

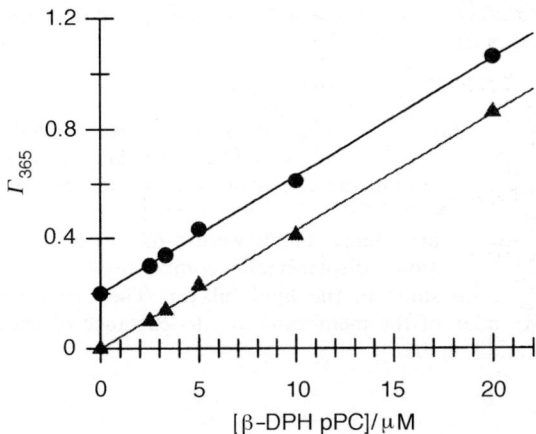

Fig. 2 Optical density Γ_{365} (D, V) at $\lambda = 365$ nm (●) of a doped PS : POPC (1 : 2) vesicle suspension as a function of the total concentration of β-DPH pPC. The absorbance A_{365} (▲) of the "reporter lipid" β-DPH pPC at $\lambda = 365$ nm is given by the difference of the doped (D, V) vesicle system and the non-doped (V) one: $A_{365} = \Delta\Gamma_{365} = \Gamma_{365}$ (D, V) $- \Gamma_{365}$(V); experimental conditions as in Fig. 1. Both Γ_{365} and A_{365} are linear in [β-DPH pPC], hence there is no optically visible interaction between the reporter molecules.

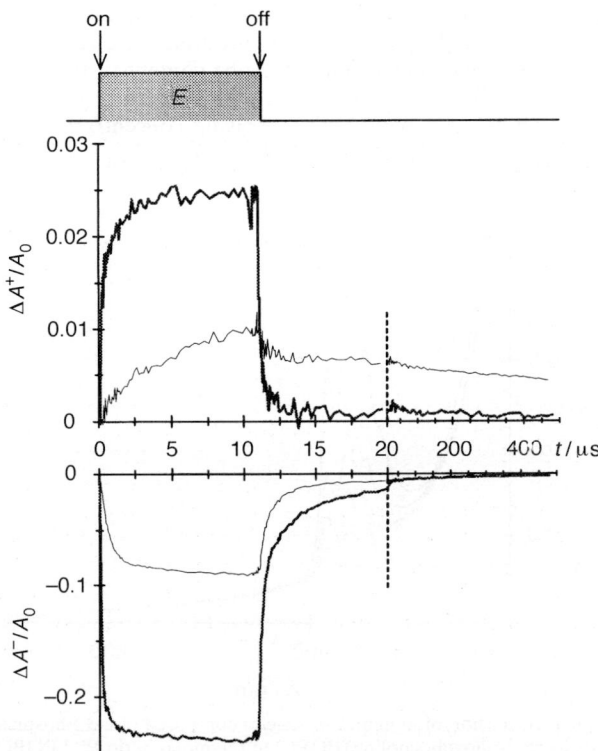

Fig. 3 Field-on and field-off relaxations of the absorbance terms $\Delta A^+/A_0$ and $\Delta A^-/A_0$, at the two extreme field strengths $E = 2$ (thin line) and $E = 8$ MV m^{-1} (thick line), [β-DPH pPC] = 5 μM. One rectangular electric pulse of field strength E and pulse duration $t_E = 10$ μs at $T = 293$ K. Note the change in the timescale for $t \geqslant 20$ μs. Experimental conditions as in Fig. 1.

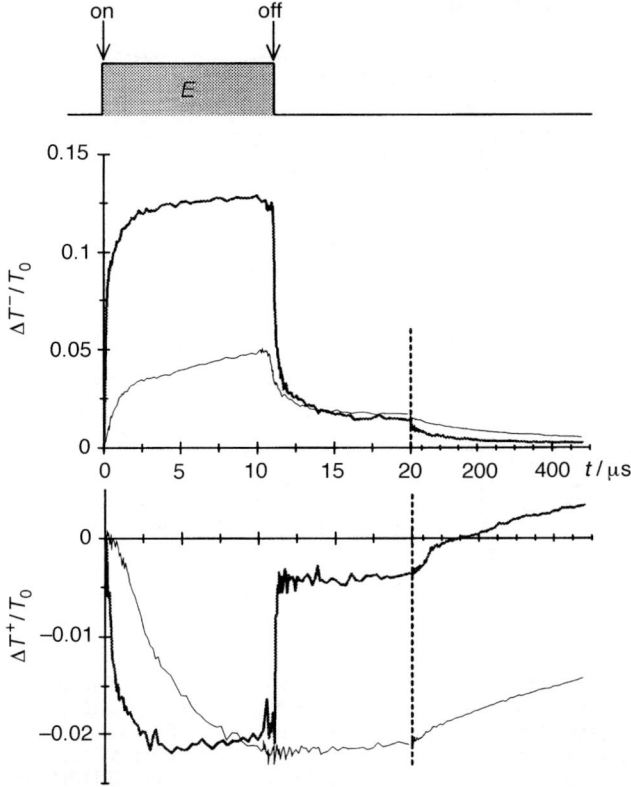

Fig. 4 Field-on and field-off relaxations of the turbidity terms $\Delta T^-/T_0$ and $\Delta T^+/T_0$ ($\lambda = 365$ nm) for non-doped vesicles, at the two extreme field strengths $E = 2$ (thin line) and $E = 8$ MV m^{-1} (thick line). Wavelength $\lambda = 365$ nm. Experimental conditions as in Fig. 3. Visual inspection shows that there are at least two kinetic modes, I and II, in the presence of E.

absorbance terms $\Delta T^+/T_0$, $\Delta A^+/A_0$ are 10-fold smaller than the dichroitic modes ($\Delta T^-/T_0$, $\Delta A^-/A_0$), justifying the approximations $\Delta T/T_0 = \Delta T^-/T_0$ and $\Delta A/A_0 = \Delta A^-/A_0$, respectively.

4 Theory and data analysis

4.1 Vesicle orientation

If the vesicles shortly after extrusion were slightly elongated,[15] they can be oriented in an electric field. However, the very rapid (ca. 1 µs) after-field relaxations of the major part of the absorbance $\Delta A^-/A_0$ (Fig. 3) and the turbidity $\Delta T^-/T_0$ (Fig. 4) modes exclude the possibility that the optical signals are due to orientation of non-spherical vesicles in an external field. After-field disorientation of slightly elongated vesicles may be described by rotational diffusion of a sphere of radius $a = 50$ nm with the rotational relaxation time given by[16] $\tau_{rot} = 4\pi\eta a^3/(3kT)$, where η is the viscosity of the solvent, here water, k the Boltzmann constant and T is the absolute temperature. With $\eta = 10.02 \times 10^{-4}$ kg m^{-1} s^{-1} at $T = 293$ (20°), $\tau_{rot} = 130$ µs. If the vesicles were modelled by ellipsoids, τ_{rot} should be even slightly larger. In any case, the time constants of the after-field dichroisms (ca. 1 µs) are appreciably smaller than τ_{rot}. Since the after-field rotational relaxation is independent of the field strength, comparison of the time constants suggests that the major parts of the $\Delta T^-/T_0$ and $\Delta A^-/A_0$ relaxations are caused by a mechanism different from the vesicle orientation, namely electric pore formation as well as membrane stretching and smoothing of thermal undulations under Maxwell stress.

4.2 Hydrophilic pore model

If the reduced absorbance mode $\Delta A^-/A_0$ is directly due to the electropores, pore formation can be

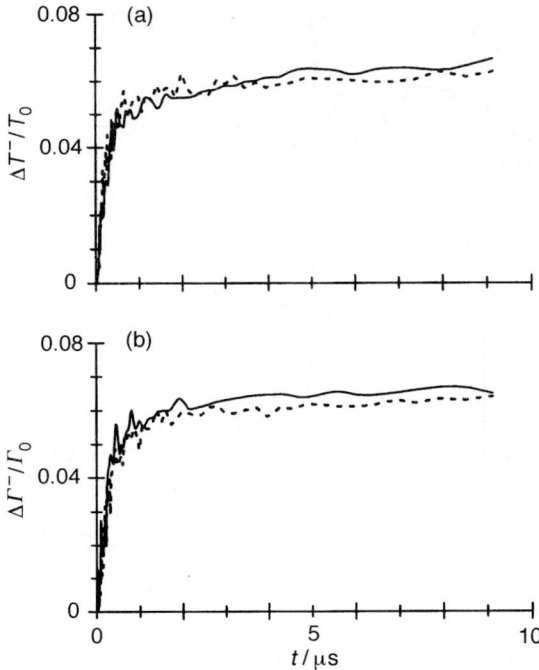

Fig. 5 Comparison of the reduced dichroitic (turbidity) modes (a) $\Delta T^-/T_0$ of an unlabelled vesicle suspension with (b) $\Delta \Gamma^-/\Gamma_0$ of a doped vesicle suspension, [β-DPH pPC] = 5 μM; at wavelengths λ = 281 (---) and 436 (——) nm, outside the β-DPH-residue absorption band (see Fig. 1). The extent and the rate of the turbidity dichroisms are independent of both the wavelength [$\Delta T^-/T_0$ (281 nm) ≈ $\Delta T^-/T_0$ (436 nm) ≈ $\Delta \Gamma^-/\Gamma_0$ (281 nm) ≈ $\Delta \Gamma^-/\Gamma_0$ (436 nm)] and the presence or absence of the reporter lipid β-DPH pPC (b). One rectangular electric pulse of field strength E = 5 MV m^{-1} and pulse duration t_E = 10 μs was applied at T = 293 K. Experimental conditions as in Fig. 1.

described in terms of local lipid phase transitions involving cooperative clusters L_n of n lipids in the pore edge. During this transition, the membrane probes, together with lipid molecules, are locally rotating (Fig. 8). The entrance of the highly polarizable aqueous solvent (water and ions) modifies the local environment of the chromophores and is described by the term $\Delta A^+/A_0$. Based

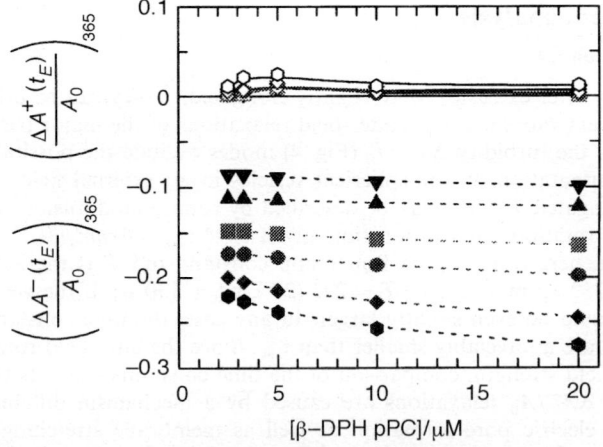

Fig. 6 Dependence of the absorbance terms $\Delta A^+(t_E)/A_0$ (open symbols) and $\Delta A^-(t_E)/A_0$ (filled symbols), at t_E = 10 μs and λ = 365 nm, on the β-DPH pPC concentration, at field strengths E/MV m^{-1} = 2.0 (▽, ▼); 2.8 (△, ▲); 4.0 (□, ■); 5.0 (○, ◉); 6.5 (◇, ◆); 8.0 (○, ●). One rectangular electric pulse field strength E and pulse duration t_E = 10 μs at T = 293 K. Experimental conditions as in Fig. 1.

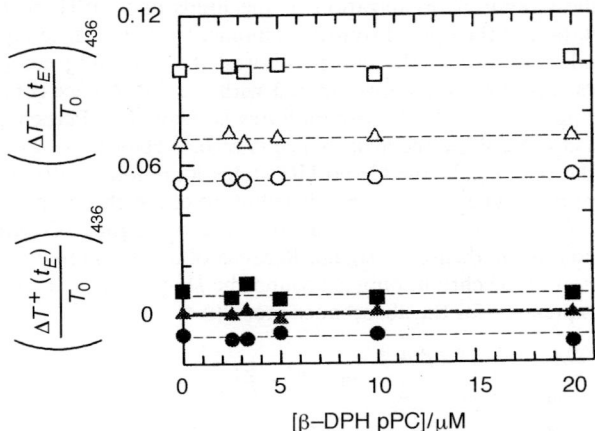

Fig. 7 Dependence of the turbidity terms $\Delta T^-(t_E)/T_0$ (open symbols) and $\Delta T^+(t_E)/T_0$ (filled symbols) at the end of the pulse t_E, at $\lambda = 436$ nm, *i.e.* outside absorbance band, on the β-DPH pPC concentration at field strengths $E/\text{MV m}^{-1} = 2.0$ (▽, ▼); 2.8 (△, ▲); 4.0 (□, ■); 5.0 (○, ◉); 6.5 (◇, ◆); 8.0 (○, ●). Experimental conditions as in Fig. 6.

on the original concept of HO and HI pores,[17] a specific scheme for electropore formation has been proposed:[5]

$$\begin{array}{c} H_2O + \text{ions} \\ \searrow \\ C \rightleftharpoons HO \rightleftharpoons HI \\ \nearrow \\ \text{Rotation of lipids and} \\ \text{DPH probes in HI-pore edge} \end{array} \qquad (5)$$

The state transitions from the closed membrane state (C) to hydrophobic (HO) and hydrophilic (HI) pore states are associated with the rotation of the lipids and chromophores in the HI-pore

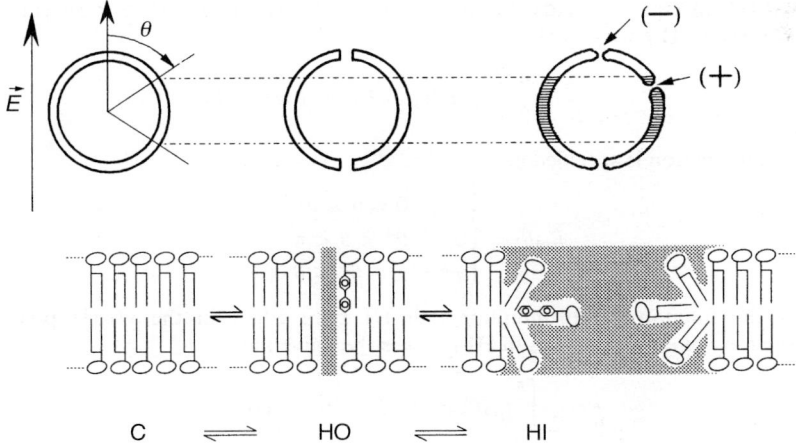

Fig. 8 HI pore model and vesicle geometry relative to the field direction E. Scheme for the molecular rearrangements of the lipids in the pore edges of the lipid vesicle membrane. C denotes the closed bilayer state. The induced membrane field leads to entrance of water into the membrane to produce pores (P), cylindrical HO pores or inverted HI pores. In the pore edge of the HI pore states the lipid molecules are rotated into the pore region to minimize the hydrophobic contact with water. In the pole caps the rotation of DPH lipids leads to a negative absorbance dichroism (−) and in the equatorial region to a positive contribution to the dichroism (+).

wall edge. The specific structural organization of the lipids and DPH in the HI pore states is modelled by a 90° rotation of the optical transition moment μ of the chromophore with respect to the membrane normal (Fig. 8). For the chromophores, the scattering component of the molar absorption coefficients is negligibly small compared with the absorbance component. Therefore, a change in the scattering of the rotated chromophores is negligible. Perpendicular to the optical transition moment of chromophore the absorbance is zero.[18] Hence, the absorbance is associated predominantly with the vector μ. In the C and HO states μ predominantly shows positions parallel to the membrane normal. The orientational distribution of the chromophores around the membrane normal in the C and HO states and around the orthogonal direction in the HI state decreases the amplitude of the dichroitic signal. Because of axial symmetry about the field direction and free orientation of the chromophores around the HI-pore axis, the Lambert–Beer formalism finally yields an expression for the absorbance mode $(\Delta A^-/A_0)$:

$$\frac{\Delta A^-}{A_0} = \frac{3}{4} \int_0^\pi \left(\frac{\beta + 3}{2} f_{HI} - \beta f_{HO}\right)(1 - 3\cos^2\theta)\sin\theta\, d\theta \tag{6}$$

and for the chemical mode $(\Delta A_{CH}/A_0 = \Delta A^+/A_0)$ accordingly:

$$\frac{\Delta A^+}{A_0} = \frac{\beta}{2} \int_0^\pi (f_{HI} + f_{HO})\sin\theta\, d\theta \tag{7}$$

where θ is the positional angle of the chromophore (Fig. 8), relative to the field direction. The relative molar absorption coefficient difference is defined by the molar absorption factor: $\beta = (\varepsilon^* - \varepsilon^C)/\varepsilon^C$. For simplicity, we assume that, in state C, the microscopic molar absorption coefficients of the chromophore in the presence (ε_E^C) and in the absence (ε^C) of the field are equal: $(\varepsilon_E^C = \varepsilon^C)$. In contrast to C, the membrane probes in HO and HI pore edges are partially exposed to water and ions; hence ε^{HI} and $\varepsilon^{HO} \neq \varepsilon^C$. We assume equal molar extinction coefficients of membrane probes in HO and HI pore edges, respectively: $\varepsilon^{HI} = \varepsilon^{HO} \equiv \varepsilon^*$. Similar to the orientation distribution function for chromophores in a solution, the angular distribution functions representing the probabilities of finding chromophores in HO or HI membrane states at an angle between θ and $\theta + d\theta$ are defined by: $f_{HO}(\theta) = [HO]_\theta/[C_0]$ and $f_{HI}(\theta) = [HI]_\theta/[C_0]$, respectively, where $[HO]_\theta$ and $[HI]_\theta$ are the DPH probe concentrations of HO and HI pores, respectively, at θ and $[C_0]$ is the total probe concentration. Mass conservation suggests: $f_{HI}(\theta) + f_{HO}(\theta) + f_C(\theta) = 1$, where $f_C(\theta) = [C]_\theta/[C_0]$. Here it is recalled that $|\Delta A^-/A_0| \gg |\Delta A^+/A_0|$ (Fig. 6 and Fig. 3), hence the molar absorption factor β is practically zero and the approximation $\Delta A^- = \Delta A$ is valid. Substitution of $\beta = 0$ in eqn. (6) yields the absorbance dichroism (due to the local rotation of the membrane probes in HI pore edges):

$$\frac{\Delta A}{A_0} = \frac{9}{8} \int_0^\pi f_{HI}(\theta)(1 - 3\cos^2\theta)\sin\theta\, d\theta \tag{8}$$

The distribution function is specified as:

$$f_{HI}(\theta) = \begin{cases} f^*, & 0 \leq \theta \leq \theta^* \\ 0, & \theta^* < \theta < \pi - \theta^* \\ f^*, & \pi - \theta^* \leq \theta \leq \pi \end{cases} \tag{9}$$

where f^* is the fraction of chromophores in HI pore edges in the vesicle pole caps. The θ-average fraction $\bar{f}$ of chromophores in HI pores is then given by:

$$\bar{f} = \frac{1}{2} \int_0^\pi f_{HI}(\theta)\sin\theta\, d\theta = f^*(1 - \cos\theta^*) \tag{10}$$

Substitution of eqn. (9) in eqn. (8) and integration yields:

$$\frac{\Delta A}{A_0} = \frac{9}{4} f^* \cos\theta^*(\cos^2\theta^* - 1) \tag{11}$$

Analysis of eqn. (11) shows that, at a given value of $\Delta A/A_0$, f^* has a minimum at $\theta^* = \arccos(\sqrt{1/3}) = 54.7°$. Actually, the rotation of chromophores in HI pores at

$54.7° < \theta < (180° - 54.7°)$ contributes to the dichroism positively, reducing the total amplitude of the negative absorbance dichroism. Substitution of $\theta^* = 54.7°$ in eqn. (11) and in eqn. (10) yields:

$$f^* = -1.155 \frac{\Delta A}{A_0} \qquad (12)$$

and

$$\bar{f} = -0.487 \frac{\Delta A}{A_0} \qquad (13)$$

respectively.

4.3 HI pore model and absorbance dichroism

At the maximum field strength $E = 8$ MV m^{-1} the amplitude value of the absorbance dichroism increases in the interval $0.225 \leq |\Delta A/\Delta A_0| \leq 0.28$ with increasing concentrations of the membrane probe $2.5 \leq [\beta\text{-DPH pPC}]/\mu M \leq 20.0$ (Fig. 6). Hence, eqn. (12) and eqn. (13) yield $0.25 \leq f^* \leq 0.32$ and $0.11 \leq \bar{f} \leq 0.135$, respectively. Note, the fractions of membrane probes in HI pores are defined as $f^* = [\text{HI}]_p/[C_0]$ and $\bar{f} = [\text{HI}]/[C_0]$, where $[\text{HI}]_p$ is the concentration of β-DPH pPC in HI pores in vesicle pole caps ($0 \leq \theta \leq \theta^*$; $\pi - \theta^* \leq \theta \leq \pi$) and [HI] is the θ-average concentration of membrane probes in HI pores. Clearly, eqn. (12) and eqn. (13) imply that $\Delta A \propto [\text{HI}]_p$, [HI] and $A_0 \propto [C_0] = [\beta\text{-DPH pPC}]$. Therefore, at constant fraction of membrane area covered by HI pores, f^* and $\bar{f}$ should not depend on [β-DPH pPC]. However, the data show that there is a 28% increase in f^* and $\bar{f}$ accompanying an eight-fold increase in [β-DPH pPC] at $E = 8$ MV m^{-1}. The reporter lipids in the membrane might decrease the bending rigidity of the membrane and, thereby, increase the degree of vesicle deformation. However, the analogous turbidity terms $\Delta T^-/T_0$ and $\Delta T^+/T_0$, measured outside the absorbance band, are independent of [β-DPH pPC] (Fig. 7), suggesting that the degree of the vesicle deformation is constant.

In any case, the fractions f^* and $\bar{f}$ are too large to be identified with the fraction of porated membrane area. Actually, $0.11 \leq \bar{f} \leq 0.135$ implies that 11–13.5% of all membrane probes are in the HI pore edges. At random distribution of the chromophores in the membrane it would mean that 11–13.5% of the membrane surface area is covered by HI pores. However, the upper limit for the fraction of membrane surface area covered by conductive pores has been estimated to be only ca. 0.002, which does not correlate with $0.11 \leq \bar{f} \leq 0.135$.[19]

Alternatively, the β-DPH pPC molecules may locally change the ability of the membrane to be electroporated, such that the pore formation is facilitated at the site of the probe. This assumption is in line with comparative fluorescence anisotropy and scanning calorimetry data, suggesting that DPH pPC "disrupts" bilayer order in its near vicinity.[20] In this case the electropores are concentrated at DPH sites. However, at the maximum concentration [β-DPH pPC] = 20 μM and at a lipid concentration of 1.0 mM, there is, on average, one β-DPH pPC molecule per 50 lipid molecules in the two membrane monolayers. The surface area per lipid molecule in the liquid crystalline state of the membrane is about 0.6 nm^2.[21] If we assume that the minimum HI pore radius is $r_p \approx d/2 = 2.5$ nm, the minimum value of the relative increase in the membrane surface area is $\Delta S/S_0 = 0.11\pi \times 2.5^2/(0.6 \times 50/2) = 0.14$ (or 14%). This fraction is again unreasonably large and should lead to membrane rupture and to vesicle elongation.

Thus, the absorbance dichroism does not directly reflect HI pores. Rather the $\Delta A^-/A_0$ mode must be predominantly due to global rotations of membrane probes caused by electroporative shape elongation of the vesicles under the electrical Maxwell stress.

4.4 Electroporative deformation model

Vesicles can be elongated if either the membrane area is increased or the intravesicular volume is decreased. However, for the short duration of the electric pulse (10 μs), solvent transport through the membrane electropores is usually very small.[2] If the optical transition moments of the membrane probes are predominantly aligned along the membrane normal, the elongation of the vesicles leads to a global orientation of the membrane chromophores apart from the external field direction, leading to negative absorbance dichroism (Fig. 9). If chromophores are distributed at finite angles around the membrane normal (Fig. 10), the amplitude of the dichroism is smaller than that for normal-parallel positions. At small deformations, the shape of the elongated vesicles may be approximated by a spheroid.[22] In view of iono-electrochromic effects or formation of electropores, the molar absorption coefficient of the chromophores in the membrane may change.[2] Formally, we can describe the modified molar absorption coefficients in terms of HO pores with the parameters f_{HO} and ε^{HO}. If the water content of the membrane in the field is increased without pores, because of field-induced orientation of the lipid head groups and isotropic penetration of water and ions into the membrane/solvent interface (Fig. 10), the radius of the HO pores is, theoretically, zero. Because of axial symmetry about the field direction, the reduced absorbance dichroism is given by:

$$\frac{\Delta A^-}{A_0} = -\frac{3}{8}\{3\cos^2\alpha - 1\}\int_0^\pi [1 + \beta f_{HO}]\{1 - 3\cos^2[\arctan(p^2\tan\theta)]\}\sin\theta\,d\theta \quad (14)$$

where $p = c/b$ is the axis ratio of the spheroid ($c > b$, Fig. 9) and α is the average angle between the membrane normal and the optical transition moment of DPH pPC (Fig. 10). It is recalled that f_{HO} is also dependent on θ. The chemical absorbance term is given by eqn. (7).

Applying eqn. (14) for $\alpha = 0$ to the data in Fig. 6 we obtain the axis ratio in the range $1.10 \leq p \leq 1.13$ (for $0.225 \leq |\Delta A^-/A_0| \leq 0.28$ at $E = 8$ MV m^{-1}). For small vesicle elongations ($p \leq 1.13$), the relative increase in the membrane surface area at constant vesicle volume is described by:

$$\frac{\Delta S}{S_0} \approx \frac{8}{45}(p-1)^2 \quad (15)$$

where $S_0 = 4\pi a^2$ is the vesicle surface area. Substituting $1.10 \leq p \leq 1.13$ in eqn. (15) we obtain $0.0018 \leq (\Delta S/S_0) \leq 0.0029$ at $2.5 \leq [\beta\text{-DPH pPC}]/\mu M \leq 20.0$ and $E = 8$ MV m^{-1}. Alternatively, without electropore formation the membrane surface area could also increase due to membrane straining by the electrical Maxwell stress.[23]

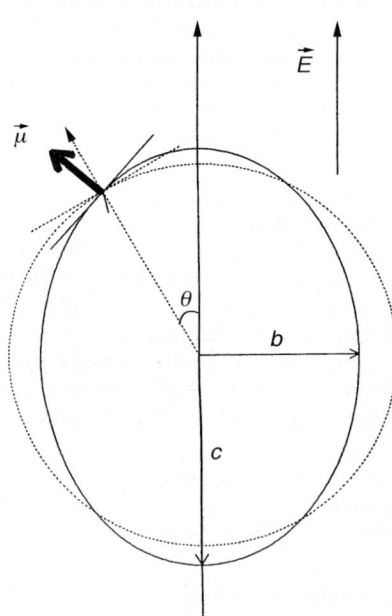

Fig. 9 Rotational displacement of the optical transition moment μ of the DPH chromophore in the membrane, caused by vesicle shape deformation in the electric field E. If μ is oriented predominantly perpendicular to the membrane surface, the vesicle elongation changes the direction of μ out of the parallel field direction, leading to negative absorbance dichroism (ΔA). The elongated vesicle is modelled by a spheroid with the principal semiaxes c and b.

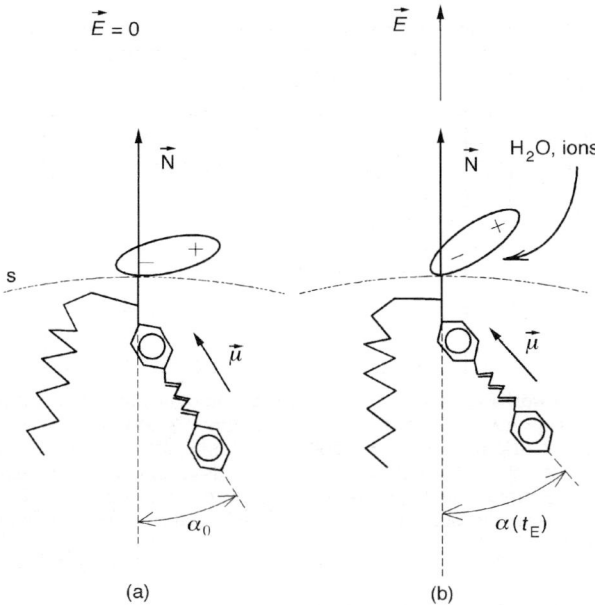

Fig. 10 Positions of the optical transition moments μ of β-DPH pPC relative to the membrane normal N (perpendicular to the membrane surface S), characterized by the average angle α, are different without electric field α_0 (a) and in a field $\alpha(t_E)$ (b). In the electric field the head-group dipole moments are partially oriented in the field direction. Entrance of water and ions in the head-group regions of the pole caps of the lipid membrane is enhanced by the electric field (b), compared with zero field (a).

In order to check the mechanism of vesicle elongation, the reduced absorbance dichroism $\Delta A^-/A_0$ is compared for vesicles of different radii a, but at a constant transmembrane potential drop $\Delta \varphi_m$. At constant $\Delta \varphi_m$, the free energy change associated with electroporation should be constant and the extent of membrane electroporation should not change with vesicle radius.[5] Moreover, vesicle elongation without pores should decrease with increasing curvature $H = 1/a$. Axis ratio and curvature are related by:[6,23]

$$p = 1 + \left\{ \frac{3}{64} \frac{\varepsilon_0 \varepsilon_w (Ea)^2}{\kappa} \right\} \frac{1}{H} \quad (16)$$

where ε_0 is the vacuum permittivity, ε_w the relative permittivity constant of the solvent and κ is the membrane bending rigidity. Note, that the product (Ea) in eqn. (16) is a constant. Insertion of eqn. (16) in eqn. (14) shows that the absorbance term $|\Delta A^-/A_0|$ should decrease with increasing H. If vesicle deformation occurs without electropore formation, controlled solely by the increase in surface area due to stretching of the membrane or reduction of membrane thermal undulations or the smoothing of a membrane super structure, the decrease in p with increasing H should be even steeper, because of the amplitude of the thermal undulations or superstructure decreases with decreasing vesicle radius.[2,24] In any case, increasing H should lead to a decrease in the vesicle elongation due to such elastic membrane stretching by the electrical Maxwell stress, and thus, to a decrease in $|\Delta A^-/A_0|$, in contrast to the experimental data (Fig. 11). On the other hand, the increase in membrane surface area by electroporation is facilitated by the increase in the packing density difference in the two membrane leaflets due to increasing membrane curvature H. The enhanced formation of conical pores with increasing H is quantified by the concept of the area difference elasticity (ADE) energy[25] and has been applied previously by Toensing et al.[9] Clearly, the experimental data suggest that vesicle elongation is rate limited by surface area increase due to pore formation. The increase in surface area $(\Delta S/S_0)$ by the electropores can be readily transformed into the vesicle elongation under Maxwell stress. Moreover, the largest increase in the membrane volume is in the pole caps of the vesicle, facilitating the vesicle elongation.

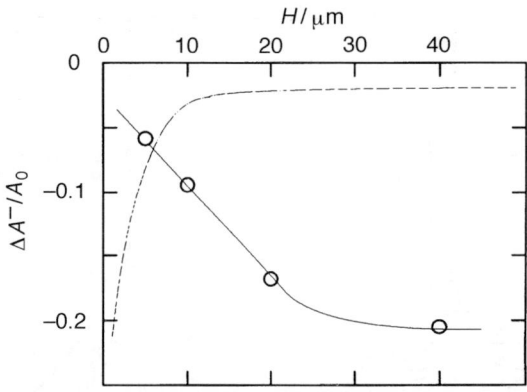

Fig. 11 Amplitude of the absorbance term $\Delta A^-/A_0$ as a function of membrane curvature $H = 1/a$ at constant transmembrane voltage drop $\Delta\varphi_m = -1.5aE = -1.5 \times 25$ nm $\times 8$ MV m^{-1} = -1.5×50 nm $\times 4$ MV m^{-1} = -1.5×100 nm $\times 2$ MV m^{-1} = -1.5×200 nm $\times 1$ MV m^{-1} = -0.3 V. The circles joined by a solid line are the experimental data. The dotted line corresponds to the vesicle elongation due to membrane stretching or smoothing of membrane undulations under the electric Maxwell stress. The increase in $|\Delta A^-/A_0|$ with H suggests that in this case the electroporative vesicle deformation is predominant. Experimental conditions as in Fig. 1 and Fig. 3.

The concept of electroporative vesicle elongation is also confirmed by the increase in the conductivity of a suspension of salt-filled vesicles exposed to an electric field pulse.[2,4] Clearly, the conductivity increase after the field pulse shows convincingly that electroporation causes permeabilization of the vesicle membrane, i.e. there are electropores. Interestingly, the reduced absorbance dichroism suggests a continuous increase in the electroporated surface area with increasing field strength, whereas massive ion transport through the membrane only occurs above a threshold value. Therefore, turbidity and absorbance electro-optics are suitable for an investigation of the very initial stages of membrane electroporation and global shape deformations.

The relaxation time of vesicle deformation due to smoothing undulations can be estimated according to:

$$\tau \approx -\frac{5}{16}\frac{\eta a^3}{\kappa}\ln\left(1 - \frac{64}{3}\frac{(p-1)\kappa}{\varepsilon_0 \varepsilon_w E^2 a^3}\right) \qquad (17)$$

Inserting $\eta = 10.02 \times 10^{-4}$ kg m^{-1} s^{-1}, the vesicle radius $a = 50$ nm, a typical value for the membrane bending rigidity $\kappa = 2.5 \times 10^{-20}$ J and $p = 1.1$ in eqn. (17), we calculate $0.8 \geqslant \tau/\mu s \geqslant 0.009$ in the field strength range $1 \leqslant E/\text{MV m}^{-1} \leqslant 8$. The time constant of stretching is also small (2.5×10^{-10} s);[26] the apparatus time constant is 20 µs and the field builds up rapidly (6×10^{-8} s). Since all the time constants are much smaller than those of electropore formation ($3 \geqslant \tau_p/\mu s \geqslant 0.2$), the slow mode (Fig. 4) of vesicle deformation is rate limited by, and rapidly coupled to, the primary electroporation process. Note, that $p = 1.1$ corresponds to a maximum change of 6.6% in the membrane curvature; therefore, the relative motion between monolayers may be neglected. Only a rapid, ca. 1000-fold increase in curvature could lead to a retardation of the shape deformation due to viscous impedance arising from relative monolayer motion.[27]

It is recalled that the absorbance dichroism contains the ratio p and the angle α; see eqn. (14). For electroporative deformation the surface area increase $\Delta S(\text{II})$ of the slow mode is proportional to the concentration [P] of electropores. In our chemical model for membrane electroporation (C $\rightleftharpoons$ P), the pore kinetics is described by [P] = [P]$_\infty$(1 − exp[$-t/\tau_p$]), where [P]$_\infty$ is the amplitude and τ_p the time constant of the poration-resealing process. Both relaxation quantities are dependent on the field strength and on the positional angle θ (Fig. 8). Hence, all measured parameters reflect θ-averages. Application of a Mie type numerical code for confocal coated spheroids[28] to the field-on time course of the turbidity terms $\Delta T^-(t)/T_0$ and $\Delta T^+(t)/T_0$ (see, e.g. Fig. 4) yields the dependence of the total increase ΔS_∞ on the field strength (Fig. 12). The increase in membrane area in Fig. 12 is certainly large, compared with 0.002 fraction of conductive pores reported by Hibino et al.[19] However, not all of ΔS_∞ is due to electropores. Membrane stretching and smooth-

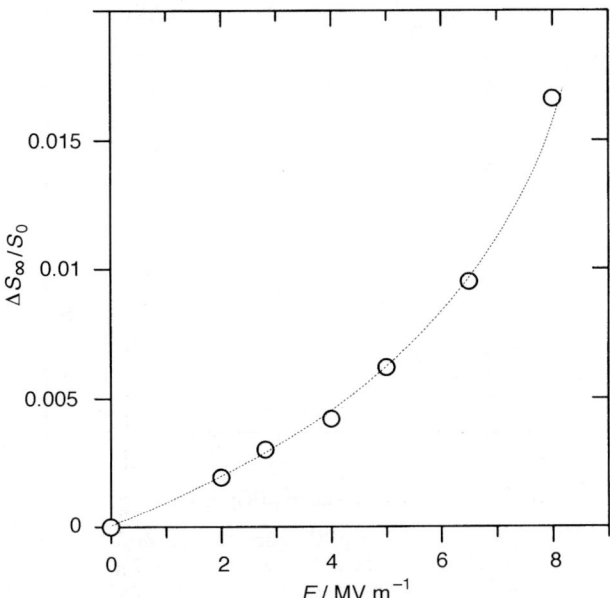

Fig. 12 Amplitude value $\Delta S_\infty/S_0$ of the relative change in the membrane surface area of a vesicle as a function of the external field strength E. S_0 is the total surface area of the vesicle membrane. ΔS_∞ is calculated from the field-on turbidity relaxations $\Delta T^-(t)/T_0$ and $\Delta T^+(t)/T_0$, assuming constant intravesicular volume. Note that $\Delta S_\infty = \Delta S_\infty(\mathrm{I}) + \Delta S_\infty(\mathrm{II})$ is the total amplitude value of the electrochemomechanical surface area increase according to $\Delta S = \Delta S_\infty(\mathrm{I})(1 - \exp[-t/\tau(\mathrm{I})]) + \Delta S_\infty(\mathrm{II})(1 - \exp[-t/\tau(\mathrm{II})])$. Experimental conditions as in Fig. 4.

ing undulations in the electric field also contribute to ΔS_∞. Fortunately, the turbidity terms $\Delta T^-/T_0$ and $\Delta T^+/T_0$ are independent of the presence of the chromophores. Comparison of the p values at different times calculated from the turbidity terms $\Delta T^-(t)/T_0$ and $\Delta T^+(t)/T_0$ and from the absorbance dichroism $\Delta A^-(t)/A_0$ using eqn. (14) with $\beta = 0$ and a finite value of α (Fig. 13)

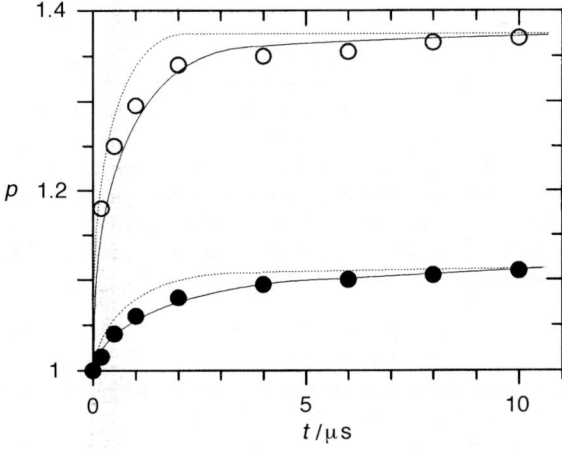

Fig. 13 Axis ratio $p = c/b$ of the elongated vesicle as a function of the pulse time at the two extreme field strengths $E/\mathrm{MV\ m^{-1}} = 2$ (●) and 8 (○). The circles are the values of p calculated from the turbidity terms $\Delta T^-/T_0$ and $\Delta T^+/T_0$ at different times. The corresponding solid curves are the theoretical simulations with $\Delta S = \Delta S(\mathrm{I}) + \Delta S(\mathrm{II})$ for the kinetics of the increase in the membrane surface area ΔS. The dotted curves are calculated from the absorbance $\Delta A^+/A_0(\approx 0)$ and $\Delta A^-/A_0$ at the concentration of the membrane probes [β-DPH pPC] = 5 μM. The values $\Delta A^+(t_\mathrm{E})/A_0$, $\Delta A^-(t_\mathrm{E})/A_0$ and $\Delta T^-(t_\mathrm{E})/T_0$, $\Delta T^+(t_\mathrm{E})/T_0$ at the end of the pulse $t_\mathrm{E} = 10$ μs were used to calculate the average angle $\alpha(t_\mathrm{E}) = 39.7°$ between the optical transition moment of β-DPH pPC and the membrane normal. Experimental conditions as in Fig. 3.

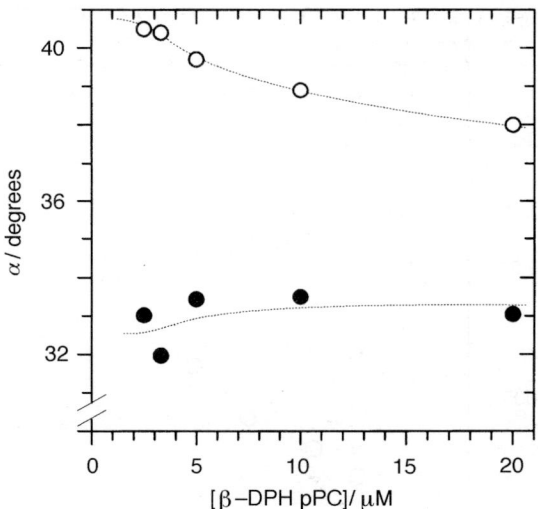

Fig. 14 Dependence of the angle α on β-DPH pPC concentration (●), the angle α_0 extrapolated to the beginning of the pulse ($t = 0$), and (○) $\alpha(t_E)$ at the pulse end $t_E = 10$ μs. Note for neither α_0 nor $\alpha(t_E)$ is there any dependence on the field strength in the range $2 \leqslant E/\text{MV m}^{-1} \leqslant 8$. Experimental conditions as in Fig. 3.

indicates that the angle α also changes, in the field strength range $E \geqslant 2$ MV m^{-1}. It is readily seen that, unlike the turbidity dichroism, the absorbance dichroism also contains information on the position of the lipid side chains, here the DPH residue, relative to the membrane normal. The field-induced increase in α and the dependence of α on the β-DPH pPC concentration (Fig. 14) cannot be rationalized unequivocally. One possible explanation is the rotational displacement of the chromophore residue caused by the alignment of the polar head group in the field direction (Fig. 10).

5 Conclusion

It appears that the concept of electrochemomechanical vesicle deformation can consistently rationalize the electro-optic data. However, the local structural changes associated with the formation of HI pore edges cannot be directly identified with electro-optics. The electropores are optically enhanced by vesicle deformation, because the slow mode $\Delta S(\text{II})$ of shape elongation under the electric Maxwell stress is caused by and rapidly coupled to the electroporative increase in the membrane area. This fundamental result of electroporative shape deformation also applies to biological cells and tissue. However, the lipid-protein plasma membrane of cells is a part of a larger envelope structure comprising extracellular matrix components and intracellular cytoskeletal elements.[29] Therefore, the extent and rate of cell deformation will be determined by the elastic properties of this network.[30]

Acknowledgements

We thank the Deutsche Forschungsgemeinschaft for the grant Ne227/9-2 to E. Neumann.

References

1 E. Neumann and S. Kakorin, *Radiol. Oncol.*, 1998, **32**, 7.
2 E. Neumann and S. Kakorin, *Curr. Opin. Colloid Interface Sci.*, 1996, **1**, 790.
3 S. Kakorin, E. Redeker and E. Neumann, *Eur. Biophys. J.*, 1998, **27**, 43.
4 E. Neumann, E. Werner, A. Sprafke and K. Krueger, in *Colloid and Molecular Electro-Optics*, ed. B. R. Jennings and S. P. Stoylov, Institute of Physics, Bristol, 1992, p. 197.
5 S. Kakorin, S. P. Stoylov and E. Neumann, *Biophys. Chem.*, 1996, **58**, 109.
6 S. Kakorin and E. Neumann, *Ber. Bunsen-Ges. Phys. Chem.*, 1998, **102**, 670.

7   S. Kakorin and E. Neumann, *Ber. Bunsen-Ges. Phys. Chem.*, 1996, **100**, 721.
8   A. Revzin and E. Neumann, *Biophys. Chem.*, 1974, **2**, 144.
9   K. Toensing, S. Kakorin, E. Neumann, S. Liemann and R. Huber, *Eur. Biophys. J.*, 1997, **26**, 307.
10 L. D. Mayer, M. J. Hope and P. R. Cullis, *Biochim. Biophys. Acta*, 1986, **885**, 161.
11 K. R. Foster and A. E. Sowers, *Biophys. J.*, 1995, **69**, 777.
12 G. K. Batchelor, *J. Fluid. Mech.*, 1976, **74**, 1.
13 M. Planck, *Berlin Ber.*, 1904, 740.
14 S. Stoylov and E. Neumann, *Bulgarian Chem. Commun.*, 1992, **25**, 445.
15 B. Mui, P. Cullis, E. Evans and T. Madden, *Biophys. J.*, 1993, **64**, 443.
16 T. Chang and H. Yu., *Comments Mol. Cell. Biophys.*, 1990, **7**, 27.
17 I. G. Abidor, V. B. Arakelyan, L. V. Chernomordik, Y. A. Chizmadzhev, V. P. Pastuchenko and M. R. Tarasevich, *Bioelectrochem. Bioenerg.*, 1979, **6**, 37.
18 E. Fredericq and C. Houssier, *Electric Dichroism and Electric Birefringence.* Clarendon Press, Oxford, 1973, p. 219.
19 M. Hibino, H. Itoh and K. Kinosita, *Biophys. J.*, 1993, **64**, 1789.
20 B. Lentz, *Chem. Phys. Lipids*, 1993, **64**, 99.
21 J. J. Lopez Cascales, M. L. Huertas and J. Garcia de la Torre, *Biophys. Chem.*, 1997, **69**, 1.
22 B. Zeks and S. Svetina, in *Springer Proceedings in Physics*, ed. R. Lipowsky, D. Richter and K. Kremer, Springer-Verlag, Berlin, 1992, vol. 66, p. 174.
23 M. Kummrow and W. Helfrich, *Phys. Rev. A*, 1991, **44**, 8356.
24 B. Kloesgen and W. Helfrich, *Eur. Biophys. J.*, 1993, **22**, 329.
25 U. Seifert and R. Lipowsky, in *Structure and Dynamics of Membranes*, ed. R. Lipowsky and E. Sackmann, Elsevier North-Holland, Amsterdam, 1995, vol. 1A, p. 519.
26 S. Komura, in *Vesicles*, ed. M. Rosoff, Marcel Dekker, New York, 1996, p. 197.
27 E. Evans and A. Yeung, *Chem. Phys. Lipids*, 1994, **73**, 39.
28 V. G. Farafornov, N. V. Voshcinnikov and V. V. Somsikov, *Appl. Opt.*, 1996, **35**, 5412.
29 R. Lipowsky, *Encyclopedia Appl. Phys.*, 1998, **23**, 199.
30 F. G. Schmidt, F. Ziemann and E. Sackmann, *Eur. Biophys. J.*, 1996, **24**, 348.

Paper 8/06461J

Lipid–protein interactions in the membrane: Studies with model peptides

Sanjay Mall, Ram P. Sharma, J. Malcolm East and Anthony G. Lee*

Division of Biochemistry and Molecular Biology, School of Biological Sciences, University of Southampton, Southampton, UK SO16 7PX

Received 27th November 1998

We have used fluorescence quenching of tryptophan-containing *trans*-membrane peptides by bromine-containing phospholipids to study the specificity of peptide–lipid interactions. We have synthesized peptides Ac-$K_2GL_mWL_nK_2A$-amide where $m = 7$ and $n = 9$ (L_{16}) and $m = 10$ and $n = 12$ (L_{22}). Binding constants of L_{22} for dioleoylphosphatidylserine [di(C18:1)PS] or dioleoylphosphatidic acid [di(C18:1)PA] relative to dieoleoylphosphatidylcholine [di(C18:1)PC] were close to 1. However, for L_{16}, whilst the bulk of the di(C18:1)PA molecules bound with a binding constant relative to di(C18:1)PC close to 1, a small number of di(C18:1)PA molecules bound much more strongly. Assuming just one high affinity binding site on L_{16} for anionic lipid, the affinity of the site for di(C18:1)PS was calculated to be *ca.* 8 times that for di(C18:1)PC. The relative binding constant was little affected by ionic strength and close contact between the anionic headgroup of di(C18:1)PS and a lysine residue on the peptide was suggested. The relative binding constant for di(C18:1)PS at this high affinity site was less than for di(C18:1)PA. Cholesterol interacts with L_{22} with an affinity about 0.7 of that of di(C18:1)PC. The structure of the peptide itself is important. The peptide Ac-KKGYL$_8$WL$_8$YKKA-amide (Y_2L_{14}) incorporated into bilayers of dinervonylphosphatidylcholine [di(C24:1)PC] whereas L_{16} did not incorporate into this lipid. It is suggested that thinning of a lipid bilayer around a peptide to give optimal hydrophobic matching is less energetically unfavourable when a Tyr residue is located in the lipid/water interfacial region.

The membrane-spanning region of an intrinsic membrane protein adopts either an α-helical or a β-sheet structure to ensure formation of the maximum number of intramolecular hydrogen bonds: of these structures, the α-helix is the most common.[1] The transmembrane α-helices of a membrane protein contain largely hydrophobic residues because the solubility of polar residues is low in a non-polar environment such as that of a lipid bilayer; since the thickness of the hydrophobic core of a bilayer composed of naturally occurring phospholipids is about 30 Å, a stretch of about 20 hydrophobic residues is required to span the bilayer. Charged residues are usually found on either side of the hydrophobic stretch of residues, the charged residues being located in the interface region of the lipid bilayer, interacting either with the polar backbone region of the lipid bilayer or with the charged headgroups or with water. The distribution of hydrophobic residues in the trans-membrane region of a membrane protein is non-random.[1] The most common amino acid is Leu, but aromatic residues, particularly Trp and Tyr are often found at the lipid/water interface in both α-helical and β-sheet proteins;[1] it has been suggested that these residues could act as 'floats' with their polar groups facing the water and their non-polar aromatic regions penetrating into the hydrocarbon region, serving to fix the helix within the lipid bilayer.

The importance of the lipid bilayer for the proper functioning of a membrane protein has been demonstrated for the Ca^{2+}-ATPase of skeletal muscle sarcoplasmic reticulum. A high rate of ATP hydrolysis by the Ca^{2+}-ATPase requires that the fatty acyl chains of the surrounding phospholipids be between C16 and C20 in length; this presumably ensures that the bilayer thickness matches the hydrophobic thickness of the Ca^{2+}-ATPase.[2,3] Outside this chain length range, the activity of the ATPase is low, and changes on the ATPase can include a decrease in the rate of phosphorylation of the ATPase by ATP and a change in the stoichiometry of Ca^{2+} binding.[4,5] High activity also requires that the phospholipids be in the liquid-crystalline, bilayer phase; low activities are observed in the gel phase and in the hexagonal H_{II} phase.[6,7] Finally, higher activities are observed in bilayers of zwitterionic phospholipids such as phosphatidylcholine (PC) or phosphatidylethanolamine (PE) than in bilayers of anionic phospholipids such as phosphatidic acid (PA) or phosphatidylserine (PS).[8]

The region of the Ca^{2+}-ATPase embedded in the lipid bilayer is a bundle made up of 10 transmembrane α-helices,[9,10] and so it must be these α-helices that are sensing the character of the bilayer. To characterize the nature of the interaction between lipid bilayers and α-helices we have turned to simple synthetic peptides of the type $Ac-K_2-G-L_m-W-L_n-K_2-A$-amide (P_{m+n}) consisting of a long sequence of hydrophobic Leu residues capped at both the N- and C-terminal ends with two polar lysine residues with a centrally located tryptophan residue to act as a fluorescence reporter group.[11] It has been shown that peptides of this type (without the tryptophan residue), when mixed with phospholipid, form stable α-helices spanning the phospholipid bilayer, the ends of the helix being anchored in the phospholipid headgroup region by the charged Lys caps.[12,13] A fluorescence quenching method has been used to measure the strength of interaction between the peptides and particular lipids, in bilayers in the fluid, liquid crystalline phase.[14] The peptide is incorporated into bilayers containing the brominated phospholipid, dibromostearoylphosphatidylcholine [di(Br_2C18:0)PC]; di(Br_2C18:0)PC behaves much like a conventional phospholipid with unsaturated fatty acyl chains because the bulky bromine atoms have similar effects on lipid packing as a *cis* double bond.[14] In mixtures of brominated and non-brominated phospholipids, the degree of quenching of the fluorescence of the tryptophan residue in the peptide is related to the fraction of the surrounding phospholipids which are brominated, and thus to the strength of binding of the non-brominated lipid to the peptide.

Experimental

Dimyristoleoylphosphatidylcholine [di(C14:1)PC], dioleoylphosphatidylcholine [di(C18:1)PC], dierucoylphosphatidylcholine [di(C22:1)PC], dinervonylphosphatidylcholine [di(C24:1)PC], dioleoylphosphatidylserine [di(C18:1)PS] and dioleoylphosphatidic acid [di(C18:1)PA] were obtained from Avanti Polar Lipids. Di(C18:1)PC was brominated to give di(Br_2C18:0)PC as described by East and Lee,[14] and the same procedure was used to brominate di(C18:1)PS to dibromostearoylphosphatidylserine [di(Br_2C18:0)PS] and di(C18:1)PA to dibromostearoylphosphatidic acid [di(Br_2C18:0)PA]. Cholesterol was brominated to give 5,6-dibromocholestan-3β-ol (dibromocholesterol) as described in Simmonds *et al.*[15] Peptides Ac-KKGL$_7$WL$_9$KKA-amide (L_{16}), Ac-KKGL$_{10}$WL$_{12}$KKA-amide (L_{22}), Ac-KKGYL$_6$WL$_8$YKKA-amide (Y_2L_{14}), and Ac-KKGYL$_9$WL$_{11}$YKKA-amide (Y_2L_{20}) were synthesised using t-Boc chemistry,[16] and purity was confirmed using electrospray and MALDI-TOF mass spectroscopy.

Peptides (20 nmol) were incorporated into phospholipid bilayers by mixing peptide and lipid at a molar ratio of peptide to phospholipid of 1:100 in chloroform–methanol (2:1 v/v). Solvent was removed under vacuum, and the mixture was resuspended in buffer (400 μl) by sonication in a bath sonicator for 5–10 min. Aliquots (100 μl) were diluted into buffer (2.5 ml; 20 mM Hepes, 1 mM EGTA, pH 7.2) and fluorescence intensities were recorded at 25 °C using an SLM-Aminco 8000C fluorimeter, with excitation and emission wavelengths of 280 and 340 nm, respectively.

Theory

Tryptophan fluorescence is quenched by bromine-containing molecules by a process of heavy atom quenching which requires contact between tryptophan and bromine. The fluorescence lifetime for tryptophan is considerably less than the time for two lipids to exchange position in a bilayer, so that quenching can be considered to be a static phenomenon.[17] Quenching can then be analyzed in either of two ways. The first is a lattice model of quenching.[14,18] The degree of quenching in the lattice model is proportional to the probability that a brominated lipid occupies a lattice site close enough to the peptide to cause quenching. For a random distribution of lipids, the probability that any lattice site is not occupied by a brominated lipid is $1 - x_{Br}$ where x_{Br} is the mole fraction of brominated lipid in the bilayer. The probability that any particular peptide will give rise to fluorescence is proportional to the probability that none of the n lattice sites close enough to the peptide to cause quenching is occupied by a brominated lipid. Thus

$$F/F_0 = F_{min} + (F_0 - F_{min})(1 - x_{Br})^n \quad (1)$$

where F_0 and F_{min} are the fluorescence intensities for the peptide in non-brominated and in brominated lipid, respectively, and F is the fluorescence intensity in the phospholipid mixture when the mole fraction of brominated lipid is x_{Br}. In a hexagonal lattice, if quenching can only be caused by immediate neighbours, n will equal 6.

An alternative description of quenching is the sphere of action model.[19] This assumes the existence of a sphere of volume around a fluorophore within which a quencher will cause quenching with a probability of unity. Adapting this model to the two-dimensional case of a biological membrane gives

$$F/F_0 = F_{min} + (F_0 - F_{min})\exp(-\pi r^2 C_{Br}) \quad (2)$$

where πr^2 is the area of the quenching circle around each tryptophan and C_{Br} is the concentration of brominated lipid in units of molecules per unit area. Here we have assumed an area of 70 Å² per lipid molecule.

The lattice model can be readily extended to describe quenching of the peptide in a mixture of two lipids of different affinities for the peptide, in the case where the sites around the peptide are equivalent. At each site, an equilibrium will exist

$$PL + Q \rightleftharpoons PQ + L$$

where PL and PQ are complexes of peptide with non-brominated lipid (L) and brominated lipid (Q), respectively. The equilibrium can be described by an equilibrium constant K given by:

$$K = [PQ][L]/[PL][Q] \quad (3)$$

where K is the binding constant of the brominated lipid relative to that of the non-brominated lipid. Fluorescence quenching then fits to the equation:

$$F/F_0 = F_{min} + (F_0 - F_{min})(1 - f_{Br})^n \quad (4)$$

where the fraction of sites at the lipid/peptide interface occupied by brominated lipid is f_{Br}.[14] The fraction of sites occupied by brominated lipid is related to x_{Br} by:

$$f_{Br} = Kx_{Br}/(Kx_{Br} + [1 - x_{Br}]) \quad (5)$$

The lattice model is not readily extended to the case where the sites around the peptide are non-equivalent. However, the quenching sphere approach can be readily extended to the case where there is a single high affinity binding site for the brominated lipid together with a large number of non-specific lipid-binding sites. Binding at the high affinity site can be described by an equation analogous to eqn. (3) giving the probability that the site is not occupied by the brominated lipid as $(1 - x_{Br})/(1 - x_{Br} + Kx_{Br})$. Quenching from the non-specific sites can be described by eqn. (2) so that

$$\frac{F}{F_0} = F_{min} + (F_0 - F_{min})\left(\frac{1 - x_{Br}}{1 - x_{Br} + Kx_{Br}}\right)\exp(-\pi r^2 C_{Br}) \quad (6)$$

Results

We synthesized two peptides Ac-KKGL$_7$WL$_9$KKA-amide (L$_{16}$) and Ac-KKGL$_{10}$WL$_{12}$KKA-amide (L$_{22}$) containing a tryptophan residue in the centre of a hydrophobic domain composed of

Leu residues and a corresponding pair of peptides Ac-KKGYL$_6$WL$_8$YKKA-amide (Y$_2$L$_{14}$) and Ac-KKGYL$_9$WYL$_{11}$KKA-amide (Y$_2$L$_{20}$) in which the Leu residues at each end of the hydrophobic stretch have been replaced by Tyr residues. Peptides were incorporated into lipid bilayers at a molar ratio of peptide : lipid of 1 : 100 by mixing peptide and lipid in organic solvent, followed by removal of the solvent and hydration of the mixture. The fluorescence emission spectrum of the tryptophan residue is environmentally sensitive, the emission maximum moving to shorter wavelength with decreasing environmental polarity.[19] The emission spectra of all the peptides incorporated into bilayers of lipids containing oleoyl chains [di(C18 : 1)PC, di(C18 : 1)PA, or di(C18 : 1)PS] are centred at about 323 nm indicating a very hydrophobic environment for the tryptophan, consistent with the expected localization in the middle of the bilayer.

Fluorescence quenching

The fluorescence intensity for the peptide L$_{16}$ incorporated into bilayers of di(Br$_2$C18 : 0)PC is about 5% of that in di(C18 : 1)PC, demonstrating highly efficient quenching of the tryptophan by the bromine-containing fatty acyl chains (Fig. 1). The fluorescence intensity for L$_{16}$ in mixtures of di(Br$_2$C18 : 0)PC and di(C18 : 1)PC decreases with increasing content of di(Br$_2$C18 : 0)PC and fits to eqn. (1) with a value of n, the number of 'sites' around the peptide where binding can result in quenching, of 3.4; the same value of n fits the quenching data in mixtures of di(Br$_2$C18 : 0)PS and di(C18 : 1)PS (Fig. 1). For the longer peptide L$_{22}$ the data in either di(Br$_2$C18 : 0)PC/di(C18 : 1)PC or di(Br$_2$C18 : 0)PS/di(C18 : 1)PS again fit to eqn. (1) but now with a value of n of 2.1. These results suggest that the peptides L$_{16}$ and L$_{22}$ adopt different structures in the lipid bilayer, and,

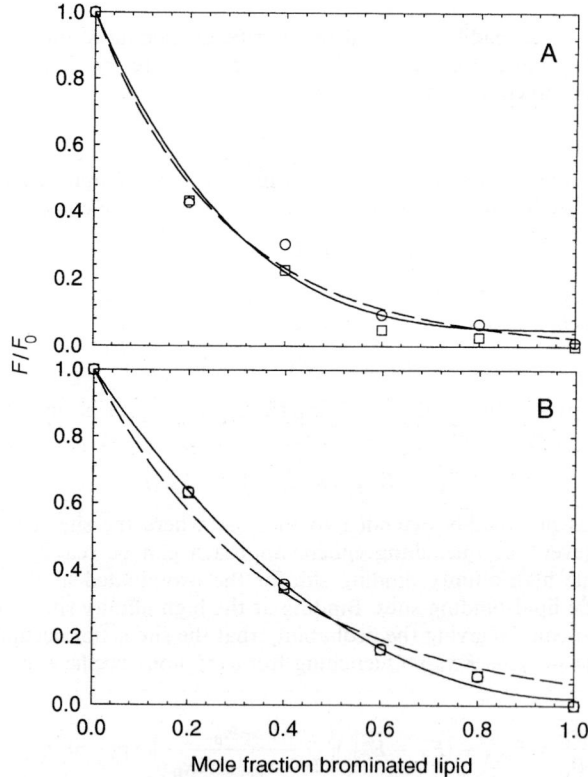

Fig. 1 Fluorescence intensities for L$_{16}$ and L$_{22}$ in mixtures containing brominated phospholipids. L$_{16}$ (A) or L$_{22}$ (B) were incorporated into mixtures of di(Br$_2$C18 : 0)PC and di(C18 : 1)PC (○) or di(Br$_2$C18 : 0)PS and di(C18 : 1)PS (□), at a molar ratio of peptide : phospholipid of 1 : 100. Fluorescence intensities are expressed as a fraction of that recorded for peptide in di(C18 : 1)PC (A) or di(C18 : 1)PS (B). The solid and broken lines show best fits of the data to eqn. (1) and (2) respectively, with the parameters given in the text.

indeed, whereas the hydrophobic length of L_{16} (27 Å) is close to the thickness of a bilayer of di(C18:1)PC (29.8 Å; ref. 20) the hydrophobic length of L_{22} is greater (36 Å) so that L_{22} will be tilted in the bilayer.[11] As shown in Fig. 1, the quenching data also fit to the circle of quenching model eqn. (2), with values for the quenching radius r of 9 and 8 Å respectively, for L_{16} and L_{22}.

Relative binding affinities of anionic and zwitterionic phospholipids

Fitting the fluorescence quenching curve for a peptide in a mixture of di(Br_2C18:0)PC and lipid X to eqn. (4) gives the peptide binding constant for lipid X relative to that for di(Br_2C18:0)PC. If all the binding sites on the peptide are equivalent, then fitting the fluorescence quenching curve for the peptide in a mixture of di(C18:1)PC and brominated lipid X should give the same relative lipid binding constant, assuming that bromination of the chains does not significantly affect interaction with the peptide. The quenching curve for L_{22} in mixtures of di(Br_2C18:0)PS and di(C18:1)PC (Fig. 2) in buffer of low ionic strength fits to eqn. (4) with a binding constant for PS relative to PC of 1.3 ± 0.4, showing no significant selectivity in binding. Consistent with this interpretation, the quenching curves for L_{22} in mixtures of di(Br_2C18:0)PC and di(C18:1)PS fit to eqn. (4) with a binding constant for PS relative to PC of 0.8 ± 0.1 (Fig. 2). Similarly, relative binding constants for PS and PC are close to 1 in media of high ionic strength, and relative binding constants for PA and PC are also close to 1 (Table 1). In contrast to these results with L_{22}, for L_{16} different relative binding constants are obtained in mixtures of PC and PS or PC and

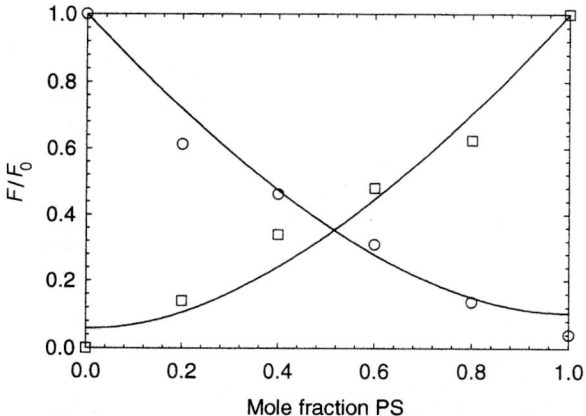

Fig. 2 Fluorescence intensities for L_{22} in mixtures of PC and PS. The experimental points show fluorescence intensities in mixtures of di(Br_2C18:0)PS and di(C18:1)PC (○) or di(Br_2C18:0)PC and di(C18:1)PS (□), as a function of the mole fraction of PS. The buffer was 20 mM Hepes, 1 mM EGTA, pH 7.2. The solid lines show fits to eqn. (4) with the relative binding constants given in Table 1.

Table 1 Relative binding constants for peptide L_{22}

Lipid mixture	Lipid binding constant relative to PC[a]	
	Low salt[b]	High salt[c]
di(C18:1)PS–di(Br_2C18:0)PC	1.3 ± 0.4	0.8 ± 0.4
di(C18:1)PC–di(Br_2C18:0)PS	0.8 ± 0.1	0.8 ± 0.2
di(C18:1)PA–di(Br_2C18:0)PC	—	1.6 ± 0.3
di(C18:1)PC–di(Br_2C18:0)PA	1.6 ± 0.1	0.8 ± 0.2

[a] Concentrations of phospholipids expressed as mole fractions.
[b] 20 mM Hepes, 1 mM EGTA, pH 7.2. [c] 20 mM Hepes, 1 mM EGTA, pH 7.2, 300 mM KCl.

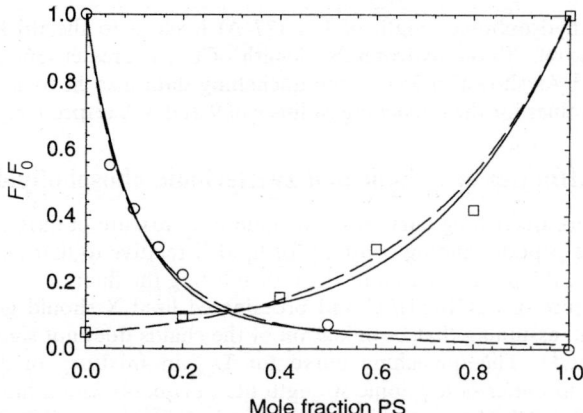

Fig. 3 Fluorescence intensity for L_{16} in mixtures of PC and PS. The experimental points show fluorescence intensities in mixtures of di(Br_2C18 : 0)PS and di(C18 : 1)PC (○) or di(Br_2C18 : 0)PC and di(C18 : 1)PS (□), as a function of the mole fraction of PS. The buffer was 20 mM Hepes, 1 mM EGTA, pH 7.2. The solid lines show fits to eqn. (4) with the relative binding constants given in Table 2 and the broken lines show fits to eqn. (6) with the parameters given in Table 3.

PA depending on whether the brominated lipid is the phosphatidylcholine or the anionic phospholipid (Fig. 3). The large relative binding constant for the anionic phospholipid derived from experiments in which the anionic phospholipid is brominated together with a relative binding constant close to 1 derived from experiments in which the phosphatidylcholine is brominated suggests that the anionic phospholipids bind to a small number of sites with high affinity on the peptide whereas the zwitterionic phospholipid binds to a large number of non-specific sites. This

Table 2 Relative binding constants for peptide L_{16}

Lipid mixture	Lipid binding constant relative to PC[a]	
	Low salt[b]	High salt[c]
di(C18 : 1)PS–di(Br_2C18 : 0)PC	0.9 ± 0.3	1.4 ± 0.1
di(C18 : 1)PC–di(Br_2C18 : 0)PS	3.0 ± 0.3	1.7 ± 0.1
di(C18 : 1)PA–di(Br_2C18 : 0)PC	1.1 ± 0.1	1.1 ± 0.2
di(C18 : 1)PC–di(Br_2C18 : 0)PA	4.2 ± 0.1	4.2 ± 0.4

[a] Concentrations of phospholipids expressed as mole fractions.
[b] 20 mM Hepes, 1 mM EGTA, pH 7.2. [c] 20 mM Hepes, 1 mM EGTA, pH 7.2, 300 mM KCl.

Table 3 Relative binding constants for anionic phospholipids for peptide L_{16}

Lipid mixture	Binding constant relative to di(C18 : 1)PC[a]	
	Low salt[b]	High salt[c]
di(Br_2C18 : 0)PS–di(C18 : 1)PC	8.6 ± 1.4	2.9 ± 0.4
di(Br_2C18 : 0)PA–di(C18 : 1)PC	23 ± 3	21 ± 4

[a] Concentrations of phospholipids expressed as mole fractions.
[b] 20 mM Hepes, 1 mM EGTA, pH 7.2. [c] 20 mM Hepes, 1 mM EGTA, pH 7.2, 300 mM KCl.

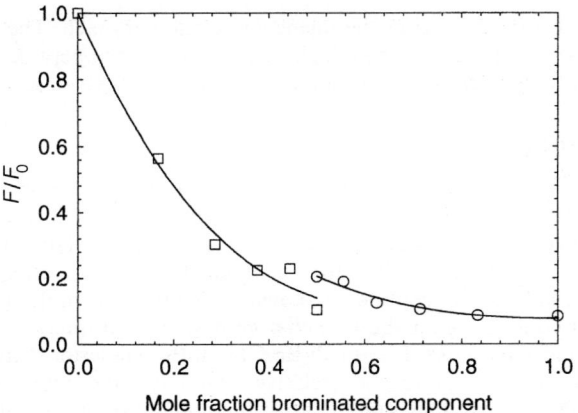

Fig. 4 Fluorescence intensities for peptide L_{22} in mixtures containing cholesterol. The peptide was mixed with di(Br_2C18:0)PC and cholesterol (○) or dibromocholesterol and di(C18:1)PC (□) and fluorescence is plotted as a function of the mole fraction of the brominated component. The solid lines show the best fit to the data with $n = 2.1$, with a relative binding constant K for cholesterol of 0.7 for mixtures of di(Br_2C18:0)PC and cholesterol and with $n = 6$ and a relative binding constant K for cholesterol of 1.4 for mixtures of di(C18:1)PC and dibromocholesterol.

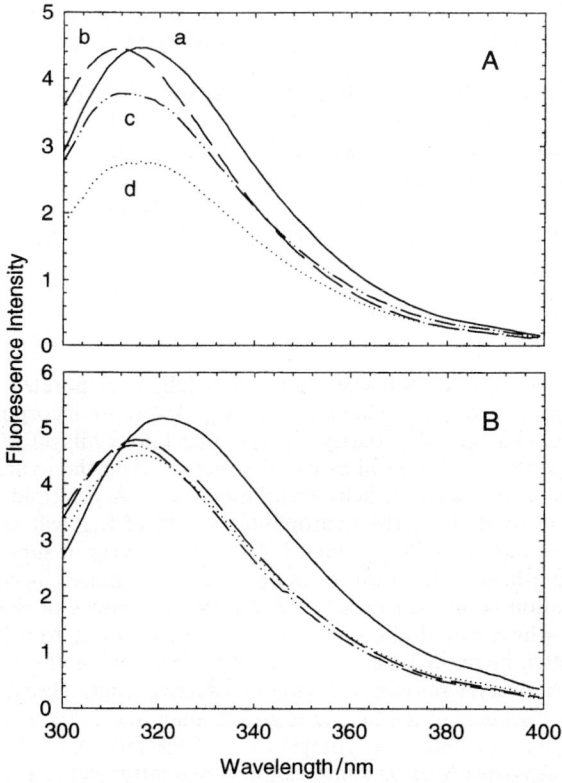

Fig. 5 Fluorescence emission spectra for Y_2L_{14} and Y_2L_{20} in lipid bilayers. (A) shows fluorescence emission spectra for Y_2L_{14} in: (a), di(C14:1)PC; (b), di(C18:1)PC; (c), di(C22:1)PC; (d) di(C24:1)PC. (B) shows fluorescence emission spectra for Y_2L_{20} in: solid line, di(C14:1)PC; broken line, di(C18:1)PC, dash-dot line, di(C22:1)PC; dotted line, di(C24:1)PC.

situation is better analyzed in terms of the quenching circle approach. The data shown in Fig. 3 can be fitted to eqn. (6), assuming a single high affinity site on the peptide for anionic phospholipid, with the relative binding constants for anionic phospholipids given in Table 3.

Interaction with cholesterol

We can also use these fluorescence quenching methods to study interactions between peptides and cholesterol in the membrane. Fig. 4 shows that the tryptophan fluorescence emission of the peptide L_{22} is quenched by dibromocholesterol in bilayers of di(C18:1)PC so that dibromocholesterol must be able to interact with the peptide. Conversely, cholesterol can displace di(Br_2C18:0)PC from around the peptide, as shown by the increase in fluorescence intensity with increasing cholesterol content, again showing that cholesterol can interact with the peptide. The experimental quenching curves for L_{22} in di(Br_2C18:0)PC–cholesterol mixtures can be fitted to eqn. (4) with $n = 2.1$ and a relative binding constant [peptide.cholesterol]/[peptide].[di(C18:1)PC] of 0.7. This same relative binding constant also fits the data for L_{22} in di(C18:1)PC–dibromocholesterol mixtures, assuming a value for n for dibromocholesterol of 6.0, suggesting a hexagonal packing arrangement for cholesterol around the peptide.

Effects of Tyr residues on hydrophobic matching

In a previous paper we showed that whereas L_{22} incorporated fully into bilayers of phosphatidylcholines with chain lengths between C14 and C24, L_{16} did not incorporate at all into a bilayer of di(C24:1)PC and only partly incorporated into a bilayer of di(C22:1)PC.[11] The level of incorporation of the peptide can be quantitated from the fluoresence spectra since tryptophan fluorescence intensities are very low for unincorporated peptide, due to the formation of non-fluorescent aggregates in water.[11] Fig. 5B shows the fluorescence spectra of Y_2L_{20} in phosphatidylcholines of chain length between C14 and C24. The very similar fluorescence intensities observed for the peptide in all bilayers indicate full insertion into these bilayers; the higher wavelength of the emission maximum in di(C14:1)PC than in the other lipids suggests that the tryptophan residue is buried less deeply in bilayers of di(C14:1)PC than in thicker lipid bilayers. For Y_2L_{14}, although the fluorescence intensity in di(C24:1)PC is less than that in di(C14:1)PC, the decrease in fluorescence intensity is only 40% (Fig. 5A), compared to an 85% decrease in intensity for L_{16}, combined with a shift to longer wavelength.[11] Thus whilst L_{16} is completely excluded from a bilayer of di(C24:1)PC, Y_2L_{14} is at least partially incorporated.

Discussion

In a previous paper[11] we showed that whereas L_{22} was fully incorporated into bilayers of phosphatidylcholines irrespective of lipid chain length, L_{16} failed to incorporate normally into a bilayer of di(C24:1)PC and was only partly incorporated into a bilayer of di(C22:1)PC. If it is assumed that the peptide adopts an ideal α-helical structure, then the hydrophobic length of L_{16} will be about 27 Å, calculated using a helix translation of 1.5 Å per residue and a hydrophobic stretch of 18 residues in total. Thus the hydrophobic length of L_{16} will match the hydrophobic thickness of a bilayer of di(C16:1)PC or di(C18:1)PC. In bilayers of sub-optimal thickness, significant changes in the α-helical structures of the peptides are unlikely because of the stability of the α-helix, an expectation confirmed in infrared studies of peptides of this type.[21,22] There are then two mechanisms which would allow matching of a thin bilayer to a long peptide: the fatty acyl chains could stretch, increasing the thickness of the bilayer, or the peptide could tilt away from the direction of the bilayer normal, reducing its effective length across the bilayer. However, when the hydrophobic thickness of the bilayer is greater than that of the peptide, there is only one way to achieve matching and that is by compression of the fatty acyl chains. The observations with L_{16} suggest that any stretching or compression of the fatty acyl chains is rather limited and becomes energetically unfavourable when the hydrophobic thickness of the bilayer exceeds the hydrophobic length of the peptide by more than about 10 Å.[11] The peptide L_{22}, however, incorporates normally into all bilayers with chain lengths in the range C14 to C24 and presumably any mismatch is accommodated by tilting the peptide. There will, of course, be an energetic cost associated with stretching or compressing the lipid fatty acyl chains, which will be reflected in values of relative lipid binding constants. Thus strongest binding of lipid to L_{16} is observed for di(C18:1)PC and for L_{22} strongest binding is observed with di(C22:1)PC, as expected for optimal matching.[11]

Here we have explored the importance of charge effects on the interaction between anionic phospholipids and the positively charged peptides. For the peptide L_{16} anionic phospholipids were found to bind more strongly than the zwitterionic phosphatidylcholine (Table 3). The binding constant for PS relative to PC in a medium of low ionic strength was found to be 8.6 (in mole fraction units), corresponding to a difference in unitary binding energies of -5.3 kJ mol^{-1}. At pH 7.2, PS bears a single negative charge.[23] The binding constant for PS changes little with ionic strength (Table 3), suggesting that the interaction with the positively charged peptide does not follow simply from a high positive potential in the vicinity of the positively charged Lys residues on the peptide, increasing the local concentration of anionic phospholipid. The energy of interaction between two ions U is given by:

$$U = z_1 z_2 e^2 / 4\Pi\varepsilon_0 \varepsilon_r r \tag{7}$$

where z_1 and z_2 are the charges on the two ions, ε_r is the relative permittivity (dielectric constant) of the medium and r is the distance between the two ions. Assuming a dielectric constant of 78.5 (water), an energy of interaction of 5.3 kJ mol^{-1} corresponds to a distance of separation between two monovalent ions of 3.3 Å. This therefore suggests that strong interaction requires the anionic headgroup of PS to be in close contact with one of the Lys residues on the peptide. Once this strong interaction with a single PS molecule has been made, other PS molecules will then interact with L_{16} relatively non-specifically, with a binding constant relative to PC close to 1. This picture is consistent with results of a molecular dynamics simulation of the individual α-helices of bacteriorhodopsin in bilayers of di(C14:0)PC carried out by Woolf.[24] This showed that a small proportion of the lipid molecules interacted with the α-helices much more strongly than the others, and that these strong interactions were dominated by electrostatic terms rather than van der Waals terms.[24]

The relative binding constants for PS are less than for PA and are more sensitive to ionic strength (Table 3). For phosphatidylserine, the presence of the positively charged ammonium group as well as the negatively charged carboxyl group in the headgroup region may reduce interaction with the positively charged peptide.

In contrast to L_{16}, the binding constants for anionic phospholipids to L_{22} are very similar to those for zwitterionic phospholipids, with a relative binding constant close to 1 (Table 1). In principle it is possible that a high affinity binding site for anionic phospholipids exists on L_{22} but that binding of a brominated anionic phospholipid to this site does not result in efficient quenching of tryptophan fluorescence. However, the considerable flexibility of the fatty acyl chains in the liquid crystalline phase make this rather unlikely. It therefore appears that tilting of L_{22} in the bilayer, necessary to match the hydrophobic length of L_{22} to the hydrophobic thickness of a bilayer of di(C18:1)PC, locates the Lys residues on the peptide too far from the lipid headgroup region to allow a strong interaction between the anionic phospholipid and the peptide.

In general, binding constants for phospholipids to membrane proteins show relatively little selectivity for anionic phospholipids. For example, binding constants for PA and PS relative to PC are close to 1 for the Ca^{2+}-ATPase[8] and for the (Na$^+$-K$^+$)-ATPase binding constants for PA and PS are about twice those for PC.[25] However, there is evidence for the presence of a small number of special phospholipids binding to some membrane proteins, acting as 'cofactors'. An example is provided by cytochrome c oxidase whose crystal structure shows the presence of a lipid molecule bound between the transmembrane α-helices.[26] Specific high affinity binding sites for anionic phospholipids on a membrane protein could involve close interaction between the anionic headgroup and a positively charged residue on the protein, as suggested by the results presented here.

The studies with cholesterol shown in Fig. 4 show that cholesterol binds to the peptide with a binding constant only a factor of about 2 less strongly than di(C18:1)PC. This is rather surprising given the relatively rigid structure of the steroid ring of cholesterol and the molecularly rough surface of the peptide. In other studies, we have shown that cholesterol binds relatively weakly at the lipid/protein interface of the ATPase;[15,27] comparison with the peptide studies reported here suggests that weak binding of cholesterol to the ATPase involves interactions in the lipid headgroup region, rather than interactions between the sterol ring and the hydrophobic transmembrane α-helices.

Aromatic residues, particularly Trp and Tyr, are often found in membrane proteins at the lipid/water interface and it has been suggested that these residues could act as 'floats' serving to fix the helix within the lipid bilayer.[1] This is consistent with the results shown in Fig. 5A for the peptide Y_2L_{14}. Whereas L_{16} does not incorporate into bilayer of di(C24:)PC,[11] Y_2L_{14} does (Fig. 5A). Incorporation of a short peptide into a thick bilayer requires dimpling of the lipid bilayer around the protein; the results shown in Fig. 5 suggest that this distortion of the lipid bilayer is energetically more favourable when Tyr residues are located close to the lipid/water interface.

Acknowledgements

We thank Dr R. J. Webb for helpful discussions and the BBSRC for financial support.

References

1 D. C. Rees, A. J. Chirino, K. H. Kim and H. Komiya in *Membrane Protein Structure*, ed. S. H. White, OUP, New York, 1994, p. 3.
2 A. G. Lee, K. A. Dalton, R. C. Duggleby, J. M. East and A. P. Starling, *Biosci. Rep.*, 1995, **15**, 289.
3 A. G. Lee, *Biochim. Biophys. Acta*, 1998, **1376**, 381.
4 F. Michelangeli, E. A. Grimes, J. M. East and A. G. Lee, *Biochemistry*, 1991, **30**, 342.
5 A. P. Starling, J. M. East and A. G. Lee, *Biochem. J.*, 1995, **310**, 875.
6 A. P. Starling, J. M. East and A. G. Lee, *Biochemistry*, 1995, **34**, 3084.
7 A. P. Starling, K. A. Dalton, J. M. East, S. Oliver and A. G. Lee, *Biochem. J.*, 1996, **320**, 309.
8 K. A. Dalton, J. M. East, S. Mall, S. Oliver, A. P. Starling and A. G. Lee, *Biochem. J.*, 1998, **329**, 637.
9 C. J. Brandl, N. M. Green, B. Korczak and D. H. MacLennan, *Cell*, 1986, **44**, 597.
10 A. G. Lee in *Biomembranes, Volume 5. The ATPases.*, ed. A. G. Lee, JAI Press, Greenwich, Connecticut, 1996, p. 1.
11 R. J. Webb, J. M. East, R. P. Sharma and A. G. Lee, *Biochemistry*, 1998, **37**, 673.
12 J. C. Huschilt, B. M. Millman and J. H. Davis, *Biochim. Biophys. Acta*, 1989, **979**, 139.
13 F. A. Nezil and M. Bloom, *Biophys. J.*, 1992, **61**, 1176.
14 J. M. East and A. G. Lee, *Biochemistry*, 1982, **21**, 4144.
15 A. C. Simmonds, J. M. East, O. T. Jones, E. K. Rooney, J. McWhirter and A. G. Lee, *Biochim. Biophys. Acta*, 1982, **693**, 398.
16 E. Atherton and R. C. Sheppard, *Solid phase peptide synthesis: a practical approach*, IRL Press, Oxford, 1989.
17 J. M. East, D. Melville and A. G. Lee, *Biochemistry*, 1985, **24**, 2615.
18 E. London and G. W. Feigenson, *Biochemistry*, 1981, **20**, 1932.
19 J. R. Lakowicz, *Principles of Fluorescence Spectroscopy*, Plenum Press, New York, 1983.
20 B. A. Lewis and D. M. Engelman, *J. Mol. Biol.*, 1983, **166**, 211.
21 Y. P. Zhang, R. N. A. H. Lewis, R. S. Hodges and R. N. McElhaney, *Biochemistry*, 1992, **31**, 11579.
22 Y. P. Zhang, R. N. A. H. Lewis, R. S. Hodges and R. N. McElhaney, *Biophys. J.*, 1995, **68**, 847.
23 G. Cevc, *Biochim. Biophys. Acta*, 1990, **1031**, 311.
24 T. B. Woolf, *Biophys. J.*, 1998, **74**, 115.
25 M. Esmann and D. Marsh, *Biochemistry*, 1985, **24**, 3572.
26 S. Iwata, C. Ostermeier, B. Ludwig and H. Michel, *Nature*, 1995, **376**, 660.
27 J. Ding, A. P. Starling, J. M. East and A. G. Lee, *Biochemistry*, 1994, **33**, 4974.

Paper 8/09299K

General Discussion

Dr Templer opened the discussion of Dr Lee's paper: Are you able to tell us whether or not the "bilayer" is moved sideways by the adhesion of the AFM tip to the top layer of DOPE in the fluid phase?

Dr Lee responded: The fact that the lipid bilayers (and monolayers) can be scanned repeatedly with the AFM probe without changing the film morphology indicates that we are not irreversibly deforming them. However, there is a significant increase in the lateral force measured by the cantilever as the probe passes from the solid lipid domains to the DOPE domains, which we associate with an increase in friction.[1] The increased friction on the DOPE phase may result from an increased probe–surface contact area or viscous dissipation.

1   Y. F. Dufrêne, W. R. Barger, J-B. D. Green and G. U. Lee, *Langmuir*, 1997, **13**, 4779.

Prof. Klein commented: MD calculations can perhaps help explain some of your observations. Unpublished calculations by Tarek *et al.*[1] have examined the behaviour of lipid monolayers and bilayers supported on SAMs. These calculations find that the supported lipid in the L_α-phase is rather mobile. If this finding is correct, the AFM tip may indeed displace the liquid-like outer film. In the gel-like phase this situation does obtain. As expected, the gel-like bilayer has crystalline alkane chains.

1   M. Tarek and D. J. Tobias, *Molecular dynamics simulation of the glassy behaviour in globular proteins*, 1999, to be published.

Dr Lee responded: Molecular dynamics simulations have provided powerful insight into the nature of the AFM probe–surface interaction in UHV. I think it would be extremely interesting to simulate a probe indenting into a supported lipid bilayer in water.

Prof. Barclay commented: When cells come into contact there is normally a clear gap between opposing lipid bilayers as proteins interact. I can see the lipid interactions being important in the membrane fusion events but are they important in adhesion events as suggested?

Dr Lee responded: You are absolutely correct in noting that the lipids we have used in this study are more biologically relevant to membrane fusion than cell adhesion. However, this work demonstrates that the AFM can be used to map the surface forces, chemistry and mechanics of multicomponent lipid systems on the nanometer scale, and thus has set the stage for studies of more complex systems. For example, we are preparing a manuscript that describes the interaction of several GM1s with hydroxy, carboxyl and methyl functionalized probes.

Prof. Tiddy asked: In Fig. 8B the chains are drawn almost straight. What is the consequence of the double bond? How much hydrophobic contact is there between the DOPE chains in the contact region and water? Do you know the state of hydration of the headgroups of chains in contact with the surface and how this differs from a normal layer in water?

Dr Lee responded: One would expect the DOPE double-bond to disorganise the lipid chains. It is clear that the unsaturated DOPE phase is thinner than the saturated phases, but we really have very little direct information about the molecular structure of the film. The weakly repulsive surface forces on DOPE suggests the film naturally has a low density of hydrophilic groups at the solid/lipid interface (or a higher density of hydrophobic groups). As the probe comes into contact with the film one would expect that the lipid headgroups would be displaced from their equilibrium conformation to present hydrophobic groups to the probe/lipid interface. The level of perturbation of the film should vary across the probe contact area. However, we really do not have any direct evidence about the molecular structure of the probe/surface interface, and this is why it would be great to have a MD simulation of this system.

Dr L. Fisher asked: Could you clarify whether your experiments were performed with the specimen and probe tip fully immersed in aqueous solution? If in aqueous solution, what were the Debye lengths? Are these significant in the context of your experimental results?

Dr Lee responded: The bilayer and alkanethiol monolayer studies were performed with both the probe and surface immersed in water or 0.15 M NaCl. The Debye length in the triply distilled water is large but electrostatic effects were ruled out on these nominally uncharged surfaces using the fairly high concentration salt solutions. The study of the mixed lipid monolayer was performed in air.

Prof. Roux said: The probe is coated by a SAM of $C_{16}OH$. Is there some degree of entanglement between the long chains of the probe and the chains of the surface you are trying to probe? Would it make a difference to use a coating with shorter chains?

Dr Lee responded: The physical properties of the SAM film are dependent on the length of the alkane chain, and for saturated alkanethiols the transition from solid to liquid behaviour takes place somewhere between C_6 and C_{12}. The $C_{16}OH$ SAM is a highly ordered, solid phase and its elastic modulus is significantly higher than that of the lipid bilayer. We have also measured the surface forces between shorter alkane thiol films (C_6 and C_{11}) and found a significant difference in the surface forces.

Prof. Evans asked: Could you explain what thickness you expect for a monolayer—including estimates for headgroup and chain dimensions—and how this compares with data from X-ray diffraction measurements?

Dr Lee responded: The thickness of a lipid monolayer is of course dependent on the deposition conditions and the specific lipids used. Theoretical estimates and ellipsometric measurements of the thickness of the lipid films used in this study are presented in Table 2 in our paper.

In order to relate molecular dimensions to monolayer thickness, one must know the orientation of each part of the lipid in the film. It has been established that the maximum chain length of saturated lipids in $0.154 + 0.1265 \times$ (number of carbons) nm.[1] For DSPE and DGDG we can expect the carbon chain length to be 2.4 nm, which would make the chain contribution to the film thickness 2.1 nm if the molecule is tilted 30° to the surface normal. The chain thickness of unsaturated DOPE is expected to be significantly less than the saturated lipids due to the disorder the *cis* double bond introduces into the film.

Experimental measurements of chain and headgroup contribution to film thickness come from X-ray or neutron diffraction studies. Unfortunately, most of this work has been done on suspensions of multiwalled vesicles so it does not exactly correlate with our work on surfaces. It should be noted, however, that there are a growing number of studies of lipid films on surfaces using the extremely powerful techniques of small angle X-ray scattering (SAXS) and grazing-incidence X-ray diffraction (GIXD).

Both phosphatidylethanolamine[2] and DGDG[3] vesicle systems have been studied by X-ray diffraction. The repeat distance of the DSPE bilayers is approximately 6.4 nm, which is in reasonable agreement with our ellipsometric DSPE monolayer thicknesses of 3.3 ± 0.1 nm. It is not possible to directly compare the AFM and X-ray results for DOPE because the liquid-like isotherm of this material makes thickness highly dependent on surface pressure.

McDaniel suggests that the DGDG headgroup is oriented parallel to the plane of a lipid bilayer forming a tightly packed 0.8 nm layer. If we add the X-ray DGDG headgroup thickness to the calculated chain thickness the total thickness of the monolayer becomes 2.9 nm, which again agrees with our ellipsometric measurements. In conclusion, I think that the thicknesses we have measured with ellipsometry are in fairly good agreement with X-ray diffraction supporting our interpretation of the AFM images and force curves.

1   C. Tanford, *The Hydrophobic Effect*, Wiley, New York, 1980; J. N. Israelachivili, D. J. Mitchell and B. W. Ninham, *J. Chem. Soc., Faraday Trans.*, 1976, **72**, 1525.
2   J. M. Seddon, G. Cevc, R. D. Kaye and D. Marsh, *Biochemistry*, 1984, **23**, 2634.
3   R. V. McDaniel, *Biochem. Biophys. Acta*, 1988, **940**, 158.

Dr Chen asked: AFM is a powerful tool to study the surface properties of the membrane, and lipid peroxidation is a big problem of the biomembrane. Have you done some work related to lipid peroxidation of the membrane with AFM? Or do you have an opinion on it?

Dr Lee responded: No. I haven't done any of this work. I do not have any opinion on this.

Prof. Holzwarth asked: Do you have any hard evidence about the smoothness of the tip used during your AFM experiments, for example an inspection by electron microscopy?
Did you try to measure at different temperatures, especially in the temperature range of a phase transition, to see different states of order coexisting as claimed by my group.[1]

1   R. Groll, A. Böttcher, J. Jäger and J. F. Holzwarth, *Biophys. Chem.*, 1996, **58**, 53.

Dr Lee responded: The AFM probes have been examined with FI-SEM and $SrTiO_3$ calibration gratings.[1] SEM confirmed that the microfabricated probes were pyramidal in shape and had a radius of curvature < 100 nm. The $SrTiO_3$ analysis was not performed on the probes used in this paper, but in subsequent studies these calibration gratings have been used to characterize the radius of curvature of each probe used. The conclusion of our recent work is that the force curves presented in our paper are consistent with the measurements made with spherical, ~50 nm radius probes.
We have not made measurements on the LB films at different temperatures. In principle, AFM can be performed at the solid/water interface at temperatures between 0 and 100 °C, but in practice it is not easy to obtain high-resolution images due to the influence of thermal drift. However, the experiment you are suggesting is tractable and would identify the different states of order.

1   S. S. Sheiko, M. Möller, E. M. C. M. Reuvekamp and H. W. Zandbergen, *Phys. Rev. B*, 1993, **48**, 5675.

Dr Pawlak asked: You investigated membrane surfaces using defined hydrophilic tip/ hydrophilic surface and hydrophobic tip/hydrophilic surface contacts on flat mica. Are there differences in the surface roughness of these two measurements? In the case of hydrophobic/hydrophilic contacts, would it be possible that hydration forces or a different water structure in close vicinity of the hydrophobic surface generate an apparent increase of surface roughness in topographical images, which is not related to the real surface roughness of the support? We have actually measured such an increase (in the range of 1 nm rms) during imaging of very hydrophobic, self-assembled layers of C_{16}-alkane on metal-oxide using Si_3N_4 tips. The roughness increase was not observed when comparing clean hydrophilic substrate surfaces and surfaces with a terminating monolayer of lipid.

Dr Lee responded: In this manuscript we describe force measurements in aqueous solutions between: (i) a gold coated, CH_{16}-OH alkanethiol functionalized probe and alkanethiol functionalized surfaces (CH_{16}-OH and CH_{15}-CH_3); (ii) a gold coated, CH_{16}-OH alkanethiol functionalized probe and lipid films (DSPE/DOPE and DGDG/DOPE). We also analyze the contact imaging mechanism of the CH_{16}-OH tip on the lipid films in water and air.
I think the issue that you are trying to get at is: Can AFM measure surface hydration at the nanometer scale and if so can this information be used to understand the chemical homogeneity of a surface? Our work on these model systems clearly shows that local hydration, mechanical and topographical properties each contribute to the AFM contrast, and changes in any one of these factors will produce a change in the observed rms roughness in an AFM image. However, I do not advocate the using of surface roughness as a means of characterizing surface homogeneity because the AFM imaging mechanism is dependent on several other factors, *e.g.*, probe radius and feedback parameters. Rather, friction, phase and force modes can be used in parallel with imaging modes to identify nanometer scale surface inhomogeneities. The power in using surface force

mapping is that the force curves can be used to semi-quantitatively interpret the contribution of each imaging factor.

Dr Sansom opened the discussion of Prof. Smith's paper: In the graph (Fig. 4) of displacement *vs.* residue, can one see: (a) correlations between residues with high displacement and structure (loop *vs.* helix) of the protein; and (b) correlations between those residues which give high displacement at ~ 150 K and those which give high displacement at ~ 250 K?

Prof. Smith responded: Fig. 4 indicates that transitions take place at ~ 150 and ~ 240 K. This result is found to be reproducible with the present simulation model and different initial conditions. However, I feel that the model may be too crude to permit meaningful investigation of the correlations you referring to. We need to repeat the calculations with bacteriorhodopsin in its trimeric form and in an explicit membrane.

Prof. Holzwarth asked: Did your simulations include the special lipids which are strongly bound to bacteriorhodopsin (BR) and are located between the helical structures? Those lipids are still mysterious in respect of their function.

Did your simulations account for lipids which are arranged outside the helical structure of BR, especially how far the influence of BR might reach into its surroundings? We tried to measure the influence of BR on its lipid environment and found that this protein influences as many as five to six lipid layers around itself.[1]

1 A. Böttcher, N. Dencher, R. Groll, F. Meyer and J. F. Holzwarth, *Reactions in Compartmentalized Liquids*, ed. W. Knoche and R. Schomäcker, Springer-Verlag, Berlin, 1989, pp. 105–115.

Prof. Smith responded: Our simulations did not explicitly include any lipids. It is something we would hope to do in the future. The influence of a protein on its lipid environment is something that molecular simulation is indeed beginning to explore, as evidenced by other papers in the Discussion. With Dan Mihailescu we are also beginning to explore this for a smaller lipid/ membrane system, gramicidin S, and see a significant change in the order parameters of peptide-associated membrane lipids.

Prof. Finney said: I would be interested in your comments on Fig. 3. (1) The transition identified at 150 K looks a little different from that measured experimentally by Réat *et al.*[1] Rather than showing an increase in $d\langle u^2\rangle/dT$ between two regions, the simulated transition depends on a previous almost horizontal plateau region. Do you have any comments on this? (2) Have you looked at the processes that give rise to this plateau region? What effects prevent the normally-expected increase of $\langle u^2 \rangle$ with T? (3) $d\langle u^2\rangle/dT$ *before* the plateau is *greater* than $d\langle u^2\rangle/dT$ *after* 150 K. What does this imply with respect to changes in the flexibility behaviour?

1 V. Réat, H. Patzelt, M. Ferrand, C. Pfister, D. Oesterhelt and G. Zaccai, *Proc. Natl. Acad. Sci. U.S.A.*, 1998, **95**, 4970.

Prof. Smith responded: The present work demonstrates anharmonic transitions in the protein simulation at ~ 150 and ~ 240 K, similar to the temperatures found experimentally by Réat *et al.*[1] Beyond that fact, it is unclear to what extent experiment and simulation resemble each other. Direct comparison of the $\langle u^2 \rangle$ data indeed show some significant differences. Further information may be obtained from a more detailed simulation model and direct experiment–theory comparison at the level of the dynamic structure factor.

To answer your second point, a clue to this might be seen in Fig. 4. At the transition, some residues increase $\langle u^2 \rangle$ and others decrease it. For the latter the transition may have involved moving into a steeper-walled potential. The run effect can conceivably average to a plateau. To answer your third point, I am not sure that the gradient difference is significant. If it were it would imply a higher 'average force constant' above 150 K such that on average the atoms move in a steeper-walled region of the protein potential surface.

1 V. Réat, H. Patzelt, M. Ferrand, C. Pfister, D. Oesterhelt and G. Zaccai, *Proc. Natl. Acad. Sci. U.S.A.*, 1998, **95**, 4970.

Prof. Klein asked: Neutron scattering data provide a powerful probe of protein dynamics. However, it is important to calculate the full scattering function from the MD simulation and to treat the MD data in the same fashion as experimentalists treat their data. In the case of the soluble protein BPTI a large "discrepancy" existed between the calculated and measured values of mean square displacements. This "problem" is at least partly due to intrinsic limitation on the instrumental resolution. With careful treatment of instrumental effects the simulation and neutron data agree rather well.[1] A similar situation may obtain for bacteriorhodopsin. So, I urge you to make more direct comparison with neutron data.

1 M. Tarek and D. J. Tobias, *Molecular dynamics simulation of the glassy behaviour in globular proteins*, 1999, to be published.

Prof. Smith responded: We have performed direct comparison of the full scattering function, including instrumental effects, many times in the past (see ref. 1 and 2 here, and references cited therein, for examples ranging from soluble proteins to molecular crystals). The present preliminary simulation data on bacteriorhodopsin are sufficiently encouraging to warrant such an investigation but I would prefer to do this with a more complete model. Concerning BPTI, our experiments with John Finney and Steve Cusack on this protein concentrated on the inelastic scattering, and were not used to derive mean-square displacements (see ref. 3 for a review). Therefore, I find the existence of new experimental mean-square displacement data on this protein of much interest.

1 N.-D. Morelon, G. R. Kneller, M. Ferrand, A. Grand, J. C. Smith and M. Bee, *J. Chem. Phys*, 1998, **109**, 2883.
2 A. Lamy, J. C. Smith, J. Yunoki, S. F. Parker and M. Kataoka, *J. Am. Chem. Soc.*, 1997, **119**, 9268.
3 J. C. Smith, *Quart. Rev. Biophys.*, 1991, **24**, 227.

Dr Smart asked: What effect, if any, does the value taken for the harmonic restraint constant (used to stabilize the structure) have on the transition temperatures? In addition, what about the value taken for the relative permittivity on the transition temperatures?

Prof. Smith responded: The force constant for the harmonic restraints would be expected to affect the dynamics if set high enough. We did not have sufficient computer time to examine this, and chose the smallest possible value that would maintain the average structure to within ~ 2 Å of experiment. In the future one would hope to do away with the restraints and include the environment explicitly.

Variation of the relative permittivity would be of interest. We have examined the effect of electrostatic model variation on the density of states of a soluble protein[1] but it would be of fundamental interest to extend this to membrane proteins and the temperature dependence.

1 J. C. Smith, S. Cusack, B. Tidor and M. Karplus, *J. Chem. Phys.*, 1990, **93**, 2974.

Mr Schuler commented: MD of protein structures in many cases has been shown to be rather sensitive to equilibration times. Would you agree that there might be a different (much larger) mean square displacement in this particular system if the equilibration times averaged were extended to, say 1 ns instead of 110 ps?

Prof. Smith responded: The question is somewhat ambiguous. What, in the present paper, is referred to as 'equilibration' is a 5 ps period in which the temperature was rescaled if necessary prior to 110 ps 'production' which was performed at constant energy. If the production periods had been extended then larger mean-squared displacements would be expected. But the finite energy resolution of the neutron instrument also limits the timescale of the experimentally-accessible mean-square displacement.[1] One also finds that protein fluctuations sampled over a given production period can depend on the equilibration length. An example of this is given in Mihailescu and Smith.[2]

1 R. Daniel, J. C. Smith, M. Ferrand, S. Héry, R. Dunn and J. L. Finney, *Biophys. J.*, 1998, **75**, 2504.
2 D. Mihailescu and J. C. Smith, *J. Phys. Chem. B*, 1999, in the press.

Prof. Tiddy said: Your simulation places the protein in an environment which does not include details of the lipid membrane and water. Is it a coincidence that the 240 K transition is close to the water melting temperature? What is the transition behaviour of the membrane lipids?

Prof. Smith responded: The 240 K transition has experimentally been shown to be strongly affected by the hydration state of the membrane, and is indeed close to the water melting temperature. However, we also see a transition at that temperature in a simulation without hydration, indicating that the protein also has an intrinsic dynamical anharmonicity that reveals itself at ~240 K.

I am not aware of data on the transition behaviour of the membrane lipids.

Dr Templer asked: Would you tell us which parts of BR are rigid and which parts flexible?

Prof. Smith responded: The loop regions are more flexible than the transmembrane helices, in our simulations. Beyond that, there is some diffraction evidence for fluctuation asymmetry across the membrane and, by combining site-directed labelling (H/O) with incoherent neutron scattering Réat and co-workers[1] were able to demonstrate relative rigidity of the retinal binding pocket. Simulation work by several groups is providing additional information in this regard.

1 V. Réat, H. Patzelt, M. Ferrand, C. Pfister, D. Oesterheltz and G. Zaccai, *Proc. Natl. Acad. Sci. U.S.A.*, 1998, **90**, 9668.

Prof. Evans asked: Are there calorimetric data for the transitions indicated by neutron scattering and implied by the MD simulations?

Prof. Smith responded: I am unaware of such calorimetric data on this system. Our experience with soluble proteins is that the thermodynamic and dynamic transitions do not necessarily coincide, and that there can be a timescale-dependence of the mean-square displacement profile. Good calorimetric data would certainly be useful, and would allow comparisons to be made with glass formation, in which the dynamical transition occurs at temperatures above the glass transition temperature, as predicted by the mode-coupling theory of non-linear coupling between density fluctuation modes.

Dr L. Fisher asked: In your 3D figure which shows mean square displacement as a function of residue number and temperature, a number of residues show comparatively large displacement at the highest temperatures examined. The residue numbers are widely separated, but it is possible that the actual residues in the folded three-dimensional structure might be close together in space. Can you tell from a comparison with the known crystalline structure of bacteriorhodopsin whether this is in fact the case?

Prof. Smith responded: Yes, one can tell from the simulation (or the crystal structure) whether residues non-local in the sequence are close together in space. It is also useful to go further than this and to examine correlations between fluctuations of different atoms. An example of this, in which the interatomic distance fluctuation matrix of crystalline lysozyme was examined and corresponding X-ray diffuse scattering intensities determined, using molecular dynamics, is given in Héry *et al.*[1] For bacteriorhodopsin, I think we need a more accurate simulation model before examining these properties.

1 S. Héry, D. Genest and J. C. Smith, *J. Mol. Biol.*, 1998, **279**, 303.

Dr P. N. Edwards opened the discussion of Prof. Needham's paper: The ionisation behaviour of polymeric acids or bases cannot be described by the Henderson–Hasselbach equation. Only titration can yield the proportions ionised at particular pH values since the multiple microscopic pK_a values are structure and chain-length dependent and typically in your system would occur over a wide range—probably from ~3 to ~10.

Prof. Needham responded: Dr Edwards is correct in his point that the only way to obtain the actual proportions of ionised polymer at a particular pH is by titration. In response to this question, the titration experiment for the polymer used in the liposome release experiments has now been performed and the methods and results are briefly described below.

PEAA was dissolved in an equimolar amount (with respect to monomer) in 0.01 M NaOH and 0.137 M NaCl was added to match the salt concentration used in our liposome release experi-

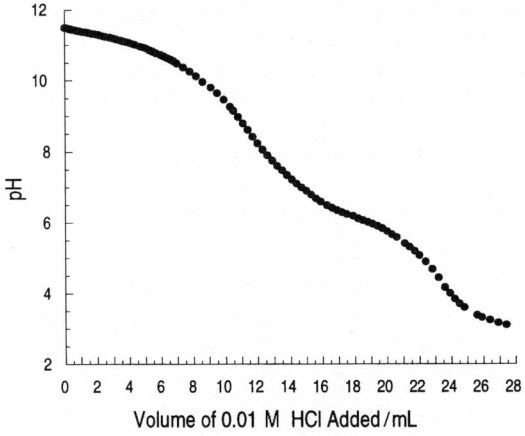

Fig. 1 Titration curve for the PEAA polymer.

ments. The polymer was then titrated with 0.01 M HCl under a nitrogen atmosphere. The pH response with added HCl is shown in Fig. 1 above.

Using the electroneutrality condition defined below[1,2] we calculated the degree of dissociation of the polymer, α.

$$\alpha = [C_{Na^+} + C_{H^+} - C_{Cl^-} - C_{OH^-}]/C_{mon}$$

Here, C_{Na^+} and C_{Cl^-} are the concentrations of Na$^+$ and Cl$^-$ ions due to initial NaOH, NaCl and added HCl. C_{H^+} and C_{OH^-} are the concentrations of OH$^-$ and H$^+$ ions and were calculated from the measured pH values. C_{mon} is the concentration of monomer units calculated as the titration proceeded. All concentrations were in mol L^{-1}.

Using this equation to calculate α, the plot shown in Fig. 2 here was constructed showing pH *vs.* percent protonation ([1 − α] × 100). Also shown in this plot is the Henderson–Hasselbach equation assuming a single pK_a of 7.9, *i.e.*, the pH at which α = 0.5 as determined from the degree of dissociation equation. This plot serves to support Dr Edwards assertion; the protonation behaviour of the polymer does cover a much wider pH range than the Henderson–Hasselbach equation would predict.

With regard to the correlation drawn in the paper between membrane cohesion and the degree of protonation of the polymer, Fig. 3 is a plot showing the corresponding change in elastic expansion modulus for each SOPC : cholesterol membrane composition (relative to pure SOPC) *vs.* the

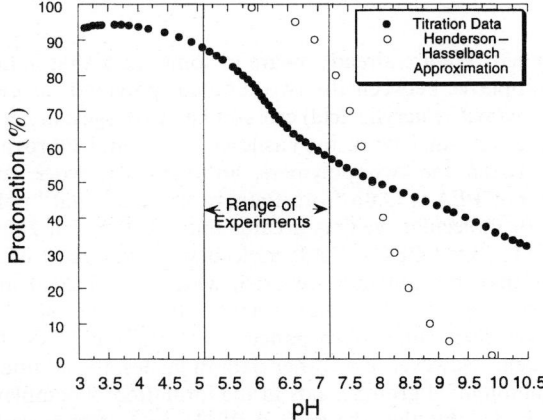

Fig. 2 Comparison between titration data and Henderson–Hasselbach approximation for the percent protonation of the polymer *versus* pH.

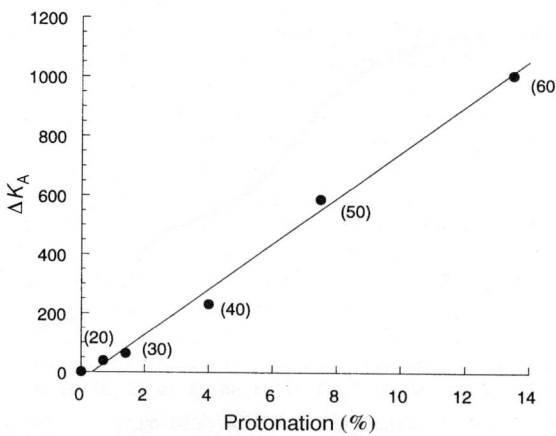

Fig. 3 Increase in the elastic area expansion modulus (over and above that for the pure SOPC bilayer) *versus* the increase in percent protonation of the polymer (over and above that which occurs at pH 6.7) at the pH where dye is released, for bilayers containing increasing amounts of cholesterol (mol% cholesterol is given in parentheses next to each data point).

change in percent protonated at the critical release pH for each composition (relative to the percent protonation required to release CF from SOPC vesicles).

Thus, although the percent protonation actually does cover a different range than we originally presented in Fig. 4 in the paper, the trend between elastic modulus and protonation of the polymer still holds. An increase in the elastic expansion modulus, K_A, is directly proportional to an increase in the percent protonation required to cause release of fluorescent dye from the vesicles.

1 U. K. O. Schröder, D. A. Tirrell and K. H. Langley, personal communication.
2 M. Mandel, *Eur. Polym. J.*, 1970, **6**, 807.

Prof. Almgren commented: The interaction of PEEA with vesicles covered with PEG-lipids is perhaps not surprising in view of the pronounced attractive interaction between polyacrylic acid, at low pH, and PEO as well as PEG-surfactants.[1] It would be of considerable interest to assess the ability of PEG lipids to hinder the approach of a polymer/polyelectrolyte that interacts only with the bilayer interface; the described experiment is not informative in this respect.

1 M. Vasilescu, D. F. Anghel, M. Almgren, P. Hansson and S. Saito, *Langmuir*, 1997, **13**, 6951 and references cited therein.

Prof. Needham responded: We are already aware of some work that indicates the formation of inter-macromolecular complexes between the two different polymers, as evidenced by the PEG-induced contraction of poly(methylacrylic acid) cross-linked hydrogels in neutral medium.

Thank you for pointing out other work by Vasilescu et al.[1] and Petrova et al.[2] that speaks of the same association between the two polymers, although this work studied the interaction between PEG and PAA not PEAA. Data from Petrova indicates that "polycomplex" formation occurs between PEG with molecular weights greater than 6000 and a 250 000 molecular weight poly(acrylic acid) (PAA), *i.e.*, at a PEG : PAA molecular weight ratio of 1 : 42. This molecular weight cut-off is greater than the polymer we used, which was a 2000 molecular weight PEG grafted to the lipid-bilayer, and the molecular weight ratio between the 2000 PEG and 30 000 PEAA is 1 : 15. There are then some discrepancies between the PEG : PAA system and our grafted-PEG : PEAA system. However, if as other data indicates, the addition of "anchor" groups to the PEG (5-nitro-8-quinolinoxyl groups), aids in the formation of complexes with 2000 molecular weight PEG, then it is possible that the grafted PEG could interact with the ethyl groups on PEAA in a similar fashion to that with PAA, thus aiding in the process of polymer–polymer association and bilayer disruption.

In the PEG:PAA system, the "polycomplexes" form between the two different polymer molecules due to hydrogen bonds between the oxygen atoms of PEG and the hydrogen atoms of the carboxyl groups of PAA. The literature indicates that this interaction is prevalent at low pH (data is shown for pHs below 5.0) where the carboxyl groups are highly protonated to allow for the hydrogen bond interaction.[2] The pH of our solutions is generally much greater than 5.0, and so the same polymer–polymer interaction is not as favoured for our PEG:PEAA system as in the lower pH PEG:PAA case. However, there still could be an interaction between the protonated groups on the PEAA at our working pHs of 7.5–6.0 and the PEG molecules.

I appreciate that it is important to evaluate the effects in isolation and finding a polymer that does not interact with PEG but does interact with the bilayer is an interesting idea. Whether the two events can be uncoupled remains to be seen, since a polymer that doesn't interact with PEG may well have structural and chemical elements that also prevent it from interacting with the bilayer interface. The point of the experiment that we reported in the paper though was to determine if PEG acted as a repulsive barrier to the PEAA polymer in much the same way that it does to more globular macromolecules. It didn't, and the PEAA did cause contents release at the same or even slightly higher pH than for the unmodified bilayer. As we discussed, PEG does not represent a repulsive barrier to PEAA and its passage through the PEG layer may also involve polymer–polymer associative interactions (in line with the above references in previous literature).

1 M. Vasilescu, D. F. Anghel, M. Almgren, P. Hansson and S. Saito, *Langmuir*, 1997, **13**, 6951 and references cited therein.
2 T. Petrova, I. Rashkov, V. Baranovsky and G. Borisov, *Eur. Polym. J.*, 1991, **27**, 189.

Prof. Robinson asked: Can you comment on the size/shape of your vesicles before and after the addition of PEAA?

Have you any information on how the PEAA is distributed in the vesicles? What is the loading? *i.e.*, the number of polymers/vesicle? (Please could you give the molecular weight of polymer in your response.) Your kinetic processes at low pH take place rather rapidly. What is the effect (on structure and dynamics) of decreasing the PEAA concentration?

Have you considered doing these experiments in a different way? For example, by adding the polymer to the vesicle at high pH (~ 7) and then suddenly decreasing the pH to say 6. Could you speculate as whether the results will be similar?

Prof. Needham responded: We have not done any size or shape determinations of the vesicles. Size analysis by light scattering is difficult because of the small amount of liposomes that are present in our experimental samples. However, Borden *et al.*[1] showed previously that at a lipid:polymer weight ratio of 1:1, the size of the measured particles in the sample (which were initially liposomes) changed in diameter from 90 to 16 nm with a drop in pH in the presence of the polymer after a transition at pH 6.5. We have far more polymer in our experiments than this and so we would suspect that the vesicles do undergo a transition to micelles when the polymer interacts with them. Since we have not yet measured the actual amount of polymer bound per vesicle, the number of polymers/vesicle at this point can only be approximated by calculating the surface area of a liposome, the cross-sectional area of the polymer and finding the number of polymers that could potentially cover an area of this size.

Polymer molecular weight	$= 30\,000$ g mol^{-1}
Radius of gyration, R_g, of polymer	$= 3.37 \times 10^{-9}$
SA of one, 100 nm liposomes	$= 3.14 \times 10^{-14}$ m^2
Cross-sectional area of polymer	$= \pi(R_g/2)^2 = 8.92 \times 10^{-10}$ m^2
Number of polymer liposome	$= 35\,201$

Decreasing PEAA concentration from 1 to 0.002 mg mL^{-1} has no effect on the critical release pH for pure SOPC membranes. A close examination of the release kinetics has not been done, but in all cases, release is noted within minutes, and so it does not seem that for these concentrations the process has reached a diffusive limit.

Yes, we have considered doing the experiment as you suggest, but controlling the pH to as fine a degree as is required here is not practical. I would speculate that the results would be similar though. Preadsorbing the polymer to the vesicles and then dropping the pH would eliminate any diffusion in the kinetics and would put the polymer in direct contact with the lipids. This is something we will try in bulk (even though the level of control over pH is not as good as the way we have so far conducted the experiment).

The best way of investigating this interaction is with the micropipet manipulation technique which allows us to take a single vesicle, with preadsorbed (or electrostatically bound) polymer on the membrane and then change the pH by either transferring the vesicle to a second chamber or using a second pipette to blow the new pH solution over the vesicle. These techniques have been developed and used by us and Prof. Evans in a series of experiments concerning molecular exchange with vesicle bilayers and are described in detail elsewhere.[2,3]

Preliminary work in collaboration with Evans, Thomas and Tirrell, that is as yet unpublished, indicated that this method can provide important and unique information including the area per molecule a_S of the polymer in the lipid bilayer. Assuming a superposition of compliances for the overall modulus of the bilayer with polymer (K_M), the pure lipid bilayer (K_L) and an "effective modulus" due to polymer partitioning (K_S), a plot of K_L/K_M against the fractional change in area of the vesicle due to polymer partitioning yields a slope of 7.5. K_S is approximately given by, $K_S \approx 2k_B T/a_S$, which gives a_S as ≈ 30 Å^2, a very reasonable number for the cross-sectional area of the polymer in the membrane.

1 K. A. Borden, K. M. Eum, K. H. Langley, J. S. Tan, D. A. Tirrell and C. L. Voycheck, *Macromolecules*, 1998, **21**, 2649.
2 E. Evans, W. Rawicz and A. F. Hoffman, *Bile Acids in Gastroenterology Basic and Clinical Advances*, ed. A. F. Hoffman, G. Paumgartner and A. Stiehl, Kluwer Academic Publishers, Dordrecht, Boston, London, 1994, pp. 59–68.
3 D. Needham, N. Stoiceva and D. V. Zhelev, *Biophys. J.*, 1997, **73**, 2615.

Dr Jones asked: In the 6-CF release experiments from vesicles (Fig. 6) below the pH where release of 6-CF occurs, you show the fluorescence for both pure SOPC and PEGylated SOPC vesicles presumably normalized to the same fluorescence. Did you in fact find that the extent of 6-CF encapsulated in the vesicles was reduced in the case of the PEGylated vesicles due to the space occupied by the PEG?

Prof. Needham responded: The fluorescence is normalised to the fluorescence equivalent to 100% release of the entrapped CF dye upon dequenching, whose fluorescence is largely quenched when inside the vesicles. The measured fluorescence therefore represents the relative percents of the total amount encapsulated not necessarily the absolute concentrations of dye either entrapped or released. The fluorescence prior to release was not of direct concern here. It also pertinent to note that a constant loading of CF was difficult to achieve from experiment to experiment, which is why we decided to control for it by using the relative percent release measurement. So, the small differences in fluorescence intensity for the liposome samples before pH-induced/polymer release of dye seen in Fig. 6, may or may not be significant, and, (without appropriate calibration) can not be used to estimate whether in fact PEG on the inside of the vesicles does reduce the available space for CF which is at a fairly high concentration. The data actually show that the fluorescence with PEG on the inside of the vesicle is in fact higher than with out it and so this would not *a priori* fit with an inference that the presence of PEG reduces the available space for this particular molecule. But to reiterate, the amount of loading per vesicle sample did vary from preparation to preparation whether PEG-lipids were present or not and so we cannot infer anything from the present data regarding available space inside the vesicles.

Prof. Holzwarth asked: Cholesterol tends to cluster in lipid bilayers and I wonder how you could have as much as 60% cholesterol in a bilayer without destroying its structure? You might inspect the literature about the influence of cholesterol on bilayers, especially the appearance of clusters of cholesterol which was observed at concentrations as low as 7%.[1]

1 A. Genz, J. F. Holzwarth and T. Y. Tsong, *Biophys. J.*, 1986, **50**, 1043.

Prof. Needham responded: Cholesterol concentration is represented as the "nominal" mol% cholesterol in the lipid sample from which the vesicles were made. In an earlier paper[1] we discussed this and showed that the saturation of cholesterol composition as measured by the elastic area expansion modulus occurred at ~58% (nominal) mol% cholesterol in SOPC vesicles. So, 60 mol% cholesterol does represent the maximum possible, bilayers simply do not form at concentrations greater than this and excess cholesterol is probably present as crystallites.

Thank you for referring us to the Genz et al. paper. We will certainly consider the issues of cholesterol clusters on polymer uptake further.

1 D. Needham and R. S. Nunn, *Biophys. J.*, 1990, **58**, 997.

Prof. Tiddy commented: There is segregation in between PC and cholesterol, with the formation of cholesterol-rich patches. Could this occur with your PEG-2000 lipid, particularly with the low poly(ethylene oxide) fraction?

Prof. Needham responded: The purchased PEG-lipid was chosen such that its hydrocarbon chains (diC18 saturated, DSPE-PEG) matched as closely as possible to the bilayer-lipid SOPC (C18:0/C18:1), and so we expect that at these low PEG-lipid mole fractions of 5 mol% the two lipids would be ideally mixed and no PEG-lipid rich domains would be formed. In all our experiments with PEG-lipids, using X-ray diffraction (with Tom McIntosh) and exchange of avidin and lysolipid micelles, which are blocked by the presence of PEG-lipids, the behavior is well modeled by a continuous, well distributed layer of PEG, irrespective of density, up to the point of saturation.

Prof. Evans asked: Does uptake of the protonated polymer into cholesterol–lipid bilayers correlate with the excess mole fraction of lipid (*i.e.* lipid–cholesterol), which would indicate partitioning predominately in the lipid component?

Prof. Needham responded: This is a very interesting idea, and goes back to what Prof. Tiddy mentioned regarding cholesterol-rich domains. To what extent cholesterol-rich patches actually occur in the SOPC–cholesterol system is, I believe, at the moment unknown. The elastic modulus varies in a fairly continuous way with increasing cholesterol[1] with the possible exception of a slight jump at around 40 mol% cholesterol, and so mechanically the bilayer behaves as a continuum. We have not actually yet measured the uptake of polymer in a quantitative fashion and so cannot answer this question directly. However, if the critical percent protonation required for release of contents can be taken to indicate the amount of polymer bound (which is a tenuous assumption at best), then the question would be, is the reciprocal of this critical protonation linearly related to the mol% cholesterol in the membrane? It turns out that it is (analysis of data carried out subsequently to the meeting). So, yes, it may be that the polymer is finding the pure lipid component and could be relatively impermeant for cholesterol-rich domains.

1 D. Needham and R. S. Nunn, *Biophys. J.*, 1990, **58**, 997.

Prof. Neumann asked: Is the release of contents from the liposomes dependent on the chain lengths of PEAA at otherwise equal conditions? Also, a comment: PEAA is an anionic polyelectrolyte. The protonation of PEAA is multiple binding, each EAA residue having its own pK value. However, at low degrees of protonation (less than or equal to 0.2, the H$^+$ binding is to separated sites, independent and of high electrostatic affinity associated with practically one high-affinity pK value. Similarly, at high degrees of protonation (greater than or equal to 0.8), the remaining unprotonated sites may be considered as localised and independent islands with one low-affinity pK value.

Dr Needham responded: All of the experiments that we and others have performed looking at release from vesicles use approximately 30 000 molecular weight PEAA, so we do not have any direct data to correlate chain length with release. However, the critical pH of the collapse transition of the polymer itself in aqueous solution has been studied as a function of molecular weight. This transition has been noted to decrease with decreasing molecular weight,[1] and so we might expect that for smaller polymers, the critical pH for release of contents might be reduced to lower pH. In answer to your comment, please see my answer to Dr Edwards (earlier), where I present new data.

1 U. K. O. Schroeder and D. A. Tirrell, *Macromolecules*, 1989, **22**, 765.

Dr P. N. Edwards said: The amazing sensitivity of your system to changes of pH seems to require a very high order of co-operativity in the disruption of membranes. Perhaps multiple polymer molecules form a channel through the membrane?

Prof. Needham responded: PEAA certainly can form channels in lipid membranes. Data from Chung *et al.*[1] shows that, at relatively low polymer : lipid weight ratios, near 1 : 100, PEAA causes the formation of cation selective pores that are often stable for many seconds. However, our experiments were carried out at much higher ratios, and it is unlikely that the mechanism of release is related to pore formation, since, as I responded to Dr Robinson earlier, at the polymer : lipid ratios we were using, the polymer induces a vesicle to micelle transition and so probably dissolves all the bilayers, making the role of channels a moot point.

1 J. C. Chung, D. J. Gross, J. L. Thomas, D. A. Tirrell and L. R. Opsahl-Ong, *Macromolecules*, **29**, 4636.

Dr Bezrukov asked: Did you try experiments on planar lipid bilayers to understand the mechanism(s) of dye release in more detail?

Prof. Needham responded: No, we have not looked at black lipid films, just extruded vesicles and we plan to examine giant lipid vesicles next. Such techniques are not presently in the laboratory tool box, although I have worked on black lipid films in the past with Dennis Haydon, and I can appreciate that this would be an interesting study to perform.

Miss Bucak communicated: You suggest a possible binding of the protein to PEG as it can make its way to the membrane in the presence of PEG whereas the same size micelles cannot. Do you think it is possible to look at the kinetics of this in the presence and absence of PEG to find out if there is an affinity between PEG and the protein?

Prof. Needham communicated in response: In the paper we do not talk about protein binding to PEG or the membrane, only PEAA interactions. Evidence to suggest that certain proteins, like avidin, do not traverse the PEG layer (that are the same size as micelles that also do not traverse the layer) have been carried out and published.[1,2] What we found, and have since modeled for proteins or particles of different size[3] is that the polymer essentially behaves as a hard sphere, and that the on rate for the polymer compared to the bare membrane can be orders of magnitude slower. Debbie Leckband seems to have data that suggests that an immobilized layer of protein will be bind to an immobilized layer of PEG in the surface force apparatus, if the force applied is large.

1 D. Noppl-Simson and D. Needham, *Biophys. J.*, 1996, **70**, 1391.
2 D. Needham, N. Stoiceva and D. V. Zhelev, *Biophys. J.*, 1997, **73**, 2615.
3 D. Needham, T. J. McIntosh and D. V. Zhelev, *Liposomes. Rational Design*, ed. A. S. Janoff, Marcel Dekker, New York, 1998.

Dr Fontana communicated: Looking at your data the role of cholesterol is to increase either the cohesiveness or the hydrophobicity of the membrane. Can you suggest a control experiment to separate these two contributions?

Prof. Needham communicated in response: This is an interesting question, and is one that we have recently been investigating with respect to water permeability. Using the micropipet method on single giant lipid vesicles has shown that water permeability through bilayers progressively decreases as the compressibility of the bilayer decreases, showing the role of cohesion as fluctuations in surface density which determine both compressibility and permeability to water. However, for the unsaturated lipids, whilst the area expansion modulus of bilayers made from lipids that contained multiple unsaturation did not change much as the unsaturation was changed from 1 to 6 double bonds per molecule, the water permeability coefficient increased from 20 $\mu m\ s^{-1}$ for 18 : 0/1 bilayers to 100 $\mu m\ s^{-1}$ for di 18 : 2 bilayers. This observation indicates that whilst membrane interface cohesion is important in restricting water transport for the more condensed membranes, such as ones containing cholesterol, it appears to be the absolute solubility of water in the membrane that dictates the rates of transport for the softer membranes, *i.e.*, this experiment does isolate hydrophobicity at constant interface cohesion, since it is known that the solubility of water in alkenes is slightly greater than in alkanes. These compositions have not yet been studied with respect to polymer insertion but it will be of interest to check the polymer (and other molecule) uptake into bilayers made from lipids with increasing unsaturation (decreasing hydrophobicity) to complement studies on bilayers of increasing cohesion.

Dr P. N. Edwards opened the discussion of Prof. Neumann's paper: To what extent does the electric field existing across plasma membranes of living cells change the expectations relative to the liposomes you describe?

Prof. Neumann responded: The electric field $E_m = -\Delta\phi_{ind}/d$ which is induced by interfacial polarization across the membrane of thickness d, is dependent on the positional angle θ relative to the direction of the external field E and is very much larger than E. The induced potential difference drop in the direction of E is given by $\Delta\phi_{ind} = -(3/2) Ea|\cos\theta|$, where a is the radius of the spherical vesicle or cell. At the pole cap facing the positive condenser electrode, $\Delta\phi_{ind}$ is directed from the outside to the vesicle inside whereas at the other pole cap, facing the negative electrode, $\Delta\phi_{ind}$ is in the direction inside to outside.

In living cells there is a natural potential difference $\Delta\phi_{nat} = \phi_{in} - \phi_{out} \approx$ from -50 to -200 mV, where the electric potential ϕ_{out} of the outside is taken as the reference $\phi_{out} = 0$. Note that $\Delta\phi_{nat}$ exists even in the absence of an external field. Generally, the total potential difference $\Delta\phi_m = -E_m d = \Delta\phi_{ind} + \Delta\phi_{nat} = -\{(3/2)Ea + \Delta\phi_{nat}/\cos\theta\}|\cos\theta|$. Therefore, for living cells at one pole cap ($\theta = 180°$, $\cos\theta = -1$) $\Delta\phi_{nat}$ is in the same direction as $\Delta\phi_{ind}$ and at the other one ($\theta = 0°$, $\cos\theta = +1$) $\Delta\phi_{nat}$ is opposite to $\Delta\phi_{ind}$. Hence $\Delta\phi_{nat}$ facilitates electric pore formation on one side and reduces it at the other side of the cell.

Prof. Klein asked: Can you comment on the effect of the applied field on the orientation of the PC headgroups? Molecular dynamics simulations indicate that the P → N dipoles have a broad distribution of angles with respect to the bilayer normal.[1] Some lipids even have P → N dipoles perpendicular to the interface. Is it possible for the field to increase the population of such lipids, which in turn could locally encourage pore formation?

1   K. Tu, D. J. Tobias and M. L. Klein, *Biophys. J.*, 1995, **69**, 2558.

Prof. Neumann responded: The chemical modes $\Delta T^+/T_0$ and $\Delta A^+/A_0$ are slower than the shape deformations indicated by the respective dichroitic modes $\Delta T^-/T_0$ and $\Delta A^-/A_0$. We associate the chemical modes with the orientation of the lipid headgroups toward the direction of the transmembrane field, allowing the entrance of additional water and ions into the surface of the membrane in the pole can regions; see Fig. 10 in our paper. We have so far not used lipids with dipolar groups perpendicular to the interface.

Prof. Bohne asked: Is there a dependence of the relaxation times (fast and slow) on the length of time that the field is applied?

Prof. Neumann responded: The relaxation times of the rapid transients are independent of the pulse length. However, at longer pulse durations additional slower deformational modes become apparent.

Prof. Robinson asked: Is there useful information to be obtained from the *amplitudes* of the perturbation? Some of your relaxations are extremely fast—of the order of 1 μs. In the paper you do not give much kinetic information based on the time resolution of these transients? What are the technical limitations to achieving good resolution? *e.g.* finite rise/decay time of electric field pulse.

Prof. Neumann responded: The turbidity relaxations $\Delta T^-/T_0$ and $\Delta T^+/T_0$ (Fig. 4 in our paper) both indicate that in the presence of the field there are at least two kinetic phases: phase I and phase II. At $E = 8$ MV m^{-1} the kinetic analysis yields the amplitudes and the time constants: $\Delta T^-(\text{I})/T_0 = 0.080$ and $\tau(\text{I}) = 0.15$ µs and $\Delta T^-(\text{II})/T_0 = 0.043$ and $\tau(\text{II}) = 1.24$ µs.

The time resolution is limited by the spark gap discharge to switch on the electric field; here this machine time is $\leqslant 20$ ns.

The kinetic analysis shows that both modes reflect shape deformations of the vesicle due to a field-induced increase ΔS in the membrane surface S at constant vesicle volume. Applying a Mie-type numerical code program by Farafornov and co-workers,[1] the turbidity data yield the axis ratio p of the ellipsoidally elongated vesicle as a function of time and field strength (Fig. 13). On the other hand, p is related to the surface fraction $f = \Delta S/S_0$ with $S_0 = 4\pi a^2$ being the total surface area of the vesicle of radius a, by (Kakorin and Neumann, unpublished data):

$$f = p^{-2/3}/2 + p^{1/3} \arcsin[(1 - p^{-2})^{1/2}]/[2(1 - p^{-2})^{1/2}] - 1 \tag{1}$$

The question is now: how can the applied electric field cause a biphasic increase in the fractional surface area?

The experimental data are quantitatively consistent with the following assignments: The rapid phase I reflects membrane stretching and smoothing of thermal undulations whereas the slower phase II is caused by electric pore formation.

Clearly, the electric field E_m across the vesicle membrane reaches its final value by the ionic interfacial polarization (Maxwell–Wagner). The time constant τ_{pol} of this polarization is given by $\tau_{pol} = aC_m (\lambda_{in}^{-1} + \lambda_{ex}^{-1}/2)$, where $C_m = 5 \times 10^{-3}$ F m^{-2} is the specific lipid membrane capacity and λ_{in} and λ_{ex} are the conductivities of the vesicle interior and of the external solution, respectively. Here $a = 50$ nm and $\lambda_{in} = \lambda_{ex} = 6.9$ mS m^{-1}. Therefore, $\tau_{pol} = 0.06$ µs, being about twofold smaller than the experimental value of $\tau(\text{I}) = 0.15$ µs.

The analysis of the relaxation modes in terms of amplitudes and time constants is rather demanding because the experimental values reflect the inhomogeneous, position-dependent field effects of the vesicle membrane: the pole caps experience a larger induced field than the equatorial regions. In addition, the field-induced membrane tension T_m has two components T_g and T_l which both are dependent on the positional angle θ relative to the field direction E. T_g refers to the global vesicle shape deformation and the sin θ-average is given by (Kakorin and Neumann, unpublished data)

$$\langle T_g \rangle = \frac{3a}{40} \{\varepsilon_0 \varepsilon_w E^2 - [64\kappa(p - 1)/3a^3]\} \tag{2}$$

where ε_0 is the vacuum permittivity, $\varepsilon_w = 80$ at $T = 293$ K the relative permittivity of water and κ is the bending rigidity. For $E = 8$ MV m^{-1}, $\kappa \approx 2 \times 10^{-20}$ J and $p = 1.37$, we obtain $\langle T_g \rangle = 1.7 \times 10^{-4}$ N m^{-1}.

The local electrostrictive contribution is given by:

$$\langle T_l \rangle = C_m \langle \Delta\phi_{ind}^2 \rangle/2, \quad \text{where } \Delta\phi_{ind} = -(3/2)Ea|\cos\theta|$$

and

$$\langle \Delta\phi_{ind}^2 \rangle = (1.5Ea)^2 \int_0^\pi \cos^2\theta \sin\theta \, d\theta/2$$

For $E = 8$ MV m^{-1} we have $\langle \Delta\phi_{ind}^2 \rangle = 0.12$ V^2 and $\langle T_l \rangle = 3 \times 10^{-4}$ N m^{-1}.

The relative surface area increase due to stretching is given by Needham and Hochmuth.[2]

$$f_T = \Delta S_T/S_0 = (\langle T_g \rangle + \langle T_l \rangle)/K \tag{3}$$

where $K \approx 0.2$ N m^{-1} is the compression modulus. The time constant for stretching is given by $\tau_T \approx \eta a/K$, where η is the viscosity. Here we obtain $f_T = 2.3 \times 10^{-3}$ and $\tau_T = 2.5 \times 10^{-10}$ s.

The smoothing of thermal undulations by the electrical Maxwell stress in the membrane with the initial lateral tension $T_0 = 1$ mN m^{-1} yields an additional increase in the projected membrane area according to Klösgen and Helfrich.[3]

$$f_{und} = \Delta S/S_0 = [kT/(8\pi\kappa)]$$
$$\times \ln\{[\pi/4a^2 + (T_g + T_l + T_0)/\kappa]/[\pi/4a^2 + T_0/\kappa]\} \quad (4)$$

where k is the Boltzmann constant and T the absolute temperature.

The time constant τ_{und} of the undulative bending can be estimated from eqn. (17) in the paper. For $E = 8$ MV m^{-1} and $T = 298$ K we obtain $f_{und} \approx 1.2 \times 10^{-3}$ and $\tau_{und} \approx 5 \times 10^{-8}$ s. Inserting $p_0 = 1.35 \pm 0.01$ from Fig. 13 in eqn. (1) we find that $f \approx 1.5 \times 10^{-2}$. The surface fraction $f(I) = f_T + f_{und} = 0.35 \times 10^{-2}$ yields the surface ratio $x_f(I) = f(I)/f$ and $x_f(II) = 1 - x_f(I)$. From $x_f(I)$ we readily obtain the ratio $x(I) = [p(I) - 1)/(p_0 - 1)] = [x_f(I)]^{1/2} \approx 0.5$. This value compares well with the experimental ratio $x = (\Delta T^-(I)/T_0)/\Delta T^-/T_0) = (p(I) - 1)/(p - 1) \approx 0.6$.

Since $(p - 1)$ is proportional to $f^{1/2}$ we obtain the 'surface' relaxation times (at $E = 8$ MV m^{-1}): $\tau_f(I) = \tau(I)/2 = 0.07$ μs, which is close to the interfacial polarization time $\tau_{pol} = 0.06$ μs and $\tau_f(II) = \tau(II)/2 = 0.62$ μs.

At higher fields the rate constant $k_{C \to P}$ of pore formation according to the scheme C$\rightleftharpoons$P is approximated by $1/\tau = k_{C \to P} + k_{C \leftarrow P} \approx k_{C \to P}$. For $E = 8$ MV m^{-1} we obtain $k_{C \to P} \approx 1/\tau_p = 1/\tau_f(II) = 1.6 \times 10^6$ s^{-1}, comparing well with previous estimates for asolectin vesicles ($1/\tau \approx 6 \times 10^5$ s^{-1}).[4]

1 V. G. Farafornov, N. V. Voshchinnikov and V. V. Somiskov, *Appl. Opt.*, 1996, **35**, 5412.
2 D. Needham and R. M. Hochmuth, *Biophys. J.*, 1989, **55**, 1001.
3 B. Klösgen and W. Helfrich, *Eur. Biophys. J.*, 1993, **22**, 329.
4 S. Kakorin, S. P. Stoylov and E. Neumann, *Biophys. Chem.*, 1996, **58**, 109.

Dr Goñi asked: Do you have an estimate of the size of the electrically-generated pores in your system? And what would be the density of pore per unit of membrane surface? What is the influence of surface potential (*i.e.* electrically charged lipids) on the number and electrical properties of the pores?

Prof. Neumann responded: To answer your first two points, we have calculated the radius of electropores in lipid membranes in three ways. First, from the field dependence of the equilibrium constant $K = [P]/[C]$ for the electroporation reaction scheme C$\rightleftharpoons$P, where C is the closed membrane state and P the electroporated membrane state of the membrane, respectively.[1] At small field strengths $E \leqslant 2$ MV m^{-1}, pulse duration $t_p = 10$ μs and for the vesicle radius $a = 160$ nm, the average pore radius is $\bar{r}_p = 0.35(\pm 0.05)$ nm of the assumed cylindrical pore of thickness $d = 5$ nm, suggesting an average cluster size of $\langle n \rangle = 12 (\pm 2)$ lipids per pore edge.[1] The percentage of membrane area $f_p \approx (\lambda_m/\lambda_i) \times 100\%$ of conductive openings filled with the intravesicular medium of conductivity $\lambda_i = 2.2$ S m^{-1} linearly increases from $f_p \approx 0$ at $E = 1.8$ MV m^{-1} to $f_p = 0.017\%$ at $E = 8.5$ MV m^{-1}. The membrane conductivity λ_m was estimated from the deviation of the linear dependence of $\ln[K(E^2)]$ on E^2. The percentage of the conductive membrane area $f_p = 0.017\%$ refers to 142 pores per vesicle or the pore density: $\rho = 4.35 \times 10^{-4}$ nm^{-2}.[1]

Second, the pore radius was obtained from the efflux of electrolyte through the electropores of salt-filled vesicles. The kinetic analysis of the conductivity increase yields, at field strength $E = 1.0$ MV m^{-1} and in the range of pulse durations $5 \leqslant t_E/\text{ms} \leqslant 60$, the number of water-permeable electropores to be $N = 35 \pm 5$ per vesicle of radius $a = 50$ nm with mean pore radius of $\bar{r}_p = 0.9 \pm 0.1$ nm under Maxwell stress.[2,3]

Third, pore radii are obtained from the kinetics of uptake of dyes and DNA by electroporated cells. For instance, the fractional surface area for the dye (SBG, M_r 854) conductive pores is $f_p = 0.035 \pm 0.003\%$ and the mean pore radius is $\bar{r}_p = 1.2 \pm 0.1$ nm. The maximum pore number is $N_p = 1.5 \pm 0.1 \times 10^5$ per average electroporated intact FcγR$^{\cdot}$ mouse B cell (cell diameter 25 μm, $E = 2.1$ kV cm^{-1} and $t_E = 200$ μs).[4,5] The kinetics of the direct transfer of plasmid DNA (YEp 351, 5.6 kbp, supercoiled, $M_r \approx 3.5 \times 10^6$) by membrane electroporation of yeast cells (*Saccharomyces cerevisiae*, strain AH 215, diameter 5.5 μm) yields the mean radius of the pores in the DNA permeable porous patches: $\bar{r}_p = 0.39 \pm 0.05$ nm at $E_0 = 4.0$ kV cm^{-1}. The corresponding mean number of pores per cell is $N_p = 2.2 \pm 0.2 \times 10^4$. The maximum membrane area which is involved in the electrodiffusive penetration of adsorbed DNA into the outer surface of the electroporated cell membrane patches is only $0.023 \pm 0.002\%$ of the total cell surface.[4,6]

To answer your third point, the surface potential acts in two main ways on the membrane electroporation. First, different surface potentials on the two membrane surfaces leads to a transmembrane potential difference, which either adds to the externally induced membrane field on one pole cap or is opposite to it on the other one. This asymmetry in the total transmembrane field leads to an increase in the extent and rate of membrane electroporation on one pole cap and to a corresponding decrease on the other.

Second, if the repulsive forces between the charged lipid molecules are larger in the outer membrane leaflet than those in the inner one, because of asymmetrical ion clouds and unequal Debye lengths at the two interfaces, the extent and rate of membrane electroporation are increased to produce the conical electropores. Thereby the distance between the charged headgroups is increased and the free energy due to the repulsive forces is reduced in the outer membrane leaflet.

1 S. Kakorin, S. P. Stoylov and E. Neumann, *Biophys. Chem.*, 1996, **58**, 109.
2 S. Kakorin and E. Neumann, *Ber. Bunsen-Ges. Phys. Chem.*, 1998, **102**, 670.
3 S. Kakorin, E. Redeker and E. Neumann, *Eur. Biophys. J.*, 1998, **27**, 43.
4 E. Neumann and S. Kakorin, *Radiol. Oncol.*, 1998, **32**, 7.
5 E. Neumann, K. Toensing, S. Kakorin, P. Budde and J. Frey, *Biophys. J.*, 1998, **74**, 98.
6 E. Neumann, S. Kakorin, I. Tsoneva, B. Nikolova and T. Tomov, *Biophys. J.*, 1996, **71**, 868.

Prof. Tiddy commented: The area per headgroup for PC is *ca.* 65–70 Å^2. How much does this increase in the pore region? Can you stabilise the pores by the inclusion of larger headgroup lipids?

Prof. Neumann responded: The surface area of a primary electropore of radius $\bar{r}_p = 0.35(\pm 0.05)$ nm, induced by 10 μs field pulse of $E = 2$ MV m^{-1} in the membrane of lipid vesicles of radius $a = 160$ nm, is $S_p = 4\pi \bar{r}_p^2 = 0.39$ nm^2. S_p compares well with the area ≈ 0.5 nm^2 of a lipid molecule in the membrane surface plane.[1] Presumably, the pore radius increases stepwise with increasing field strength and pulse duration, because of a discrete number of lipid molecules composing the pore and because of the hydrophobic force oscillating with distance between pore walls.[2,3] We have so far no information about the potential stabilisation of pores by the inclusion of larger headgroup lipids.

1 S. Kakorin, S. P. Stoylov and E. Neumann, *Biophys. Chem.*, 1996, **58**, 109.
2 E. Neumann, K. Toensing, S. Kakorin, P. Budde and J. Frey, *Biophys. J.*, 1998, **74**, 98.
3 E. Neumann, S. Kakorin, I. Tsoneva, B. Nikolova and T. Tomov, *Biophys. J.*, 1996, **71**, 868.

Prof. Morantz asked: Can you comment on the charge distribution within the pore?

Prof. Neumann responded: We expect that in the curved surface of a hydrophilic pore the charge density due to lipid headgroups is smaller than the planar membrane part.

Prof. Laggner commented: One would expect the pore formation property to depend on the lipid composition—*e.g.* the presence of PEs—and also on the presence of additional solutes in the vesicles solution, *e.g.* salts or alcohol. How far has this been explored?

Prof. Neumann responded: Actually, the pore formation properties do depend on the lipid composition and on the concentration of salt and other substances in the external and internal vesicle media. We have found that the salt concentration gradient across the membrane has a profound influence on the electroporation of membranes containing charged lipids, if the membrane surface charge density is $\sigma \geqslant 10^{-2}$ C m^{-2}. The difference in the Debye lengths of the inner and outer compartments of the vesicles and cells has a considerable effect on the spontaneous curvature. The spontaneous curvature caused by different chemical composition of the internal and external membrane leaflets appreciably affects membrane electroporation. Analysis of electro-optical data obtained with asolectin vesicles doped with 1,6-diphenyl-1,3,5-hexatrien (DPH) in terms of Gibbs area difference elasticity energy suggests that the fraction of electroporated membrane area increases 2-fold, if the concentration of NaCl in vesicles inside increases from $c_{in} = c_{out} = 0.2$ mM up to $c_{in} = 0.2$ M $> c_{out} = 0.2$ mM, where c_{in} and c_{out} are the concentrations of salt in the inner and outer vesicle compartments, respectively.

Prof. Holzwarth asked: How large is the temperature jump resulting from the decay of the electric field? Can you distinguish between the electric field effect and the temperature effect?

Prof. Neumann responded: The temperature increase due to Joule heating is up to 3°, depending on the applied field jump lasting only 10 μs. The additional electrooptical signal is negligibly small compared to the signal change of the direct field effect.

Prof. Svetina commented: An estimate presented for the relaxation time for the orientation of non-spherical vesicles in an external field is in the order of 100 μs. Couldn't this time be shorter if the vesicle aligned itself by appropriate shape transformations rather than by a rotation?

Prof. Neumann responded: Actually, if the electrooptical signal were due to the orientation of initially elongated vesicles (*e.g.*, resulting from the extrusion procedure), the time constant should be equal to the rotational relaxation time $\tau_{rot} = 130$ μs at field zero. The measured time constants are in the range $0.1 \leq \tau/\mu s \leq 3$. Therefore, we deal with initially spherical vesicles which are rapidly elongated by the electrical field and rapidly relax to the spherical shape after the electric pulse. However, the vesicle elongation either requires an increase in the membrane area or a decrease in the internal vesicle volume. The decrease in the vesicle volume by efflux of intravesicular medium through electropores is very slow; the characteristic time constant is about 10 ms.[1,2] The increase in the surface membrane area due to membrane stretching and smoothing of thermal undulations as well as membrane electroporation in the electrical field must be much faster than orientation of non-spherical vesicles (<1 μs). However, membrane stretching and smoothing of thermal undulations can only rationalise about 23% of the measured increase in the membrane area. The residual increase (of 77%) in the membrane surface area is due to electroporation with a time constant of about 1 μs.

1 S. Kakorin and E. Neumann, *Ber. Bunsen-Ges. Phys. Chem.*, 1998, **102**, 670.
2 S. Kakorin, E. Redeker and E. Neumann, *Eur. Biophys. J.*, 1998, **27**, 43.

Prof. Roux opened the discussion of Prof. Lee's paper: How much are the conclusions about the affinity of DOPC *vs.* DOPS model-dependent?

Prof. Lee responded: The key observation shown in Fig. 3 in the paper is that quenching is more efficient when the bromine is on phosphatidylserine than when it is on phosphatidylcholine. This lack of symmetry means that there must be at least two types of site on the ATPase. The observation that displacement of brominated phosphatidylcholine by phosphatidylserine fits to a relative binding constant close to 1 suggests that there are a large number of sites at which phosphatidylserine and phosphatidylcholine bind with equal affinity. The observation of more efficient quenching by brominated phosphatidylserine than phosphatidylcholine then means that there must be a small number of sites at which phosphatidylserine can bind with greater affinity than phosphatidylcholine. The experiments do not show what the number of sites is, but assuming there is just one such site we get the relative binding affinities at this site given in Table 3. If we were to assume two sites, we could fit the data equally well, but with a lower relative binding constant for phosphatidylserine. We picture phosphatidylserines bound to these 'sites' as phosphatidylserine molecules interacting strongly with the positively charged lysines in the peptides. There is likely therefore to be only one of these sites per peptide on each side of the bilayer, and, since the peptide contains a spacer Gly residue on one side of the membrane and not the other, it may be that there is only one such site and not two. We should be able to distinguish between these possibilities by suitable design of new peptides.

Prof. Tiddy asked: What are the error bars for data in Fig. 2 and 3?

Prof. Lee responded: Errors in measuring fluorescence intensity are very small, and will be smaller than the data points in the figures. The greatest problem in these experiments is ensuring total inclusion of the peptide into the lipid bilayer. This seems to be largely a matter of ensuring that the peptide is kept totally dry.

Dr Dijkstra commented: An alternative to the electrostatic interpretation of the lipid protein interaction presented in the paper could be a non-specific, bending energy induced effect. Arguments to support this are:

(1) The data shown in Fig. 3 appear to fit more or less as well to eqn. (4) as to eqn. (6) indicating all 'binding' sites would be of equal importance, rather than implying a single high affinity site on the peptide.

(2) From the results presented in Table 2 it is observed that in general only slight preferences at most are found from the peptide L_{16} for the anionic lipids *versus* the phosphatidylcholine (PC) if the PC is brominated rather than the anionic lipid. This is in line with the lack of a preference from the L_{22} for one lipid over the other.

(3) This is in contrast to the data where the anionic lipid has been brominated (Table 3). This difference would imply the bromination technique does not quite leave the saturated brominated lipid functionally equivalent to the unsaturated, non-brominated lipid. Intuitively one would also not expect these species to be the same, *e.g.* one could expect bromination to induce a dipole at that site which a double bond would not.

(4) The 5 kJ mol^{-1} difference in interaction energy observed between PA relative to PC, could also be explained by a difference in the hydrophobic length of the one lipid *vs.* the other. Calculations show the energy difference associated with hydrophobic mismatches of about 1 Å to be of the same order of magnitude.[1] It can be expected that anionic headgroups will experience more headgroup–headgroup repulsion than zwitterionic headgroups, and therefore result in large hydration and headgroup areas per molecule and thus shorter hydrophobic lengths.

(5) A hydrophobic mismatch effect rather than specific binding would also explain why no difference in the relative affinities for either lipid from the L_{22} was observed. Here the mismatch between the peptide and the lipids is so large, 1 Å more or less would, relatively speaking, make less difference here than close to the 'perfect', as with L_{16}. The paper suggests no difference between the lipids affinities is seen due to the tilting the L_{22} peptide locating the Lys too far from the lipid headgroup region to allow strong interactions, thereby minimising the difference between the anionic and zwitterionic lipids. However, tilting will, out of geometrical necessity, bring all parts of the peptide outside the membrane closer to the membrane surface. This would lead to *increased* interactions with the lipid headgroups, unless tilting goes so far as to place the Lys-residues inside the hydrophobic region. In this case the efficiency of incorporation into thinner bilayers would be expected to decrease, which is not the case based on earlier reports from the authors that full incorporation has been observed for L_{22} into PC-bilayers with chain lengths between C_{14} and C_{24}.

(6) A distance of 3.3 Å is calculated between the anionic headgroup and the Lys-residue using the relative permittivity of water. Given that this distance is the same order of magnitude as the size of a water molecule, the validity of the calculation, using a bulk relative permittivity value for water in the calculation might be questioned.

1 C. Nielsen, M. Goulian and O. S. Andersen, *Biophys. J.*, 1998, **74**, 1966; D. R. Fattal and A. Ben-Shaul, *Biophys. J.*, 1993, **65**, 1795.

Prof. Lee responded: Since anionic phospholipids show stronger binding to the positively charged peptides than do zwitterionic phospholipids, the obvious interpretation is in terms of charge–charge interactions. The simple calculation included in the paper shows that relative binding constants of the kind we find can be obtained in this way, and that they require close contact between the negatively charged headgroup of the phospholipid and the positively charged Lys on the peptide. The observation that a small number of anionic phospholipids interact with the peptide with high affinity, with a large number interacting with the same affinity as phosphatidylcholine, is easy to explain if charge–charge interactions are important; I do not see any obvious explanation for this in terms of bending energies. However, we could test the idea of an involvement of bending energies by measuring binding constants to phosphatidylethanolamines, but this we have not yet done.

(1) Indeed, the data in Fig. 3 fit equally well to eqn. (3) (assuming a single class of binding site) with the parameters in Table 2 or eqn. (6) (assuming two classes of binding site) with the parameters in Table 3. However, the point is that the fit to a single class of binding site gives different relative affinities for the anionic and zwitterionic lipids, depending on which one is brominated;

this is physically impossible—if there was truly only a single class of binding site, quenching curves such as those shown in Fig. 3 would be symmetrical, being the same, independent of which lipid was brominated. The fact that the quenching curves are not symmetrical shows that the anionic lipids can bind to sites where they have high affinity but the zwitterionic lipids do not. That is the basic for the two-class of model presented in the paper, and the basis for eqn. (6).

(2) I am not sure that I understand the question. The data in Tables 2 and 3 show that for the short peptide L_{16} there are a small number of 'sites' where anionic phospholipids bind with relatively high affinity, whereas, as shown by the data in Table 1, L_{22} such 'sites' do not appear to be present.

(3) As described above, the physical properties of brominated and non-brominated phospholipids are very similar. Quenching observed in mixtures of brominated and non-brominated phosphatidylcholine, or in mixtures of brominated and non-brominated phosphatidylserine are the same and all fit assuming that the brominated and non-brominated lipids bind with equal affinity to the peptide (Table 1). I know of no information that suggests that a dipole in the chain region would affect interaction with the hydrophobic region of a peptide or protein.

(4) We know the effects of hydrophobic mismatch in these systems from our previous studies (ref. 11). When the hydrophobic thickness of the bilayer is more than 10 Å greater than the hydrophobic length of the peptide, the peptide does not incorporate. For hydrophobic mismatch less than this, the peptide does incorporate, but with a weaker binding to the mismatched lipid. However, the difference in binding is no greater than a factor of 2, much less than the differences we observed here. Any differences in hydrophobic lengths between dioleoylphosphatidylcholine and, for example, dioleoylphosphatidic acid will be very small, and will certainly be much too small to account for the effects of anionic phospholipids we see here. Further, of course, it is not at all obvious how a difference in effective chain length could give rise to two classes of binding site of the type necessary to explain our data, whereas two classes of site fit naturally with what would be expected for binding of a charge lipid to a charged peptide.

(5) It is hard at this stage to be certain about what structure will be adopted by a tilted peptide in a lipid bilayer. There might well be some unfolding of the α-helix at the ends of the tilted peptide. In a tilted peptide it is also possible that the charged residues will be located higher above the lipid/water interface than for a peptide oriented parallel to the bilayer normal. I think what we want here is more experimental data: what is the orientation of the tilted peptide in a bilayer?; does the α-helix unfold at all?; how does the charged lipid headgroup interact with the charged groups on the peptide?, and is the structure of the charge residue important? We do not yet know the answer to any of these questions.

(6) Yes, indeed. The calculation simply shows that the interaction is likely to be a close one. The reason for using the relative permittivity of water in these calculations is that, necessarily, the anionic headgroup–Lys pair will be surrounded by water, and that this surrounding water will have a dominating effect, whatever the nature of the 'medium' actually between the headgroup and the Lys.

Prof. Laggner commented: A quasi-ternary system (two lipid species and one peptide) should be described in terms of a phase triangle which implies that the interactions also depend on the peptide : lipid ratio. How is that situation in your systems?

Prof. Lee responded: In these experiments we have to have a lipid : peptide molar ratio of at least 15 : 1 to ensure total incorporation of the peptide into the lipid bilayer. We can then increase the lipid : peptide molar ratio up to 1000 : 1 with no further change in the system. For convenience all the experiments shown here were performed with a molar ratio of 100 : 1. Because of the positive charge on the peptides we assume that the peptides will be simply dispersed in monomeric form in the bilayers, although we have no direct evidence for this.

Prof. Klein asked: Do you have evidence on the state of aggregation of the peptides, particularly those with Trp and Tyr?

Prof. Lee responded: No, we have no direct evidence. However, we assume that because of the positive charge on the peptides they will be present in monomeric form.

Dr Goñi commented: Your results with Tyr residues at the ends of the peptide remind me of the "mystery" of Trp residues at the ends of many transmembrane peptides. Could you share with us your ideas on this point? Have you detected any signs of hexagonal phase formation with the short peptides/long lipids? H_{II} phase would mean a solution for the problem.

Prof. Lee responded: The idea of the experiments with the Tyr containing peptides was to see if we could get any information about the role of these residues in membrane proteins. What we found, as shown in Fig. 5, was that Tyr-containing peptides could incorporate into thicker bilayer than corresponding peptides containing only Leu as the hydrophobic residues. The explanation could be that the Tyr residue simply acts as a 'larger' residue than Leu so that the effective hydrophobic length of the Tyr-containing peptides is greater than for the peptides lacking Tyr. If this were the case, Phe should have the same effect as Tyr; this is something we will test. The alternative possibility is that insertion of the Tyr residues in the hydrophobic region of the bilayer with the OH group oriented to take part in hydrogen bonding gives Tyr (and by analogy, Trp) unique properties at the lipid/water interface.

To answer your second question, we have not looked specifically at this. However, at the high lipid : protein ratio we use we expect the lipids to be purely bilayer; other groups have observed hexagonal and other non-bilayer phases in mixtures with peptides, but only at very low molar ratios of lipid to peptide.

Prof. Barclay said: I have two queries concerning the implications of your data on the association of specific lipids to proteins. First, if for example a protein associates preferentially with a long chain fatty acid would you expect these fatty acids to preferentially associate to build up an enriched zone or domain? Secondly, I realise you haven't any data on the other common method of membrane integration namely through glycosylphosphatidylinositol (GPI)/anchors. However there are a large number of data on GPI anchored proteins and association with lipids and formation of "rafts"—would you like to comment on these?

Prof. Lee responded: To answer your first point, in previous studies we have shown that phosphatidylcholines interact most strongly with peptides when the thickness of the lipid bilayer matches the hydrophobic length of the peptide (ref. 11). However, the effects are relatively small, at most a factor of 2. Thus whilst, for example, there will be more phospholipids with long chains than short chains around a long peptide, the effect will not be sufficient to build up a macroscopic domain or 'raft' of the type thought to be important in cell membranes.

In response to your second point, we have done no work on lipid-modified peptides and so cannot usefully comment on this.

Prof. Tiddy asked: How does bromination modify lipid behaviour, *e.g.* phase transition temperature. What is the phase transition temperature of the C_{24} lipid?

Prof. Lee responded: Bromination of an unsaturated lipid such as dioleoylphosphatidylcholine has very little effect on its physical properties. The low transition temperature of dioleoylphosphatidylcholine follows from the difficulty in packing the 'bent' *cis* unsaturated chains into the gel phase. Similarly, the two bulky bromines in each chain give a low transition temperature for the brominated analogue; we have shown that the transition temperature is below 5 °C. In that paper we also showed that order parameters for a variety of spin-labelled fatty acids were the same in bilayers of brominated and non-brominated phosphatidylcholines. We also showed that enzyme activities for a transport protein (Ca^{2+}-ATPase) were the same in bilayers of brominated and non-brominated phosphatidylcholines.

1   J. M. East and A. G. Lee, *Biochemistry*, 1982, **21**, 4144.

Prof. Bayerl asked: Could the morphology of the different vesicles preparation lead to different curvatures which may explain part of the differences you observed regarding peptide–lipid interaction? Did you have any control on morphology in your experiments?

Prof. Lee responded: All the experiments we reported here were with multilamellar vesicles prepared by hand-shaking. The large size of the multilamellar vesicles means that bilayers will be essentially flat. We have also carried out some studies in which lipid and peptide were mixed in cholate as detergent, followed by dilution into buffer; using this procedure, membrane fragments are formed, containing the peptide. Using both procedures we obtained the same results. I might add that in experiments with long peptides which did not incorporate into thin lipid bilayers, the peptide and lipid clearly formed separate phases, as the lipid could be separated from the peptide on Sephadex columns.

Dr P. N. Edwards commented: Anionic phosphate di-esters have very strong hydrogen bond acceptor properties and would wish to be solvated by many more water molecules (acting as H-bond donors) than can physically be accommodated at a lipid interface. Thus protein residues (Tyr; Trp) with H-bond donor ability and much weaker acceptor ability are especially favoured at the polar/apolar interface.

Prof. Lee responded: I am not sure about that. Anionic phospholipids will form very stable lipid bilayers under the right conditions. So the phosphatidylserines and phosphatidic acids used here at neutral pH in the absence of divalent cations all form bilayers in which, presumably, all hydrogen bonding needs are met, either by other lipid headgroups or by water. Indeed, crystal structures of some of these molecules show extensive hydrogen bonding networks in the headgroup region. However, it is certainly true that Tyr and Trp have unique properties amongst the 'hydrophobic' amino acids and their hydrogen bonding requirements are best met by a location at the lipid/water interface, where the polar groups can be involved in hydrogen bonding with the hydrophobic rings still immersed in a hydrophobic environment.

Prof. Deber communicated: Your Tyr-containing peptide suggests interfacial involvement of Tyr residues. Can you speculate on the nature of the interactions that attract Tyr specifically to the interfacial location? If certain peptides adopt a tilted or slanted orientation as they cross the lipid bilayer, what are the implications for local lipid packing and fluidity? Can such effects be measured?

Prof. Lee communicated in response: Your first comment was answered in reply to the previous question. With regard to your second point, it is a very good question to which I have no clear answer. One would imagine that the effect of a tilted peptide would be significantly different to one oriented parallel to the bilayer normal. For a peptide oriented parallel to the bilayer normal it would seem likely that most lipids whose fatty acyl chains are in significant contact with the peptide are also in contact in the headgroup region; this is the classic idea of a shell of 'boundary' or 'annular' lipids surrounding a peptide or protein. For a peptide with a large tilt, the arrangement will be more complex, and lipids whose headgroups interact strongly with the charged groups on the peptide may well have chains not interacting with the hydrophobic part of the peptide. Conversely, lipids whose chains interact with the hydrophobic part of the peptide may not have headgroups interacting with the charged part of the peptide. I suspect that this kind of level of detail may only be obtainable by molecular dynamics simulations. The experiments would then hopefully give sufficient thermodynamic data on strengths of binding, *etc.*, to provide suitable tests of the models.

Phospholipid chain length alters the equilibrium between pore and channel forms of gramicidin

Toby P. Galbraith and B. A. Wallace*

Department of Crystallography, Birkbeck College, University of London, London, UK WC1E 7HX

Received 26th October 1998

Gramicidin is an excellent model system for studying the passage of ions through biological membranes. The conformation of gramicidin is well defined in many different solvent and lipid systems, as are its conductance and spectroscopic properties. It is a polymorphic molecule that can adopt two different types of structure, the double helical "pore" and the helical dimer "channel". This study investigated the influence of the acyl chain length of membrane phospholipids on the conformations adopted by gramicidin. We used circular dichroism spectroscopy to examine the conformational equilibrium between the pore and channel forms in small unilamellar vesicles of phosphatidylcholine with acyl chain lengths of 18, 20 and 22 carbons. Our results show that in C_{18} and C_{20} lipids almost all the gramicidin is in the channel form, while in the longer C_{22} lipids the equilibrium shifts in favour of pore conformations, such that they form up to 43% of the total population. This change is attributed to the ability of the double helical conformation to tolerate more hydrophobic mismatch than the helical dimer, perhaps due to the greater number of stabilising intermolecular hydrogen bonds.

Introduction

Gramicidin is a linear polypeptide antibiotic, synthesised by the soil bacterium *Bacillus brevis* during sporulation.[1] It is hydrophobic and forms ion channels specific for passage of monovalent cations across biological membranes.[2] The primary structure of gramicidin includes an unusual sequence of alternating L- and D-amino acids.[3] The N- and C-termini are capped with formyl (HCO) and ethanolamine ($-NHCH_2CH_2OH$) groups, respectively, and its sequence is as follows: formyl-L-Val-Gly-L-Ala-D-Leu-L-Ala-D-Val-L-Val-D-Val-L-Trp-D-Leu-L-Trp-D-Leu-L-Trp-D-Leu-L-Trp-ethanolamine.

The peculiarities of its primary structure have important implications for its three-dimensional structure and function. Firstly, all the residues are hydrophobic, with no charged or hydrophilic side chains. Secondly, because both the N- and C-termini are blocked, this prevents the formation of a zwitterion or net charge at any pH, reinforcing the hydrophobic nature of the peptide and causing it to partition strongly into the hydrophobic core of the phospholipid bilayer. Thirdly, the sequence of alternating L- and D-amino acids allows gramicidin to adopt conformations which would be unacceptable for an all L-peptide. In this case, the β-sheet-like motif present in gramicidin means that all the side chains protrude from one side of the sheet, rather than alternate sides as is the case in all L-peptides. When the sheet is rolled up to form a helix the hydrophobic side chains project from the outside of the helix to interact with the surrounding solvent or lipid, while the central lumen is left unobstructed to allow the passage of ions.

Physical characteristics of pores and channels

Gramicidin is highly polymorphic, being able to adopt a wide range of structures with different topologies, orientations and hands. The two most common forms of gramicidin are the double helix (DH), a dimer in which two monomers are intertwined [Fig. 1(a)],[4] and the helical dimer (HD), a dimer in which two helical monomers are joined end-to-end [Fig. 1(b)].[5–7] Both the double helix and helical dimer forms have β-sheet-like hydrogen bonding patterns, differing in the number of intra- and inter-molecular hydrogen bonds, and the helical rise per residue. The HD form is commonly known as the "channel", whereas the DH structures are often referred to as "pores".

In ion-free methanol solution, gramicidin has been shown to exist as four interconvertible double helical conformers, referred to as species 1, 2, 3 and 4, differing in helical sense and chain orientation.[8] The components of this equilibrium mixture can be separated using chromatographic techniques, allowing characterisation of the individual species by circular dichroism (CD) spectroscopy. Veatch and co-workers[8] proposed that species 1 and 2 are left-handed parallel DHs differing in the staggers between the ends of their chains, species 3 is a left-handed antiparallel DH, and species 4 is a right-handed parallel DH. The relative abundance of these four species in a given solution depends on the nature of the solvent.[9]

Influences on the equilibrium between DH and HD forms

The equilibrium between the DH and HD forms of gramicidin can be shifted by a number of physical conditions. The principle influence appears to be the chemical nature of the environment. Whilst in organic solvents the DH forms are almost exclusively found (with the relative proportions of the four DH species present determined by the nature of the solvent),[9] in most lipids the HD form predominates. However, if the membranes are prepared from lipids with polyunsaturated fatty acids,[10] this has been shown to bias the equilibrium towards the DH forms in small unilamellar vesicles (SUVs), and conformational transitions between the DH and HD forms appear to occur faster in bilayers with polyunsaturated acyl chains.[11] In addition, in SUVs the balance between DH and HD forms can be affected by the chirality of the lipid molecules[12] used to produce the bilayers. It has also been shown that the relative proportions of DHs and HDs present depend on the solvent and temperature used during the preparation of the SUVs (the so-called "solvent-history dependence"). The conformation which is initially adopted when gramicidin is incorporated into membranes is determined by the solvent in which it was dissolved during sample preparation.[13–16] This is not necessarily the conformation that is finally adopted at thermodynamic equilibrium, however. Sychev et al.[10] assert that while the initial conformational state of gramicidin in hydrated lipid bilayers is affected by the solvent, the equilibrium conformational state (after heat incubation) is determined by the lipid structure rather than the nature of the solvent. When taken altogether, these studies show that considerable variation in the relative proportions of the HD and DH forms can be achieved in SUVs.

Quite a different picture is seen for gramicidin in bilayer lipid membranes (BLMs) used for conductance studies. Under ordinary conditions, essentially only HDs are detected in these

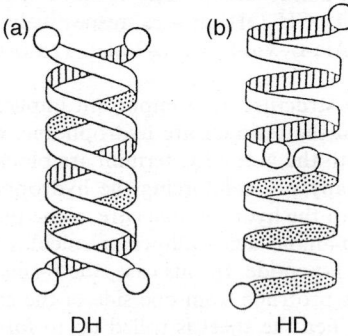

Fig. 1 Schematic representations of the (a) double helical (DH) and (b) helical dimer (HD) forms of gramicidin.

samples.[17] While occasionally channels with very long mean channel lifetimes and lower conductance are seen—which are properties attributed to DHs—the proportion of them is so low that it is difficult to quantitate. It has been estimated that approximately one in 10 000 conducting species found in GMO (glycerol monooleate) membranes may be DHs. This very low proportion suggests that the HD form is considerably more stable in these types of bilayers, perhaps by about 6 kcal mol^{-1}. In actuality, this would correspond to a very small difference in interaction energies, on the order of, say, one additional hydrogen bond formed. It has been difficult to find conditions in BLMs that shift the equilibrium towards the DH forms. Unlike in SUVs, neither the lipid chirality[18] nor the degree of fatty acid unsaturation[19] appears to have any effect—with only HDs detected in either type of lipid. Furthermore, there appears to be no solvent dependence—under similar conditions which favour DH formation in SUVs, only HDs are detected.[20] Although supposed DH forms can be produced using certain mixtures of gramicidin analogues,[17] it has otherwise been difficult to produce conditions in BLMs which even slightly shift the equilibrium towards DH formation.

The most striking aspect of these results is the difference seen in the two types of membrane samples, SUVs and BLMs, examined by circular dichroism spectroscopy and conductance measurements, respectively. It is interesting to consider what properties of the bilayers (or of the methods used to examine them) may have resulted in the different observed behaviours. Lipid vesicles and planar bilayers differ, amongst other properties, in their radii of curvature, the ratio of polypeptide-to-lipid molecules present, the packing of lipid molecules, and solvent content. Hence these features must also be taken into account when considering influences on the DH–HD propensities. An alternative explanation may be that conductance measurements only detect open channels, which may represent only a small proportion of the total population in some environments, or that the CD experiments cannot distinguish between membrane-associated and membrane-spanning conformers.

The question then arises whether conditions can be found that will influence the equilibrium in both SUVs and BLMS. The one membrane property that may do this is the chain length of the phospholipid molecules. Very long unsaturated lipids,[21] which result in a mismatch of lengths between gramicidin and the bilayer, seem to produce some (but not exclusively) DH forms in BLMs. An early study in SUVs[22] had also suggested that although the lipid phase state in short lipids did not have an influence on the balance between the forms produced, in lipids composed of saturated fatty acids, hydrocarbon chain length did influence the balance between DH and HD forms.

Therefore in order to examine for a parallel influence in both environments, this study used the same type of unsaturated lipids as were used in the conductance studies. We present data on CD studies of gramicidin incorporated into mono-unsaturated phosphatidylcholine vesicles of varying acyl chain length, and analyse the spectra using a reference database derived from the individual DH and HD structures to estimate the proportion of molecules in each sample in the DH and HD forms.

Materials and methods

Gramicidin D (the commercially available mixture of ~80% gramicidin A, 5% gramicidin B, and 15% gramicidin C) was obtained from ICN Biochemicals (Aurora, OH, USA). The phospholipids di-oleoylphosphatidylcholine (di-$C_{18:1}$-PC), di-eicosenoylphosphatidylcholine (di-$C_{20:1}$-PC) and di-erucoylphosphatidylcholine (di-$C_{22:1}$-PC) were obtained from Avanti Polar Lipids (Alabaster, AL, USA). Gramicidin and lipid (molar ratio 1 : 20) were initially dissolved in 300 μl of a 2 : 1 (v/v) $CHCl_3$–CH_3OH mixture, dried by rotary evaporation, and hydrated in 300 μl de-ionised Milli-Q water (Millipore Corp., Bedford, MA, USA). The samples were mixed thoroughly and sonicated for 10 min at around 50 °C, and then incubated overnight (approximately 16 h) at 55 °C. Prior to examination, the samples were sonicated using a Soniprep probe sonicator for 6–10 bursts of 10 s. The resulting samples were spun at 12 000g for 4 min, and the supernatant then loaded into a 0.02 cm pathlength Suprasil cell. Gramicidin in acetic acid samples were prepared using 15 mg ml^{-1} in glacial acetic acid (BDH, Analar grade). Spectra for these samples were obtained using a 0.001 cm pathlength cell. CD spectra were collected at 0.2 nm intervals over the wavelength range from 190 to 300 nm at 50 °C on an Aviv 62DS spectropolarimeter. Five repeat scans were collected for each sample preparation.

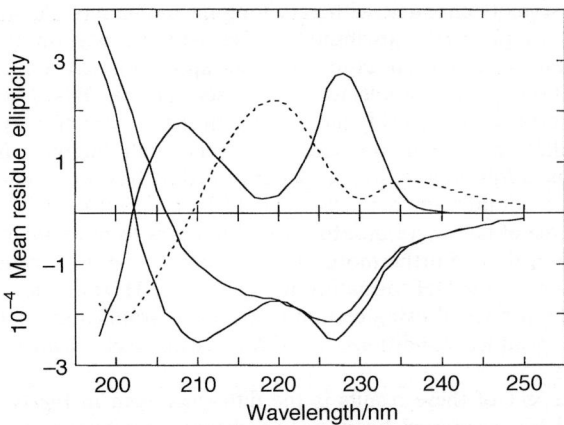

Fig. 2 Reference circular dichroism spectra for the three DH species (———) and the HD form (- - -) of gramicidin.

The spectra were averaged and baseline corrected, using similarly prepared samples without gramicidin present as the baselines. Individual spectra derived from experiments in propan-2-ol[8] provided the reference data for the double helical species 1, 3 and 4 (species 1 and 2 were represented by a single spectrum as their spectra are virtually identical), as described in Chen and Wallace,[9] and a spectrum from the di-$C_{18:1}$-PC preparation in this study was used as the reference spectrum for the helical dimer form (Fig. 2). The sample spectra were analysed using a least-squares fitting algorithm[23] over the wavelength range from 198 to 250 nm. A fit parameter, the normalised root mean square deviation (NRMSD), which is a measure of the correspondence between the calculated results and the experimental data,[23] was determined for each analysis.

Results

In a range of phospholipids with saturated acyl chains of lengths ranging from C_{12} to C_{18}, gramicidin was found to have a distinct CD spectrum, which is attributable to the $\beta^{6.3}$ HD structure.[13,14,22,24] In this study, in di-$C_{18:1}$-PC the CD spectrum of gramicidin obtained was essentially identical to the spectra previously reported in saturated lipids and thus indicates it, too, is of a HD form (Fig. 3). In di-$C_{20:1}$-PC, the spectrum obtained was very similar (Fig. 3) to the C_{18} spectrum, and analyses indicated that 99% of the molecules were in the HD form, with only 1% in DH conformations (Table 1). This was consistent with conductance measurements, which showed only one short-lived channel type in this lipid, corresponding to the $\beta^{6.3}$ helical dimer.[21]

In di-$C_{22:1}$-PC, however, a significantly different spectrum was obtained (Fig. 3). It shows a marked decrease in the peak at approximately 219 nm, and a slightly negative peak at ~230 nm, where the HD spectrum exhibits a positive ellipticity. According to the analysis (Table 1), the HD conformation only contributes 57% of the components present, with DH conformers making up the rest: 27% was found to correspond to species 1, with the remaining 16% being attributable to species 4. This is consistent with conductance studies that detected both HD and other (presumably DH) structures in the C_{22} lipids.[21] It has recently been suggested[25] that acetic acid is a solvent which can produce the "channel" form of gramicidin, in contrast to all other organic

Table 1 Estimated contents of double helical and helical dimer conformers in lipids of different acyl chain lengths

Acyl Chain	Proportion of conformer (%)				
	DH Species 1	DH Species 3	DH Species 4	HD	NRMSD
C_{18}	0.00	0.00	0.00	1.00	0.00
C_{20}	0.01	0.00	0.00	0.99	0.09
C_{22}	0.27	0.00	0.16	0.57	0.17

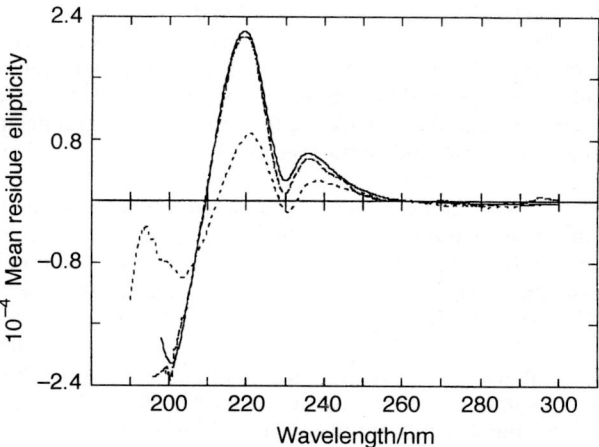

Fig. 3 Circular dichroism spectra of gramicidin in SUVs composed of lipids with fatty acid chains of different lengths: (———) di-$C_{18:1}$-PC, (– – – –) di-$C_{20:1}$-PC, and (······) di-$C_{22:1}$-PC.

solvents which have been found to produce "pore" forms. To examine this, we obtained the CD spectrum of gramicidin in this solvent (Fig. 4). The spectrum in acetic acid clearly differs from the signature channel spectrum, in both shape and peak positions. When calculations were done to determine the proportions of channel and pore components present, the very high NRMSD fit parameter (0.30) suggested that not only was it not the channel structure, but also that a different conformation was present in this solvent. If the reference database was then altered to also include the spectrum of an ion-containing pore form[26] the fit improved slightly (to 0.20), but was still poor. Therefore, the structure in this solvent is not well represented as any one of the standard helical dimer or double helical structures. Clearly, then, acetic acid is not a membrane-mimetic solvent for gramicidin, and thus, despite the authors' claims,[25] the crystal structure in this solvent is not of the channel form. To date, no organic solvent has been found to give rise to the "channel" structure.

Discussion

In this study, we investigated the acyl chain length dependence of the various DH and HD conformations of gramicidin in SUVs formed from lipids with long fatty acid chains which result in a

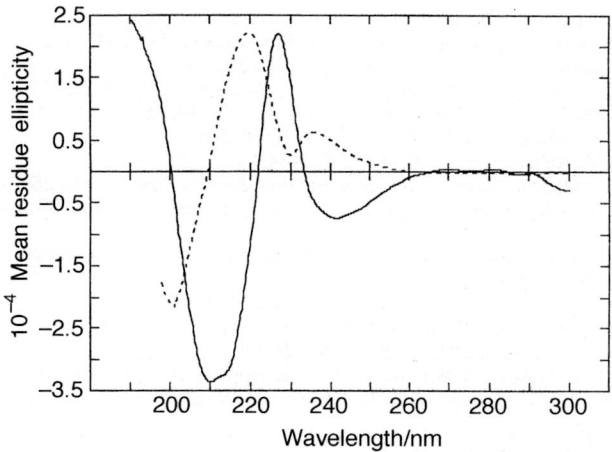

Fig. 4 Circular dichroism spectrum of gramicidin (15 mg ml^{-1}) in glacial acetic acid (———) compared with the channel spectrum of gramicidin in C_{18} lipids (– – –).

mismatch between the length of the gramicidin dimer and its surrounding bilayer. These studies can be directly compared with conductance studies on the same types of lipids.

Previous studies with other lipid types have not quantified the proportions of the various forms present under different conditions in SUVs. In this study we were able to do so by augmenting the existing spectral databases produced for analysing the proportion of gramicidin DH species with a gramicidin HD reference spectrum and using these with an established algorithm for secondary structure analysis.

Previously Mobashery et al.[21] had carried out conductance measurements to assess the effect of bilayer thickness on the conformation of gramicidin. They observed that under most experimental conditions gramicidin A adopts the $\beta^{6.3}$ HD conformation, and that normally DH structures do not form membrane-spanning channels unless there is a considerable mismatch of lengths. In di-$C_{18:1}$-PC and di-$C_{20:1}$-PC BLMs the conductance was shown to be of a single type, corresponding to the $\beta^{6.3}$ HD. However, in di-$C_{22:1}$-PC BLMs, significant numbers (although still a minority of the total) of conducting forms were of a different type assumed to be DH structures. The present CD study suggests the same situation is true in SUVs. In these samples, the HD is still the predominant form, but a significant proportion of DHs is also formed.

Thus it appears that in di-$C_{18:1}$-PC and di-$C_{20:1}$-PC the energetically favourable conformation for gramicidin is the $\beta^{6.3}$ helical dimer channel. However, in the thicker bilayer formed by di-$C_{22:1}$-PC, the hydrophobic mismatch created by the difference in length of the $\beta^{6.3}$ channel and the membrane means that the equilibrium is shifted in favour of double helical conformations. This has been explained[21] in terms of the contributions to the free energy difference between gramicidin conformations from gramicidin itself and from its interactions with its environment: the free energy difference between the DH and HD forms increases as the hydrophobic mismatch increases such that the equilibrium shifts significantly for DH species to be observed, both by CD (this study) and conductance studies.[21] This may be attributable to the greater dimeric stability of the double helices, which are held together by 28 intermolecular hydrogen bonds, compared to only 6 intermolecular bonds in the channel conformation.

References

1   R. D. Hotchkiss and R. J. Dubos, *J. Biol. Chem.*, 1940, **132**, 791.
2   S. B. Hladky and D. A. Haydon, *Biochim. Biophys. Acta*, 1972, **274**, 294.
3   R. Sarges and B. Witkop, *J. Am. Chem. Soc.*, 1965, **87**, 2011.
4   W. R. Veatch and E. R. Blout, *Biochemistry*, 1974, **13**, 5257.
5   D. W. Urry, *Proc. Natl. Acad. Sci. U.S.A.*, 1972, **68**, 672.
6   G. N. Ramachandran and R. Chandrasekaran, *Indian J. Biochem.*, 1972, **9**, 1.
7   A. S. Arseniev, I. L. Barsukov, V. F. Bystrov, A. L. Lomize and Y. A. Ovchinnikov, *FEBS Lett.*, 1985, **186**, 168.
8   W. R. Veatch, E. T. Fossel and E. R. Blout, *Biochemistry*, 1974, **13**, 5249.
9   Y. Chen and B. A. Wallace, *Biopolymers*, 1997, **42**, 771.
10 S. V. Sychev, L. Barsukov and V. Y. Ivanov, *Eur. Biophys. J.*, 1993, **22**, 279.
11 K. J. Cox, C. Ho, J. V. Lombardi and O. D. Stubbs, *Biochemistry*, 1992, **31**, 1112.
12 J. D. Callahan, R. Bittman and B. A. Wallace, unpublished results.
13 J. A. Killian, K. U. Prasad, D. Hains, and D. W. Urry, *Biochemistry*, 1988, **27**, 4848.
14 P. V. LoGrasso, F. Moll III and T. A. Cross, *Biophys. J.*, 1988, **54**, 259.
15 M. C. Bano, L. Braco and C. Abad, *Biochemistry*, 1991, **30**, 886.
16 M. Bouchard and M. Auger, *Biophys. J.*, 1993, **65**, 2484.
17 R. E. Koeppe II and O. S. Andersen, *Annu. Rev. Biophys. Biomol. Struct.*, 1996, **25**, 231.
18 L. L. Providence, O. S. Andersen, D. V. Greathouse, R. E. Koeppe II and R. Bittman, *Biochemistry*, 1995, **34**, 16404.
19 J. Girshman, D. V. Greathouse, R. E. Koeppe and O. S. Andersen, *Biophys. J.*, 1997, **3**, 1310.
20 D. B. Sawyer, R. E. Koeppe II and O. S. Andersen, *Biophys. J.*, 1990, **57**, 515.
21 N. Mobashery, C. Nielsen and O. S. Andersen, *FEBS Lett.*, 1997, **412**, 15.
22 B. A. Wallace, W. R. Veatch and E. R. Blout, *Biochemistry*, 1981, **20**, 5754.
23 B. A. Wallace and C. L. Teeters, *Biochemistry*, 1987, **26**, 65.
24 D. V. Greathouse, J. F. Hinton, K. S. Kim and R. E. Koeppe II, *Biochemistry*, 1994, **33**, 4291.
25 B. M. Burkhart, N. Li, D. A. Langs, W. A. Pangborn and W. L. Duax, *Proc. Natl. Acad. Sci. U.S.A.*, 1998, **95**, 12950.
26 B. A. Wallace, *Biophys. J.*, 1984, **45**, 114.

Paper 8/08270G

Protein inclusion in lipid membranes: A theory based on the hypernetted chain integral equation

Patrick Lagüe,[a] **Martin J. Zuckermann**[b] **and Benoît Roux**[a]

[a] *Departments of Physics and Chemistry, Université de Montréal, C.P. 6128, succursale Centre-Ville, Montréal (Québec), H3C 3J7 Canada*
[b] *Physics Department, McGill University, Montréal (Québec), Canada*

Received 11th September 1998

A theory for describing the structure of the hydrocarbon chains around a protein inclusion embedded in a lipid bilayer is developed on the basis of the hypernetted chain integral equation formalism for liquids. The exact lateral density–density response function of the hydrocarbon core, which is extracted from a molecular dynamics simulation of a pure lipid bilayer, is used as input to the theory. Numerical calculations show that the average lipid order is perturbed over a distance of 25 to 30 Å around a hard repulsive cylinder of 5 Å radius representing an α-helical polyleucine protein inclusion. The lipid-mediated protein–protein interaction is calculated and is shown to be non-monotonic, being repulsive at an intermediate range but attractive at short range. It is found that the lipid matrix contributes a free energy well of 8 $k_B T$ to the association of two cylindrical inclusions.

1 Introduction

The microscopic factors which drive the assembly of membrane-bound proteins are of considerable interest to both biophysicists and biologists since the energetics of this process is a key element in understanding membrane protein stability. It has been demonstrated that specific protein–protein interactions play a crucial role in the stability of protein aggregates and that the physical state of the bilayer modulates the activity of membrane-bound proteins and affects the lateral distribution of protein in the membrane surface.[1–3] For example, the energetics of inclusion-induced bilayer deformation has been investigated using the gramicidin A channel as a molecular probe.[4] The membrane spanning portions of many integral membrane proteins consist of one or a number of transmembrane α-helices. Interactions between transmembrane helices contribute to the energetics of folding and oligomerization in a manner that could be highly specific in some cases, but relatively non-specific in others.[3] Since the presence of proteins perturbs the structure of bilayer membranes (see ref. 5), it seems reasonable that transmembrane helices may also interact with one another *via* some non-specific lipid-packing effects. This concept raises important questions concerning the nature of lipid-mediated driving forces for protein aggregation and assembly. In other words, what is the influence of the average structure of the core of the membrane, a liquid-crystalline fluid environment of partially ordered hydrocarbon chains, on protein–protein association?

Theoretical investigations of this problem were initiated by Marcelja in 1976, who proposed a mean-field model of a lipid bilayer based on order parameters related to lipid chain conformational states.[6] His theory described the effects resulting from a non-specific interaction between membrane integral proteins and the surrounding lipids. The model assumed that the most important change in lipid structure was restricted to the annulus of those lipid chains which are in direct contact with the protein. However, at temperatures above the main gel/liquid crystal phase transition, it was found that the disturbance caused by the protein inclusion extended to the second or third neighboring chains. Marcelja showed that the change in lipid order gave rise to an indirect lipid-mediated interaction between membrane integral proteins, leading to a monotonically attractive potential between two proteins embedded in a membrane in the fluid state with a free energy well of 2–3 $k_B T$ at contact.

Different approaches, again using mean-field theories, were due to Schröder,[7] Owicki et al.[8,9] and Pearson et al.[10] In these theories, the state of the lipid bilayer is characterized by several spatially inhomogeneous "order parameters" which are directly related to fluctuations in the lateral density of the lipid chains. The equation for the spatial variation of the order parameter field is derived from a Landau–de Gennes free energy functional with a limited gradient expansion.[11] Interaction strengths and correlation lengths for the pure bilayer are described in terms of phenomenological parameters in these theories. Similar ideas were used in more recent theoretical work which incorporates the influence of membrane stretching, bending moduli, and the spontaneous curvature.[12–14]

These earlier theoretical treatments generally predicted an attractive lipid-mediated protein–protein interaction which decays monotonically as a function of distance for two proteins embedded in a membrane in the fluid state (though the interaction was often found to depend on the state of the bilayer, see ref. 12, 14). Given that the influence of a protein inclusion is to perturb the natural state of the bilayer, the indirect interaction between membrane proteins was obtained under the assumption that fluctuations are suppressed in the vicinity of the proteins. The lipid-mediated protein–protein interaction is then caused by the overlap of the annuli and its range was found to increase with increasing correlation length. Typically, the magnitude of this interaction was on the order of a few $k_B T$ at protein contact.

One important limitation of these previous studies is the significant amount of simplification that has to be introduced in order to construct tractable analytical theories. The bilayer membrane, a complex macromolecular assembly of amphiphilic phospholipid molecules, is thus described in terms of a necessarily limited phenomenological free energy functional. In contrast a realistic approach for the study of the structure and dynamics of biological membranes is the use of molecular dynamics simulations based on detailed inter-atomic potentials.[15] In the last few years, studies of pure lipid bilayers[16,17] and protein–membrane systems[18–20] have demonstrated the feasibility and success of such detailed simulations (see also ref. 15 and references therein). In principle, free energy molecular simulations and perturbation techniques could be used to calculate the solvation free energy of inclusions.[21] However, molecular dynamics simulations are computationally intensive and cannot be used to address such aspects of membrane structure. It is at present not feasible to examine long-range protein–protein interactions embedded in a membrane with molecular dynamics simulations due to the very long time scales involved in the relaxation of the lipids. Progress on general questions about protein–protein interactions clearly requires a different approach. Recently Sintes and Baumgärtner[22] examined the problem of lipid-mediated protein interactions by using Monte Carlo computer simulations based on a simple model of the lipid membrane. The model represented the bilayer with 2 × 500 lipid molecules and each molecule was modeled by a flexible chain with five monomers; two hard cylinders represented the proteins. The simulations gave a depletion-induced attraction between proteins lying closer than one lipid diameter and a fluctuation-induced attraction for larger inter-protein displacements. The latter has a correlation length of about three lipid diameters. However the main limitation of the approach was again that the description of the membrane lipids was necessarily simplified to decrease the computational cost.

In this short paper we propose and develop a different approach to examine the influence of lipid chains on protein–protein interactions. The present approach is based on statistical mechanical theories involving integral equations which were developed for the study of liquids[23,24] and which use recent computational advances for multi-dimensional systems.[25,26] The integral equation theory presented here is constructed as a hypernetted chain (HNC) equation projected in the two-dimensional space of the lipid membrane plane. The exact lateral density–density response function of the hydrocarbon core is used as an input to this theory. This response function is closely related to the membrane structure factor which can be extracted from X-ray or neutron scattering measurements.[23] However, since such experimental data are not presently available the response function was calculated from the configurations of a molecular dynamics simulation of a lipid membrane.[17] Our HNC integral equations then allowed us to calculate the perturbation of the hydrocarbon core around a protein inclusion and the lipid-mediated protein–protein interaction based on the information extracted from the molecular dynamics simulation of the lipid membrane. From this point of view, the present theory offers an intermediate approach, combining aspects of both mean-field theories and fully detailed atomic simulations. Our integral equation for the average structure of membranes under the influence of protein inclusion is similar, in spirit, to the theory developed by Pratt and Chandler to describe the hydrophobic effect.[27] In their theory, the experimental oxygen–oxygen pair correlation function was used as an input to calculate the free energy of non-polar solutes in water. To illustrate the present approach, the lateral perturbations on the membrane structure as well as the lipid-mediated protein–protein interaction were calculated for a 5 Å radius cylinder representing an α-helical polyleucine. In Section II, we develop the theoretical formulation of our approach and preliminary results for a dipalmitoyl phosphatidylcholine (DPPC) bilayer are described in Section III. The paper is concluded with a brief summary and a discussion of future work.

2 Integral equation theory

We consider isolated protein inclusions embedded in a uniform membrane in the fluid state. It is assumed that the carbons along the lipid acyl chain are affected by a repulsive potential $u(r)$ due to the presence of the protein inclusions. For the sake of simplicity, we assume that the protein inclusions are hard repulsive objects which only interact with the hydrocarbon chains and that the polar headgroups are not directly affected by the protein. Furthermore, we assume that the perturbation acts only on the lateral positions of the lipids. In this case we require an integral equation which allows us to calculate the average carbon density projected in the two dimensional membrane plane $\langle \rho(r) \rangle$, where $r \equiv (x, y)$. We begin by writing an expression for the free energy density functional theory for non-uniform liquids in the HNC approximation,[23,24]

$$\mathscr{A}[\langle \rho(r) \rangle] = \int dr \, \langle \rho(r) \rangle \ln\left[\frac{\langle \rho(r) \rangle}{\bar{\rho}}\right] - \Delta\rho(r) + \beta u(r)\langle \rho(r) \rangle$$
$$- \frac{1}{2} \int dr \int dr' \, \Delta\rho(r) C_m(|r - r'|) \Delta\rho(r') \quad (1)$$

where $\bar{\rho}$ is the density of the hydrocarbon chains in the uniform two-dimensional membrane plane, $\Delta\rho(r) = \langle \rho(r) \rangle - \bar{\rho}$ is the deviation from the uniform density $\bar{\rho}$. The direct lipid–lipid correlation function $C_m(|r - r'|)$ is defined in terms of $\chi_m(r)$, the equilibrium carbon–carbon density susceptibility of the uniform unperturbed membrane

$$C_m(r) = (\bar{\rho})^{-1}\delta(r) - \chi_m^{-1}(r) \quad (2)$$

$\chi_m(r)$ is a response function which is related to the density fluctuations of carbon pairs in the unperturbed bilayer (see below). According to the free energy variational principle,[24] the average density is obtained by minimization of the functional $\mathscr{A}$ with respect to the functions $\langle \rho(r) \rangle$. This leads to the HNC-like integral equation

$$\langle \rho(r) \rangle = \bar{\rho} \exp\left[-\beta u(r) + \int dr' \, C_m(|r - r'|) \Delta\rho(r')\right] \quad (3)$$

which must be solved self-consistently. The integral equation can be rewritten in a form more suitable for numerical calculations as a pair of coupled 2d-HNC equations,

$$c(r) = \exp[-\beta u(r) + h(r) - c(r)] - h(r) + c(r) - 1 \quad (4)$$

and

$$\bar{\rho} h(r) = \int dr' \, c(|r - r'|)\chi_m(r) \quad (5)$$

where $h(r) \equiv \Delta\rho(r)/\bar{\rho}$ is the protein–lipid correlation function, and $c(r)$ is the direct protein–lipid correlation function. Eqn. (5) is the well-known Ornstein–Zernike equation[23] for an isolated impurity in an infinite bulk system. The excess Helmholtz free energy due to the protein inclusion is obtained by substituting the self-consistent solution to eqn. (3) into the free energy functional $\mathscr{A}$ of eqn. (1). This leads to the closed form expression

$$\beta\Delta\mathscr{A} = \bar{\rho}\int dr\{\tfrac{1}{2}[h(r)]^2 - \tfrac{1}{2}h(r)c(r) - c(r)\} \qquad (6)$$

This expression is used to calculate the lipid-mediated protein–protein interaction.

Pair correlation function of the unperturbed membrane

A central quantity in the present theory is the response function of the uniform unperturbed membrane, $\chi_m(r)$ defined in eqn. (2). Functions such as $\chi_m(r)$ play a central role in the response of the average structure of an equilibrium system to a small perturbation.[23] The response function χ_m is related to lipid density–density fluctuations of carbon pairs in the unperturbed bilayer at equilibrium,

$$\chi_m(|r - r'|) = \langle(\rho(r) - \langle\rho(r)\rangle)(\rho(r') - \langle\rho(r')\rangle)\rangle$$
$$= \langle\rho(r)\rho(r')\rangle - \langle\rho(r)\rangle\langle\rho(r')\rangle \qquad (7)$$

where $\rho(r)$ is the density of the ensemble of carbon atoms comprising the lipid chains,

$$\rho(r) = \sum_i \sum_{\alpha=1}^{n} \delta(r_\alpha^{(i)} - r) \qquad (8)$$

Here i is the index of the lipid molecules and α, which goes from 1 to n, is the index of the carbon atom along the lipid chains. In the uniform unperturbed system, the average of $\rho(r)$ is the average carbon density per unit area,

$$\langle\rho(r)\rangle = \sum_i \sum_{\alpha=1}^{n} \langle\delta(r_\alpha^{(i)} - r)\rangle$$
$$= \bar{\rho} \qquad (9)$$

The average density $\bar{\rho}$ is equal to $2 \times n \times$ the surface density of lipid molecule per leaflet, where n is the number of carbon atoms in the hydrophobic moiety of one DPPC molecule (the factor 2 appears because of the upper and lower leaflets of the bilayer).

Density–density fluctuations of carbon pairs can also be expressed in terms of the radial intramolecular and intermolecular pair correlation functions of the pure unperturbed membrane,

$$\chi_m(r - r') = \left\langle\sum_i \sum_{\alpha=1}^{n} \sum_j \sum_{\gamma=1}^{n} \delta(r_\alpha^{(i)} - r)\delta(r_\gamma^{(j)} - r')\right\rangle - \bar{\rho}\bar{\rho}$$
$$= \bar{\rho}[\delta(r - r') + S_m(r - r') + H_m(r - r')\bar{\rho}] \qquad (10)$$

where the function $S_m(r - r')$ represents the carbon–carbon intramolecular pair correlation within a given lipid molecule ($i = j$), while the function $H_m(r - r')$ represents the carbon–carbon intermolecular pair correlation between distinct lipids ($i \neq j$). By symmetry, the pair correlation functions depend only on the distance r projected in the x,y plane, with $r = \sqrt{(x - x')^2 + (y - y')^2}$.

In the present study, the pair correlation functions were calculated from the configurations generated by molecular dynamics simulations of a detailed atomic model of a pure DPPC bilayer at 323.15 K performed by Feller et al.[17] The function $S_m(r)$ was calculated as

$$S_m(r) = \left\langle\frac{1}{n}\sum_\alpha \frac{N_\alpha^{\text{intra}}(r; r + \Delta r)}{a(r; r + \Delta r)}\right\rangle \qquad (11)$$

where $N_\alpha^{\text{intra}}(r; r + \Delta r)$ is the total number of carbons from a given lipid found within the two-dimensional annulus going from r to $r + \Delta r$ centered around carbon α, and

$a(r; r + \Delta r) = 2\pi[(r + \Delta r)^2 - r^2]$ is the area of the annulus. The function $H_m(r)$ was calculated as

$$H_m(r) = \left\langle \frac{1}{n} \sum_\alpha \frac{N_\alpha^{\text{inter}}(r; r + \Delta r)}{a(r; r + \Delta r)\bar{\rho}} \right\rangle - 1 \qquad (12)$$

where $N_\alpha^{\text{inter}}(r; r + \Delta r)$ is the number of carbons from the other lipids found within the two-dimensional annulus going from r to $r + \Delta r$ centered around carbon α of a lipid molecule. At large values of r, intermolecular correlations vanish and the function $H_m(r) \to 0$.

All the heavy atoms from the acyl chains and glycerol backbone were counted in the calculation of the pair correlation function for a total 39 particles per lipid. The polar headgroup was not included. Since the average cross-sectional area is 62.9 Å^2 per DPPC,[17] the average density per unit area $\bar{\rho}$ is equal to 1.24 Å^{-2}.

Computational details

The 2d-HNC eqn. (4) and (5) were solved for two different systems. First, a single isolated α-helical protein inclusion modeled as a hard repulsive cylinder was considered. Secondly, two identical cylindrical protein inclusions were examined at various separations and the lipid-mediated protein–protein free energy was calculated using eqn. (6). The radius of the hard cylinder was chosen to be 5 Å, corresponding closely to that of a polyleucine α-helix.

Eqn. (4) and (5) were solved numerically by a method used previously to solve HNC integral equations.[25,26] It involves a mapping of all functions onto a two-dimensional discrete grid, e.g., $u(x, y) \to u(i, j)$, and $h(x, y) \to h(i, j)$. Two grid dimensions were used, depending on the number of proteins inserted in the system. A discrete grid with $N = 1024 \times 1024$ with a spacing d of 0.10 Å was used with single protein systems, and $N = 2048 \times 1024$ with the same d spacing was used for two protein systems. The two-dimensional convolution in eqn. (5) was calculated using a numerical two-dimensional fast Fourier transform (FFT) procedure.[28] The convolution was calculated directly, without zero-padding. This corresponds to a periodic system in the x and y directions. An iterative scheme with simple mixing was used to solve eqn. (4) and (5) self-consistently. In this scheme, the mth iteration is obtained from

$$c^{(m+1)} = \lambda\{\exp[-\beta u + h^{(m)} - c^{(m)}] - 1 - h^{(m)} + c^{(m)}\} + (1 - \lambda)c^{(m)} \qquad (13)$$

Approximately 50 iterations were necessary for convergence. Numerical solution takes 5 min on a Pentium II 400 mHz.

3 Results and discussion

Fig. 1 gives the results for the carbon-carbon distribution function as extracted from the molecular dynamics simulations of Feller et al.[17] The figure shows that the carbon–carbon intramolecular correlation function, $S_m(r)$, is positive definite. This correlation function has a large contribution followed by a peak up to 3 Å and then a slow decay over a distance of 10-15 Å. The short range contribution to the intramolecular correlation arises from nearest neighbor carbons along the acyl chains (i.e., carbon i with carbon $i - 1$ and $i + 1$). Such a peak is an indication of a significant amount of short range order in the lipid chains perpendicular to the plane of the bilayer. The long range contribution, which extends to 15 Å, is due to carbons located in different acyl chains of a single lipid molecule. The carbon–carbon intermolecular correlation function $H_m(r)$ has a strong negative contribution at short distances due to the lipid–lipid core repulsion. Interestingly, the negative contributions are much less important at distances greater than 5 Å. As a result, the response function $\chi_m(r)$ exhibits a strong peak for distances up to 3 Å, arising from the intramolecular correlations and a second strong peak at a distance of 4–5 Å arising from the intermolecular correlations. The response function decays in an oscillatory manner, with small positive peaks appearing around 9 and 14 Å.

The observed structure of the correlation functions of a pure DPPC membrane suggests that the expected response to perturbations may be quite complex. To assess the lateral response of the bilayer quantitatively, the average density around an α-helical protein inclusion modeled as a hard cylinder was examined. The pair correlation $h(r)$ for the distribution of carbon atoms around the cylinder as calculated using the integral eqn. (4) and (5) of Section II is shown in Fig. 2. It can be

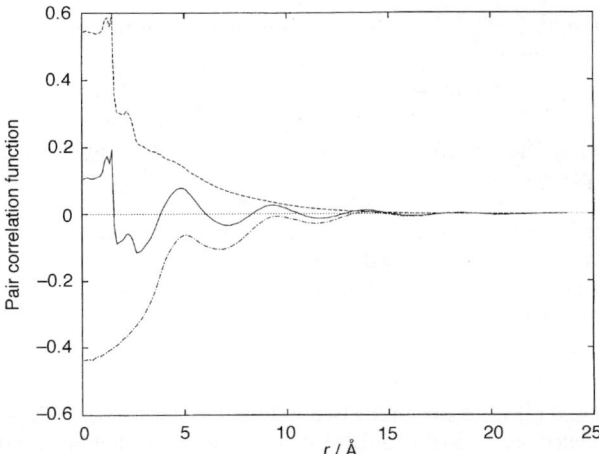

Fig. 1 Carbon–carbon density–density response function $\chi_m(r)$ (solid line) extracted from molecular dynamics simulations and used as input in the integral equation. The intramolecular $S_m(r)$ (dashed line) and intermolecular, $H_m(r)$ (dash-dotted line) pair correlation functions extracted from the molecular dynamics simulations of Feller *et al.*[17] are shown. (To match the dimensions of Å^{-2} of $S_m(r)$, the function $\chi_m/\bar{\rho}$ and the function $H_m(r) \times \bar{\rho}$ are shown.)

seen from the figure that the perturbation of the membrane structure extends from 5 to 25 Å from the center of the cylinder with strong oscillations in the correlation function separated by approximately 4 Å. Furthermore the correlation function is negative between 5 and 15 Å from the center of the cylinder. This indicates that the average lipid density next to the cylinder is lower than its bulk value. This region next to the cylinder can thus be regarded as a depletion layer with respect to the lipid molecules. It should be noted that no lipid can be closer than 5 Å to the center of the cylinder because of the hard core repulsive potential and the protein–lipid correlation function is therefore -1 inside the cylinder. The pair correlation function is positive from 15 to 25 Å which indicates that the density of carbon atoms is higher than the uniform bulk value in this interval.

The integral equation theory and the free energy expression of eqn. (6) can be used to calculate the lipid-mediated free energy between two protein inclusions as a function of their separation. The results are given in Fig. 3. The figure shows that there is a free energy barrier at a distance of 20 Å between the two proteins, followed by an attractive free energy well for distances less than

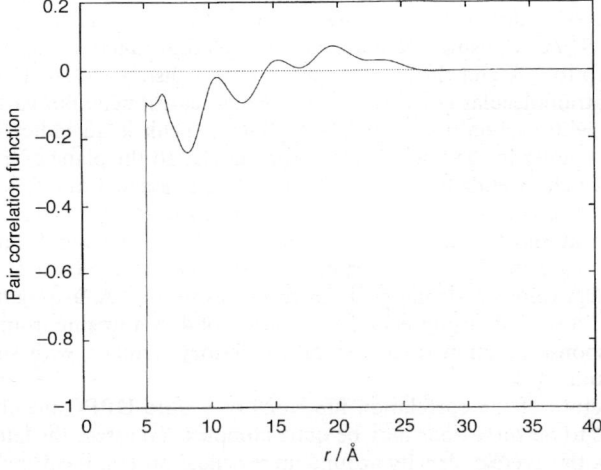

Fig. 2 Perturbation of the average structure of the hydrocarbon core around a hard repulsive cylinder of 5 Å radius.

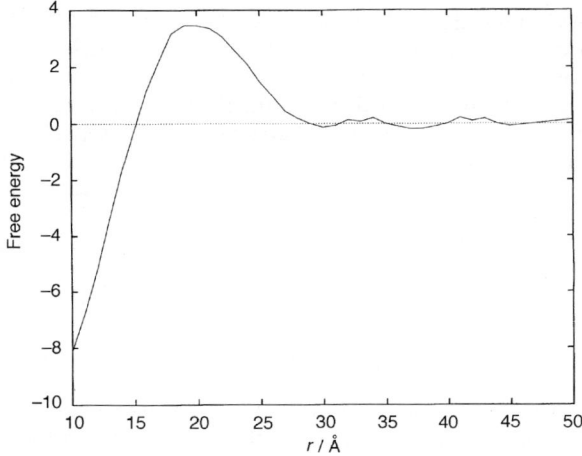

Fig. 3 Lipid-mediated free energy (in units of $k_B T$) between two hard repulsive cylinders of 5 Å radius as a function of their separation distance.

15 Å. The repulsive barrier is approximately 3.5 $k_B T$ and extends from 15 to 30 Å. At separations greater than 30 Å, the protein–protein potential is very small and oscillatory. Finally, at protein–protein contact, the magnitude of the lipid-mediated potential is -8 $k_B T$. This value is in good agreement with (though somewhat more negative than) previous estimates.[6–10,22] However, in contrast to previous studies, the results based on the integral equation theory show that the lipid-mediated protein–protein interaction is both attractive at short distances between proteins and repulsive at large distances. The attractive part of the interaction is clearly due to the presence of a depletion layer of lipid molecules close to the embedded protein. This is in good agreement with the results of Sintes and Baumgärtner[22] mentioned above. The repulsive part of the interaction has not been observed previously and has important consequences for protein association in biological membranes. The free energy maximum at 20 Å corresponds to the distance at which it is increasingly more difficult to fit one last lipid molecule between the cylinders. At 20 Å separation, there is a 10 Å free space between the two 5 Å radius cylinders. According to the cross-sectional area per lipid (62.9 Å^2), the radius of one DPPC is approximately 4.5 Å.[17] This is basically the spatial range of the dominant contribution to the membrane response function $\chi_m(r)$ observed in Fig. 1. Thus, a reasonable lengthscale for protein–protein interactions arises naturally in the present integral equation theory.

The present model accounts only for steric excluded volume interactions of the protein with the hydrocarbon chains. In particular, electrostatic interactions were ignored. Ben-Tal and Honig[29] calculated the electrostatic contribution to helix–helix interactions using a continuum electrostatic model.[29] Their results led to the conclusion that there is a non-specific attractive interaction on the order of 1 to 2 kcal mol^{-1} between transmembrane helices in an anti-parallel configuration. This suggests that the magnitude of our lipid-mediated potential is thus on the order of the electrostatic interaction between α-helices embedded in a membrane.

4 Conclusion

A theory for examining the structure of the hydrocarbon chains around a protein inclusion embedded in a lipid bilayer was developed on the basis of the hypernetted chain integral equation theory for liquids. The exact lateral density–density response function of the hydrocarbon core, which is extracted from a molecular dynamics simulation of a pure lipid bilayer, was used as an input in the calculations. Although the integral equation resembles that of a simple HNC theory for a two-dimensional liquid, both intramolecular and intermolecular correlations of the lipid molecules were included through the response function χ_m which was calculated using configurations taken from the molecular dynamics simulation of a DPPC lipid bilayer membrane by Feller et al.[17]

To illustrate our new approach, we first examined the structure of the DPPC bilayer in the neighborhood of a hard 5 Å radius cylindrical inclusion which modeled an α-helical polyleucine protein. The results showed that the average lipid order is perturbed over a distance of 25 to 30 Å from the center of the protein. We then calculated the lipid-mediated protein–protein interaction and the results suggest that there is a free energy barrier for protein separations between 15 and 30 Å which inhibits protein–protein association. In contrast, the interaction is attractive at distances between proteins which are less than 15 Å. In particular the calculations indicate that the lipid matrix gives rise to a free energy well of 8 $k_B T$ for the association of two hard cylindrical inclusions of 5 Å radius. It should be stressed that this result was obtained using a two-dimensional discrete grid of approximately 204 Å by 102 Å. More than 600 DPPC molecules with 24 000 water molecules would have been required to assemble an atomic model of a bilayer system of corresponding size for molecular dynamics simulations (*i.e.*, more than 150 000 atoms). Therefore, it is clear that the present approach permits the investigation of factors that are not readily accessible by straight molecular dynamics simulations.

In the future, the dependence of the lipid-mediated interaction upon the protein size will be investigated. In addition other familiar closures, such as the Percus–Yevick (PY) integral equation,[23] will be examined. Finally, further developments of the integral equation theory to include the lateral and transversal response of the bilayer are in progress.

Acknowledgements

We are grateful to S. Feller, R. M. Venable and R. W. Pastor for making their trajectory of a DPPC bilayer membrane available. Financial support from NSERC (Canada) and FCAR (Québec) is acknowledged. B.R. is a research fellow of the Medical Research Council of Canada. M.J.Z. is an associate of the Canadian Institute of Advanced Research.

References

1. W. Kleeman and H. M. McConnell, *BBA*, 1974, **345**, 220.
2. B. A. Lewis and D. M. Engelman, *J. Mol. Biol.*, 1984, **166**, 203.
3. M. A. Lemmon and D. M. Engelman, *Quart. Rev. Biophys.*, 1994, **27**, 157.
4. C. Nielsen, M. Goulian and O. S. Andersen, *Biophys. J.*, 1998, **74**, 1966.
5. R. B. Gennis, *Biomembranes Molecular Structure and Function*, Springer-Verlag, New York, 1989.
6. S. Marcelja, *Biochim. Biophys. Acta*, 1976, **455**, 1.
7. H. Schröder, *J. Chem. Phys.*, 1977, **67**, 1617.
8. J. C. Owicki, M. W. Springgate and H. M. McConnell, *PNAS*, 1978, **75**, 1616.
9. J. C. Owicki and H. M. McConnell, *PNAS*, 1979, **76**, 4750.
10. L. T. Pearson, J. Edelman and S. I. Chan, *Biophys. J.*, 1984, **45**, 863.
11. P. G. deGennes, *The Physics of Liquid Crystals*, Oxford University Press, London, 1974.
12. M. Goulian, R. Bruinsma and P. Pincus, *Europhys. Lett.*, 1993, **22**, 145.
13. P. A. Kralchevsky, V. N. Paunov and N. D. Denkov, *J. Chem. Soc., Faraday Trans.*, 1995, **91**, 3415.
14. H. Aranda-Espinoza, A. Berman, N. Dan, P. Pincus and S. Safran, *Biophys. J.*, 1996, **71**, 648.
15. *Biological Membranes. A molecular perspective from computation and experiment*, ed. K. M. In Merz and B. Roux, Birkhauser, Boston, 1996.
16. E. Egberts and H. J. C. Berendsen, *J. Chem. Phys.*, 1988, **89**, 3718.
17. S. E. Feller, R. M. Venable and R. W. Pastor, *Langmuir*, 1997, **13**, 6555.
18. T. B. Woolf and B. Roux, *Proteins: Struct. Funct. Genet.*, 1996, **24**, 92.
19. L. Shen, D. Bassolino and T. Stouch, *Biophys. J.*, 1997, **73**, 3.
20. D. P. Tieleman and H. J. C. Berendsen, *Biophys. J.*, 1998, **74**, 2786.
21. J. P. Postma, H. C. Berendsen and J. R. Haak, *Faraday Symp. Chem. Soc.*, 1982, **17**, 55.
22. T. Sintes and A. Baumgärtner, *Biophys. J.*, 1997, **73**, 2251.
23. J. P. Hansen and I. R. McDonald, *Theory of Simple Liquids*, 2nd edn., Academic Press Inc., San Diego, 1986.
24. D. Chandler, J. D. McCoy and S. J. Singer, *J. Chem. Phys.*, 1986, **85**, 5971.
25. D. Beglov and B. Roux, *J. Chem. Phys.*, 1995, **103**, 360.
26. D. Beglov and B. Roux, *J. Phys. Chem.*, 1997, **101**, 7821.
27. L. W. Pratt and D. Chandler, *J. Chem. Phys.*, 1977, **67**, 3683.
28. M. Frigo and S. G. Johnson, *ICASSP Conference Proceedings*, 1998, **3**, 1382.
29. N. Ben-Tal and B. Honig, *Biophys. J.*, 1996, **71**, 3046.

Paper 8/07109H

Lipid packing stress and polypeptide aggregation: alamethicin channel probed by proton titration of lipid charge

Sergey M. Bezrukov,[a] R. Peter Rand,[b] Igor Vodyanoy[c] and V. Adrian Parsegian[d]

[a] *NICHD, National Institutes of Health, Bethesda, MD 20892-0924, USA and St. Petersburg Nuclear Physics Institute, Gatchina, 188350 Russia*
[b] *Biological Sciences, Brock University, St. Catharines, Ontario L2S 3A1, Canada*
[c] *Office of Naval Research Europe, London, UK NW1 5TH*
[d] *NICHD, National Institutes of Health, Bethesda, MD 20892-5626, USA*

Received 20th August 1998

Lipid membranes are not passive, neutral scaffolds to hold membrane proteins. In order to examine the influence of lipid packing energetics on ion channel expression, we study the relative probabilities of alamethicin channel formation in dioleoylphosphatidylserine (DOPS) bilayers as a function of pH. The rationale for this strategy is our earlier finding that the higher-conductance states, corresponding to larger polypeptide aggregates, are more likely to occur in the presence of lipids prone to hexagonal H_{II}-phase formation (specifically DOPE), than in the presence of lamellar L_α-forming lipids (DOPC). In low ionic strength NaCl solutions at neutral pH, the open channel in DOPS membranes spends most of its time in states of lower conductance and resembles alamethicin channels in DOPC; at lower pH, where the lipid polar groups are neutralized, the channel probability distribution resembles that in DOPE. X-Ray diffraction studies on DOPS show a progressive decrease in the intrinsic curvature of the constituent monolayers as well as a decreased probability of H_{II}-phase formation when the charged lipid fraction is increased. We explore how proton titration of DOPS affects lipid packing energetics, and how these energetics couple titration to channel formation.

Introduction

The evidence mounts. Membrane lipids are not just a filler or an inert solvent for membrane proteins; they are functionally involved. Interactions between lipids and embedded proteins control conformational equilibrium between different functional states of proteins. Among natural membrane proteins one clear example is the shift between the Meta-I and Meta-II forms of Rhodopsin with varied lipid species in the host membrane. The amount of the Meta-II form increases with the increase in phosphatidylcholine acyl chain unsaturation,[1] thus demonstrating the important role of lipids in modulating membrane-signaling systems. Many other examples of the crucial role of lipid–protein interactions in enzymatic reactions and receptor regulation can be found in a recent review.[2]

It is well known that the conductance, lifetime, and formation 'on-rate' of the channel-forming drug gramicidin A[3] depend on host-lipid species. In addition to electrostatic effects of surface charge[4,5] and poorly understood effects of neutral lipids[6–10] on channel conductance, extensive systematic studies of the gramicidin channel in different host-lipid compositions have shown that its properties can be controlled by purely mechanical parameters.[11] It was shown that gramicidin channel life-time and free energy of dimerization are modified by bilayer curvature stress[12] and membrane tension[13] in a quantitatively predictable way.

Channels formed by the 20-amino acid peptide alamethicin[14–16] also show properties that depend on membrane lipid composition[17–20] or on tension applied to the bilayer.[21] A clear correlation between the tendency of lipids to form the inverted hexagonal phase and the expression of higher-conductance states of alamethicin peptide channels in those lipids was demonstrated in studies with dioleoylphosphatidylethanolamine (DOPE)/dioleoylphosphatidylcholine (DOPC) mixtures.[22] Alamethicin was inserted into bilayer membranes composed of lipids of empirically determined inverted hexagonal phase spontaneous radii. Lipids with different spontaneous radii form planar membranes with expectedly different degrees of stress of forcing lipids into a planar structure. It was found that this mechanical parameter of the host-lipid bilayer plays a crucial role in alamethicin channel formation. In particular, states of higher conductance were found to be much more probable in DOPE, a lipid of high curvature, than in DOPC, lipid of low curvature. In the case of mixtures, the relative probability of states was a monotonic function of the DOPE/DOPC ratio.[22]

In this paper, whose preliminary version was reported elsewhere,[23] we demonstrate that a continuous variation of factors that stress membrane structure can direct the conformational equilibrium of channel-forming peptides. Specifically, we insert alamethicin into DOPS bilayers and change the bilayer surface charge and lipid head group electrostatic interactions with varied pH and varied salt concentration. Dramatic changes in relative probabilities of channel conductance states observed as a result of such manipulations provide further evidence of the importance of host-membrane mechanical parameters for channel protein function.

By X-ray diffraction we show that the decreased electrostatic energy of the polar surface (that goes with the decreased charge at low pH values) shifts DOPS from the purely lamellar form seen at neutral pH to an H_{II} phase of ever-higher spontaneous curvature. This shift agrees well with our transport measurements performed on the same lipids. In 0.1–0.3 M sodium chloride solutions at neutral pH, alamethicin channels exhibit the DOPC-like pattern expected for lamellar lipids; in acidic solutions they show the DOPE-like pattern expected for H_{II}-prone lipids of high spontaneous curvature.[22] Specifically, we find that the higher conductance states of the channel are expressed much more at pH 2.0 than at pH 6.0. The corresponding observed 50-fold change in the relative probability of a particular state *vs.* the adjacent state suggests that there is a change of ~ 4 kT in the difference between free energies of adjacent states. As expected, a qualitatively similar change was observed when salt concentration was increased to 1 or 2 M.

The energies of the different functional states are the most important factors in protein regulation.[24] For membrane proteins these energies depend on the lipid molecules outside the protein. We see now that it is possible to regulate the energies of these lipids themselves to modulate their influence on proteins.

Materials and methods

Alamethicin channels were inserted into 'solvent-free' planar lipid bilayer membranes that had been formed by apposition of two phospholipid monolayers spread on aqueous solutions of sodium chloride (Baker Analyzed grade, Baker, Phillipsburg, NJ, USA). The monolayers were prepared from 10% DOPS or DOPE (Avanti Polar Lipids, Alabaster, AL, USA) in pentane (HPLC grade, Burdick and Jackson, Muskegon, MI, USA). The Teflon chamber[25] (after Montal and Mueller[26]) with two compartments of 1 ml was divided by 15 µm thick Teflon partition (CHEMFAB, Merrimack, NH) with a 60 µm diameter aperture. The aperture was pretreated with 1% solution of hexadecane (Aldrich, Milwaukee, WI, USA) in pentane and dried during 10 min prior to monolayer opposition. The same partition was used throughout all measurements reported in this paper.

Natural alamethicin (Sigma, St. Louis, MO, USA) was added only to one side of a membrane from 10^{-5} M stock solution in ethanol to a final concentration of $(1-3) \times 10^{-8}$ M. All experiments were done at 150 mV, positive from the side of alamethicin addition, and at a room temperature of $(23 \pm 1)\,°C$. Alamethicin concentration was adjusted to a concentration that gave first current bursts in about 20 min after peptide addition; in this way we were able to monitor single-channel activity (no channel overlapping) for about 10 min. Ion currents, amplified with an Axopatch 200A integrating patch clamp amplifier (Axon Instruments, Foster City, CA, USA), were recorded with a sampling rate of 50 kHz into computer memory and, simultaneously, onto recordable compact discs.

Statistical analysis of state probabilities was performed using direct comparison of the time spent by a channel at different conductance states (levels). First, current histograms were plotted and appropriate windows around each state were determined. Second, the total numbers of points within each such window were calculated and their ratios were taken to represent the relative probabilities of corresponding states. Each point in a relative probability graph represents averaging over more than 100 channels that were obtained, typically, from one membrane. A new membrane was formed for every pH or salt concentration.

Results

Typical recordings of alamethicin-induced currents in DOPS bilayers at different pH are shown in Fig. 1. They demonstrate that the probabilistic character of a conductance burst, corresponding to a single alamethicin channel, is very sensitive to membrane-bathing solution acidity. At relatively high acidity (pH 2.0), when lipid charge of the membrane is mostly neutralized by protons, a typical channel undergoes many transitions between different conductance states. Higher conductance states (labeled 4, 5) are well-expressed and are typically observed in every current burst. Increased pH and, presumably, increased lipid charge progressively suppress higher states. At pH 5.0 a typical channel goes only to Level 0 and back to the closed state. Note that the current burst at pH 2.0 represents a single ion channel of fluctuating size. Conductance increments, corresponding to channel transition to the next higher-conductive state increase with the level number from 0.104 nS (background to Level 0 transition) to 0.51 nS (Level 4 to Level 5 transition). If the burst were representing several identical channels occurring at the same time, the increments would be equal or would decrease with level number due to interference of ion currents in access areas.

While the probabilistic character of a single-channel burst changes dramatically with pH (Fig. 1), the acidity of the medium only slightly influences the channel conductance itself. Fig. 2 shows that at pH 2.0 all levels exhibit a conductance increase. This conductance increase is several times higher than the corresponding increase in solution specific conductivity (data not shown). The

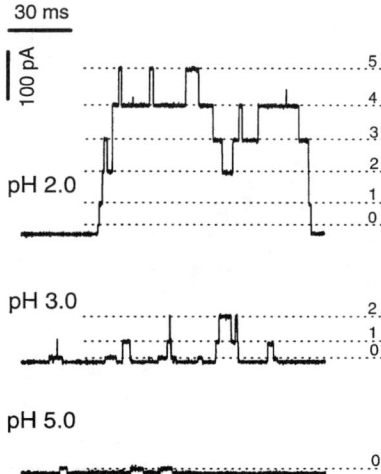

Fig. 1 Typical current bursts representing alamethicin channels in DOPS membranes bathed by 0.3 M NaCl at three different pH values. Current is displayed with a 50 μs resolution. Horizontal dotted lines with numbers show conductance states' (levels') notation used in this paper. The pH-dependent character of current bursts corresponding to single alamethicin channels is clearly seen. At pH 2.0, the current burst always appears through the lowest conductance state, Level 0, fluctuates between several higher conductance states (Level 1 to Level 5, or even higher), and then disappears. At pH 5.0, a typical channel is seen at Level 0 only.

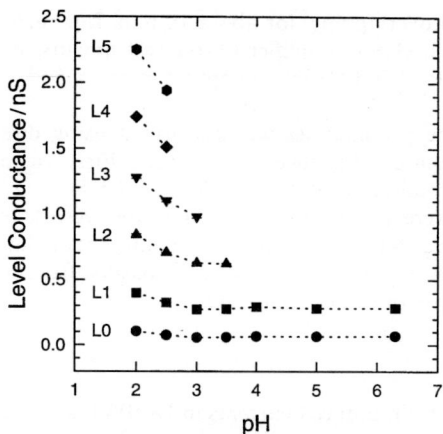

Fig. 2 Conductance of alamethicin channel levels in DOPS membranes bathed by 0.3 M NaCl as a function of pH. Due to the strong dependence of level probability on acidity of the medium (Fig. 1), conductance of higher levels could be measured only low pH. At pH above pH 3.5, levels higher than Level 1 were virtually nonexistent. Attempts to increase the number of channels per unit time to resolve these levels led to channel overlapping and smearing of current histograms.

disparity is probably related to preferential transport of protons *vs.* sodium cations. At pH 3.0, lower levels show a small dip that reflects titration of the membrane surface charge and corresponding depletion in counterion concentration. This effect is similar to the recently reported titration of gramicidin channel conductance.[5]

The 'smooth' dependence of channel conductance on the acidity of the medium is helpful for statistical analysis of relative probabilities. Fig. 3 demonstrates the results of such an analysis of relative probabilities to quantify the pH-dependence clearly seen 'by naked eye' in Fig. 1. The change in pH from 2.0 to 6.3 changes the relative probability of Level 1 *vs.* Level 0 observation (filled symbols) by a factor of $e^4 \approx 50$. Most of the probability change occurs between pH 2.0 and 4.0, that is, within the range that includes the lipid's hydroxy group pK_a.[5] Relative probability of Level 2 *vs.* Level 1 (open symbols) changes similarly and, for some reason, is very close to that of Level 1 *vs.* Level 0 in its absolute value.

Fig. 4 shows that substituting 0.1 M NaCl for 0.3 M NaCl does not have any statistically significant effect on the structure of channel probabilities. Within error bars the quantitative

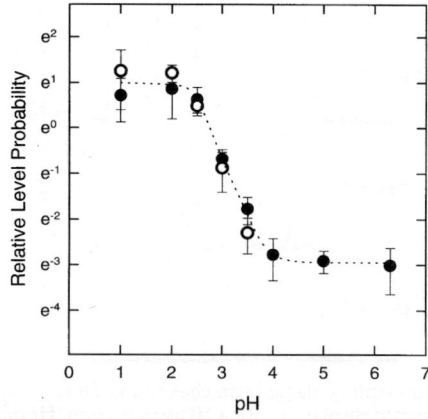

Fig. 3 Relative probability of channel levels in DOPS membranes bathed by 0.3 M NaCl as a function of pH. Filled symbols, probability of Level 1 *vs.* Level 0; open symbols, probability of Level 2 *vs.* Level 1. Higher-level relative probabilities decrease with pH increase; most of the probability change (about four orders of natural logarithm base) occurs between pH 2.0 and 4.0.

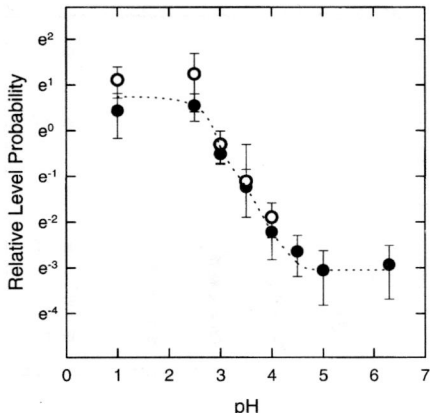

Fig. 4 Relative probability of channel levels in DOPS membranes in 0.1 M NaCl shows behavior much like that in 0.3 M NaCl (Fig. 3). Filled symbols, probability of Level 1 vs. Level 0; open symbols, probability of Level 2 vs. Level 1.

behaviour of levels' relative probability as a function of pH is the same at these salt concentrations. As shown, the relative probability of Level 2 vs. Level 1 (open symbols) in 0.1 M NaCl closely follows the relative probability of Level 1 vs. Level 0 (closed symbols).

To check for a possible direct influence of acidity or salt concentration on the probabilistic character of the alamethicin channel, we ran a series of control measurements with neutral DOPE bilayers. Fig. 5 displays a typical channel in 2 M NaCl exhibiting well-defined conductance levels. It is seen that the higher levels are quite probable. This result agrees with earlier observations[22] though in the present study we use a neutral form of alamethicin that, by Glu[18] to Gln[18] substitution,[15,16] differs from peptide used earlier.

Fig. 6 shows that conductance of channel levels is a monotonic function of salt concentration. Comparison to solution specific conductivity (solid line) indicates that channel conductance grows slower than does the solution conductivity. Similar to alamethicin channels in DOPS (Figs. 1 and 2), differences between conductance levels diverge with level number (although the levels themselves are sublinear in salt concentration). In particular, for 2 M NaCl these increments (Level 0 through Level 5, measured in nS) are 0.27, 1.10, 1.59, 1.80, 2.13. This means that the current burst, shown in Fig. 5, does represent a single ion channel of fluctuating size and not a random overlap of several identical channels.

Fig. 7 demonstrates that the probabilistic character of alamethicin channel reconstituted into uncharged lipid does not depend on salt concentration or on the shift of pH from a neutral to acidic value. Within experimental error, changing sodium chloride concentration from 0.1 to 2.0 M or acidity from pH 6.2 to 2.5 does not influence levels' relative probabilities. This suggests that

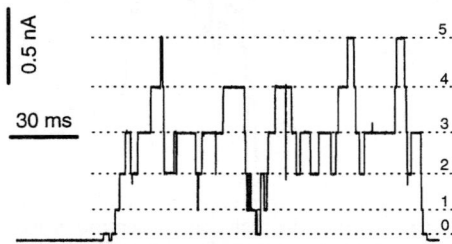

Fig. 5 Typical current burst of alamethicin-induced conductance obtained from a DOPE bilayer bathed by 2.0 M NaCl at pH 6.3. Channel 'switches on' through Level 0 and then fluctuates between different well-defined conductance states reaching Level 5 several times during its lifetime. The probabilistic character of the channel is very close to that of the alamethicin channel in DOPS at pH 2.0 (Fig. 1).

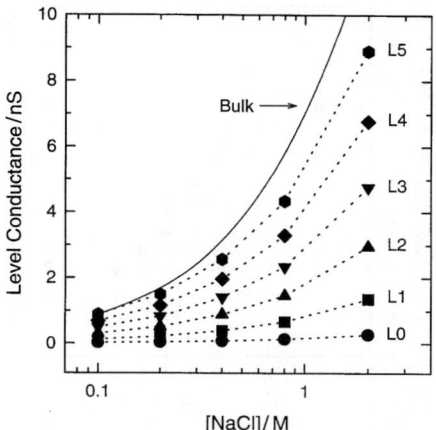

Fig. 6 Conductance of alamethicin channel states as a function of sodium chloride concentration measured on DOPE bilayers at pH 6.3. The solid line gives bulk solution specific conductivity scaled in such a way that Level 5 conductance and solution conductivity coincide at 0.1 M concentration to facilitate comparison. The comparison shows that conductance of Level 5 is a weaker function of salt concentration than bulk conductivity.

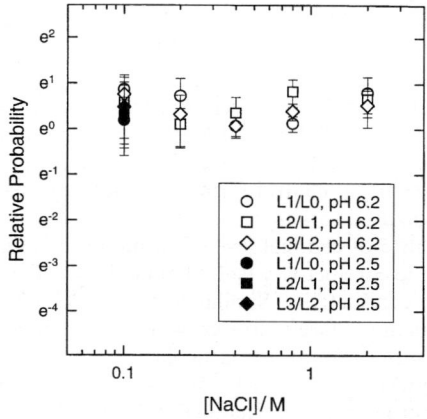

Fig. 7 Relative probability of channel conductance states in DOPE bilayers at different pH and sodium chloride concentrations. Within experimental error, the relative probability of higher states does not depend on salt concentration or acidity of the medium if the channel is inserted into neutral lipid. Whatever pH or salt concentration, relative probabilities in DOPE stay close to those in DOPS at pH 2.0–2.5 (Figs. 3 and 4).

Fig. 8 Typical current burst of alamethicin-induced conductance obtained from a DOPS bilayer bathed by 2.0 M NaCl at pH 6.2. It is clearly seen that the high salt concentration increases the probability of higher conductance states (compare to the current recording of Fig. 1 for pH 5.0). However, due to the appearance of strange conductance substates (shown by tilted arrows), it was impossible to quantify the salt effect in the alamethicin/DOPS system as done for DOPE (Fig. 7).

the dramatic pH effects shown in Figs. 1, 3, and 4 are related to lipid charge titration that affects channel behavior *via* lipid–channel interactions.

In addition to lipid charge titration by proton one can think about the analogous action of high salt concentration. As clearly seen in Fig. 8, illustrating a typical current burst obtained from a DOPS bilayer bathed by 2 M NaCl at neutral pH, we do observe a strong increase in higher level probabilities at high sodium chloride concentrations (compare to Fig. 1 recording for pH 5.0). Unfortunately, the effect of salt could not be quantified reliably because of additional, strange (compared to the channel in Fig. 5) sublevels in the alamethicin/DOPS system at high salt concentrations. Three of these sublevels are marked by tilted arrows (Fig. 8). Appearance of sublevels prohibited clear determination of level positions that is crucial to the subsequent statistical analysis. It should be noted that the bursts, an example of which is presented in Fig. 8, were rare enough (separated, on average, by 10-fold longer periods of 'silent' background recording) to exclude a trivial reason of several channels overlap.

Discussion

Many parameters influence the conformational equilibrium of membrane proteins. It is well known that a conformational transition can be triggered by a pH shift or by ligand and multivalent cation binding. The number of appropriate examples is overwhelming since these reactions constitute a basis for diverse physiological regulation at the cellular level.[24,27]

Much less studied are mechanisms of protein regulation by membrane lipids that act either directly or *via* the mechanical properties of host bilayers. Recent progress in this field demonstrates the possible ubiquity of such regulation. It has been shown that lipids modulate catalytic activity and binding properties of integral membrane proteins that include Insulin-R, Na^+/K^+-ATPase, Ca^{2+}-ATPase, GABA transporter, acetylcholine receptor (Table 1, in ref. 2), and Rhodopsin,[1] to name just a few. Nevertheless, the nature of the physical forces underlying lipid action is still a subject of considerable controversy.[28]

Membrane protein function can be modified by changes in the mechanical properties of a host membrane, *e.g.*, by changes in spontaneous curvature of membrane lipids.[12,22,29–33] There has been corresponding theoretical effort (*e.g.*, ref. 11, 34, 35). The conformational equilibrium of a protein between different functional states is governed by the total free energy differences between these states. Clearly, if a conformational transition between states involves a change in the shape or length of the protein surface that is exposed to lipids, this transition has to be sensitive to membrane mechanics. Although the idea is general, the particular approaches permitting a quantitative description are model-specific.

A correlation between packing stress and conformational equilibrium of a *single* membrane-bound polypeptide structure was first observed with alamethicin channels in planar bilayer membranes made from DOPE/DOPC mixtures.[22] It was shown that an increase in the mole fraction of DOPE, which favors a highly curved H_{II}-phase, shifts the distribution of conductance levels towards those of higher conductance. The following relationship between spontaneous curvature of the lipid and polypeptide aggregation in the membrane was established: higher curvature stress promotes larger alamethicin aggregates.

Now we demonstrate that with respect to the probabilistic character of the alamethicin channel, the same charged lipid species can be made equivalent to DOPC or DOPE by changing bathing solution pH or salt concentration. Again, we correlate ion channel function with the stress of forcing lipids of a given spontaneous curvature into a planar membrane form. The cartoon in Fig. 9 illustrates how a change in the charge of lipid head groups is able to change lipid spontaneous curvature. A useful notion in description of mechanical properties of a bilayer is the effective 'shape' of the membrane molecules. A cylindrical molecular shape (when the cross-sectional area of the polar head group is similar to the cross-sectional area of the acyl chains) will correspond to lamellar phases and stress-free packing into the bilayer form. At pH 6 and small enough salt concentration, DOPS molecules have an approximately cylindrical effective shape because of repulsion of fully charged neighboring head groups. At pH 2 proton binding titrates out the head group charge; lipid shape is conical. Correspondingly, DOPS hexagonal phase can go from a rather low spontaneous curvature at neutral pH to a high spontaneous curvature in an acidic environment.

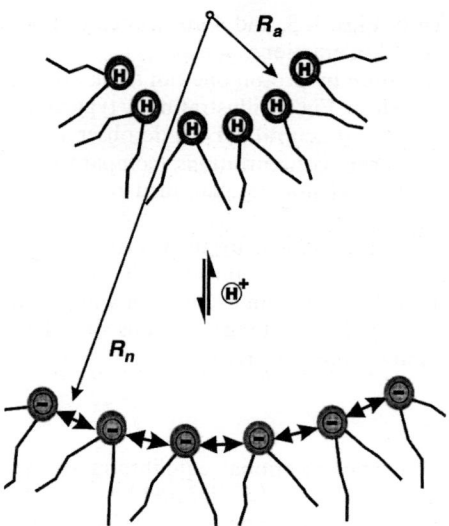

Fig. 9 Cartoon illustrating the proton-induced increase in spontaneous curvature of DOPS. Neutralization of the lipid head group charge by proton binding reduces head–head repulsion; it thus effectively changes lipid molecule 'shape'. At a fully deprotonated, charged state, repulsion between DOPS head groups drives system into a lamellar structure; however, at high proton concentrations this repulsion is 'switched off' so that a preferred packing is an H_{II}-phase of high curvature.

Strong X-ray diffraction evidence, Fig. 10, supports this interpretation. Samples with an excess of water solution were used in diffraction measurements and solution pH was adjusted before, and checked after, sample equilibration. It is seen that the Bragg repeat spacing for the DOPS hexagonal phase changes sharply between pH 2.5 and 4.0, in excellent agreement with transport measurements (Figs. 3 and 4). For samples above pH 4, X-ray scattering showed disorder characteristic of highly and irregularly separated bilayer membranes. The common but puzzling coexistence of hexagonal and lamellar phases at low pH was probably related to the small free energy difference between these two lipid assemblies.[36,37]

A general thermodynamic analysis permits us to quantify the energetics of alamethicin channel regulation. Indeed, because the different conductance levels of the channel are well defined states,

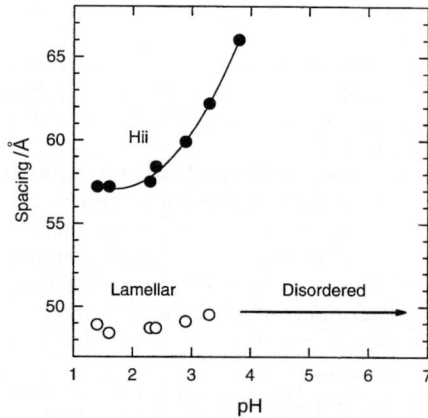

Fig. 10 Bragg repeat spacing of hexagonal and lamellar DOPS phases in excess solution. A sharp decrease in hexagonal spacing (increase in spontaneous curvature) occurring between pH 2.0 and 4.0 is clearly seen. The structure above pH 4.0 (that is designated as 'disordered') most probably corresponds to charged bilayers with irregular separation.[40] The reasons for coexistence of hexagonal and lamellar phases at low pH are not yet clear.

it is possible to speak of the chemical potentials (or free energies) of the individual states, $\mu_i = \mu_i(\text{pH}, \mu_{\text{NaCl}})$. The relative probability of two adjacent levels, $p(i, i-1) = P_i/P_{i-1}$, is given then by

$$p(i, i-1) = \exp[-(\mu_i - \mu_{i-1})/kT] \tag{1}$$

where k and T have their usual meaning of Boltzmann constant and absolute temperature. A change in the relative probability induced by a pH shift can be expressed as the ratio of relative probabilities at the two pH values and equals $\exp[-\Delta(\mu_i - \mu_{i-1})/kT]$. Here $\Delta(\mu_i - \mu_{i-1})$ is the pH-induced change in the states' chemical potential difference. This ratio, reflecting total change in relative probability caused by acidity shift from pH 2 to 6 (Figs. 3 and 4), is measured to be close to 50. Thus

$$\Delta(\mu_i - \mu_{i-1})|_{\text{pH 2} \Rightarrow \text{hpH 6}} \cong 4 \, kT \tag{2}$$

The rate of change in relative probability with pH is an indication of difference in the number of protons associated with the surface. By standard Gibbs–Duhem reasoning, assuming that the active factors are proton and sodium activity, the changes in state energy go as

$$d\mu_i = -n^i_{\text{NaCl}} d\mu_{\text{NaCl}} - n^i_{\text{H}} d\mu_{\text{H}} \tag{3}$$

where the functions n^i_{NaCl} and n^i_{H} themselves depend on NaCl and H activity. These are the Gibbs excess numbers of sodiums or protons that are associated with the membrane–channel system when the channel is in state i. The change in the relative probability of two conductance levels i and $i - 1$ depends on the *difference* in these ns

$$d(\mu_i - \mu_{i-1}) = -(n^i_{\text{NaCl}} - n^{i-1}_{\text{NaCl}})d\mu_{\text{NaCl}} - (n^i_{\text{H}} - n^{i-1}_{\text{H}})d\mu_{\text{H}} \tag{4}$$

When salt concentration is kept fixed but pH varied, the change in relative probabilities gives us $n^i_{\text{H}} - n^{i-1}_{\text{H}}$, a difference in the number of protons associated with the system upon a transition from conductance state $i - 1$ to conductance state i. From these equations and $\mu_{\text{H}} = kT \ln[\text{H}]$, $\text{pH} = -\log[\text{H}]$ we obtain:

$$n^i_{\text{H}} - n^{i-1}_{\text{H}} = -\frac{\partial(\mu_i - \mu_{i-1})}{\partial \mu_{\text{H}}} = \frac{\partial \ln p(i, i-1)}{\partial \ln[\text{H}]} = -\frac{\partial \ln p(i, i-1)}{\ln 10 \, \partial(\text{pH})} \tag{5}$$

Analysis of the data for the Level 0 to Level 1 and Level 1 to Level 2 transitions around pH 3.0 (which corresponds to the maximum slope of the relative probability dependence on pH, Fig. 3) gives $n^i_{\text{H}} - n^{i-1}_{\text{H}} = 1.1 \pm 0.1$. The change in the number of associated protons upon transition to a higher conductance state is thus positive. Higher conductance states by rearranging the whole channel/bilayer system accommodate more protons so that the increase in proton chemical potential makes these states more favorable. As a result, higher conductance states are more expressed at high proton concentrations.

The thermodynamic description shown above is very useful for quantifying the effect of pH on channel function and for restricting the number of possible models. However, as usual, thermodynamics does not elucidate a specific physical mechanism. To approach the mechanism of channel regulation by pH, more structural knowledge of the conformation transformations of channel opening/closing and on lipid–peptide interactions is needed. Such knowledge is necessary to discriminate between stress of packing, surface tension, or other mechanical factor contributions.

As recently pointed out on the basis of a careful study of different mechanical contributions to the energetics of protein inclusions into lipid bilayers,[11] the increase in hydrophobic mismatch between the protein and the lipid by a mere 0.3 Å can change the equilibrium distribution between the corresponding protein states by a factor of 10. Also, the energy of the protein-induced bilayer deformation can be as high as $2–3 \, kT$ per one lipid molecule. In this case, the $4 \, kT$ effect reported in the present study can be explained by perturbation of a few lipid molecules only.

It is worthwhile to compare the pH-shift-induced energy change found in our study to other characteristic energies in the system. The electrostatic energy of recharging of a single lipid head group in the fully charged DOPS bilayer (*e.g.*, at pH 6.0) can be easily found as a product of the membrane surface potential ψ_0 (*e.g.*, ref. 38) and the elementary charge e

$$e\psi_0 = 2 \, kT \, \sinh^{-1}\{[8 \, kT\varepsilon\varepsilon_0(\text{Na})]^{-1/2}\sigma\} \tag{6}$$

Here σ is the lipid charge surface density, ε is the dielectric constant, and ε_0 is the permittivity of free space. Taking $\sigma = 0.25$ C m^{-2} (one elementary charge per 64 Å^2) and [Na] = 6.02×10^{25} m^{-3} (0.1 M NaCl), we obtain $e\psi_0 = 5.2$ kT. This energy compares well with the 4 kT change found for the energy of the alamethicin state to state transitions [eqn. (2)].

The work of forcing one lipid molecule from the hexagonal H_{II}-phase into the lamellar phase can be estimated as[36,39]

$$E = \frac{ak_c}{2R_0^2} \qquad (7)$$

where k_c is the monolayer bending modulus (about 10 kT[37]), a is the area per lipid molecule, and R_0 is the radius of spontaneous curvature. Taking $a = 64$ Å^2 and deducing the radius of spontaneous curvature from the Bragg spacing for the hexagonal phase at pH 2.0 (Fig. 10), we get $E \approx 0.3$ kT. This energy is about an order of magnitude less than the acidity-induced change in the energy of alamethicin state-to-state transitions. However, many lipid molecules are in direct contact with the alamethicin aggregate. If, in addition, the lipids that are to be perturbed by protein conformational change extend over distances of many lipid molecules,[11] this estimate is reasonable.

To conclude, varying the pH changes the probabilistic character of alamethicin channels in a way that correlates with the pH-induced changes in the nonlamellar tendency of the host-lipid. This correlation, made for the charged lipid DOPS, agrees with the logic of our conclusions drawn from earlier observations on alamethicin in mixtures of neutral lipids.[22] The channel's response to pH is all the more impressive given the fact that the present measurements were made on a neutral form of alamethicin that does not possess any residues that can be titrated in this pH range. Still, there is a pronounced, 50-fold effect on channel state-to-state transitions from a pH shift of about two units.

Whatever the mechanism, lipid charge titration modifies channel structural equilibrium. This finding probably unveils an additional, previously unrecognized way of pH regulation in membrane transport.

R.P.R. acknowledges the financial support of the Natural Sciences and Engineering Research Council of Canada and the expert assistance of Mrs. Nola Fuller.

References

1 B. J. Litman and D. C. Mitchell, *Lipids*, 1996, **31**, S193.
2 A. Bienvenue and J. S. Marie, *Curr. Top. Membr.*, 1994, **40**, 319.
3 R. E. Koeppe, II and O. S. Andersen, *Annu. Rev. Biophys. Biomol. Struct.*, 1996, **25**, 231.
4 H. J. Apell, E. Bamberg and P. Lauger, *Biochim. Biophys. Acta*, 1979, **552**, 369.
5 T. K. Rostovtseva, V. M. Aguilella, I. Vodyanoy, S. M. Bezrukov and V. A. Parsegian, *Biophys. J.*, 1998, **75**, 1783.
6 E. Bamberg and P. Lauger, *Biochim. Biophys. Acta*, 1974, **367**, 127.
7 E. Neher and H. Eibl, *Biochim. Biophys. Acta*, 1977, **464**, 37.
8 V. Fonseca, P. Daumas, L. Ranjalahy-Rasoloarijao, F. Heitz, R. Lazaro, Y. Trudelle and O. S. Andersen, *Biochemistry*, 1992, **31**, 5340.
9 J. A. Killian, *Biochim. Biophys. Acta*, 1992, **1113**, 391.
10 J. Girshman, D. V. Greathouse, R. E. Koeppe II and O. S. Andersen, *Biophys. J.*, 1997, **73**, 1310.
11 C. Nielsen, M Goulian and O. S. Andersen, *Biophys. J.*, 1998, **74**, 1966.
12 J. A. Lundbaek and O. S. Andersen, *J. Gen. Physiol.*, 1994, **104**, 645.
13 M. Goulian, O. N. Mesquita, D. K. Fygenson, C. Nielsen, O. S. Andersen and A. Libchaber, *Biophys. J.*, 1998, **74**, 328.
14 M. S. P. Sansom, *Prog. Biophys. Mol. Biol.*, 1991, **55**, 139.
15 G. A. Wooley and B. A. Wallace, *J. Membrane Biol.*, 1992, **129**, 109.
16 D. S. Cafiso, *Annu. Rev. Biophys. Biomol. Struct.*, 1994, **23**, 141.
17 G. Boheim, *J. Membrane Biol.*, 1974, **19**, 277.
18 R. Latorre and J. J. Donovan, *Acta Physiol. Scand. (Suppl.)*, 1980, **481**, 37.
19 J. E. Hall, I. Vodyanoy, T. M. Balasubramanian and G. Marshall, *Biophys. J.*, 1984, **45**, 233.
20 S. Stankowski, U. D. Schwarz and G. Schwarz, *Biochim. Biophys. Acta*, 1988, **941**, 11.
21 L. R. Opsahl and W. W. Webb, *Biophys. J.*, 1994, **66**, 71.
22 S. L. Keller, S. M. Bezrukov, S. M. Gruner, M. W. Tate, I. Vodyanoy and V. A. Parsegian, *Biophys. J*, 1993, **65**, 23.

23 S. M. Bezrukov, I. Vodyanoy, P. Rand and V. A. Parsegian, *Biophys. J.*, 1995, **68**, A341.
24 B. Alberts, D. Bray, J. Lewis, M. Raff, K. Roberts and J. D. Watson, *Molecular Biology of the Cell*, Garland Publishers, New York, 1994.
25 S. M. Bezrukov and I. Vodyanoy, *Biophys. J.*, 1993, **64**, 16.
26 M. Montal and P. Mueller, *Proc. Natl. Acad. Sci. USA*, 1972, **65**, 3561.
27 *Cell Physiology*, ed. N. Sperelakis, Academic Press, San Diego, 1988.
28 O. G. Mouritsen and M. Bloom, *Annu. Rev. Biophys. Biomol. Struct.*, 1993, **22**, 145.
29 S. M. Gruner, *Proc. Natl. Acad. Sci. USA*, 1985, **82**, 3665.
30 C. D. McCallum and R. M. Epand, *Biochemistry*, 1995, **34**, 1815.
31 C. D. Stubbs and S. J. Slater, *Chem. Phys. Lipids*, 1996, **81**, 185.
32 R. M. Epand, *Chem. Phys. Lipids*, 1996, **81**, 101.
33 J. A. Lundbaek, A. M. Maer and O. S. Andersen, *Biochemistry*, 1997, **36**, 5695.
34 N. Dan, A. Berman, P. Pincus and S. A. Safran, *J. Phys. II France*, 1994, **4**, 1713.
35 N. Dan and S. A. Safran, *Isr. J. Chem.*, 1995, **35**, 37.
36 M. M. Kozlov, S. Leikin and R. P. Rand, *Biophys. J.*, 1994, **67**, 1603.
37 S. Leikin, M. M. Kozlov, N. L. Fuller and R. P. Rand, *Biophys. J.*, 1996, **71**, 2623.
38 S. McLaughlin, *Curr. Top. Membr. Transport.*, 1977, **9**, 71.
39 W. Helfrich, *Z. Naturforsch.*, 1973, **28C**, 693.
40 N. Fuller, personal communication.

Paper 8/06579I

Structure-based prediction of the conductance properties of ion channels

Oliver S. Smart,*[a] Guy M. P. Coates,[a] Mark S. P. Sansom,[b] Glenn M. Alder[c] and C. Lindsay Bashford[c]

[a] *School of Biochemistry, The University of Birmingham, Edgbaston, Birmingham, UK B15 2TT. E-mail: o.s.smart@bham.ac.uk*
[b] *Laboratory of Molecular Biophysics, The Rex Richards Building, University of Oxford, Oxford, UK OX1 3QU*
[c] *Division of Biochemistry, St. George's Hospital Medical School, Cranmer Terrace, London, UK SW17 0RE*

Received 1st September 1998

The HOLE procedure allows the prediction of the absolute conductance of an ion channel model from its structure. The original prediction method uses an empirically corrected Ohmic method. It is most successful, with predictions being reliable to within a factor of two. A new modification of the procedure is presented in which the self-diffusion coefficients of water molecules from molecular dynamics simulation are used to replace the empirical correction factor. A "prediction" of the conductance for the porin OmpF by the new method is made and shown to be very close to the experimental value. HOLE also allows the prediction of the effect that the addition of non-electrolyte polymers will have on channel conductance. The method has great potential to yield structural information from data provided by single channel recordings but needs further validation by making measurements on channels of known structure. Preliminary results are given of single channel records establishing the effects of non-electrolytes on the conductance of gramicidin D channels. As an example of the potential uses of the procedure application is made to examine the oligomerization of α-toxin (α-hemolysin) channels. A model for the α-toxin hexamer, based on the crystal structure for the heptamer, is generated using molecular mechanics methods. The compatibility of the structures with single channel conductance data is assessed using HOLE.

1 Introduction

Ion channels are an important class of proteins, allowing the passage of ions through the hydrophobic barrier presented by lipid bilayers.[1] They are commonly involved in communication and regulation processes, and are the target for many drugs. Because of the difficulties in experimentally determining the three dimensional structures of membrane proteins it is common to propose channels models on the basis of low resolution data, based on techniques such as mutagenesis[2] or electron microscopy data.[3] An important part of such a modelling procedure is to validate the final models. Even where a high-resolution structure for a channel is available it is not always immediately obvious as to whether this represents the conducting form.

A reliable tool for the prediction of the conductance properties of a channel model on the basis of its structure would be invaluable in addressing these questions. This paper describes, in Sections

2 and 3, steps taken in the development of such a tool. Section 4 describes attempts to extend the method to allow the prediction that the addition of non-electrolyte molecules has on channel conductance, including preliminary results for gramicidin D. Finally the method is applied to examine the complex question of the stoichiometry of α-toxins in lipid bilayers.

2 The HOLE method

The HOLE method[4] has become widely used in the analysis of ion channel structures. It uses Monte Carlo simulated annealing to maximize the radius of a sphere that is forced to squeeze through the van der Waals surface of a channel. The user defines a channel direction vector that is used to define a stack of planes with a uniform separation s. A separate optimization is conducted for each plane, maximizing the radius of the probe sphere, while avoiding any overlap with the atoms of the channel and keeping the centre of the sphere on the plane. Although the use of a spherical probe is reasonable for narrow channels it becomes problematic for larger, anisotropic pores. For this reason an adaptation was made,[5] in which the probe sphere is replaced by a capsule (more accurately termed a sphereocylinder). Although this feature allows the measurement of anisotropy the routine suffers from stability problems in the irregular channels that result from molecular dynamics simulations. For this reason, work in progress aims to adapt the Connolly procedure[6] to more accurately delimit larger channels.

3 Predicting absolute conductance properties using HOLE

The method[5] is based on simple Ohmic considerations. The HOLE program can be thought of as measuring the cross-sectional area $A(z)$ of a pore as a function of distance along the channel direction vector z. Consider the pore to be filled with an electrolytic solution of resistivity ρ. A reasonable approximation of the conductance (inverse resistance) of such a column of fluid is given by:

$$G_{\text{macro}}^{-1} = \sum_{z=\text{low}}^{z=\text{high}} \rho s / A(z) \tag{1}$$

where s is the width between parallel planes used in the HOLE method. This assumes that the conductivity of an ionic solution within the channel is equal to that of bulk solution. The assumption would be true if ion channels had macroscopic dimensions (much larger than a water molecule). In practice it is found that real channels have a conductance which is around five times lower than that expected from the above equation giving the macroscopic limit G_{macro}. In order to make a reasonable estimate of the conductance of a channel, an empirically based correction factor is used:

$$G_{\text{pred}} = G_{\text{macro}}/C(\xi) \tag{2}$$

where ξ is a characteristic parameter for the channel, e.g. minimum radius. Although a number of characteristic parameters have been tested,[5] a constant correction equal to 5.6 performs well. The prediction routine was tested on all channel-forming proteins and peptides where both a high-resolution structure and conductance data are available. Results are given in Fig. 1. Overall, the algorithm yielded good results with predictions accurate to within an average factor of 1.8 to the experimental values. This accuracy is sufficient to make the method a useful part in validating model structures.

3.1 Using molecular dynamics simulation data to improve predictions

The use of empirical correction factors to Ohmic predictions of absolute conductance, although remarkably successful is limited: reducing the physical insights[5] available and is unlikely to be able to account for the subtle differences between different channels. Analysis of the motion of water molecules during molecular dynamics (MD) simulations provide an alternative approach, while avoiding the necessity of adopting a full scale microscopic simulation of the complete ion translocation process inherent in an *ab initio* prediction of conductance.[7]

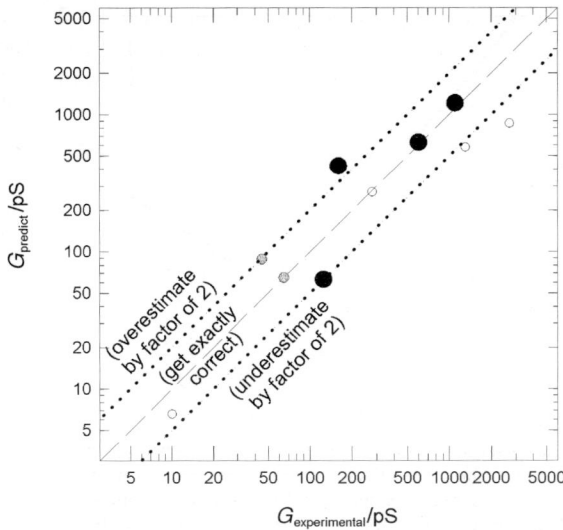

Fig. 1 Prediction of channel conductance using the 1st generation empirical correction function (adapted from Smart et al.[5]). Filled circles mark results for experimentally determined models. Open circles mark results for model channels.

MD methods use the numerical solution of Newton's equations of motion to follow the course of the motion of a protein and its surroundings as a function of time, under physically realistic conditions. Breed et al.[8] used methods to analyse the restriction on the motion of water molecules within simplified ion channel models. The self-diffusion coefficient of water molecules from within the pore was reduced by around a factor of five compared to the value in bulk.[8] The similarity of this reduction to the empirical correction factor used in HOLE led to the proposal that data from MD could be used instead.[5]

We replace the assumption that the resistivity of the ionic solution within the pore is equal to that of bulk solution (used to date with the consequent necessity of empirical correction functions). Instead the resistivity is made to be a function of z, the coordinate along the channel vector and is assumed to vary in inverse proportion to the change in the water self-diffusion coefficient derived by MD simulation:

$$\rho_{MD}(z) = \rho_{bulk} \frac{D_{bulk}}{D_z(z)} \quad (3)$$

where D_z is the self-diffusion coefficient in the direction z, averaged over all water molecules at a particular z and over the MD run. D_{bulk} is the water self-diffusion coefficient derived from bulk phase using the same water model as the ion channel run. The approximation in this procedure is that the reduction in the self-diffusion coefficients of ions within the channel will be equal to that of water molecules.

A MD-corrected conductance prediction G_{MD} can then be made:

$$G_{MD}^{-1} = \left\{ \sum_{z=z_{low}}^{z=z_{high}} \frac{\rho_{MD}(z)s}{A(z)} \right\} + \frac{\rho_{MD}(z_{low})}{4r'(z_{low})} + \frac{\rho_{MD}(z_{high})}{4r'(z_{high})} \quad (4)$$

where s is the width of the slabs being considered. The two last terms represent access resistance[9] where $r(z_{low})$ and $r(z_{high})$ are effective pore radii at the ends of the channel. The access resistance, which typically contributes around 10% of the total resistance, was previously ignored in eqn. (1), being effectively absorbed by the large empirical correction factor.[5]

To gauge the effectiveness of the modified procedure a "prediction" is made for the E. coli porin OmfP based on a MD simulation performed by Tieleman and Berendsen,[10] using the X-ray

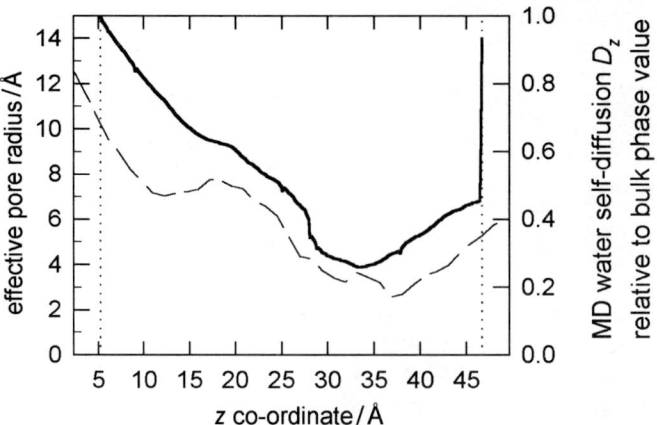

Fig. 2 Using MD simulation data to make a prediction of the absolute conductance of porin OmpF. The solid line and left-hand ordinate show the effective pore radius profile of the crystal structure[11] found by the capsule option of HOLE. The dashed line and right-hand ordinate show the reduction in the water self-diffusion coefficient (D_z/D_{bulk}) for this structure, obtained by Tieleman and Berendsen.[10]

crystal structure determined by Cowan and co-workers.[11] The simulation was performed on the complete OmpF trimer in POPE bilayer/water box, with data being collected over a 1 ns period.[10] Peter Tieleman kindly supplied the self-diffusion coefficient of the water molecules in the direction of the channel vector averaged over the trimer. Fig. 2 shows the marked reduction in the mobility of water molecules within the pore.

The data on the reduction in the mobility of water molecules were combined with the effective pore radius profile for the crystal structure measured by the capsule option of HOLE (Fig. 2) and used in eqn. (3) and (4). This results in an expected conductance of 850 pS for 1 M KCl solution which is encouragingly close to the experimental value[5,12,13] of 700 pS.

Work in progress aims at extending the technique to be generally applicable. The method will be tested by application to all the systems illustrated in Fig. 1. Short molecular dynamics runs are desirable to allow general applicability. At present a conductance prediction can be made in a few tens of minutes processing time on a modern workstation. Although including molecular dynamics in the process will by necessity increase the requirement, a reasonable limit is a few tens of hours of processing time. The use of an explicit lipid environment in the simulation is excluded by this limit. Although the explicit modelling of lipid increases the realism of the simulation it can be difficult to set up the initial ensemble, as demonstrated by the variety of approaches adopted by different workers.[10,14,15] As the simulations only require the water dynamics to be realistic it is expected that excluding the, relatively distant, lipid will not be too severe an approximation. This is borne out by a comparison of results obtained for water behaviour from alamethicin simulations using an explicit lipid representation with earlier applications without.[16]

Without an explicit lipid representation restraints on protein backbone atoms will be necessary.[17] An issue to be explored is the effect of the imposition of restraints during simulation. Restraints are desirable in maintaining the structure close to the starting conformation as all MD simulations tend to suffer from problems of drift. However, restraints on the protein may seriously affect the dynamic mobility of the water within the pore, particularly for narrow channels, such as gramicidin.

4 Structure-based prediction of the effect of adding non-electrolytes on conductance

Measurements of the effect of the addition of uncharged polymers to the medium in single channel conductance measurements can be used to derive information about the pore dimensions of a channel.[5,18,19] The standard experimental protocol is to add 15 or 20% w/v polyethylene glycol

(PEG) to the conducting medium. The variation in channel conductance is measured as a function of the molecular weight of the PEG used. The bulk conductivity of an ionic solution is reduced in the presence of PEG, the magnitude of the reduction being broadly independent of the molecular weight, provided a constant weight fraction is maintained.[18] A recent development of the procedure allows the determination of the effect of adding different molecular weights on either side of the lipid bilayer,[20] which allows more information to be derived.

When a low molecular weight PEG is added to a wide channel, the conductance of the channel will drop by the same factor as the change in bulk conductivity. But as the molecular weight of the PEG is increased the polymer will be progressively excluded from the channel. This will result in a progressive increase in the conductance measured. Results are normally interpreted using the hydrodynamic radii of the PEG or other polymer used.

In the limit of very large molecular weight PEGs it is possible to observe an *increase* in conductance.[21] This has been proposed to be because PEG is dehydrating and increases the water activity with the pore from which it is excluded. The increase in conductance depends on the ratio between the channel and access resistance, and will be larger for thinner channels.[5]

The HOLE conductance prediction method has been adapted to predict the effect that PEGs will have on conductance. Full details are described in another publication,[5] but broadly a conductance prediction is made using HOLE and following eqn. (1), with the modification that the channel is divided into areas which are accessible to probe spheres of a radius, R_{PEG} which is set equal to the hydrodynamic radius of the polymer being considered, and areas which are not. The former are assigned a resistivity of the PEG bulk solution and the latter a lower resistivity. The conductance prediction is then made as before but no empirical correction factor is required because results are expressed as a ratio of the polymer free case.

An application[5] of the procedure has been made to assess the effect of PEG on the conductance of cholera toxin B_5 by comparing the value computed from the X-ray structure,[22] with that obtained experimentally.[23] The results are most encouraging with the prediction falling within experimental error (Fig. 3). A further application was comparing results found for models of different alamethicin conducting states,[24] with experimental observations[21] arriving at the conclusion that the experiments were consistent with the barrel-stave hypothesis.[5] Encouraged by these two successes, we have decided to make a systematic study. The aim is to collect polymer-addition data for every ion channel for which a high resolution structure is available to validate the techniques and give real confidence in the interpretation of data for biologically important channels. The next section describes preliminary results for the first study undertaken in this scheme.

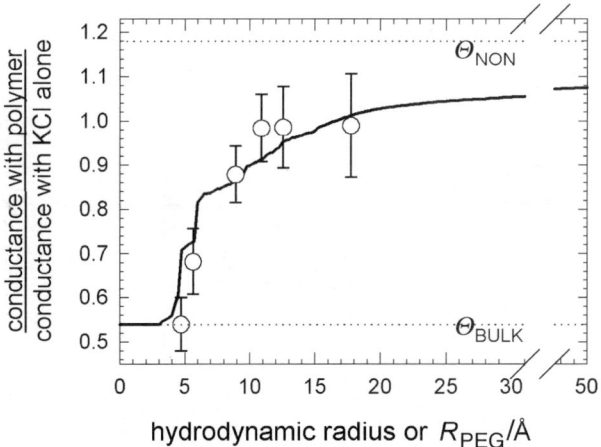

Fig. 3 Comparing the effect of non-electrolytes on experimental conductance[23] of cholera toxin B_5, with the expected result using HOLE (solid line) based on the X-ray crystal structure.[22]

4.1 The effect of non-electrolytes on the conductance of gramicidin D

An important step in the validation of the non-electrolyte addition technique is to apply the method to a narrow channel of known structure. Gramicidin D is the archetypal model ion channel[25] with a high resolution solid-state NMR structure available for the channel forming conformation.[26] The conductance pathway is narrow, with a minimum HOLE pore radius of 1.15 Å,[4] which contains a single file of water molecules. We have started to make measurements of the effect of non-electrolyte on gramicidin conductance. Preliminary results are reported here.

Gramicidin D (ICN) was incorporated into bilayers of diphytanoylphosphatidylcholine (Avanti Polar Lipids) across a 10–20 μm aperture separating two Teflon chambers as described previously.[19] Current was recorded in voltage-clamp mode *via* Ag/AgCl electrodes in each chamber. Fig. 4 shows typical traces obtained from membranes containing a few gramicidin channels in the absence and presence of 20% v/v PEG 10 000.

Fig. 5 shows the results found and provides a comparison with the results expected using the prediction routine, described above, with the experimental channel structure.[26] The results are most interesting.

It is encouraging to observe that the addition of high molecular weight PEG results in an increase in channel conductance. This counterintuitive effect can be ascribed to an increase in the water activity within the channel[5,21] and has previously been observed for the low conductance states of alamethicin.[21] The value found is in approximate accordance with that expected on the basis of the structure. This early result encourages our forecast[5] that the asymptotic value may be the most useful parameter in obtaining structural information. More detailed experiments are required to test this assertion.

The effect that low molecular weight polymers have on conductance (Fig. 5) is more surprising. Both glucose and PEG300 reduce channel conductance by around half the maximum (bulk) effect. On the basis of the "resting" structure of gramicidin in lipid it can be expected that a molecule as large as glucose would not be able to penetrate the channel at all. We know that gramicidin readily conducts cations up to the size of formamidinium[27] but is blocked by guanidinium.[27] However, these ions are still considerably smaller than glucose or PEG300. This makes it most unlikely that gramicidin would be flexible enough to allow the passive diffusion of glucose into the channel.

If the effect of glucose and PEG300 is not due to non-electrolyte permeation, what is its cause? A possibility is that it may be due to a much larger than expected access resistance at the channel mouths. Gramicidin is known to readily form channels in lipids with thicknesses of up to 40 Å,[28] despite having a length of only 26 Å.[4] It has been proposed that the channel may cause "crimping" of the lipid.[29] This local thinning would create "vestibules" which would be accessible

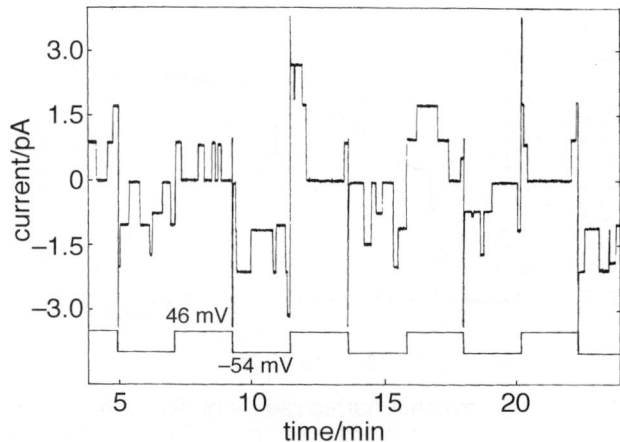

Fig. 4 Gramicidin channels in diphytanoylphosphatidylcholine bilayers, in the presence of PEG. Gramicidin D ($<10^{-12}$ M) treated bilayers in 1 M KCl, 0.005 M Hepes containing 20% v/v PEG10000, pH 7.4 at room temperature. The voltage applied is indicated by the lower trace.

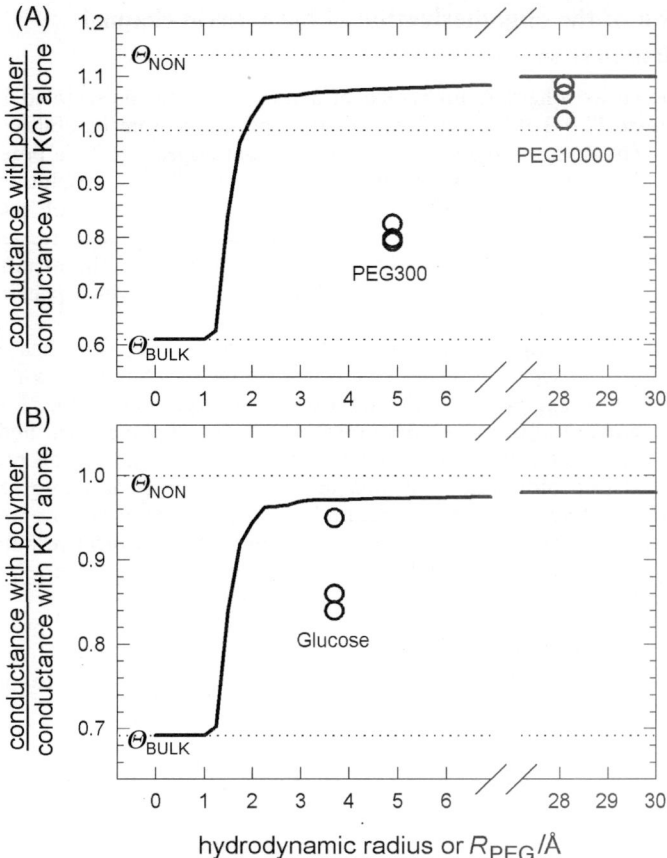

Fig. 5 The effect of non-electrolytes on the single channel conductance of gramicidin. The results are preliminary. Panel (A) shows the effect of adding PEGs and panel B sugars. Circles mark individual experimental results and the solid lines show the effect expected using the HOLE prediction routine[5] on the basis of the solid state NMR structure of Ketchem et al.[26] In each case, Θ_{PEG} and Θ_{NON} are the theoretical limits to the ratio.[5]

to small polymers but not large ones. The magnitude of this effect is testable by making measurements with lipid bilayers of different thickness. This effect may contribute to the large impact of smaller polymers but is unlikely to be the principal cause. This can be seen by noting that PEG300 has approximately 50% of the result expected if it were able to permeate the whole channel. It is difficult to imagine access resistance ever having such a significant impact for a channel as narrow as gramicidin.

The likely source is the breakdown of the simplistic model of PEG-accessible regions of high resistivity switching instantly to inaccessible regions with low resistivity. It is likely that non-electrolytic polymers have significant effect in the region around them. Following the philosophy described in Section 3, we plan to undertake molecular dynamics simulations to gauge this effect. We hope this will provide data to produce a reliable prediction method, which would provide a very valuable aid to modelling studies. In the meantime we urge caution in the interpretation of polymer addition experiments, particularly for narrow channels. Using the simple interpretation used by most authors to date, the data presented here would likely be interpreted as representing a channel with a characteristic radius above 4 Å. This clearly contradicts a wealth of evidence on gramicidin from other sources.

5 Examination of the oligomerization of the α-toxin channel

5.1 Background: seven vs. six

α-Toxin (also known as α-hemolysin) is one of a number of toxins secreted by the bacterium *Staphylococcus aureus*.[30] α-Toxin's cytotoxicity arises primarily from its ability to form pores in cell membranes.[30] The crystal structure of α-toxin, in what appears to be a pore forming conformation, has recently been solved by X-ray crystallography,[29] at a resolution at 1.9 Å. This structure[31] consists of seven identical subunits, each of which consists of a large globular β-sandwich domain and a 42 amino acid β-hairpin loop. The subunits are arranged in a "bunch of tulips" pattern, where the head domains lie on the outer surface of the membrane, and the β-hairpin loops bunch together to form a 14 strand β-barrel "stem" which penetrates through the membrane (Fig. 6).

The mechanism of α-toxin channel assembly and pore formation has been extensively studied.[32–35] A number of separate stages have been identified.[32,34] Initially α-toxin monomers in a pre-pore conformation aggregate on the surface of the target membrane.[34] A number of the monomers then assemble to form a pre-pore oligomer.[34] The oligomer then undergoes a conformational change to the active channel form, with little change in the overall proportions of secondary structure.[32] This conformational change involves the loop of 42 amino acids inserting into the membrane and adopting a β-hairpin structure.[31] It is likely that in the pre-pore oligomer these residues adopt Greek-key motif similar to that found in anthrax protective antigen.[36,37] The β-hairpins from each monomer come together to form a transmembrane β-barrel, which provides a path for ion conductance in a similar manner to the porins.

Although the crystal structure of α-toxin showed it to be heptameric,[31,33] previous biochemical studies including cross-linking studies[32] and electron microscopy (EM)[38] indicated that the assembled conformation in lipid vesicles was hexameric. A recent atomic force microscopy (AFM)

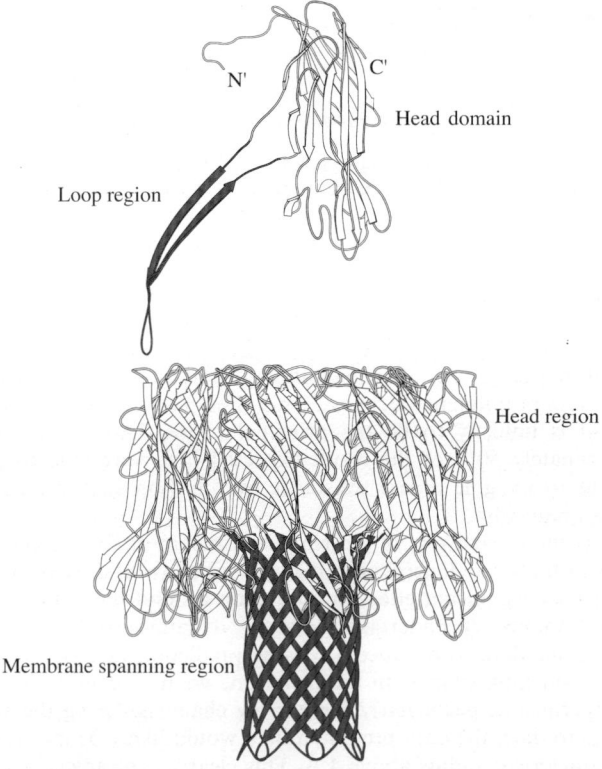

Fig. 6 The construction principles of the α-toxin heptamer structure.[31] Produced using the molscript program.[52]

study[39] of α-toxin in lipid bilayers under physiological conditions has revealed that α-toxin can form long lifetime hexamers but no evidence was found of the existence of heptamers. This is in direct contrast to Yang and co-workers[40] who applied AFM to a mutant form of the protein, which is trapped in the pre-pore oligomer, and found it to be heptameric.

In view of the current debate on the stoichiometry of α-toxin, it was decided to construct a model for a hexameric form, based on the "design principles" revealed by heptameric crystal structure. The compatibility of the experimental heptamer structure and the model hexamer with the single channel conductance data available is then assessed, using the HOLE techniques described above.

5.2 Producing a model for α-toxin hexamer

5.2.1 The pore domain.
In the heptameric structure the transmembrane domain is formed by each subunit donating an amino acid loop to form a 14 stranded β barrel. It was assumed that a hexameric toxin would form a pore along similar principles, forming a 12-stranded barrel. A great deal is known about the geometric factors that constrain the structures of β barrels.[41,42] The overall structure of a β barrel can be described by two parameters, the number of strands (N) and the stagger of the barrel (S).[41] The stagger number S is a measure of how far the β barrel strands tilt relative to axis of the barrel. Certain combinations of S and N are favoured in β barrels as they maximize inter-strands hydrogen bonding and favourable sidechain contacts on the inside of the barrel but do not unduly strain the backbone conformation. Murzin et al.[41] showed theoretically that for large β barrels with a central water-filled cavity $S = N + 4$ resulted in the best geometry. This combination has been found for the E. coli porins OmpF,[11] PhoE[11] and *Rhodobacter capsulatas* porin,[43] which crystal structures reveal to have $N = 16$ and $S = 20$. The structure of maltoporin[44] has a larger barrel with $N = 18$ and $S = 22$. The α-toxin heptamer structure is an exception to this rule with $N = 14$ and $S = 14$.[31]

To explore the possibility that the hexamer may adopt a different stagger from the heptamer two separate 12 stranded β barrels were constructed. One was made to have $S = 12$, following the example of the heptamer and the other was made to adopt a great stagger with $S = 16$. Initial coordinates for each conformation were generated using existing models of alanine decamers β barrels with ($N = 12$, $S = 12$) and ($N = 12$, $S = 16$) produced by Sansom and Kerr[45] by molecular dynamics simulated annealing. To produce barrels of the required length (20 amino acids) the model barrels were duplicated and joined end to end. Loops of three amino acids, in a β turn conformation, were added between alternate strands. Individual amino acids were then "mutated" from alanine to the α-toxin pore domain sequence, using the QUANTA package (Molecular Simulations, Inc).

This resulted in starting conformations for regular ($N = 12$, $S = 12$) and ($N = 12$, $S = 16$) β barrels, with good hydrogen bond geometry. The models consist of six separate polypeptide strands with the same sequence and broad topology as the α-toxin pore domain of the heptamer crystal structure. Each of the models was then refined using the X-PLOR package.[46] The PARAM19 extended atom parameter set was used. Non-crystallographic symmetry was imposed between the six subunits. This forces the individual monomers to adopt a similar conformation resulting in a regular geometry. The models were subjected to 10 000 steps of Powell energy minimization,[47] to remove bad contacts. Following this the non-crystallographic restraints were removed and a further 5000 steps of minimization were undertaken.

The ($N = 12$, $S = 16$) more staggered conformation was found to have a larger potential energy than the ($N = 12$, $S = 12$) (Table 1). A more detailed comparison between the different terms contributing to the potential energy function shows that of the bonded terms only the bond angle is markedly different. The absolute values of the energy show that both conformations are relatively strained and the difference in bond angle energy that the more staggered conformation is more strained than the less.

Most of the potential energy difference lies in the non-bonded terms. In particular a large difference in the van der Waals term reveals poorer side chain packing. The smaller difference in electrostatic energy shows that the hydrogen bonds between main chain groups within the barrel have a worse geometry in the more staggered conformation.

Table 1 XPLOR potential energy in kcal mol^{-1} for α-toxin hexamer models

Model	Total energy	Bonds and torsions	Bond angle term	Van der Waals	Electrostatic
($N = 12, S = 12$)	−13 104	465	308	−1759	−12 119
($N = 12, S = 16$)	−12 424	466	379	−1590	−11 681

The relative stability of the two different staggers is in contrast to the results found for the model polyalanine barrels[45] used in constructing the domain. This effect can be seen to arise from two principal sources.

An important difference is that the polyalanine barrel is constructed from individual unconnected strands, whereas the α-toxin barrel is composed of two-stranded covalently linked loops. For the barrel to be regular (i.e. the monomers to be quasi-equivalent), geometrical considerations necessitate that $S = \gamma N$, where γ is an even integer. Other geometries, such as $S = N + 4$ can only be achieved if at least one of the loops (i.e. two strands of the barrel) are translated along the barrel axis relative to the other strands. Such a distortion is observed in the ($N = 12$, $S = 16$) structure modelled here, and leads to the strain discussed above. While such an arrangement may be allowed in soluble proteins, such a distortion would be highly unfavourable in the case of α-toxin. The ends of the barrel are clearly delimited by a ring of charged residues at one end (Asp127, Asp128 and Lys131) and a ring of aromatic residues at the other (Tyr118 and Phe120). Any distortion of the protein loops would mean that the charged residues would penetrate into the core of the membrane or the aromatic residues would be exposed to the extra-membrane environment. It is possible to exclude the $S = 2N$ and $S = 4N$ as the barrels would be too short to span the membrane.

The sequence of the pore domain provides an additional feature which favours the $S = 12$ form. One of the major factors in restricting the possible geometries of β barrels is the packing of the amino acids sidechains on the inside of the barrel. The $S = N + 4$ stability rule arises partly from this effect.[41] Unusually the pore domain is rich in glycine residues, which are generally positioned towards the barrel centre[31] (Fig. 7). This in effect releases the geometric constraints on the barrel allowing it to adopt other conformations. By these considerations the ($N = 12$, $S = 12$) β barrel was taken as the model for the pore domain of the putative α-toxin hexamer and attention turned to constructing the full oligomer from this core.

5.2.2 Adding the head group. The starting assumption was that the individual head group domains in the hexamer would adopt similar conformations to those of the heptameric crystal structure. The six individual head domains were then packed together using a geometrical procedure, coded as a CHARMm[48] script. The starting point was the structure for a single head domain (residues 1–108 and 152–293), taken from the crystal structure.[31] Five further copies of this structure were generated on top of each other; with each copy rotated 60° with respect to the previous copy around their common centroid. The six head groups were progressively moved outwards from the centroid, in 0.5 Å steps. The direction of movement was such that the orientation of the monomers with respect to the centre of the channel was kept the same as that found in the crystal structure. The potential energy of complex was evaluated, using the CHARMm potential energy function at each step during the transformation.

The process resulted in a clear potential energy minimum at channel centre–centroid distance of 30 Å (data not shown). This conformation was adopted as a starting structure for the hexameric head group. To remove local bad contacts between the monomers the structure was subjected to energy minimization using the steepest descents and adopted basis set Newton–Raphson techniques implemented with CHARMm.[48]

Visual comparison between the model for the hexameric head group model, developed here, and the model hexameric β barrel stem, described in Section 5.2.1, showed that they were compatible. The head and stem were joined together manually using the quanta package. The complete molecule was then subject to energy minimization. The local geometry of the resulting molecule was then checked using the PROCHECK package.[49] A few improbable or unphysical side chain

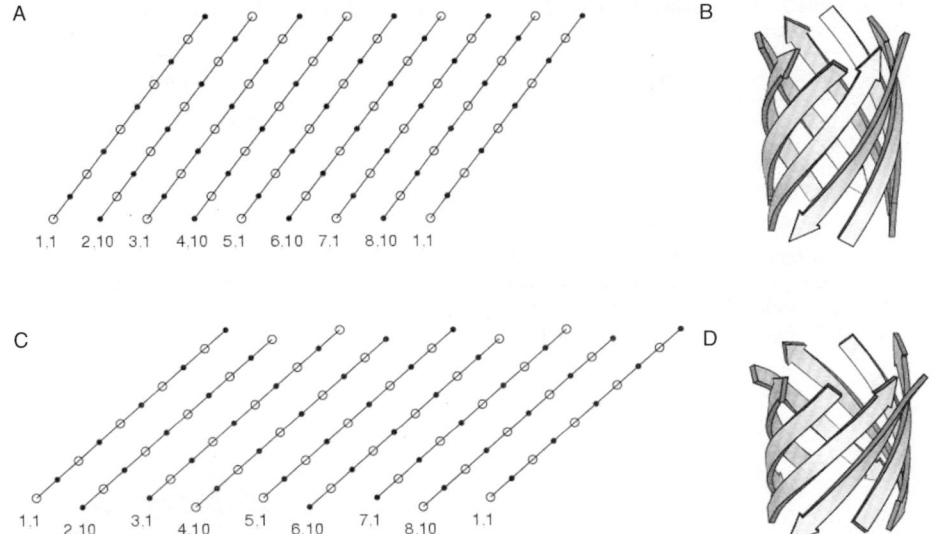

Fig. 7 The effects of changing the stagger number for $N = 8$ β barrel. The diagrams on the left show the topology of the barrel when it is unrolled. In each case strand β1 is shown twice. Empty circles mark residues whose side chains lie above the page whereas filled dots show side chains below the page. Panels A and B show the ($N = 8$, $S = 8$) structure. Panels C and D show the more staggered ($N = 8$, $S = 12$) structure. Figure adapted from Sansom and Kerr.[45]

geometries were found. There were corrected using manual repositioning (with the exception of those that were present in the heptameric crystal structure).

5.3 Analysis of α-toxin structures

5.3.1 Local geometry. An important issue is the compatibility of this model structure with the images of α-toxin hexamers obtained by EM[38] and AFM.[39] The direct comparison of the images with the structure is the best way of resolving this. However, in the absence of this information some idea can be gained by examination of dimensions quoted in the works (although these must be treated with caution, as they are dependent on interpretation). Table 2 shows that the hexamer model is compatible with the outer diameter for hexamers measured by EM and AFM. Interestingly the quoted outer diameter for heptameric pre-pores obtained from AFM[40] is larger and in accord with the crystal structure[31] (Table 2).

An analysis of the protein geometry revealed that there were no overall changes in the structure of the protein. However, several residues have poor stereochemistry. Ser 262 has a forbidden conformation in all of the subunits in the hexameric model, but also appears strained in the original heptameric crystal structure. Ser 262 lies at the apex of a β-turn at the end of the barrel domain. The hydrogen-bonding network between residues in the turn, especially Thr 261 and Asn 264, forces Ser 262 into its poor conformation. It is interesting to note that these residues are located at the membrane/protein interface. It is likely that the polar surface of the membrane stabilizes the turn by interactions with the polar residues in the loop.

5.3.2 Analysis using HOLE. The dimensions of pore in the hexamer are similar to the heptamer (Fig. 8 and 9). The overall profile of the channel is similar to that of the heptamer (Fig. 9). The length of the channel is the same for the hexamer and the heptamer, while the diameter of the channel is narrower for the hexamer. The minimum radius of the hexamer pore is 2.5 Å compared to 5.4 Å for the heptamer. This minimum occurs at a constriction caused by an annulus of charged residues that lie towards the distal end of the β barrel region of the pore.

Table 2 Comparing the outer diameter for the α-toxin head group measured by imaging methods and from structural models

Ref.	Method	Stoichiometry	Outer diameter/Å
38	EM	6mer	70–80
39	AFM	6mer	76
40	AFM	7mer (pre-pore)	89
31	X-ray crystallography	7mer	90[a]
This work	Modelling	6mer	75[b]

[a] Distance between atoms identified in the experimental structure as C_α A : 71 and C_α D : 270 (from the point outer point of one subunit to the flat face of the opposite). [b] Distance between the C_α C : 276 and $C_{2\alpha}$ F : 276, between the opposing flat faces of the hexamer.

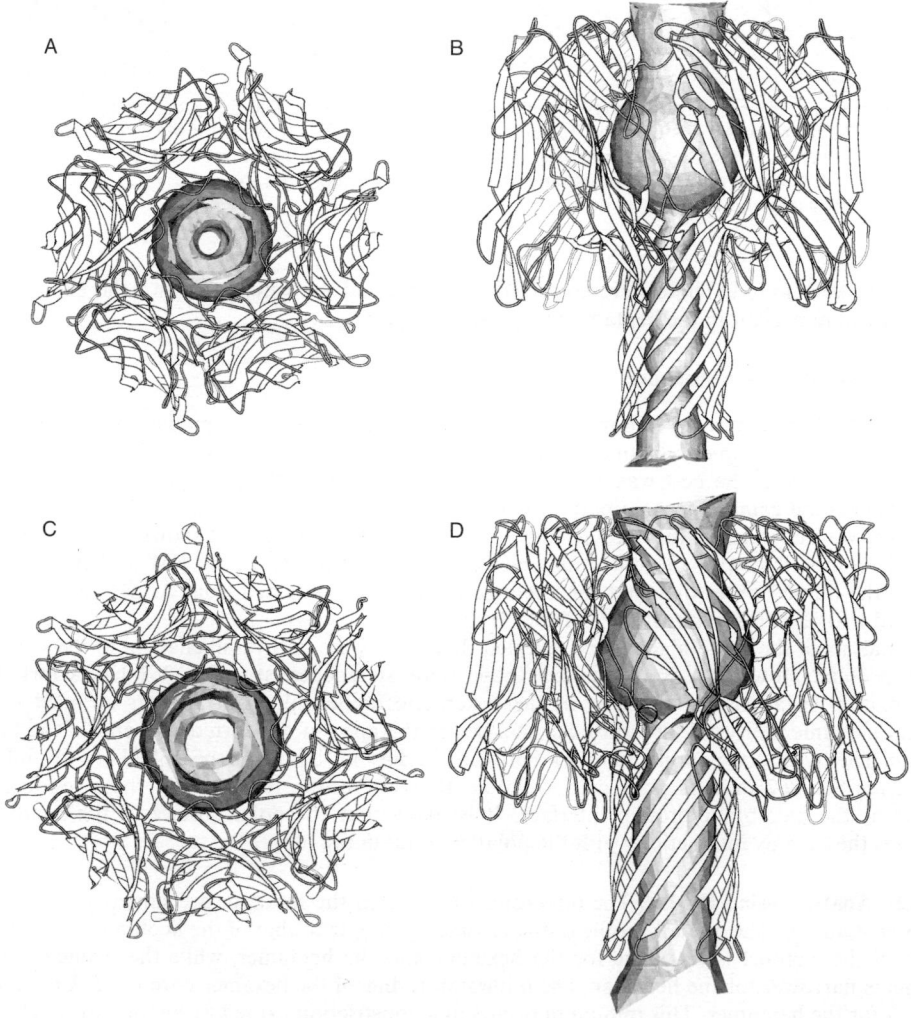

Fig. 8 Orthogonal views of the HOLE surface of the heptamer (panels A and B) and the hexamer (panels C and D). Produced using the molscript program.[52]

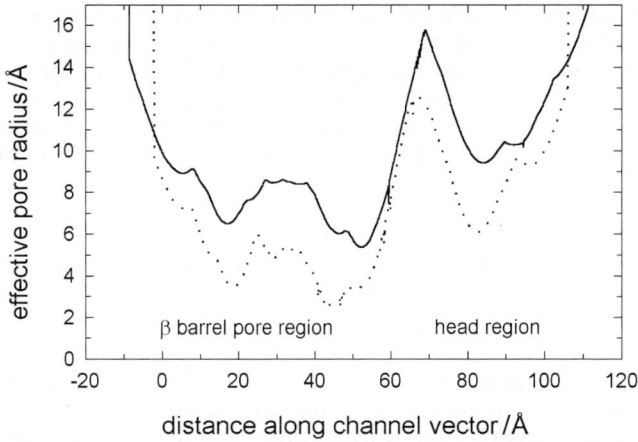

Fig. 9 Comparing the pore radius for the X-ray crystal structure[31] for α-toxin heptamer (solid line) and model hexamer (dotted line).

The HOLE procedures, described in Sections 3 and 4, were used to predict both the absolute conductance and expected effect of the addition of PEG for both the experimental heptamer and model hexamer structures. Table 3 compares the absolute conductances and Fig. 10 shows the effect of PEG.

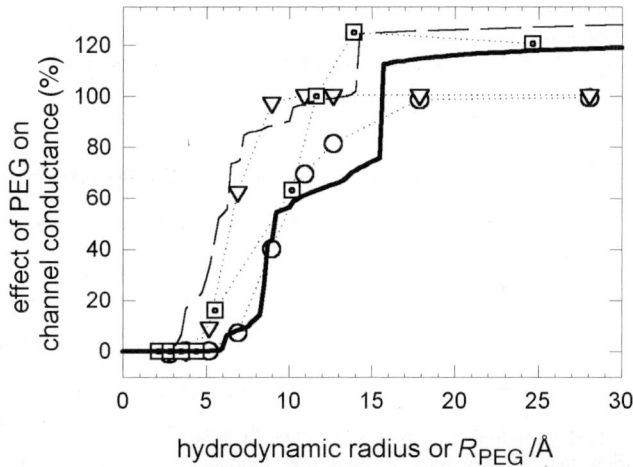

Fig. 10 Effect of PEG on the conductance of α-toxin. Points marked by circles and triangles show the effect found experimentally by Korchev *et al.*[19] for the high and low conductance states, respectively. Points marked by dotted squares show the experimental observations for the high conductance state by Krasilnikov *et al.*[18] The thick solid line shows the effect predicted for the heptameric crystal structure[31] and the dashed line for the model hexamer.

Table 3 HOLE prediction of the conductance of α-toxin

	Conductance/pS (0.1 M KCl)
Experimental—high conductance state[a]	90
Experimental—low conductance state[b]	<10
Prediction heptamer crystal structure[c]	52, 44[d]
Prediction for model hexamer structure	18, 18[d]

[a] Ref. 51. [b] Ref. 19. [c] Ref. 31. [d] 1st and 2nd generation predictions.[5]

The data suggest that it is possible that both hexameric and heptameric forms may be seen in conductance measurements. The low conductance state would correspond to the hexamer and the high conductance state to a heptamer. This provides an explanation why the two forms have been observed in imaging techniques. However, it leaves a rather large problem, namely, how can the channel rapidly switch between the two forms. This switching would entail the making and breaking of a large number of hydrogen bonds. However, the insertion and pore-forming processes will also involve such changes. The hypothesis is worthy of further consideration.

Work in progress seeks to establish the compatibility of the hexamer with AFM and EM data. We also plan to undertake electrostatics calculations to explain the pH sensitivity of the channel and the difference in ion selectivity observed between the high and low conductance states.[50]

6 Conclusion

This work shows the importance of systematically linking the structure of ion channels to their structure. The approach has great promise in aiding both modelling studies and the interpretation of experimental data. An important aim is to provide useful tools for the general scientific community working on ion channels. For this reason the program suite HOLE2 is freely available to all non-profit organizations; for further details see http://www.biochemistry.bham.ac.uk/hole, or e-mail: o.s.smart@bham.ac.uk.

This work was supported by: the Wellcome Trust through the provision of a Career Development Fellowship (042889 to O.S.S.), the UK Medical Research Council (grant G4600017), and the Cell Surface Research Fund. We are most grateful to Drs Peter Tieleman and Herman Berendsen, of the University of Groningen, for the provision of data,[10] used in Section 3.1. We thank Joe Neduvelil and Xiaonan Wang for their contributions to the HOLE package. HOLE was originally developed at Birkbeck College with the support of Dr Bonnie Wallace and Prof. Julia Goodfellow.

References

1 B. Hille, *Ionic Channels of Excitable Membranes*, 2nd edn., Sinauer Associates, Inc., Sunderland, MA, 1992.
2 P. D. Adams, I. T. Arkin, D. M. Engelman and A. T. Brunger, *Nature (London), Struct. Biol.*, 1995, **2** 154.
3 M. S. P. Sansom, C. Adcock and G. R. Smith, *J. Struct. Biol.*, 1998, **121**, 246.
4 O. S. Smart, J. M. Goodfellow and B. A. Wallace, *Biophys. J.*, 1993, **65**, 2455.
5 O. S. Smart, J. Breed, G. R. Smith and M. S. P. Sansom, *Biophys. J.*, 1997, **72**, 1109.
6 M. L. Connolly, *Science*, 1983, **221**, 709.
7 B. Roux, in *Theory of transport in ion channels: from molecular dynamics simulations to experiments*, ed. J. M. Goodfellow, Weinham, 1995.
8 J. Breed, R. Sankararamakrishnan, I. D. Kerr and M. S. P. Sansom, *Biophys. J.*, 1996, **70**, 1643.
9 J. E. Hall, *J. Gen. Physiol.*, 1975, **66**, 531.
10 D. P. Tieleman and H. J. C. Berendsen, *Biophys. J.*, 1998, **74**, 2786.
11 S. W. Cowan, T. Schirmer, G. Rummel, M. Steiert, R. Ghosh, R. A. Pauptit, J. N. Jansonius and J. P. Rosenbusch, *Nature (London)*, 1992, **358**, 727.
12 R. Benz, A. Schmid and R. E. W. Hancock, *J. Bacteriol.*, 1985, **162**, 722.
13 B. K. Jap and P. J. Walian, *Q. Rev. Biophys.*, 1990, **23**, 367.
14 T. B. Woolf and B. Roux, *Proteins*, 1996, **24**, 92.
15 L. Y. Shen, D. Bassolino and T. Stouch, *Biophys. J.*, 1997, **73**, 3.
16 M. S. P. Sansom, personal communication.
17 M. Watanabe, J. Rosenbusch, T. Schirmer and M. Karplus, *Biophys. J.*, 1997, **72**, 2094.
18 O. V. Krasilnikov, R. Z. Sabirov, V. I. Ternovsky, P. G. Merzliak and J. N. Muratkhodjaev, *Fems Microbiology Immunology*, 1992, **105**, 93.
19 Y. E. Korchev, C. L. Bashford, G. M. Alder, J. J. Kasianowicz and C. A. Pasternak, *J. Membr. Biol.*, 1995, **147**, 233.
20 O. V. Krasilnikov, J. B. DaCruz, L. N. Yuldasheva, W. A. Varanda and R. A. Nogueira, *J. Membr. Biol.*, 1998, **161**, 83.
21 S. M. Bezrukov and I. Vodyanoy, *Biophys. J.*, 1993, **64**, 16.
22 E. A. Merritt, S. Sarfaty, F. Vandenakker, C. Lhoir, J. A. Martial and W. G. J. Hol, *Protein Sci.*, 1994, **3**, 166.
23 O. V. Krasilnikov, J. N. Muratkhodjaev, S. E. Voronov and Y. V. Yezepchuk, *Biochim. Biophys. Acta*, 1991, **1067**, 166.

24 J. Breed, P. C. Biggin, I. D. Kerr, O. S. Smart and M. S. P. Sansom, *Biochim. Biophys. Acta Biomembranes*, 1997, **1325**, 235.
25 B. A. Wallace, *J. Struct. Biol.*, 1998, **121**, 123.
26 R. R. Ketchem, W. Hu and T. A. Cross, *Science*, 1993, **261**, 1457.
27 B. Turano, M. Pear and D. Busath, *Biophys. J.*, 1992, **63**, 152.
28 V. B. Myers and D. A. Haydon, *Biochim. Biophys. Acta*, 1972, **274**, 313.
29 J. A. Killian, *Biochim. Biophys. Acta*, 1992, **1113**, 391.
30 S. Bhakdi and J. Tranumjensen, *Microbiol. Rev.*, 1991, **55**, 733.
31 L. Z. Song, M. R. Hobaugh, C. Shustak, S. Cheley, H. Bayley and J. E. Gouaux, *Science*, 1996, **274**, 1859.
32 N. Tobkes, B. A. Wallace and H. Bayley, *Biochemistry*, 1985, **24**, 1915.
33 J. A. Gouaux, G. Braha, M. R. Hobaugh, L. Z. Song, S. Cheley, C. Shustak and H. Bayley, *Proc. Natl. Acad. Sci. USA*, 1994, **91**, 12828.
34 A. Valeva, M. Palmer and S. Bhakdi, *Biochemistry*, 1997, **36**, 13298.
35 A. Valeva, J. Pongs, S. Bhakdi and M. Palmer, *Biochim. Biophys. Acta Biomembranes*, 1997, **1325**, 281.
36 C. Petosa, R. J. Collier, K. R. Klimpel, S. H. Leppla and R. C. Liddington, *Nature (London)*, 1997, **385**, 833.
37 E. Gouaux, *Curr. Opinion Struct. Biol.*, 1997, **7**, 566.
38 R. J. Ward and K. Leonard, *J. Struct. Biol.*, 1992, **109**, 129.
39 D. M. Czajkowsky, S. Sheng and Z. Shao, *J. Mol. Biol.*, 1998, **276**, 325.
40 Y. Fang, S. Cheley, H. Bayley and J. Yang, *Biochemistry*, 1997, **36**, 9518.
41 A. G. Murzin, A. M. Lesk and C. Chothia, *J. Mol. Biol.*, 1994, **236**, 1382.
42 A. G. Murzin, A. M Lesk and C. Chothia, *J. Mol. Biol.*, 1994, **236**, 1369.
43 M. S. Weiss and G. E. Schulz, *J. Mol. Biol.*, 1992, **227**, 493.
44 T. Schirmer, T. A. Keller, Y. F. Wang and J. P. Rosenbusch, *Science*, 1995, **267**, 512.
45 M. S. P. Sansom and I. D. Kerr, *Biophys. J.*, 1995, **69**, 1334.
46 A. T. Brunger, A. Krukowski and J. W. Erickson, *Acta Crystallogr., Sect. A*, 1990, **46**, 585.
47 R. Fletcher, *Practical Methods of Optimization: Volume 1 Unconstrained Optimization*, John Wiley & Sons, Chichester, 1980.
48 B. R. Brooks, R. E. Bruccoleri, B. D. Olafson, D. J. States, S. Swaminathan and M. Karplus, *J. Comput. Chem.*, 1983, **4**, 187.
49 R. A. Laskowski, M. W. Macarthur, D. S. Moss and J. M. Thornton, *J. Appl. Crystallogr.*, 1993, **26**, 283.
50 Y. E. Korchev. C. L. Bashford, G. M. Alder, P. Y. Apel, D. T. Edmonds, A. A. Lev, K. Nandi, A. V. Zima and C. A. Pasternak, *Faseb. J.*, 1997, **11**, 600.
51 G. Menestrina, *J. Membr. Biol.*, 1986, **90**, 177.
52 P. J. Kraulis, *J. Appl. Crystallogr.*, 1991, **24**, 946.

Paper 8/06771F

Molecular dynamics simulation of a hydrated diphytanol phosphatidylcholine lipid bilayer containing an alpha-helical bundle of four transmembrane domains of the Influenza A virus M2 protein

Thomas Husslein,[a,b] Preston B. Moore,[a] Qingfong Zhong,[a] Dennis M. Newns,[b] Pratap C. Pattnaik[b] and Michael L. Klein[a]

[a] *Centre for Molecular Modelling and Department of Chemistry, University of Pennsylvania, Philadelphia, Pennsylvania 19104-6323, USA*
[b] *T. J. Watson Research Center, International Business Corporation, P.O. Box 218, Yorktown Heights, NY 10598, USA*

Received 25th August 1998

An α-helical bundle composed of four transmembrane portions of the M2 protein from the Influenza A virus has been studied in a hydrated diphytanol phosphatidylcholine bilayer using molecular dynamics (MD) calculations. Experimentally, the sequence utilized is known to aggregate as a four-helix bundle and act as a pH-gated proton-selective ion channel, which is blocked by the drug amantadine hydrochloride. In the presented simulation, the ion channel was initially set up as a parallel four-helix bundle. The all-atom simulation consisted of almost 16 000 atoms, described classically, using a forcefield from the CHARMM22 database. Bilayers with and without the bundle were shown to be stable throughout the nanosecond timescale of the MD simulation. Structural and dynamical properties of the bilayer both with and without the transmembrane protein are reported.

I. Introduction

The ion channel is understood to play a key role in a wide range of biological phenomena such as nerve conduction, heartbeat, toxic response, and disease, and it can be a site for attack by drug molecules. Due to the difficulty in crystallizing membrane proteins relatively few ion channel structures are known. Thus, in spite of their importance, many key aspects of ion channels are ill understood at the present time. However, just recently, the structure of the potassium channel KscA has been reported.[1,2] The main structural motif in this channel consists of an alpha helix bundle. The alpha helix bundle is an important structural element in other channel forming proteins such as acetylcholine receptors.[3,4] Thus the system of an alpha-helical bundle in a lipid bilayer is of general importance in the context of ion-channel function.

The present article is concerned with an alpha helix bundle whose structure is not yet characterized at the atomic level. Specifically, we focus on the membrane protein termed M2, which is associated with the Influenza A virus.[5] The M2 protein is known to form an ion channel and play an essential role in infection. It is gated by hydrogen ion concentration (pH). The channel is the site of attack by anti-flu drugs.[5] As already mentioned, the structure of the channel formed by the 96-residue M2 protein is not yet well characterized. Accordingly, we focus on a simpler system,

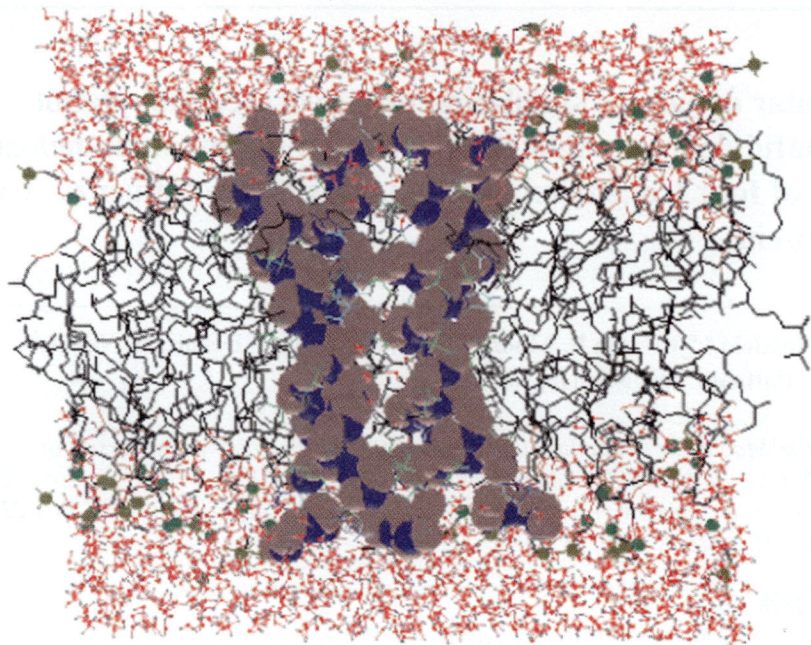

Fig. 1 Snapshots of the peptide bundle in a lipid bilayer taken from MD simulation at 300 ps. The four peptide backbones atoms have been drawn with Van der Waals radii. The lipids have been rendered in a ball and stick representation. The lipid head groups nitrogen and phosphorus atoms are drawn as balls. The hydrogens have been omitted for clarity.

namely the fully functional channel formed by a bundle of four 25-residue peptides with the M2 transmembrane sequences.[6–8] This model ion channel is also blocked by the drug amantadine hydrochloride.[6,7] The restricted objective in this paper is to understand how the insertion of this transmembrane peptide bundle affects the structure and the dynamics of a typical lipid bilayer.

The lipid we have chosen for study is diphytanol phosphatidylcholine (DPhPC), which has been successfully used in many investigations of protein–bilayer interactions, especially in experiments on channel-forming proteins. Its high resistance to proton and other ion flow account for the popularity of DPhPC in ion-conduction measurements.[9] Additionally, there is a general belief that this particular lipid has a high bilayer stability, and acyl chain packing comparable to that of naturally occurring lipid bilayers. Furthermore, it provides a well defined single-component lipid for experiments, whereas the above-mentioned characteristics are traditionally obtained by a using a mixture of different lipids.

In the next section we outline the simulation techniques and the way we set up the system. Then we describe the results of the MD simulations and address the impact of the insertion of the peptide bundle to the lipid bilayer. Finally, we present our conclusions.

II. Methods and materials

Diphytanol phosphatidylcholine is a rather typical phospholipid. It consists of a zwitterion phosphatidylcholine head group, linked to a glycerol ester and two hydrocarbon chains. In the case of DPhPC the hydrocarbon chains are branched. The main chain contains 16 carbons, but at the positions 15, 11, 7, and 3 there are methyl groups attached, so each chain consists of 20 carbon atoms. Technically, the atoms 11, 7, and 3 are chiral centres. These have been treated as having an all R configuration, in agreement with ref. 10. For the peptide bundle, we used the truncated sequence of the M2 protein (M2-DA), described by Duff and Ashley,[7] which was shown to function as a pH gated ion channel, that is also blocked by the drug amantadine hydrochloride. The sequence is from Ser22-Leu46 with standard capping residues Ac and NH_2: Ac-SSDPLVVAASIIGILHLILWILDRL-NH_2.

Both the DPhPC bilayer and the M2-DA protein were simulated using an all atom description, specifically the recently developed PARAM22b4b database of the CHARMM22 program package.[11] For the histidine (His37) in the M2-DA protein only the NE nitrogen in its aromatic ring was protonated, which is consistent with a pH of 7. The water was simulated with the flexible TIP3P model.[12]

The simulation system consisted of 4 M2-DA proteins, 50 DPhPC lipids, and 1996 water molecules. The M2-DA proteins were set up as straight α-helices, not tilted against each other, to form a four-fold symmetric bundle with the residues Ser31, His37 and Trp41 pointing towards the centre of the pore. This bundle was then embedded in the DPhPC lipid bilayer. To do so, we used a well-equilibrated structure from a simulation on the pure lipid, in which we created a hole by deleting 14 lipid molecules. The water of hydration in the pure lipid simulation was retained and an additional 170 waters were added to fill the void in the head group region, a further 36 waters were used to fill the interior of the bundle. Fig. 1 shows a typical configuration taken from the MD simulation.

The phase behaviour and dimensions of lipid bilayers are well known to be sensitive to temperature, pressure, hydration and ionic concentration.[9,13,14] We therefore employed a constant NPT simulation of a fully flexible box using the MTK extended system approach as described in ref. 15 and 16. We used a chain length of 3 for the thermostats. The cut-off for the interactions was set to 10 Å. Additionally, we employed a RESPA scheme to perform the calculation of the forces on their appropriate timescales.[17] The time-step we used for the update of the long-range Coulombic and non-bonded forces was 3 fs, the update of the short-range non-bonded forces was 1.5 fs, whereas the update of the bond stretch, bend, and dihedral interactions occurred every 0.375 fs. We used a Ewald summation technique for the long range forces. The reciprocal space summation was truncated at $k_x = k_y = k_z = 10$.

III. Structural effects

In a recent publication we compared an MD simulation on a fully hydrated DPhPC bilayer with available experimental data[18] and with MD results for DPPC.[19] Here, we investigate the influence of the inserted peptide bundle on the DPhPC bilayer structure. Specifically, we focus on the structure of the lipid, both with and without the M2-DA peptide bundle.

Fig. 2 shows the average separation along the bilayer normal between the peaks associated with the distributions of the phosphate atoms on each side of the bilayer. This measure of the bilayer thickness reveals no significant drift suggesting that the overall system dimensions have converged.

The lipid phosphate group separation across the bilayer for the case of the embedded peptide bundle is about 33 Å, which is 5 Å less than in the pure lipid case. Assuming a constant packing density for the lipid molecules, the surface area per lipid should be increased correspondingly. Accordingly, we estimate that the surface area per lipid should likely increase from 74.5 to ca. 85 Å^2. This increase in area per head group manifests itself in the changes seen in Fig. 3, where the maximum for the P-P in-plane radial distribution function is at ca. 6.1 Å in the case of the pure DPhPC, compared to a broad peak at ca. 6.7 Å for the bilayer with the peptide bundle.

In a more detailed analysis of the structure we carried out a Voronoi-type tessellation, in which individual lipid phosphate groups were assigned their own area.[20] To do so, we first construct a Wigner-type cell around each lipid, to encompass the area that is closest to it. This construction is space filling and takes into account the periodic boundaries of the simulation box. The analysis was performed on systems both with and without the M2-DA peptide bundle in the lipid bilayer. The surface roughness of the bilayer was ignored.

The surface area covered by the peptide bundle is not estimated easily. In particular, the fact that the profile of the bundle can be convex demands a rather refined method. In the present case, for the calculation of the area of the bundle, we have only taken into account those atoms that are submerged in the leaflet of the bilayer that is under consideration. Their positions are projected onto the plane of the bilayer with each atom assigned its van der Waals radius. The area occupied by the bundle is then measured by dividing the bilayer plane into a dense grid. For each element

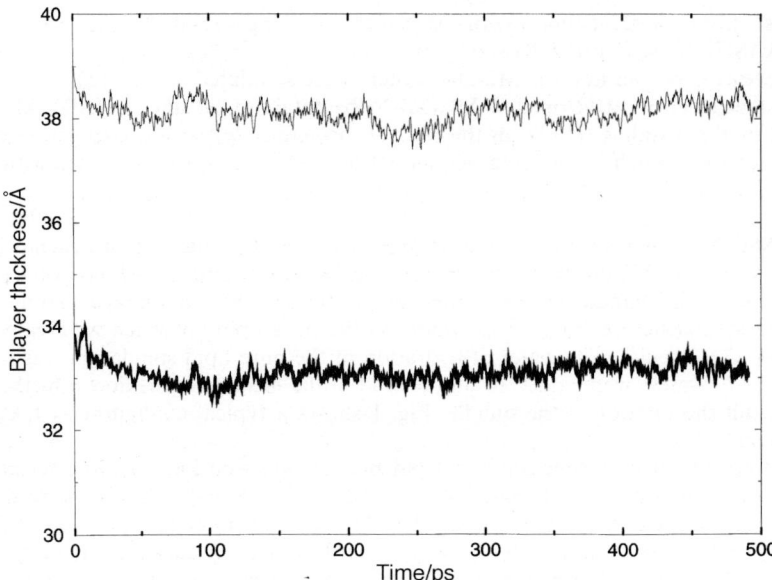

Fig. 2 Time evolution of the average peak-to-peak separation of phosphate group distributions across the bilayer for DPhPC without (thin line) and with the M2-DA peptide bundle (bold line).

of the grid we then decide whether it is within the outline of the protein or outside. When it is within, we find the corresponding Wigner-type cell, which was previously constructed, and subtract the area from there.

We utilized a cylindrical coordinate system for this analysis, slicing the area like a pie. We move the origin of the coordinate system to the centre of mass. Radiating out from the origin, we draw a line, at an angle θ from the x-axis. Starting from well outside the boundaries of the peptide bundle we follow the line until we find a peptide atom, defined by its van der Waals radius. After incrementing the θ by dθ, we draw another line and repeat the above procedure. The two radii and the origin define a triangle that gives an estimate of the area covered by the peptide in this

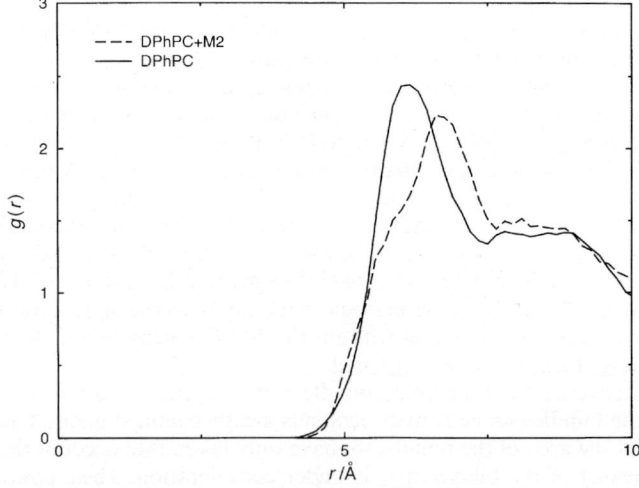

Fig. 3 The in-plane phosphorus-phosphorus radial distribution functions $g(r)$ for DPhPC without (thin line) and with (dashed line) the M2-DA peptide bundle.

segment. We now sub-grid this triangle and find the Wigner-type cell of the lipid it falls into. We checked this procedure for several sizes of the segments $d\theta$ and different sub-grids for the triangle, until the area became independent of the choice of these parameters.

For our simulation of the M2-DA peptide bundle in the lipid bilayer we find an area per lipid molecule of 85 Å^2 and without a peptide bundle we reproduce the area per lipid as 74.6 Å^2. Fig. 4 shows snapshots of the lipid bilayer at different times. It illustrates the behaviour of the Wigner-type cells for a lipid with and without a peptide bundle. In the case without the peptide, the smaller area per lipid is clearly evident.

The comparison of the NMR order parameter S_{CD} for DPhPC with and without the peptide bundle is shown in Fig. 5. In contrast to the terrace-like structure that is characteristic for the pure DPhPC case, the order parameter for the system with the peptide bundle falls off smoothly with increasing position along the chain, indicating a more fluid-like behaviour. Furthermore, the overall magnitude is also reduced, hence the chain order is decreased.

Additionally, we have investigated the orientation of the head-groups and the lipid hydrocarbon chains with respect to the bilayer normal. We measured the angle Φ, the angle between the vector connecting the first (C-1) and the last chain carbon atom (C-15) and the bilayer normal. Fig. 6 compares the probability distribution for Φ with and without the bundle in the bilayer. Upon insertion of the peptide bundle, the peak value for Φ increases from *ca.* 20° to larger values. This behaviour is consistent with the observation that the bilayer thickness is reduced (recall Fig. 2) when the peptide bundle is inserted and, hence, the chains are not as extended as before. We observe that the probability for a chain to be almost parallel to the bilayer surface is also reduced. We have also investigated the probability distribution of the angle Ψ, which is the angle between a

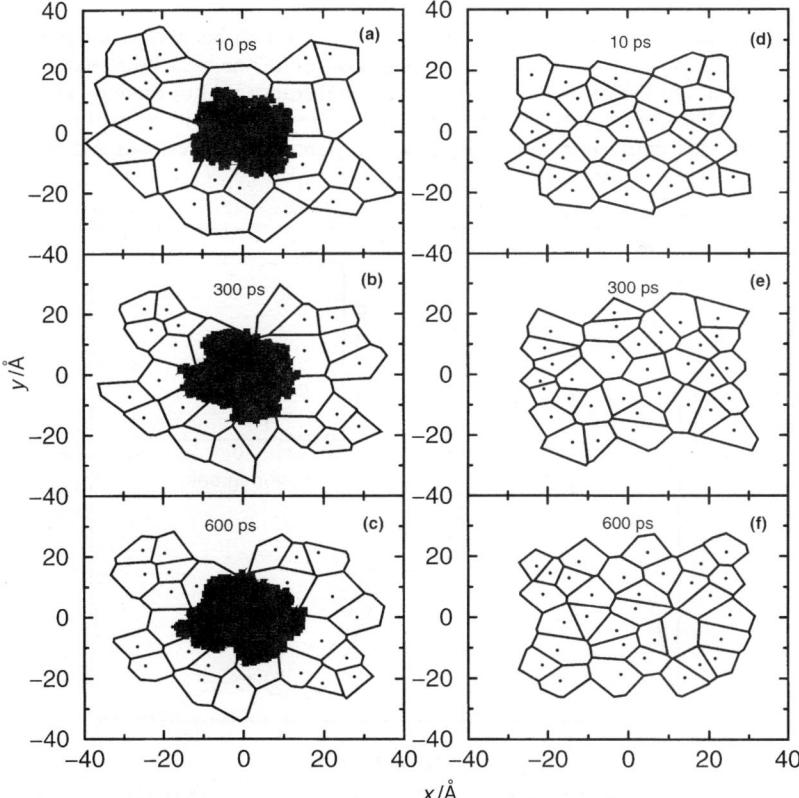

Fig. 4 Snapshots of one leaf of the lipid bilayer, where we indicate the area per lipid by a Wigner-type cell construction for different simulation times. (a)–(c) show the bilayer with the M2-DA peptide bundle within, whereas (d)–(f) show the pure lipid.

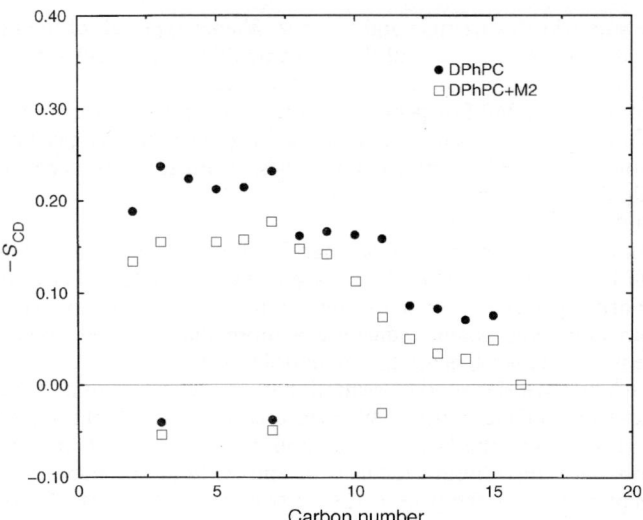

Fig. 5 Hydrocarbon chain deuterium order parameters S_{CD} as a function of the carbon position along the main chain for DPhPC both with (□) and without (●) the peptide bundle.

vector connecting P and N of the head-group and the plane of the lipid bilayer. Here, we find no major differences between the two cases.

Fig. 7(a) gives the average projected structure of the α-carbon backbone of the M2-DA peptide in the lipid. On the abscissa we show the average position of the α-carbon atoms with respect to the bilayer normal. On the ordinate we plot the mean distance $r(Å)$ of these α-carbon atoms from the channel axis. The data points are an average over the four individual peptides, and the whole trajectory. From this figure one can easily pick the pore-forming residues Ala-30,Ser-31,Gly-34,

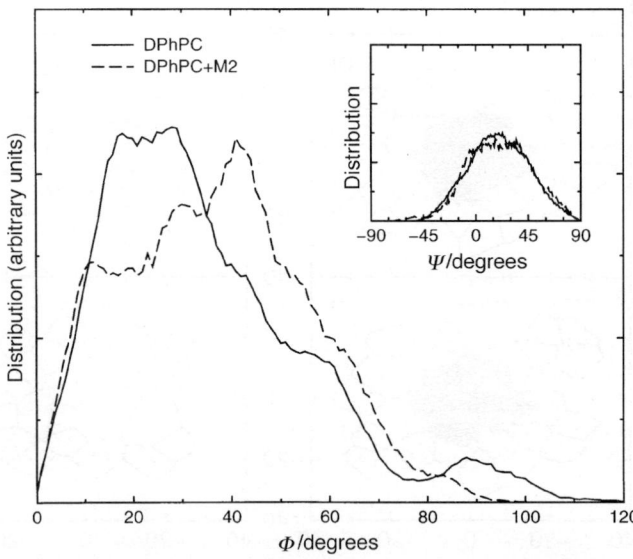

Fig. 6 Distribution of the angle Φ between the vector connecting the first (C-1) and the last (C-15) main chain carbon atom and the bilayer normal for pure DPhPC (———) and DPhPC with the M2-DA peptide (– – – –). The inset shows the angle Ψ of the vector connecting the head-group P and N atoms and the plane of the lipid bilayer for both cases. (The scale of the y-axis of both graphs is identical.)

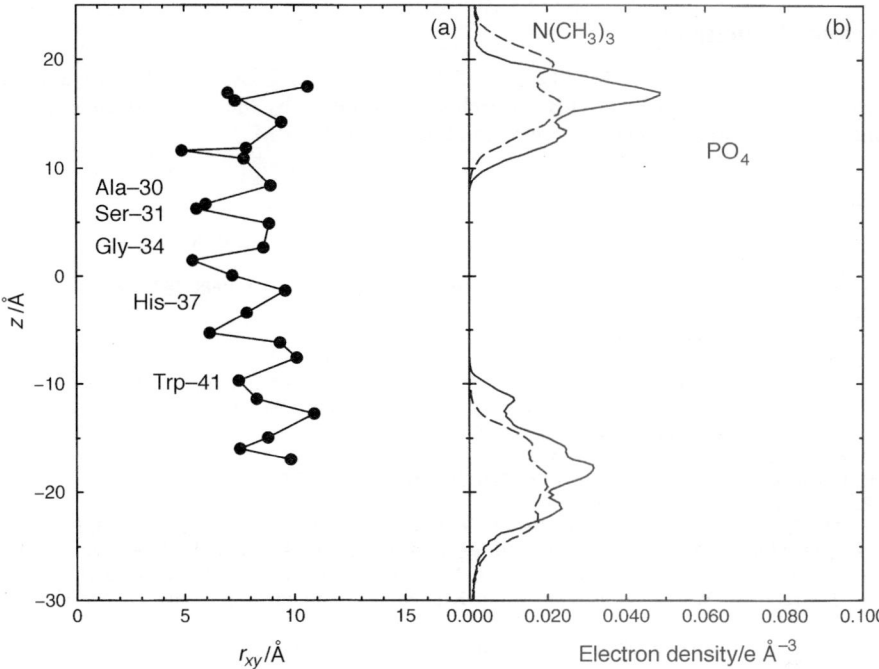

Fig. 7 Positions of important parts of the system with respect to the bilayer normal: (a) shows the average separation of the individual α-carbon atoms from the channel axis, only the pore facing residues are indicated; (b) shows the contributions to the electron density distributions associated with phosphorus (———) and nitrogen (– – – –) of the lipid bilayer head-groups.

His-37 and Trp-41. Both ends of the peptide are not α-helical but are more extended. The profile of the averaged peptide backbone reveals a twist. The position where the backbone is closest to the pore axis is at the Gly-34 position. The profile shape should not misinterpreted. Individual peptides from the bundle do not have a kink. The curved profile arises solely from the fact that the individual peptides are tilted and coiled with respect to each other, giving rise to this hourglass shape.[21]

In Fig. 7(b) we show the average positions of the lipid head groups with respect to the channel axis. Here, the channel does not seem to reside symmetrically in the bilayer. However, this only reflects the fact that at the top of the bilayer the polar Ser residues are rather short, whereas the Arg residues at the bottom have rather long side chains, with charges at the very end.

IV. Conclusion

We have reported on the structural consequences of inserting a four-helix peptide bundle into a DPhPC bilayer and compared our findings with a previous simulation of a pure DPhPC bilayer. Both simulations were performed in the NPT ensemble and covered the nanosecond timescale. We did not apply any artificial structural constraints to these systems. The focus of our investigation was on the impact of the insertion of the M2-DA peptide bundle on a DPhPC bilayer. The most striking effect of the insertion is a decrease for the bilayer thickness and a concomitant increase in the surface area per lipid. Furthermore, we observe a more fluid-like behaviour of the hydrocarbon chains indicated by a lower deuterium order parameter S_{CD}. Such observations should be amenable to experimental verification.[8] The overall stability of our all-atom simulation reinforces the notion that the study of small membrane proteins within lipid bilayers is a promising field for further studies.[14] Unfortunately, much longer MD simulations will likely be required to probe the structure and function of the ion channel itself.[22]

V. Acknowledgements

We thank Tarek Mounir, Alex MacKerell, Bill DeGrado, Tim Cross, Larry Pinto, and Jim Lear for many stimulating discussions. This research was supported by the National Institute of Health *via* grant GM 40712 and a collaborative research project with IBM.

References

1. D. A. Doyle, J. M. Cabral, R. A. Pfuetzner, A. L. Gulbis, S. L. Cohen, B. R. Chait and R. MacKinnon, *Science*, 1998, **280**, 69.
2. R. MacKinnon, S. L. Cohen, A. L. Kuo, A. Lee and B. T. Chait, *Science*, 1998, **280**, 106.
3. J. Changeux, *Sci. Am.*, 1993, 58–62.
4. N. Unwin, *J. Struct. Biol.*, 1998, **121**, 181.
5. L. H. Pinto, L. J. Holsinger and R. A. Lamb, *Cell*, 1992, **69**, 517.
6. L. H. Pinto, G. R. Dieckmann, C. S. Gandhi, C. G. Papworth, J. Braman, M. A. Shaughnessy, J. D. Lear, R. A. Lamb and W. F. DeGrado, *Proc. Natl. Acad. Sci. USA*, 1997, **94**, 11301.
7. K. C. Duff and R. H. Ashley, *J. Virol.*, 1992, **190**, 485.
8. F. A. Kovacs and T. A. Cross, *Biophys. J.*, 1997, **73**, 2511.
9. C-H. Hsieh, S-C. Sue, P-C. Lyu and W-G. Wu, *Biophys. J.*, 1997, **73**, 870.
10. L. R. Sita, *J. Org. Chem.*, 1993, **58**, 5285.
11. A. D. MacKerall Jr., D. Bashford, M. Bellott, R. L. Dunbrack Jr., J. D. Evanseck, M. J. Field, S. Fischer, J. Gao, H. Guo, S. Ha, D. Joseph-McCarthy, L. Kuchnir, K. Kuczera, F. T. K. Lau, C. Mattos, S. Michnick, T. Ngo, D. T. Nguyen, B. Prodhom, W. E. Reiher III, B. Roux, M. Schlenkrich, J. C. Smith, R. Stote, J. Straub, M. Watanabe, J. Wiorkiewicz-Kuczera, D. Yin and M. Karplus, *J. Phys. Chem. B.*, 1998, **102**, 3586.
12. W. L. Jorgensen, J. Chandesekhar, J. D. Madura, R. W. Impey and M. L. Klein, *J. Chem. Phys.*, 1983, **79**, 926.
13. R. W. Pastor, *Curr. Opin. Struct. Biol.*, 1994, **4**, 486.
14. D. P. Tieleman, S. J. Marrink and H. J. C. Berendsen, *Biochim. Biophys. Acta*, 1997, **1331**, 235.
15. G. J. Martyna, M. L. Klein and M. Tuckerman, *J. Chem. Phys.*, 1992, **97**, 2635.
16. G. J. Martyna, M. E. Tuckerman, D. J. Tobias and M. L. Klein, *Mol. Phys.*, 1996, **87**, 1117.
17. M. Tuckerman, B. J. Berne and G. J. Martyna, *J. Chem. Phys.*, 1992, **97**, 1990.
18. T. Husslein, D. M. Newns, P. C. Pattnaik, Q. F. Zhong, P. B. Moore and M. L. Klein, *J. Chem. Phys.*, 1998, **109**, 2826.
19. K. Tu, D. J. Tobias, J. K. Blasie and M. L. Klein, *Biophys. J.*, 1996, **70**, 595.
20. W. Shinoda and S. Okazaki, *J. Chem. Phys.*, 1998, **109**, 1517.
21. Q. Zhong, Q. Jiang, P. B. Moore, D. M. Newns and M. L. Klein, *Biophys. J.*, 1998, **74**, 3.
22. P. B. Moore, Q. Zhong, T. Husslein and M. L. Klein, *FEBS Lett.*, 1998, **431**, 143.

Paper 8/06675B

Alamethicin channels in a membrane: molecular dynamics simulations

D. Peter Tieleman,[a] Jason Breed,[b,†] Herman J. C. Berendsen[a] and Mark S. P. Sansom[*b]

[a] *BIOSON Research Institute and Department of Biophysical Chemistry, University of Groningen, Nijenborgh 4, 9747 AG Groningen, The Netherlands*
[b] *Laboratory of Molecular Biophysics, The Rex Richards Building, Department of Biochemistry, University of Oxford, South Parks Road, Oxford, UK OX1 3QU. E-mail: mark@biop.ox.ac.uk*

Received 7th August 1998

Alamethicin (Alm) is a 20 residue peptide which forms a kinked α-helix in membrane and membrane-mimetic environments. Ion channels formed by intramembraneous aggregates of Alm are thought to be formed by bundles of approximately parallel Alm helices surrounding a central bilayer pore. Different channel conductance levels correspond to different numbers of helices per bundle, ranging from $N = 5$ to $N > 8$. Calculation of the predicted pK_A values of the ring of Glu18 sidechains at the C-terminal mouth of the pore suggests that at neutral pH most or all of these sidechains will remain protonated. Nanosecond molecular dynamics (MD) simulations of $N = 5, 6, 7$ and 8 bundles of Alm helices in a POPC bilayer have been run, corresponding to a total simulation time of 4 ns. These simulations explore the stability and conformational dynamics of these helix bundle channels when embedded in a full phospholipid bilayer in an aqueous environment. The structural and dynamic properties of water in these model channels are examined. As in earlier *in vacuo* simulations (J. Breed, R. Sankararamakrishnan, I. D. Kerr and M. S. P. Sansom, *Biophys. J.*, 1996, **70**, 1643) the dipole moments of water molecules within the pores are aligned antiparallel to the helix dipoles. This helps to contribute to the stability of the helix bundles.

Introduction

Ion channels are formed in lipid bilayers by integral membrane proteins. Channels enable selected ions to move rapidly (*ca.* 10^7 ions s^{-1} channel^{-1}) and passively (*i.e.* down their electrochemical gradients) across membranes. Ion channels are important in numerous cellular processes, principally electrical signalling,[1] but also include diverse processes such as facilitating the uncoating of viral genomes.[2] In order to understand the physical events underlying the biological properties of channels one must characterise both their structures and their dynamic behaviour. This is far from easy. Because ion channels are membrane proteins, we remain relatively ignorant of their three dimensional structures. Indeed, a high resolution structure is known for only one ion channel, a bacterial K$^+$ channel.[3] This reflects a more general problem for membrane proteins. Although integral membrane proteins comprise *ca.* 20 to 30% of most genomes,[4,5] high resolution structures

† Present address: Fakultät für Biologie, Universität Konstanz, Postfach 5560, M656, 78434 Konstanz, Germany.

have been solved for only a small handful of these. In this context, model membrane proteins have much to offer to studies of membrane protein structure and dynamics. For example, studies of a channel-forming peptide, gramicidin A, have provided profound insights into the structural basis of channel function.[6-9] However, the structural idiosyncrasies of gramicidin (which is made up of alternating L- and D-amino acids) direct one's attention to other peptide models, which more closely mimic ion channel proteins. Many channel proteins contain a central pore lined by a bundle of approximately parallel α-helices.[10] Such channels range in complexity from the M2 protein of influenza A (ca. 100 amino acids per subunit)[11] to the nicotinic acetylcholine receptor (ca. 500 amino acids per subunit).[12] Given the importance of this structural motif in a number of channel proteins, it is important to have a simple yet detailed model system for channels formed by α-helices. Such a system is provided by alamethicin.

Alamethicin (Alm) is a largely hydrophobic 20 residue peptide which forms ion channels in lipid bilayers. The channel, structural and spectroscopic properties of Alm have been studied in considerable detail.[7,13,14] The structure of Alm in a non-aqueous environment has been determined by X-ray diffraction[15] and by NMR.[16,17] The crystal and solution structures are strikingly similar. The largely α-helical conformation of Alm is stabilised by the presence of a large number of Aib (α-amino isobutyric acid, i.e. α-methyl alanine) residues in its sequence:

Ac-Aib-Pro-Aib-Ala-Aib-Ala-Gln^7-Aib-Val-Aib-Gly-Leu-Aib-Pro^{14}-Val-Aib-Aib-Glu^{18}-Gln-Phol

The presence of Pro14 induces a central kink in the helix. NMR amide exchange data demonstrate that the largely α-helical conformation of Alm when dissolved in methanol[18] is retained when it interacts with lipid bilayers.[19] Amide exchange data also suggest that Alm in methanol undergoes hinge-bending motion about the central proline-induced kink.[20]

Alm forms *multi-conductance* channels (Fig. 1) in a voltage-dependent manner. The multi-conductance behaviour of Alm channels is generally explained in terms of the barrel stave (*i.e.* helix bundle) model[15,21-23] in which several (from *ca.* 5 to *ca.* 10) Alm helices form a bundle surrounding a central pore. Different conductance levels correspond to different numbers (N) of helices per bundle. This model is in accord with a large body of experimental data (reviewed in ref. 13) including neutron scattering studies from Alm in lipid bilayers.[24] A key feature of this model is that the helices are oriented parallel (rather than antiparallel) to one another, their helix dipole repulsions being overcome by their favourable interaction with the electrostatic field across the bilayer, and with the water inside the transbilayer pore.[25,26] A parallel orientation of the helices is supported by the asymmetry of voltage-activation of Alm channels. Thus, when Alm is added to

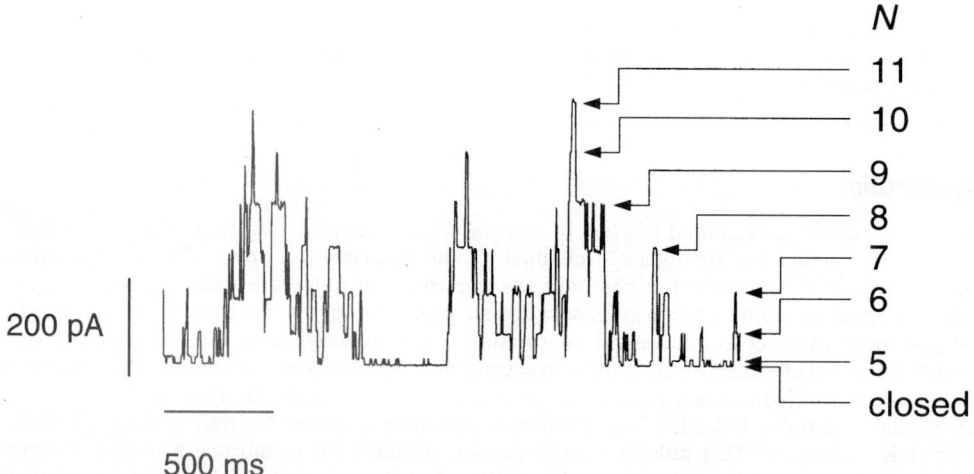

Fig. 1 Recording of ionic currents flowing through a single Alm channel. This recording[32] was obtained from a diphytanoyl phosphatidylcholine bilayer exposed to 0.25 μM Alm. The electrolyte was 0.5 M KCl and the transbilayer potential was 125 mV. The arrows and numbers on the right give the presumed number of helices per bundle (N) for each conductance level.

one face (the *cis* face) of a bilayer, only *cis*-positive voltages will induce channel formation. This argues for a structure which is asymmetric with respect to the bilayer plane, *i.e.* a parallel helix bundle. Such a structure is supported by the work of Woolley and colleagues[27,28] who have shown that channels formed by covalently coupled pairs of Alm helices (which are physically constrained to be parallel to one another) strongly resemble those formed by unmodified Alm.

Molecular models of channels formed by Alm helix bundles have been generated by restrained molecular dynamics (MD) simulations *in vacuo* and these models have been refined by short MD simulations with water molecules included within and at each mouth of the pore.[29] These models can explain the change in stability of Alm channels when the Gln7 sidechain is replaced by a smaller polar residue.[30,31] The pore dimensions of such models correlate well with experimentally observed conductance values of Alm.[13,32,33] Similar models of channels formed by Alm analogues when used in continuum electrostatics calculations predict the ionic-strength dependent non-linear current–voltage curves observed experimentally.[28] Models of Alm helix bundles have been used in MD simulations to demonstrate that water within channels formed by parallel helix bundles is ordered and shows reduced translational and rotational mobility relative to bulk water.[25] Finally, models of Alm channels have been used to explore changes in translational mobility of Na^+ ions when within narrow pores.[34] Thus, α-helix bundle models of Alm channels have been studied in some detail and have been shown to correspond well with experimental data. So, it is reasonable to use them as the basis of prolonged MD simulations of α-helix bundles in a lipid bilayer.

It has recently become feasible to run multi-nanosecond duration MD simulations of Alm channel models with full bilayer models in which the lipid molecules and the water on either face of a membrane are represented explicitly. Such simulations exploit the considerable progress of the past few years in simulations of lipid bilayers *per se*.[35–38] In particular, simulations on pure bilayers have explored the dependence of simulation behaviour on the methods employed, and have established rules-of-thumb for physically realistic simulations, both in terms of suitable parameters for inter-atomic interactions, and in terms of optimal simulation protocols. Alm channel simulations also build upon full bilayer simulations of single transmembrane α-helices,[39–41] which have shown that such simulations enable one to characterise the structural dynamics of membrane-spanning α-helices in a realistic model of their environment. In previous simulations we have investigated isolated Alm helices in a bilayer and in solution[42] and an $N = 6$ model of an Alm helix bundle in a bilayer.[26] Here, we extend the latter simulations to Alm helix bundles containing from $N = 5$ to 8 helices, paying particular attention in the initial setup of the simulation to the ionisation state of the Glu18 sidechains.

Methods

Generation of starting models

Initial models of Alm helix bundles with from $N = 5$ to 8 helices per bundle were generated using restrained MD *in vacuo* as described in ref. 29 and 43. During the final stage of the simulated annealing protocol used to build the model the restraints applied were: (i) intra-helix restraints, to maintain H-bonding of the backbone of each Alm monomer; (ii) inter-helix restraints, between the N-terminal segments of adjacent monomers of the bundle, in order to maintain the integrity of the bundle; and (iii) non-crystallographic symmetry restraints, to maintain the approximate rotational symmetry of the helix bundles. For each value of N, an ensemble of 25 structures was generated. The most symmetrical member of each ensemble was used in the bilayer MD simulations.

pK_A calculations

The ionisation states of the Glu18 sidechains in the Alm bundle models were estimated *via* pK_A calculations, as described in detail in ref. 44. This proceeded in two stages. Intrinsic pK_A values were calculated using:

$$pK_{A,\,\text{INTRINSIC}} = pK_{A,\,\text{MODEL}} - \frac{1}{2.303}[\Delta\Delta G_{\text{BORN}} + \Delta\Delta G_{\text{BACK}}]$$

where $pK_{A,\,\text{MODEL}}$ is the pK_A of an isolated amino acid, $\Delta\Delta G_{\text{BORN}}$ is the solvation contribution to the pK_A shift and $\Delta\Delta G_{\text{BACK}}$ is the contribution due to the interaction of the residue with non-titrating charges.[45,46] Absolute pK_A values were obtained *via* calculation of titration curves. The latter were obtained *via* calculation of:

$$p(x) \propto \exp[-\ln 10 \sum_i \gamma_i(pK_{A, \text{INTRINSIC}, i} - pH) - \beta \sum_i \sum_{k<i} \Delta\Delta G_{i,k}]$$

where $p(x)$ is the probability of a residue existing in its ionised state and x is an N-element state vector whose elements are 0 or 1 depending on whether the residue is un-ionised or ionised respectively, $\gamma = -1$ for a basic residue, $\gamma = +1$ for an acidic residue, and $\Delta\Delta G_{i,k}$ is the screened Coulombic interaction energy between pairs of ionisable residues i and k.[45,47]

System setup for bilayer MD

The setup of the simulation systems was essentially as described in ref. 26. An equilibrated POPC bilayer with 128 lipid molecules was used. For each bundle model a cylindrical hole was made in the centre of the bilayer by removing selected lipid molecules and running a short MD simulation with a radially acting repulsive force to drive any remaining atoms out of the cylinder into the bilayer. The Alm bundle models (with a single ionised Glu18, see below) were inserted into the cavities thus created. These systems (helix bundle plus POPC) were solvated with SPC waters (*ca.* 30 waters per lipid molecule), and a single Na$^+$ ion replacing a water molecule at the position of lowest Coulomb potential. The resultant systems (see Table 1) were simulated for 25 ps (50 ps for the $N = 8$ bundle) with positional restraints on the peptide atoms relative to the starting bundle structure, with constant surface area, and with a constant pressure of 1 bar in the z (*i.e.* bilayer normal) direction. The resultant systems were used as the starting point for the 1 ns production runs.

MD simulation details

Molecular dynamics simulations were run using GROMACS.[48] A twin range cut-off was used for longer-range interactions: 1.0 nm for van der Waals interactions, and 1.7 nm for electrostatic interactions. The timestep was 2 fs, using LINCS[49] to constrain bond lengths. NPT conditions (*i.e.* constant number of particles, pressure and temperature) were used in the simulation. A constant pressure, of 1 bar independently in all three directions, was used, with a coupling constant of $\tau_P = 1.0$ ps.[50] This allowed the bilayer/peptide area to adjust to its optimum value for the force-field employed. Water, lipid and protein were coupled separately to a temperature bath at 300 K, using a coupling constant $\tau_T = 0.1$ ps.

Lipid parameters were as in previous MD studies of lipid bilayers,[51,52] and as in our previous papers on MD simulations of Alm.[26,42] These lipid parameters give good reproduction of the experimental properties of a DPPC bilayer. The water model used was SPC,[53] which has been shown to be a reasonable choice for lipid bilayer simulations.[54]

Longitudinal diffusion coefficients of water molecules within the pore were determined from their mean square displacement along the pore (z) axis over a period of 5 ps, as described by.[55] The diffusion coefficient was assigned to the local region on the pore axis corresponding to the position of the water molecule at the start of the 5 ps period.

Computational details

Simulations were carried out on a 195 MHz R1000 Origin 2000, and took *ca.* 8 days per processor per 1 ns of simulation time (for about 17 000 atoms). Analysis was performed using facilities within

Table 1 Simulation details

N	Number of phospholipids	Number of waters	Total number of atoms	Duration of simulation/ps	Final Cα RMSD/ nm
5	103	3511	16 729	1015	0.22
6	102	3527	16 893	1015	0.20
7	96	3524	16 740	1025	0.24
8	95	3548	16 928	1010	0.22

GROMACS and with code written specifically for this project. Secondary structure analysis employed the DSSP algorithm.[56] Initial models were generated using X-PLOR.[57] Structures were examined using Quanta (Biosym/MSI) and Rasmol, and the diagrams were drawn using MolScript.[58] Electrostatics calculations (for pK_A calculations) were performed using UHBD version 5.1[59] (with some local modifications) and partial atomic charges from the Quanta/Charmm22 parameter set. Pore radius profiles were determined using HOLE.[60]

Results

Ionisation state of Glu18 sidechains

Examination of the *in vacuo* generated models of Alm helix bundles reveals a ring of Glu18 sidechains at the C-terminal mouth of the channel [see Fig. 2A]. In this respect, the Alm channel model is similar to that of the pore-lining M2 helix bundle of the α7 nicotinic receptor (nAChR) channel.[44] Studies of the ionisation state of the ring of glutamate residues at the C-terminal mouth of the nAChR suggest that their location at the C-terminus of the α-helix dipole and in proximity to one another results in a shift of their pK_As such that at neutral pH they are not fully ionised. Previous MD simulations of $N = 6$ Alm helix bundle models have suggested that the integrity of the helix bundle during the simulation is dependent upon the ionisation state of the Glu18 sidechains. Thus, it was judged to be important to determine the average ionisation state of the Glu18 sidechains before running the bilayer MD simulations. Note that this follows the approach of

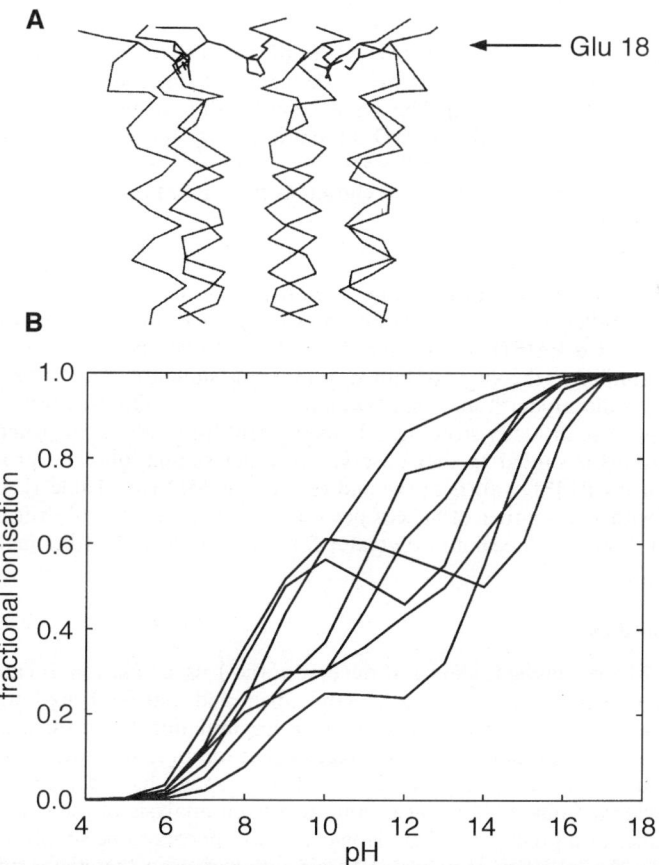

Fig. 2 A, model of the $N = 7$ Alm bundle, generated by restrained MD *in vacuo*. The ring of Glu18 sidechains is indicated. B, calculated titration curves for the seven Glu18 residues of the model shown in A.

Table 2 Calculated pK_As of Glu18 residues in Alm helix bundles

N	pK_As of Glu18 residues	Net charge at pH 7
5	6.5, 7.6, 11.3, >14.0, >14.0	−0.80
6	6.7, >10.0, 13.0, >14.0, >14.0, >14.0	−0.66
7	9.4, 10.7, 11.3, 12.2, 13.0, 13.8, 14.0	−0.69
8	5.6, 7.4, 7.5, 8.3, 8.5, 9.2, 10.3, >14.0	−1.69

Tieleman and Berendsen[55] who employed the ionisation states estimated by Karshikoff et al.[46] in their bilayer simulations of the porin OmpF.

The results of the estimation of pK_A values for the Glu18 residues are summarised in Table 2, and an example of the calculated titration curves on which these are based is provided in Fig. 2B. Note that the calculated pK_A values for the N Glu18 sidechains of a bundle are not identical. This is because the bundle models do not have exact rotational symmetry. Thus, in Fig. 2B one may see that three of the Glu18s titrate with pK_As of 9.4 to 11.3, whereas the remaining four titrate with somewhat higher pK_As. However, a consistent picture does emerge across the different N values. Overall, the Glu18 residues exhibit much higher pK_A values than that of an isolated glutamate sidechain (pK_A = 4.4). Model calculations in which the pK_A value of a Glu sidechain as a function of its position along an Ala$_{21}$ α-helix indicate that this shift is a result of the unfavourable interaction of the C-terminal end of the α-helix dipole with the ionised state of the sidechain (Adcock, Smith and Sansom, unpublished results), a result which is also supported by experimental data.[61,62]

One may estimate the consequence of this shift in pK_A values for Glu18 by calculating the net charge expected on each helix bundle at pH 7. As can be seen from Table 2, this varies from −0.66 (for N = 6) to −1.7 (for N = 8). The average across all four models is −0.96. Thus, choosing the nearest integer, for the bilayer MD simulations each bundle was modelled as having (N − 1) Glu18 residues in their protonated state and one Glu18 ionised. In each case, that Glu18 sidechain with the lowest calculated pK_A was chosen as the ionised one.

Progress of the simulations

The progress of the simulations was monitored via inspection of the Cα RMSD trajectories (not shown) of the Alm bundles in order to ascertain how far these drifted from their initial structures. For each N value, the Cα RMSD rose during the first 250 to 500 ps, reaching a plateau value of ca. 0.2 nm. This is similar to the value obtained in previous simulations of an (electrically neutral) Alm N = 6 helix bundle model[26] and in MD simulations of an OmpF porin trimer (the latter simulations starting with an X-ray structure). Thus the drift from the initial structure in all four of the current simulations is similar to that observed in other simulations of a protein in a bilayer. Examination of the Cα RMSD values at the end of the 1 ns MD run (Table 1) shows no trend in the RMSD value with the number of helices per bundle. Thus the overall drift in structure from the starting model does not appear to be greater for the large helix bundles (at least, not over a period of 1 ns).

Integrity of helix bundles

In Fig. 3 the N = 7 helix bundle is shown at the end of the 1 ns simulation. It can be seen that the helix bundle is intact, that the central pore remains open, and that the helix bundle remains with its pore axis perpendicular to the plane of the bilayer. Examination of the superimposed Cα traces at the start and end of the simulation (Fig. 4) suggests that for all four N values the integrity of the helix bundle is maintained during the simulation. There is some slight degree of expansion of the bundles. This is evident from the Cα traces, and also from analysis of the radii of gyration (not shown) of the bundles as a function of time. However, this appears to be simply a relaxation of the bundle geometry in the presence of water and lipid (remembering that these simulations started from in vacuo models), and does not appear to disrupt helix–helix packing. Such bundle expansion appears to be a little greater at the C-terminal mouth of the pore, which would suggest a limited

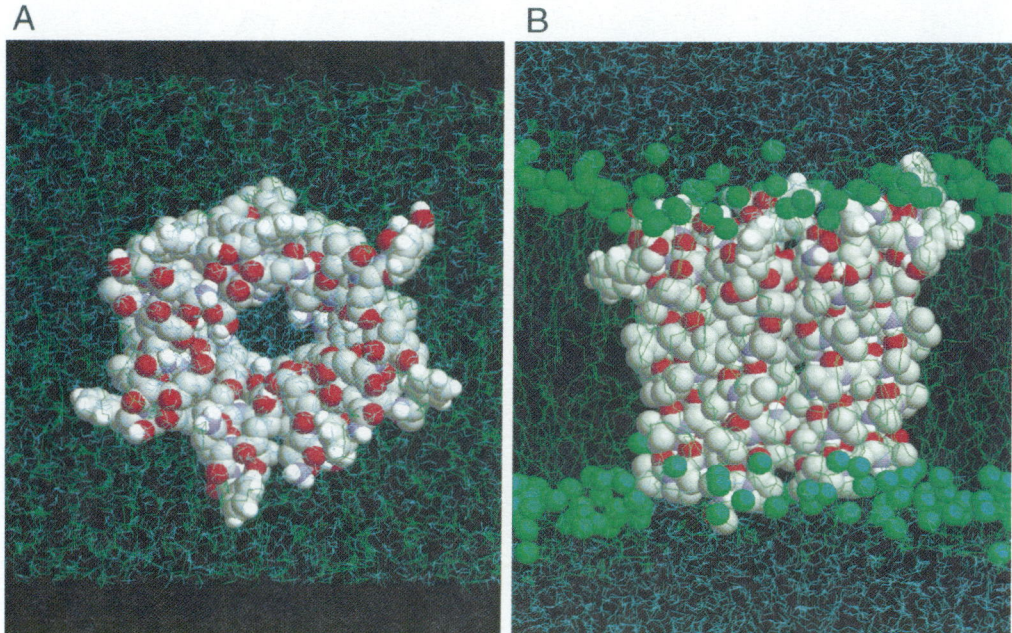

Fig. 3 Images of the $N = 7$ Alm simulation system, taken at $t = 1$ ns. Water molecules are shown in cyan, phospholipid molecules in green and the model pore in standard CPK colours. The Alm molecules and the carbonyl oxygens of the fatty acids are shown in space-filling format. A, View down the pore (z) axis, with the C-terminal mouth of the pore (z ca. $+4$ nm) towards the reader. B, View perpendicular to the pore axis, with the C-terminal mouth of the pore uppermost.

degree of conformational change of the constituent helices (see below). Inspection of the helix bundles plus the water molecules within the pore at the start (not shown) and end (Fig. 5) of the simulation suggests that such expansion corresponds to entry of additional water molecules into the pore. This is most evident for the $N = 5$ helix bundle, for which at $t = 0$ ps there were ca. 55 water molecules within the pore, whereas after 1 ns this number had risen to 94.

Internal motions of the α-helices

It is also of interest to examine the internal motions of the helices, as these will in turn influence the dynamic properties of the transbilayer pore. Experimental[20] and simulation[42] studies of Alm helices suggest that the Gly-X-X-Pro sequence motif allows for flexibility in the central hinge region of the molecule. Furthermore, comparison of the dynamic behaviour of isolated Alm molecules in solution in water and in methanol, and in a transbilayer environment,[42] suggests that the C-terminal half of the molecule becomes disordered when in an aqueous environment, but not when in a non-aqueous solvent or within a bilayer. Analysis of the RMS fluctuations from the average structure during the MD trajectories (Fig. 6) indicates that the C-terminal half of the Alm molecule *within a bundle* undergoes greater fluctuations than the N-terminal half, although these fluctuations are less marked than those for a single Alm molecule in water. This confirms the impression arrived at from visual comparison of the initial and final Cα traces (see above). Thus, it seems that the N-terminal halves of the Alm helices, which are more tightly packed together in the bundles, fluctuate less than the corresponding C-terminal halves.

One may examine the consequences of these fluctuations in terms of the secondary structure of the helices, analysed using DSSP[56] (Fig. 7). This reveals that: (i) the majority of residues remain in an α-helical conformation throughout the simulation; (ii) loss of helicity is transient, and tends to be in the C-terminal half of the Alm molecules; and (iii) patterns of loss of helicity vary between the different constituent Alm molecules of a bundle. As in earlier $N = 6$ simulations[26] the overall

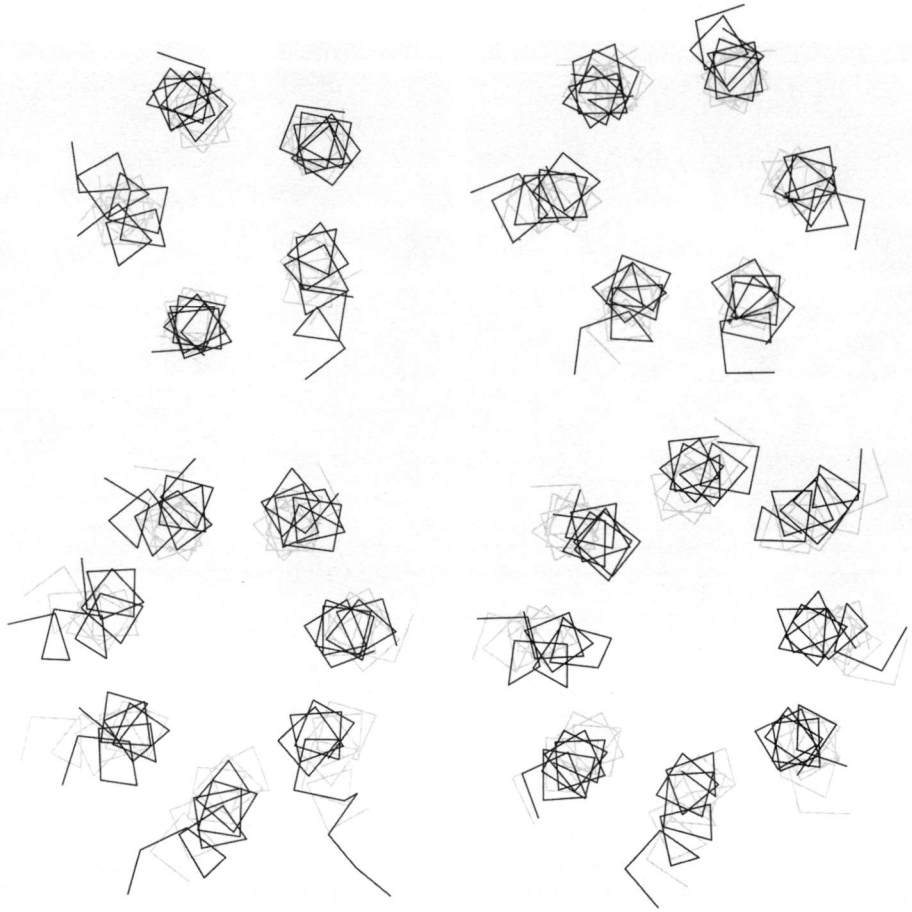

Fig. 4 Cα traces of the four Alm helix bundles, taken at the start ($t = 0$ ns; grey lines) and end ($t = 1$ ns; black lines) of each simulation. In each case the view is down the pore axis.

degree of helicity is slightly less than that of an isolated Alm molecule in a hydrophobic environment, but significantly greater than that of an isolated Alm molecule in water.[42] This presumably reflects the intermediate environment in which an Alm molecule within a bundle finds itself, with its more apolar face towards the lipid but its more polar face towards the water molecules within the pore.

Water within the pore

As seen above (Fig. 5) there is a well defined column of water within the lumen of the each of the four Alm helix bundles. As the water within a pore plays an important role in the permeation of ions through a pore, the nature of the pores and their water has been examined in some detail. Pore radius profiles were determined every 50 ps. The resultant time-averaged radius profiles are compared in Fig. 8. The $N = 5$ bundle is more constricted at its C-terminal mouth at z *ca.* 4.5 nm, *i.e.* in the vicinity of the Glu18 sidechain ring, yielding a minimum pore radius of *ca.* 0.15 nm, with a lesser degree of constriction in the vicinity of the ring of Gln7 sidechains at z *ca.* 2.8 nm. This pattern is switched in the $N = 6$, 7 and 8 bundles, with the narrower constriction in the region of the Gln7 ring, and a lesser degree of constriction at the C-terminal mouth. Thus, the smallest helix bundle, which is thought to correspond to the lowest conductance level of the channel (Fig. 1), differs somewhat in its pore geometry from the others.

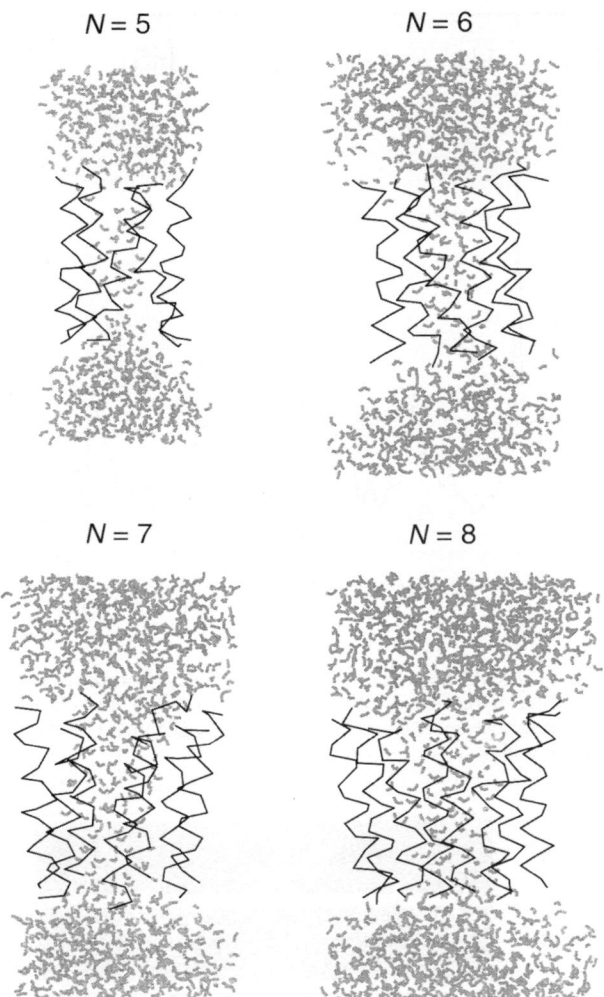

Fig. 5 Cα traces of the four Alm helix bundles, taken at the end ($t = 1$ ns; black lines) of each simulation, superimposed upon those water molecules within and close to either mouth of the pore (grey bonds).

On the basis of these pore radius profiles one may obtain approximate predictions of the pore conductances using the methods of Smart *et al.*[33] For 1 M KCl this yields a predicted single channel conductance of 170, 250, 310 and 400 pS for $N = 5, 6, 7$ and 8 helices per bundle, respectively. The $N = 6$ value is in agreement with the corresponding experimental value of 280 pS.[33,63] The predicted conductance values for $N = 7$ and 8 are too low. This may reflect difficulties in applying a simple geometry based prediction method to these large helix bundles (see below).

The longitudinal (*i.e.* along the pore, z, axis) diffusion coefficient of water molecules has been examined for all four simulation systems, plotting water diffusion coefficients as functions of their z coordinates [Fig. 8B]. Note that those waters within the pore are located between $z = 2$ and $z = 5$ nm. The diffusion coefficients of waters within the pore are markedly reduced relative to those of waters in the bulk region. Note that the bulk diffusion coefficient of SPC water is *ca.* 5×10^{-9} m^2 s^{-1}. In the narrowest regions of all four pores the water diffusion coefficients fall to less than 0.5×10^{-9} m^2 s^{-1}, *i.e.* less than a tenth of the bulk water value. Thus the substantial reduction in the translational motion of water within narrow pores seen in simpler (no bilayer) simulations[25] is reproduced in the current, more realistic study. Interestingly, although there is some dependence of the local diffusion coefficient on the pore radius, this is not as strong as might

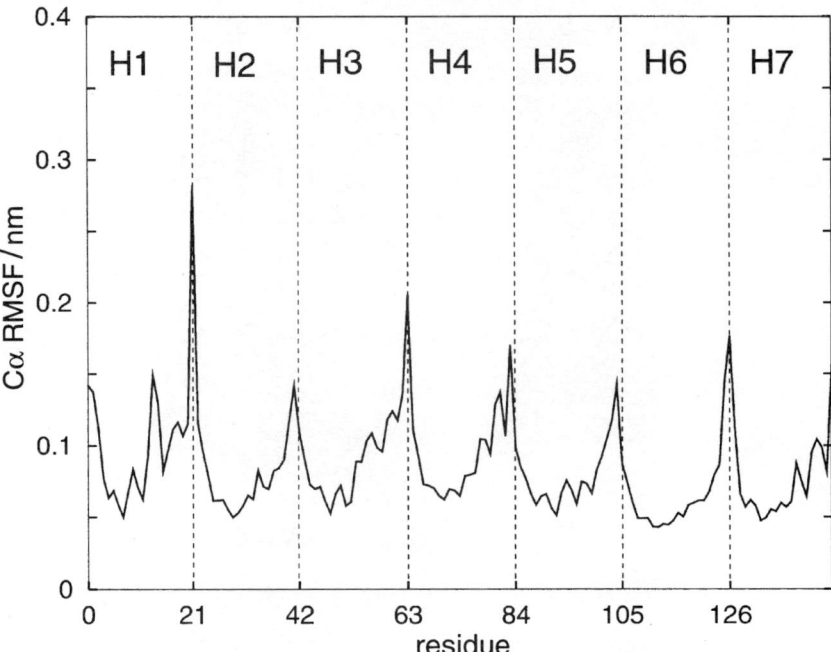

Fig. 6 Residue-by-residue Cα RMS fluctuations about their average coordinates for the Alm $N = 7$ model. The vertical broken lines delineate the extents of helices H1 (residues 0 to 20) to H7 (residues 126 to 146). Note that in our residue numbering scheme, residues 0, 21, 42, 64, 84, 105 and 126 correspond to the N-terminal acetyl groups of the Alm molecules.

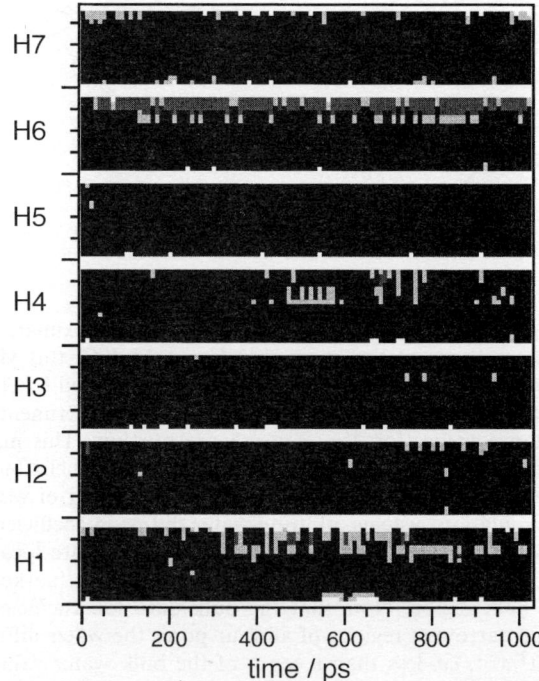

Fig. 7 Secondary structure, as defined by DSSP,[56] as a function of time for the Alm $N = 7$ model. The greyscale is: black = α-helix; dark grey = 3_{10}-helix; pale grey = turn; and white = coil. H1 to H7 denote the constituent helices of the bundles.

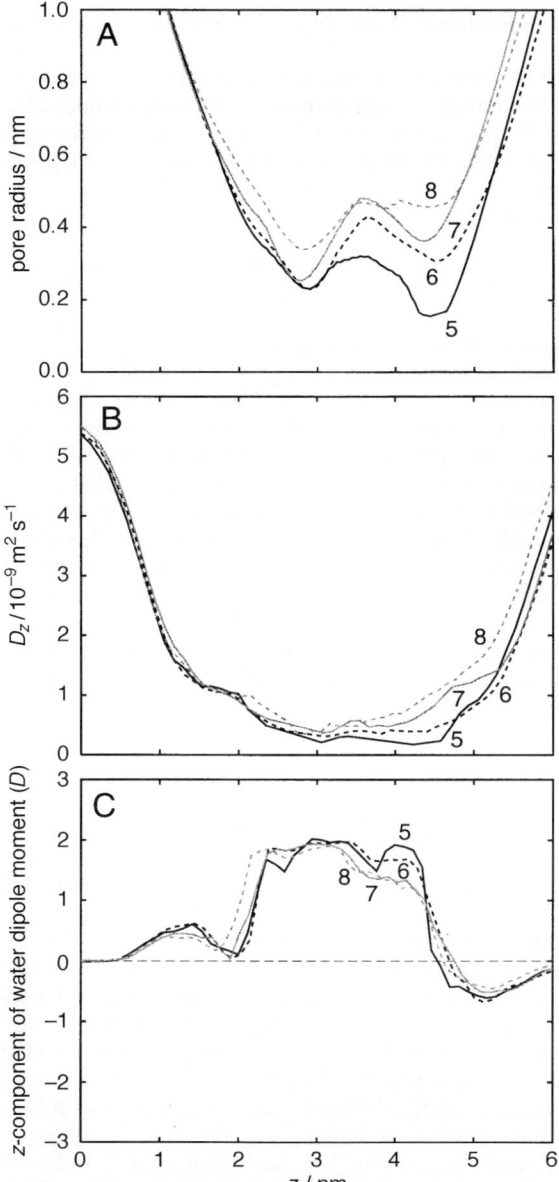

Fig. 8 A, Pore radius profiles, evaluated using HOLE,[60] for the four Alm simulations. The profiles were averaged from structures saved every 50 ps. The N-termini of the helices are at z ca. $+2$ nm and the C-termini at z ca. $+5$ nm. B, Water diffusion coefficients (D_z) as a function of position along the pore (z) axes. C, Projection of water dipole moments onto the pore axes. For each graph, the following convention is used: solid, black line for $N = 5$; broken, black line for $N = 6$; solid, grey line for $N = 7$; and broken, grey line for $N = 8$.

have been expected (compare the coefficients for the four bundles close to the Glu18 rings at z ca. 4.5 nm).

The dipoles of the water molecules within all four pores are oriented by the surrounding parallel helix dipoles [Fig. 8C]. Thus, within the pore the mean z component of the water dipoles is nearly 1.8 Debye, which should be compared with a dipole moment of 2.3 Debye for a single SPC water. There is some dependence of the degree of water dipole orientation on the number of helices per bundle.

The observed degree of orientation of the water dipoles may be employed to calculate the local field experienced by water molecules within the pore. Thus, the values of $\mu_z = 1.79, 1.77, 1.61, 1.60$ Debye averaged along the pore for $N = 5$ to 8 respectively, when compared with $\mu_0 = 2.3$ Debye for SPC water [where μ_0 is the dipole moment of water and μ_z is its projection along the z (pore) axis], can be used to estimate the field strengths within each pore due to the aligned helix dipoles. Using the Langevin equation:

$$\mu_z = \mu_0 \left[\coth\left(\frac{\mu_0 E_z}{k_B T}\right) - \left(\frac{k_B T}{\mu_0 E_z}\right) \right]$$

where E_z is the z-component of the electrostatic field due to the helix dipoles, and k_B and T are the Boltzmann constant and temperature, respectively; this yields fields of $E_z = 2.4, 2.3, 1.8$ and 1.8×10^9 V m^{-1} respectively for the $N = 5, 6, 7$ and 8 pores. (See ref. 64 for a more detailed description of the theory of interaction of α-helix bundle dipoles with pore water molecules.) These 'observed' fields may be compared with approximate predictions of the electrostatic fields generated by the peptide backbones of the corresponding bundles of aligned α-helices. Estimates of the latter fields were obtained by numerical differentiation of the electrostatic potential energy along the pore axes of initial Alm helix bundles *in vacuo*. Note that this approximate prediction is a simple Coulombic field, and does not take into account the environment surrounding the helix bundle. The fields estimated in this fashion had maximum values in the centres of the pores of *ca.* 3.6, 3.0, 2.1 and 1.2×10^9 V m^{-1}, respectively, for $N = 5, 6, 7$ and 8 helices per bundle. These values are, given the approximations involved, in good agreement with the 'observed' fields calculated from the Langevin equation. Such agreement confirms that there are strong interactions between water dipoles and aligned helix dipoles, which will contribute to the stability of the helix bundles.

Discussion

Simulation methodology

The simulations in this paper employ a similar methodology to that employed in a number of recent MD simulations of membrane proteins.[26,42,55,65] However, as all simulations remain an approximation to the true properties of a system, it is useful to consider possible limitations of the approach. In particular, 1 ns is a relatively short period of time, both in the context of lipid motions[37] and of the mean time it takes an ion to move through a channel (*ca.* 10 to 100 ns). Extension of the simulations is in progress. Furthermore, our simulations have been conducted in the *absence* of a transbilayer voltage difference. This is perhaps not too problematic, as the work of He *et al.*[24] suggests that Alm helix bundles may form in the absence of a voltage difference. However, a number of recent simulations[66-68] and theoretical[69] studies have shown that a transbilayer voltage difference may be included in MD simulations. It will be interesting to perform Alm bundle simulations in the presence of an external electrostatic field. A further limitation to our simulations is the absence of electrolyte. Experimental studies of Alm channels are conducted using *e.g.* 1 M electrolyte, corresponding to *ca.* 130 ions in the simulation box used in our simulations. This is an area which future studies will have to address. In particular, extended simulation times will be required to allow efficient equilibration of such a system. Another limitation is the relatively simple treatment of long-range electrostatic interactions in our current study. The simulation protocol we have used gives reasonable agreement with experimental results for pure bilayer simulations[37] and yields stable simulations for the porin OmpF in a POPE bilayer.[55] However, a number of studies have been concerned with the effects of different treatments of long-range electrostatic interactions.[37,38] It is likely that in the presence of explicit charges long-range interactions should be included by proper lattice sums so as to avoid artefacts. Finally, although we have tried to take into account the most likely ionisation state of the Glu18 side-chains, one must recognise that this ionisation state may be influenced by the headgroups of the lipid molecules, and also that the pK_As of the Glu18 sidechains may change dynamically with respect to time.[70] Such effects will also have to be taken into account in future simulation studies.

Biological significance

Despite the limitations of the simulations, the dynamic behaviour of the Alm $N = 5$ to 8 bundles and their interactions with water and lipids are of biological significance. In particular, if Alm helix bundles are thought of as paradigms of other, more complex, ion channels formed by helix bundles, the dynamic behaviour of Alm bundles may have relevance to channels in general. In this respect it is of interest to note that the Alm bundle simulations reveal that a stable helix bundle within a bilayer may exhibit dynamic fluctuations in its constituent helices. This is of interest in the context of K^+ channels. The recently determined X-ray structure of a bacterial K^+ channel[4] reveals a pore based upon an α-helix bundle motif, into which the pore-lining P-domain is inserted in order to confer greater ion selectivity. Site-directed spin labelling studies indicate that subtle changes in TM helix packing of the KcsA channel may be related to its gating. Our studies suggest that MD simulations in a lipid bilayer may be able to reveal details of such changes. This is also of relevance in the context of K^+ channels. Interestingly, recent MD simulations of isolated pore-lining (S6 and M2) helices of K^+ channels spanning a POPC bilayer[71] suggest that hinge-bending motions occur within these helices. Thus, K^+ channels may exhibit fluctuations in their pore dimensions similar to those seen in Alm channels.

A second significant result which emerges is that, for Alm, bundles with different numbers of helices are approximately equally stable in MD simulations within a phospholipid bilayer. This correlates nicely with the experimental observation of multiple conductance levels for the Alm channel. Other channel-forming peptides do not show such a wide multiplicity of conductance levels.[32,72] It would be of interest to see whether for such peptides helix bundle models with different N values showed similar stability in bilayer simulations.

A general property which is revealed in the current study is the altered dynamics of water molecules within Alm pores. This is seen regardless of the number of helices per bundle. Altered water dynamics have also been seen in comparable simulations of the bacterial porin OmpF.[55] Almost identical effects on dynamics of water with Alm and other model pores were observed in simulations of a pore plus water system in the absence of a bilayer model, *i.e.* essentially *in vacuo*,[25] and in simulations of peptide channels in which the bilayer was mimicked by an octane slab.[73] This is important in that it suggests that simplified simulations may capture the essence of the dynamics of water (and, by extension, of ions[34,74]) in models of transbilayer pores. The other unusual property of water within Alm channels is the high degree of alignment of water dipoles along the pore axis. This was also observed in the earlier simulations.[25,75] The current study demonstrates that water dipoles are aligned by the electrostatic field created by the parallel dipoles of the constituent α-helices of the bundle. Such alignment of intra-pore water dipoles has important consequences. Alignment of water dipoles by parallel helix dipoles means that if one should attempt to model the water within a pore as a continuum a lower dielectric (relative permittivity) than that of bulk water should be used.[64] However, there are rather more fundamental problems in treating nearly dielectrically saturated water as a simple dielectric continuum. Furthermore, as discussed above, helix–water dipolar interactions contribute to the stabilisation of parallel α-helix bundles.

Future directions

These studies have shown that MD simulations of Alm helix bundles including a lipid bilayer plus water reveal important aspects of the dynamic behaviour of both the peptide and water molecules. There are a number of ways in which this work may be extended. Firstly free energy profiles[9,76–78] for the ion as it moves along the pore should be calculated, in order to understand better the energetics of pore–ion interactions. Secondly, longer MD simulations may reveal details of slower changes in conformation of Alm helix bundles, and slower movements of helix bundles relative to their bilayer environment. Thirdly, the effects on bundle stability of changing the environment, *e.g.* by including a transbilayer voltage difference[67,68] or by using other phospholipids, should be explored. In this way it will be possible to build up a more complete picture of the behaviour of a simple ion channel at atomic resolution.

Acknowledgements

Work in MSPS's laboratory is supported by The Wellcome Trust. DPT was supported by the European Union under contract CT94-0124.

References

1. B. Hille, *Ionic Channels of Excitable Membranes*, Sinauer Associates Inc., Sunderland, MA 2nd edn., 1992.
2. M. S. P. Sansom, L. R. Forrest and R. Bull, *Bioessays*, 1998, in press.
3. D. A. Doyle, J. M. Cabral, R. A. Pfuetzner, A. Kuo, J. M. Gulbis, S. L. Cohen, B. T. Cahit and R. MacKinnon, *Science*, 1998, **280**, 69.
4. D. Boyd, C. Schierle and J. Beckwith, *Protein Sci.*, 1998, **7**, 201.
5. E. Wallin and G. von Heijne, *Protein Sci.*, 1998, **7**, 1029.
6. O. S. Andersen, *Annu. Rev. Physiol.*, 1984, **46**, 531.
7. G. A. Woolley and B. A. Wallace, *J. Membr. Biol.*, 1992, **129**, 109.
8. B. A. Wallace, *Prog. Biophys. Mol. Biol.*, 1992, **57**, 59.
9. B. Roux and M. Karplus, *Annu. Rev. Biophys. Biomol. Struct.*, 1994, **23**, 731.
10. S. Oiki, V. Madison and M. Montal, *Proteins: Struct. Func. Genet.*, 1990, **8**, 226.
11. M. S. P. Sansom, I. D. Kerr, G. R. Smith and H. S. Son, *Virology*, 1997, **233**, 163.
12. M. S. P. Sansom, C. Adcock and G. R. Smith, *J. Struct. Biol.*, 1998, **121**, 246.
13. M. S. P. Sansom, *Qt. Rev. Biophys.*, 1993, **26**, 365.
14. D. S. Cafiso, *Annu. Rev. Biophys. Biomol. Struct.*, 1994, 23, 141.
15. R. O. Fox and F. M. Richards, *Nature (London)*, 1982, **300**, 325.
16. G. Esposito, J. A. Carver, J. Boyd and I. D. Campbell, *Biochemistry*, 1987, **26**, 1043.
17. J. C. Franklin, J. F. Ellena, S. Jayasinghe, L. P. Kelsh and D. S. Cafiso, *Biochemistry*, 1994, **33**, 4036.
18. C. E. Dempsey, *J. Am. Chem. Soc.*, 1995, **117**, 7526.
19. C. E. Dempsey and L. J. Handcock, *Biophys. J.*, 1996, **70**, 1777.
20. N. Gibbs, R. B. Sessions, P. B. Williams and C. E. Dempsey, *Biophys. J.*, 1997, **72**, 2490.
21. G. Baumann and P. Mueller, *J. Supramol. Struct.*, 1974, **2**, 538.
22. M. K. Mathew and P. Balaram, *FEBS Lett.*, 1983, **157**, 1.
23. G. Boheim, W. Hanke and G. Jung, *Biophys. Struct. Mech.*, 1983, **9**, 181.
24. K. He, S. J. Ludtke, H. W. Huang and D. L. Worcester, *Biochemistry*, 1995, **34**, 15 614.
25. J. Breed, R. Sankararamakrishnan, I. D. Kerr and M. S. P. Sansom, *Biophys. J.*, 1996, **70**, 1643.
26. D. P. Tieleman, H. J. C. Berendsen and M. S. P. Sansom, *Biophys. J.*, 1999, in press.
27. S. You, S. Peng, L. Lien, J. Breed, M. S. P. Sansom and G. A. Woolley, *Biochemistry*, 1996, **35**, 6225.
28. G. A. Woolley, P. C. Biggin, A. Schultz, L. Lien, D. C. J. Jaikaran, J. Breed, K. Cowhurst and M. S. P. Sansom, *Biophys. J.*, 1997, **73**, 770.
29. J. Breed, P. C. Biggin, I. D. Kerr, O. S. Smart and M. S. P. Sansom, *Biochim. Biophys. Acta*, 1997, **1325**, 235.
30. G. Molle, J. Y. Dugast, G. Spach and H. Duclohier, *Biophys. J.*, 1996, **70**, 1669.
31. J. Breed, I. D. Kerr, G. Molle, H. Duclohier and M. S. P. Sansom, *Biochim. Biophys. Acta*, 1997, **1330**, 103.
32. M. S. P. Sansom, *Prog. Biophys. Mol. Biol.*, 1991, **55**, 139.
33. O. S. Smart, J. Breed, G. R. Smith and M. S. P. Sansom, *Biophys. J.*, 1997, **72**, 1109.
34. G. R. Smith and M. S. P. Sansom, *Biophys. J.*, 1998, **75**, 2767.
35. K. M. Merz and B. Roux, *Biological Membranes: A Molecular Perspective from Computation and Experiment*, Birkhäuser, Boston, 1996.
36. K. M. Merz, *Curr. Opin. Struct. Biol.*, 1997, **7**, 511.
37. D. P. Tieleman, S. J. Marrink and H. J. C. Berendsen, *Biochim. Biophys. Acta*, 1997, **1331**, 235.
38. D. J. Tobias, K. C. Tu and M. L. Klein, *Curr. Opin. Coll. Interface Sci.*, 1997, **2**, 15.
39. K. Belohorcova, J. H. Davis, T. B. Woolf and B. Roux, *Biophys. J.*, 1997, **73**, 3039.
40. L. Shen, D. Bassolino and T. Stouch, *Biophys. J.*, 1997, **73**, 3.
41. T. B. Woolf, *Biophys. J.*, 1997, **73**, 2376.
42. D. P. Tieleman, M. S. P. Sansom and H. J. C. Berendsen, *Biophys. J.*, 1999, in press.
43. I. D. Kerr, R. Sankararamakrishnan, O. S. Smart and M. S. P. Sansom, *Biophys. J.*, 1994, **67**, 1501.
44. C. Adcock, G. R. Smith and M. S. P. Sansom, *Biophys. J.*, 1998, **75**, 1211.
45. D. Bashford and M. Karplus, *J. Phys. Chem.*, 1991, **95**, 9556.
46. A. Karshikoff, V. Spassov, S. W. Cowan, R. Ladenstein and T. Schirmer, *J. Mol. Biol.*, 1994, **240**, 372.
47. C. Lim, D. Bashford and M. Karplus, *J. Phys. Chem.*, 1991, **95**, 5610.
48. H. J. C. Berendsen, D. van der Spoel and R. van Drunen, *Comput. Phys. Commun.*, 1995, **95**, 43.
49. B. Hess, H. Bekker, H. J. C. Berendsen and J. G. E. M. Fraaije, *J. Comput. Chem.*, 1997, **18**, 1463.
50. H. J. C. Berendsen, J. P. M. Postma, W. F. van Gunsteren, A. DiNola and J. R. Haak, *J. Chem. Phys.*, 1984, **81**, 3684.
51. O. Berger, O. Edholm and F. Jahnig, *Biophys. J.*, 1997, **72**, 2002.
52. S. J. Marrink, O. Berger, D. P. Tieleman and F. Jahnig, *Biophys. J.*, 1998, **74**, 931.

53 J. Hermans, H. J. C. Berendsen, W. F. van Gunsteren and J. P. M. Postma, *Biopolymers*, 1984, **23**, 1513.
54 D. P. Tieleman and H. J. C. Berendsen, *J. Chem. Phys.*, 1996, **105**, 4871.
55 D. P. Tieleman and H. J. C. Berendsen, *Biophys. J.*, 1998, **74**, 2786.
56 W. Kabsch and C. Sander, *Biopolymers*, 1983, **22**, 2577.
57 A. T. Brünger, *X-PLOR Version 3.1. A System for X-Ray Crystallography and NMR*, Yale University Press, New Haven, CT, 1992.
58 P. J. Kraulis, *J. Appl. Crystallogr.*, 1991, **24**, 946.
59 M. E. Davis, J. D. Madura, B. A. Luty and J. A. McCammon, *Comput. Phys. Commun.*, 1991, **62**, 187.
60 O. S. Smart, J. M. Goodfellow and B. A. Wallace, *Biophys. J.*, 1993, **65**, 2455.
61 D. Sitkoff, D. J. Lockhart, K. A. Sharp and B. Honig, *Biophys. J.*, 1994, **67**, 2251.
62 B. M. P. Huyghues-Despointes, J. M. Scholtz and R. L. Baldwin, *Protein Sci.*, 1993, **2**, 1604.
63 W. Hanke and G. Boheim, *Biochim. Biophys. Acta*, 1980, **596**, 456.
64 M. S. P. Sansom, G. R. Smith, C. Adcock and P. C. Biggin, *Biophys. J.*, 1997, **73**, 2404.
65 L. R. Forrest and M. S. P. Sansom, *Biochem. Soc. Trans.*, 1998, **26**, S303.
66 P. C. Biggin and M. S. P. Sansom, *Biophys. Chem.*, 1996, **60**, 99.
67 P. Biggin, J. Breed, H. S. Son and M. S. P. Sansom, *Biophys. J.*, 1997, **72**, 627.
68 Q. Zhong, P. B. Moore, D. M. Newns and M. L. Klein, *FEBS Lett.*, 1998, **427**, 267.
69 B. Roux, *Biophys. J.*, 1997, **73**, 2980.
70 K. R. Ranatunga, I. D. Kerr, C. Adcock, G. R. Smith and M. S. P. Sansom, *Biochim. Biophys. Acta*, 1998, **1370**, 1.
71 I. H. Shrivastava, C. Capener, L. R. Forrest and M. S. P. Sansom, unpublished work.
72 K. S. Åkerfeldt, J. D. Lear, Z. R. Wasserman, L. A. Chung and W. F. DeGrado, *Acc. Chem. Res.*, 1993, **26**, 191.
73 Q. Zhong, Q. Jiang, P. B. Moore, D. M. Newns and M. L. Klein, *Biophys. J.*, 1998, **74**, 3.
74 G. R. Smith and M. S. P. Sansom, *Biophys. J.*, 1997, **73**, 1364.
75 P. Mitton and M. S. P. Sansom, *Eur. Biophys. J.*, 1996, **25**, 139.
76 B. Roux and M. Karplus, *Biophys. J.*, 1991, **59**, 961.
77 B. Roux, *Biophys. J.*, 1996, **71**, 3177.
78 V. Dorman, M. B. Partenskii and P. C. Jordan, *Biophys. J.*, 1996, **70**, 121.

Paper 8/06266H

General Discussion

Dr Duax opened the discussion of Prof. Wallace's paper: It is not clear whether your references to the CD spectra of gramicidin in acetic acid in your contribution to this meeting are meant to question the accuracy of the X-ray structure determinations reported in our *PNAS* paper.[1]

One criticism of X-ray studies of gramicidin has been that crystals were grown from alcohols which have relative permittivities near 50. This is very high compared to the relative permittivities of lipids of 2 to 4. To address this problem we grew crystals from glacial acetic acid which has a relative permittivity much closer to that of lipids (5–6). We did not claim that acetic acid resembled lipids in other ways.

Our crystal structures of the Cs^+ complex of gramicidin from methanol and the hydronium ion complex of gramicidin from glacial acetic acid reveal the same right handed antiparallel double helix. These structures agree in all details with the model first proposed by Arseniev *et al.*[2] to interpret the NMR Noesy spectra of the Cs^+ complex of gramicidin in a methanol–chloroform mixture. Our structure is also compatible with ^{15}N NMR data on gramicidin in planar lipid bilayers.[3] Our structure is consistent with the standard characterization of the membrane active channel of gramicidin based upon CD spectra as being right handed.

While the significance and implications of an X-ray structure may be open to debate, the accuracy of a high resolution X-ray structure determination can and should be unambiguous. As you know the results of our X-ray analysis of Cs^+ and K^+ complexes of gramicidin are completely different from yours, different in molecular hand, ion content, nature of ion coordination, channel shape, and every single one of the 30 hydrogen bonds present in the antiparallel dimers. On the basis of published results we conclude that your reported structural models have chemical and stereochemical anomalies and crystallographic limitations not present in our refined structures.

The disparities between our structures and yours are all the more puzzling because the crystallization conditions were very similar and all the cell dimensions of the complexes agree to within 1%. In light of the controversy concerning the relevance of the X-ray results to solution spectra, transport properties and the mechanism of membrane transport of gramicidin we consider it vitally important to resolve the questions generated by the profound differences between our X-ray crystal structures and yours. There are three possibilities. (1) Our structures are wrong. (2) Your structures are wrong. (3) Both structures are right. Which of these possibilities is in fact the case can be quickly resolved by comparing the diffraction intensities that were used in our determinations and yours. We deposited the atomic coordinates and all of the intensities for our Cs^+ complex in the Protein Data Bank (PDB) when the paper was accepted in *PNAS* (two months before it was published in November 1998). You told me this morning that you attempted to access our intensities files through the PDB yesterday and could not. I am sorry to hear that the PDB is unable to cope with deposition and distribution in a timely fashion. We will gladly provide you with the intensity data on our Cs^+ complex directly. In the spirit of fairness and in the interest of science we would request that you reciprocate by sending us the intensity data for your Cs^+ complex within 10 days of receipt of our intensities data. We will be happy to do the same with our data on the K^+, Rb^+ and other complexes. Will you do this?

We hope that you will agree that it is important to resolve this ambiguity. If your structures and ours are correct it would be the first example that I know of where isomorphous forms of a structure with identical cell dimensions have entirely different conformations. Given the existence of 160 000 structures in the Cambridge Structural Database such an unusual phenomena is certainly worth characterizing carefully.

1 B. M. Burkhart, N. Li, D. A. Langs, W. A. Pangborn and W. L. Duax, *Proc. Natl. Acad. Sci. USA*, 1988, **95**, 12950.
2 A. S. Arseniev, I. L. Barsukov and V. F. Bystrov, *FEBS Lett.*, 1985, **180**, 33.

3   L. K. Nicholson, F. Moll, T. E. Mixon, P. V. LoGrasso, A. L. Lay and T. A. Cross, *Biochem.*, 1987, **26**, 6621.

Dr Wallace replied: You have raised a number of points in your comments, so I will try to answer them one by one: (1) The CD studies in acetic acid were done to address the statements in your *PNAS* paper that this form represented the conducting form in lipid membranes. They clearly show that the acetic acid form is not the conducting form found in phospholipids, but they in no way question the accuracy of your X-ray structure, merely the interpretation of it. (2) Some of the other evidences you cite for this being the conducting form are also controversial. While I am not an expert in solid state NMR, I understand that a paper disputing your claims that this structure is compatible with the spectroscopic data, has been submitted[1] that will address that issue better than I can in this limited space. (3) I agree that our X-ray structure of the CsCl form is completely different from yours. However, the crystals examined in each case are different, and given the polymorphic nature of this molecule, which can demonstrably form structures of different handedness, different stagger between the chains, and different hydrogen bonds in solution, it does not seem particularly surprising that a similar polymorphism is seen in the crystals.

1   T. A. Cross, A. Arseniev, B. A. Cornell, J. H. Davis, J. A. Killian, R. E. Koeppe, L. K. Nicholson, F. Separovic and B. A. Wallace, submitted.

Dr Duax further commented: You did not respond to the most important question. Will you release the intensities from your published structures to the PDB as we have?

Dr Wallace responded: We are currently re-refining our CsCl structure[1] using improved geometric parameters, as you have recently done[2] with the earlier Buffalo structure.[3] Once done, we intend to publicly release our structure *via* the PDB, in the standard fashion.

1   B. A. Wallace and K. Ravikumar, *Science*, 1988, **241**, 182.
2   B. M. Burkhart, R. M. Gassman, D. A. Langs, W. A. Pangborn and W. L. Duax, *Biophys. J.*, 1998, **75**, 2135.
3   D. A. Langs, *Science*, 1988, **241**, 188.

Dr Burkhart asked: How much interpretation can you make about the effect of side chain conformational change on the CD spectrum differences?

Dr Wallace responded: The CD spectra[1] do not show significant differences between gramicidin in different solvents. I know that your crystal structures[2] do show a difference in the orientations of the tryptophans between the methanol and ethanol or propanol structures. This may be due to the differences between solution and crystal structures, especially any constraints on the tryptophan side chains, which are located at the periphery of the moecule, due to packing in the crystals. That the two crystal structures crystallised in the same space group (ethanol and propanol) have similar tryptophan orientations and the one crystallised in another space group (methanol) is most different might tend to support the crystal packing as a source of these differences between solution and solid state studies.

1   W. R. Veatch, E. T. Fossel and E. R. Blout, *Biochemistry*, 1974, **13**, 5249; Y. Chen and B. A. Wallace, *Biopolymers*, 1997, **42**, 771.
2   B. M. Burkhart, R. M. Gassman, D. A. Langs, W. A. Pangborn and W. L. Duax, *Biophys. J.*, 1998, **75**, 2135.

Prof. Holzwarth asked: Why did you use a mixture of gramicidin A, B and C and not a pure sample? I would expect differences between a mixture and pure samples of gramicidin.

Dr Wallace responded: We used the commercially available mixture of gramicidin A, B, and C. Veatch had previously shown[1] that the mixture behaves spectroscopically in a very similar manner to pure gramicidin A.

1   W. R. Veatch, PhD Thesis, Harvard University, 1974.

Prof. Laggner commented: Could it be that both structures (Wallace and Duax) are right, but not really relevant to the structure in the bilayer system. Current ideas in this field suggest that the structure of peptides in the membranes may not be unique but rather dependent on various variables, *e.g.* the amount of peptide incorporated to the bilayer. In BLMs, monomer, it is rather difficult to control the amount of peptide actually present in the "black" part of the film.

Dr Wallace replied: I agree entirely. From all the physical and chemical studies done on gramicidin over the past 20 years or more, it is clear that the principle conducting form in membranes is the helical dimer, not any of the double helices seen in either Duax's and our structures. The studies presented in my paper at this meeting confirm this, and suggest there are only very specific conditions under which double helices may form conducting molecules, such as when there is a very severe mis-match of the lipid fatty acid chain length with the gramicidin size.

I should point out that the Duax CsCl crystals and ours are prepared under very different conditions: different peptide and ion concentrations and ratios, and different temperatures (which our recent CD studies[1] have shown produce very different spectra, and by implication, structures, in solution). The crystals have slightly different unit cell dimensions (but small variations in unit cell dimensions have significant consequences for this molecule.[2] Given the polymorphism seen for this molecule in so many different environments (solution, membranes, *etc.*) it is not surprising to find polymorphic crystal structures. It is very clear from our anomalous Patterson maps, for instance, that the Duax structure is not consistent with our data, and so must represent a different crystal form.

1   T. P. Galbraith and B. Wallace, unpublished results.
2   D. A. Doyle and B. A. Wallace, *J. Mol. Biol.*, 1997, **266**, 963.

Prof. Lee said: My question is about the effect of membrane thickness on the state of gramicidin. In di-$C_{22:1}$ PC, is gramicidin present as a $\beta^{6.3}$ dimer or monomer, or will the CD spectra of these two forms be indistinguishable?

Second, I thought that the thickness of a solvent-free bilayer and the equivalent solvent containing bilayer were very different, so that it seems surprising that chain length effects are the same in the two systems.

Dr Wallace responded: We cannot clearly distinguish $\beta^{6.3}$ monomers from $\beta^{6.3}$ dimers by CD spectroscopy. Both have similar backbone folds (which is what CD detects). CD studies we did some time ago[1] on gramicidin in which the N-terminal formyl group was replaced with an N-terminal acetyl group, which tends to destabilise the dimer and thus produce monomers, showed very little difference between the spectra, so I don't think we can say for certain whether the structures in C_{22} lipids are monomers or dimers from our work. However, the conductance studies of Mobashery *et al.*[2] suggest that a significant proportion of the population in this lipid is dimeric and conducting.

I agree with you that the bulk thicknesses of bilayers with and without solvent are quite different. But what I don't think we know is the actual thickness of the bilayer immediately surrounding and in contact with the gramicidin. In fact, I have to say that I was surprised that the "chain length effect" we are seeing only starts at C_{22}. I would have thought that much shorter lipids (even as short as C_{18}) would be a mis-match for gramicidin. But that is, again, based on thicknesses measured for bulk lipids in the absence of peptide, and the lipid surrounding the gramicidin may have substantially different properties.

1   B. A. Wallace, W. R. Veatch and E. R. Blout, *Biochemistry*, 1981, **20**, 5754.
2   N. Mobashery, C. Nielsen and O. S. Andersen, *FEBS Lett.*, 1997, **412**, 15.

Prof. Roux asked: It is not clear to my why the DH form gets stabilized in thick membrane relative to the HD form. Which one is stabilized, or, perhaps more justly, which one is destabilized?

Dr Wallace responded: I would think that really we are talking about the absence of stabilizing effects as much as anything. Numerous studies have suggested that interactions between tryptophan side chains and the interfacial region of the bilayer may be important structurally. In thin lipids, the HD, with its tryptophans near the ends of the molecule, has them in a position that could interact favourably with the lipid headgroup region, which would have a stabilising influence; in the DHs, the tryptophans are spread along the length of the molecule, and so some of them would have to be buried deep in the bilayer, a much less favourable disposition. Thus in this case the equilibrium may be shifted towards the HD. However, in thick lipids, even the HD would have to bury some of its tryptophans below the bilayer interface, therefore this factor would be less likely to shift the equilibrium towards the HD. The other factor in the mismatch case could be the number of intermolecular hydrogen bonds holding the dimers together, which is 6 for the HD and between 26 and 30 for the DH.

Dr Burkhart commented: Given your own admission that gramicidin adopts many different structural forms and that the influences of Trp residue side chain conformation cannot be discerned, is it not dangerous to choose only four structural forms to describe each CD spectrum.

Dr Wallace responded: The control we have for whether the CD data is well represented by the reference data spectra used in the analyses is the NRMSD parameter. It is much like an R-factor in crystallography, in that it is a measure of the correspondence between the experimental data and the best fit to the reference data. A low value for the NRMSD indicates the reference data set reflects well the population of conformers present in the sample.

Dr Deber communicated: What are the linear (vertical) dimensions of the DH $vs.$ the HD forms of gramicidin?

Dr Wallace communicated in response: The ion-containing DH is ~ 26 Å long, whilst the HD form is ~ 32 Å (see ref. 1).

1 O. S. Smart, J. M. Goodfellow and B. A. Wallace, *Biophysical J.*, 1993, **65**, 2455.

Dr Okazaki opened the discussion of Prof. Roux's paper: Within my knowledge, the interaction potential between protein cylinder and lipid chain carbon $u(r)$ is essential for this kind of calculation. The results must be very sensitive to the potential function between unlike particles. If you assume the strongly attractive potential between them, homogeneous mixing will be obtained but if you set the weakly attractive interaction, you will obtain a kind of phase separation, *i.e.* protein aggregation, as you show in your work. You might get even a double-well-like free energy profile for a particular potential function. The results including the free energy profile must change dramatically as a function of the protein–lipid interaction.

Now, my question is what kind of protein–lipid potential function did you assume or, in other words, how did you determine the interaction function?

Prof. Roux responded: The protein–protein potential of mean force $W(r)$ depends on several factors. One usually writes that $W(r) = U_{pp}(r) + \Delta W(r)$, where $U_{pp}(r)$ is the microscopic protein–protein potential energy and $\Delta W(r)$ is the lipid-mediated free energy potential. What you are saying is that the latter depends on the protein–lipid interactions, which is absolutely correct. In this preliminary study, we were mostly interested in exploring the magnitude of the forces arising from the influence of excluded volume of the hydrocarbon chains by the protein inclusion. Therefore, in the present calculation the protein–lipid potential was simply chosen as a repulsive hard cylinder. Of course more realistic interactions could be used (see for instance our answer to Prof. Smith below), but they would not allow an investigation of the excluded volume effect on the protein–protein lipid-mediated forces.

Dr Smart asked: Do your results have similarities to the results of Huang and coworkers[1] who have determined the 2D radial distribution function for peptides in lipid bilayers.

1 L. Yang, T. A. Harroun, W. T. Heller, T. M. Weiss and H. W. Huang, *Biophys. J.*, 1998, **75**, 641.

Prof. Roux responded: Huang has measured the in-plane distribution pair correlation function for the gramicidin channel. We intend to compare these data to the calculated lateral packing of cylinders corresponding to the size of a gramicidin channel.

Prof. Haymet commented: The level of theory seems nicely appropriate for the problem. The difficulty arises with the new calculations which you have discussed now and are not in the printed copy of the paper. The approximation for hydrophobicity which inspired your approach has long been known to break down the size of the solute increases beyond the solvent diameter. Hence calculation of the interaction of two cylinders are likely to be similarly qualitatively incorrect as the cylinder radium increases beyond the characteristic length-scale of the medium in which they are immersed. This is due to the fact that the solvent response-function is not allowed to relax in the presence of the solute.

Prof. Roux responded: In the paper were described preliminary results obtained for a hard repulsive cylinder of 5 Å diameter. We observe that the lipid-mediated potential of mean force has a complex structure, with an attractive well at contact and a repulsive barrier at a cylinder–cylinder separation distance of 20 Å. It is of interest to examine what is the dependence of this result on the size of the protein inclusion and that is the reason why we are investigating this matter. Nevertheless, I agree with you that the HNC integral equation theory for simple liquids formed by spherical particles has problems in dealing with very large differences in particle size, though I would add that the present case is more complex since there are several lengthscales in the lipid–lipid pair correlation function $\chi_{mm}(r)$ whereas the situation with monoatomic liquids is much simpler. Comparison with molecular dynamics simulations of atomic models will be done in order to assess the range of validity of the current theory for lipid bilayers. The extensive simulations performed in Prof. Klein's group will provide a good basis for that. Nevertheless, whether the theory is yielding semi-quantitative or qualitative results is secondary at this point. One must realize that an integral equation theory such as described here provides a unique route to gain some insights into lipid-mediated potential of mean force between protein inclusions and the influence of the hydrocarbon chains on the protein–protein interactions in bilayer membranes. There is presently no other theoretical approach to gain such information.

Prof. Holzwarth asked: Are you in a position to predict how far the influence of proteins could reach into the membrane; especially how many lipid layers around the protein will be influenced? We found experimentally for bacteriorhodopsin that as many as five to six layers of surrounding lipids are influenced.[1]

1 A. Böttcher, N. Dencher, R. Groll, F. Meyer and J. F. Holzwarth, *Reactions in Compartmentalized Liquids*, ed. W. Knoche and R. Schomäcker, Springer Verlag, Berlin, 1989, pp. 105–115.

Prof. Roux responded: In Fig. 2 we show that the density of the hydrocarbon core is perturbed over a distance of 30 Å around a 5 Å cylinder. The radius of one DPPC being roughly 4.5 Å based on a surface area per molecule of 64 Å^2 (using πr^2), this implies that a layer of 2 to 3 lipids are perturbed. The difference may be due in part to the fact that bacteriorhodopsin is much larger (it is a bundle of 7 transmembrane helices). Furthermore, in reality there is a direct attractive dispersion interaction between the protein and the lipids hydrocarbon chains whereas here we consider only the excluded volume effect. This may amplify the effect. Lastly, one should keep in mind that our theory offers an equilibrium statistical mechanical view of the perturbed density around an impurity. Ultimately, the number of observed lipids may depend upon the experimental method used to characterize the system (*e.g.*, magnetic resonance, fluorescence, infrared spectroscopy, differential calorimetry, *etc.*).

Prof. Klein commented: A liquid hydrocarbon against a hard wall will exhibit layering of methylene groups that extends 4 layers or so into the bulk liquid. Thus, the oscillation you observe in the hydrocarbon core around the 5 Å cylinder is reminiscent of this effect. It would be interesting to compare the DPPC lipid with, for example, hexadecane bulk liquid to see if there is any specific effect of the lipid. The main input in your calculations is the simulation data for the carbon–carbon density–density response function. These data were generated with a relatively small sample. How confident are you that the values are reliable in the range 15–20 Å and will uncertainties in this asymptotic region influence your results?

Prof. Roux responded: Fig. 1 shows that the carbon–carbon intramolecular correlation function $S_{mm}(r)$ has a large contribution followed by a peak up to 3 Å and then a slow decay over a distance of 10–15 Å. Clearly, the intramolecular correlation contributes significantly to the response function of the lipid bilayer: the short range structure in the intramolecular correlation arises from nearest neighbor carbons along the acyl chains. The peak at $r = 0$ is indicative of the significant amount of short range order in the lipid chains perpendicular to the plane of the bilayer. The oscillations you are referring to arise from the intermolecular contributions in the response function. Those correspond partly to carbon–carbon contacts, and are reminiscent of liquid hydrocarbon. You are raising a very interesting question: to what extent does the response of the hydrocarbon differ from that of an isotropic liquid hydrocarbon? We will try to address that in the future by using a response function extracted from a liquid hydrocarbon simulation. Concerning your second question, the pair correlation function was calculated from a molecular dynamics trajectory of a lipid bilayer generated by Feller *et al.*[1] The atomic system that they simulated consisted of 72 DPPC molecules (36 in each leaflet). The physical length of the periodic box is approximately 48 Å. Thus, since the the correlation function decays almost to zero over a distance of 10 to 15 Å, it seems reasonable to assume that the dominant packing structure was captured by the molecular dynamics simulation.

1   S. E. Feller, R. M. Venable and R. W. Pastor, *Langmuir*, 1997, **13**, 6555.

Prof. Petersen asked: Please speculate on the interactions among several cylinders. Will there be an optimum size? Will a bundle of 3–5 cylinders of 5 Å radius behave as a single cylinder of ∼9 Å and hence there might be no further aggregation?

Prof. Roux responded: The current calculations correspond to the infinite diluted limit of protein inclusion in a lipid bilayer. This assumption is necessary since we possess only the response function $\chi_{mm}(r)$ of the unperturbed bilayer from the simulation of Feller *et al.*[1] Nevertheless, one could try to investigate finite concentration and aggregation effects with the current theory, keeping in mind its limitations due to the response function.

1   S. E. Feller, R. M. Venable and R. W. Pastor, *Langmuir*, 1997, **13**, 6555.

Prof. Smith asked: What are the prospects for evolution of this approach towards an all-atom description of the helices?

Prof. Roux responded: The current theory is designed to address the influence of the lateral packing of the lipid hydrocarbon chains on the protein–protein potential of mean force. For this purpose the theory was kept as simple as possible. In particular, we considered only hard cylindrical protein inclusions with no details. Nevertheless, it is possible to construct a more sophisticated integral equation theory in which the detailed atomic structure of a membrane-bound protein will be used (*e.g.*, transmembrane helix or an amphipatic helix associated at the membrane/solution interface). This extended integral equation theory would require a more complete response function than the simple $\chi_{mm}(r)$ used in the present theory and which characterizes the lateral fluctuations. The extended approach would be the equivalent of integral equation theories such as those described in ref. 26. We are currently working on the development of this extended theory.

Dr Gilbert opened the discussion of Dr Bezrukov's paper. In the experiments on alamethicin reported larger pores appear to be favoured in their formation by the presence of lipids possessing a propensity for the formation of H_{II} phases (induced by pH modification). Does alamethicin itself promote H_{II} formation? Is this relevant to consideration of these data? Secondly, do the temporally co-existing hexagonal (H_{II}) and lamellar phases also coexist within a single lipid body—*i.e.* within the surface of a liposome? If so, what form would the interface between the lamellar and H_{II} phases take?

Dr Bezrukov responded: Yes, not only lipid monolayer spontaneous curvature modifies alamethicin channel behavior, but, in turn, alamethicin itself influences lipid curvature properties. According to an X-ray and NMR study by Keller *et al.*,[1] addition of as little as 1% of alamethicin to 1,2-dielaidoyl-*sn*-lycero-3-phosphoetanolamine introduces a large region of cubic phase into the thermal phase diagram. This observation could be important for the molecular model of the phenomenon. To answer your second question, in the X-ray measurements that are reported in our paper the samples were prepared in excess water. Both lamellar and hexagonal phases are independent, three-dimensional structures. So necessarily they exist separately, and not on a two-dimensional surface, but within the same bulk sample. We do not know the structural nature of the interface between them.

1 S. L. Keller, S. M. Gruner and K. Gawrisch, *Biochem. Biophys. Acta*, 1996, **1278**, 241.

Dr Goñi asked: To what extent do your "change in pH" experiments answer the criticism, raised in relation to the PC/PE data, that you were changing the chemistry of the system? Your results look reasonable to me, but in my view protonation is also changing the chemistry of the host lipid towards alamethicin. Have you thought of changing curvature of PC bilayers by adding lyso PC? Or, conversely, adding polyunsaturated PC to saturated PC? Very little changes in the "chemistry" would take place in such an experiment.

Dr Bezrukov responded: The answer probably depends very much on what you mean by "chemistry". If changing head group interaction is chemistry, then yes, we are changing the chemistry of host lipid. In our opinion, however, changing the pH means less chemical modification of the system than admixing/substituting one lipid species by a distinct second lipid species. In response to your second point, in principle, this is a very good suggestion. In practice, it may be difficult to design a reliable experiment and to rationalize obtained results due to relatively high water solubility of lyso PC. Besides, an additional worry in experiments with the lipid mixtures is a possible "demixing" of lipids in the vicinity of the channel.

Prof. Laggner asked: Two related questions: What are the relative populations of lamellar and H_{II} phases in the coexistence region between pH 1.5 and 3.5. Why do you refer to these two states as "interstable"?

Dr Bezrukov responded: We can only estimate the relative quantities based on relative X-ray intensities. Qualitatively, the lamellar phase appears to be maximum between pH 2.3 and 2.9, where it coexists with the hexagonal phase, and appears to involve less than 50% of the lipid. The lamellar phase is nearly, but not quite, absent at pHs above and below 2.3 and 2.9. As for your second question, the X-ray scattering patterns are stable over periods of at least weeks.

Prof. Neumann asked: Does the dependence of the current intensity $\Delta i(t)$ on the electrolyte concentration reflect increasing charge screening of the phosphatidylserine (PtdSer) groups and thereby affect the spontaneous curvature?

Dr Bezrukov responded: The increase of single-channel current with the electrolyte concentration is mostly related to the increase in the average channel occupancy by ions. As seen in Fig. 6 of our paper, this increase mostly follows the bulk solution conductivity. The effect of increasing lipid charge screening that you mention is very small, because the initial contribution from lipid charge to the open channel conductance is tiny. This is not always the case though. For "small" channels it can be quite pronounced. Recently we published a study where the lipid charge effect on channel conductance was studied over varied charge densities.

1 T. K. Rostovtseva, V. M. Aguilella, I. Vodyanoy, S. M. Bezrukov and V. A. Parsegian, *Biophys. J.*, 1998, **75**, 1783.

Dr L. Fisher commented: I totally agree that membrane lipids must be functionally involved and have a controlling role in membrane protein behaviour. My question is whether we are looking at the right end of the lipid. Lipid packing studies to date have tended to focus on non-glycosylated lipids. It is a fact of nature, though, that the outer membrane lipids of mammalian cells are heavily glycosylated, and it may be that interactions between the headgroups, rather than the chains, of such lipids dominate their function. Would you care to comment on this, and on whether chain and charged headgroup interaction studies are of relevance for such lipids and the membranes containing them?

Dr Bezrukov responded: Yes, this is very interesting question, though I do not think that we, or anybody else for that matter, have enough evidence to state that interactions between the headgroups are more important than chain interactions. Many factors contribute to the balance of energies in channel conformational equilibrium. Glycosylation is likely to be important; at least the cells seem to think so. Whether head or tail "dominates" I would rather not decide in general.

Prof. Roux said: I fail to see a relationship between the propensity of forming the H_{II} phase and the stabilization of alamethicin multimers, which are a bundle of transmembrane helices. Could you comment on that please?

Dr Bezrukov responded: If you ask about the empirical relationship, it is in Fig. 3 *vs.* Fig. 10 of our paper. As for a detailed molecular mechanism, we are still far from any good idea. One possibility could be the "shape" of the channel to explain the increase in the number of stress-relieved lipid molecules as the channel goes to a state of higher conductance.

Prof. Lee commented: The difficulty you have, as you have said, is in separating effects of the chemical structure changes from effects on curvature. The transition temperature into the hexagonal H_{II} phase is higher for POPE than for DOPE. Have you compared the effects of POPE and DOPE on alamethicin channel formation?

Dr Bezrukov responded: No we have not. But this is definitely a good suggestion.

Dr Smart asked: Can your results be explained if the alamethicin molecule adopts a wedge-shaped bundle (and that this varies with bundle size)? Can Dr Sansom comment whether the models he has generated support this?

Dr Bezrukov responded: Yes, the shape of the channel exterior facing lipid matrix could be a key factor. I will leave this question for Dr Sansom who did extensive work modeling alamethicin channel.

Dr Sansom responded: The models of alamethicin helix bundles we have generated are, to a limited extent, wedge-shaped (if one takes a cross section down the pore *i.e.* bundle axis). This may provide a partial explanation of the result of Bezrukov and colleagues. However, to be certain of this probably requires a systematic series of simulations with *e.g.* different lipid headgroups or headgroup protonation states. I don't think "back of an envelope" approaches are going to work here.

Dr Kakorin asked: Did you change the pH value and NaCl concentration equally on the two sides of the lipid membrane?

Dr Bezrukov responded: Yes, we have varied the pH value and NaCl concentration equally on both sides of the membrane. In the future we are going to change the pH value on one side.

Dr Kakorin responded: If there were not a difference in pH value or NaCl concentration on the two sides of the membrane, then you cannot rationalise the variation of the conductance level of alamethicin channels with pH value or [NaCl] by the proton-induced increase in spontaneous curvature. The cartoon illustration of the curved monolayer (Fig. 9) is not relevant to the data. Actually, according to Helfrich's definition the spontaneous curvature reflects a possible asymmetry in the bilayer,[1] not in a monolayer as in Fig. 9. The physical–chemical origin of the asymmetry could be either a different chemical environment on both sides of membrane or a different chemical composition of the two monolayers.[2] Indeed, the variation of pH value and NaCl concentration may change only the Debye screening length and thereby the electrostatical interaction between charged lipid headgroups in two monolayers. This can lead to a change in the lateral membrane tension which may be operative to modulate the conductance level of alamethicin channels.

1   W. Helfrich, *Z. Naturforsch. C*, 1973, **28**, 693.
2   U. Seifert, *Adv. Phys.*, 1997, **46**, 13.

Dr Bezrukov responded: While we cannot rule out other factors, we correlate our findings exactly with what you suggest: with the pH-induced change in lateral pressure. As it is well known, lateral pressure in a monolayer of a symmetric bilayer is distributed differentially and changes along monolayer depth (*e.g.* see Fig. 1 of the paper by Templer *et al.* in this volume). Therefore, every attempt to reduce actual distribution with a single number is a simplification of the actual situation. We correlate probabilities of alamethicin channel states with the spontaneous monolayer curvature[1] which is related to the first moment of the lateral pressure.[2] The pH variation changes lipid charge thus changing lipid head–head interactions; the resulting modification in lateral pressure distribution is manifested by the change in lipid spontaneous curvature (Fig. 10). In our earlier work with alamethicin channel in PE–PC mixtures we were first to relate changes in the single-channel expression with lipid spontaneous curvature.[3] Recently Olaf Andersen and his colleagues[4] have shown that this parameter is also very important for the gramicidin A channel activity. Moreover, they showed that, at least in the case of gramicidin A and solvent-free bilayers, the channel expression is much more sensitive to changes in the first moment of lateral tension than to changes in its average value at lipid substitution.[5]

Thus, to say that the cartoon in Fig. 9 is not relevant to our channel data is, to put it mildly, an exaggeration. For further reading I would recommend the special issue of *Chemistry and Physics of Lipids*,[6] devoted to the functional role of non-lamellar lipids.

1 S. M. Gruner, *Adv. Chem. Ser.*, 1994, **235**, 129.
2 R. H. Templer, S. J. Castle, A. R. Curran, G. Rumbles and D. R. Klug, *Faraday Discuss.*, 1998, **111**, 41; M. M. Kozlov and V. S. Markin, *J. Chem. Soc. Faraday Trans. 2*, 1989, **85**, 261.
3 S. L. Keller, S. M. Bezrukov, S. M. Gruner, M. W. Tate, I. Vodyanoy and V. A. Parsegian, *Biophys. J.*, 1993, **65**, 23.
4 J. A. Lundbaek, A. M. Maer and S. O. Andersen, *Biochemistry*, 1997, **36**, 5695 and references cited therein.
5 C. Nielsen, M. Goulian and O. S. Andersen, *Biophys. J.*, 1998, **74**, 1966.
6 *Chem. Phys. Lipids*, Special Issue, 1996, **81**(2).

Dr Templer commented: The curvature elastic stress that is stored in a lipid bilayer is in general non-zero. For a flat bilayer, which has the same lipid composition on either side, the Helfrich Hamiltonian, as expressed in terms of the bilayer, would indeed be zero. However, if it is expressed in terms of each constituent monolayer it becomes evident that each monolayer wishes to bend to the same degree, but in opposite directions. Helfrich has called this the torque tension, and the frustrated curvature energy stored in each monolayer per unit area is proportional to the square of the spontaneous curvature of the monolayer and the bending modulus of the monolayer. Since the monolayer is held flat this curvature elastic energy must be stored in some way. This is done by increasing the area dilation at the interface.

Dr Sansom commented: Alamethicin helices are linked (by the glycine-X-X-proline motif) and thus, to a crude approximation, helix bundles are somewhat "hourglass" shaped in cross-section. Thus, lipids which prefer the H_{II} phase may pack better around the helix bundle.

Dr Bezrukov responded: Yes, I agree. This could be one of the possibilities for explaining the positive correlation between probabilities of higher conductance channel states and H_{II} lipid propensity.

Prof. Evans asked: Is the effect of 'spontaneous curvature' on channel activity altered by organic solvents that partition in the bilayer interior (*e.g.* as your group has shown for the H_{II} phase transition of DOPC)?

Dr Bezrukov responded: We did not study that. The present paper reports results for "dry" membranes obtained by Montal–Mueller monolayer opposition technique. No solvents were used in the X-ray liquid-crystal phase preparations.

Dr Bezrukov opened the discussion of Dr Smart's paper: What you are trying to do is very important for the central issue in molecular biophysics, the issue of structure–function relationship. For ionic channels their transport properties, in particular conductance, are of prime interest. However, the predictive power of your approach is not clear. Consider the following example. It is well known (structural data, molecular models, permeation results) that the pore radius of the channel formed by gramicidin A is about three times smaller than the pore radius of amphotericin B channel. Using your approach we would expect about ten times higher conductance for amphotericin B channels. The single-channel measurements show that the actual situation is reversed: amphotericin B channel conductance is about three times *smaller* than the conductance of gramicidin A channel. Are you not worried about this 3000% discrepancy with your model prediction?

Dr Smart responded: The main point to emphasize is that the predictive power of the technique has been tested on all ion channels where an experimental high resolution structure was available.[1] In these tests prediction to within an average factor of 1.6 with a predictive r^2 (ref. 2) of 0.9 (6 systems tested). Extending the test set to include model structures with a reasonable certainty produced results to within a factor of 1.8 and a predictive r^2 of 0.46. In comparison to methods of forecasting binding affinities to enzymes on the basis of structure these figures are good, particularly for a first attempt.

The method has not yet been tested on amphotericin B, as the structure of the channel conformation has not been experimentally determined. Although models have been proposed[3,4] they are by no means certain, in particular the stoichiometry and role of sterol molecules is still unclear.[4] The pore dimensions of the models have been based on the fact that the antibiotic allows the diffusion of xylose or ribose, but not larger sugars through cell walls.[5] Experience has shown that such an identification is good for rigid molecules such as porins (see ref. 5 of our paper) but it is quite possible that the flexibility and/or changes of stoichiometry of amphotericin B may be involved in xylose transport. If the discrepancy is real then it may be due to the fact that amphotericin B is a polyene rather than peptide and is proposed to have a channel lumen lined by hydroxy groups, which makes it different in character from the channels used to parameterize the HOLE conductance prediction. In conclusion, the discrepancy may not exist but if it does then this is worthy of further investigation as this would mean that the amphotericin B has a structure activity relation which is 30-fold different from the many channels which fit within the HOLE procedure.

1 G. R. Smith and M. S. P. Sansom, *Biophys. J.*, 1998, **75**, 2767.
2 R. D. Cramer, D. E. Patterson and J. D. Bunce, *J. Am. Chem. Soc.*, 1988, **110**, 5959.
3 M. Bonilla-Marín, M. Moreno-Bello and I. Ortega-Blake, *Biochim. Biophys. Acta*, 1991, **1061**, 65.
4 M. Baginski, H. Resat and J. A. McCammon, *Mol. Pharmacol.*, 1997, **52**, 560.
5 B. de Kruihff, W. J. Gerritsen, A. Oerlemans, R. A. Demel and L. L. M. van Deenen, *Biochim. Biophys. Acta*, 1974, **339**, 30.

Prof. Roux asked: Ion channels generally exhibit saturation properties as a function of permeant ion concentration, *e.g.*, the conductance increases linearly at low concentration and then reaches a plateau at higher concentration. Sometimes it even decreases as the concentration is raised further. How is this Ohm's law approximation dealing with such phenomena?

Dr Smart responded: In short, HOLE does not attempt to deal with these phenomena. To date concentration effects have been ignored. In the original paper we state the desirability of making predictions in the low concentration range you mention. However, we are at present limited to the concentrations at which experimental data is available. The method is successful despite the problem of the diversity of concentration (for details see my earlier reply to Dr Bezrukov). Hopefully, in the future we will acquire data under consistent conditions for all channels with known structure. In many respects, although a prediction of absolute conductance in the expected range can provide a useful guide in model validation, the PEG addition experiment has the greatest potential to be of use yielding much more direct information. However, at present there is not a sufficiently large amount of data to make confident interpretations. This is why we are extending the method by collecting data for channels of known structure.

In passing, the situation is even more complex than you state. The surface conductance effects can cause higher than expected conductances at low ion concentrations (see ref. 50 of our paper) as the effective concentration of ions within a pore is increased with respect to bulk.

Dr Sansom asked: As you suggest the 6-meric and 7-meric α-toxin pores correspond to low and high conductance states, what is the timescale for switching between these two forms?

Are you suggesting that even for more complex membrane proteins, there is a similar problem to that discussed earlier for gramicidin *i.e.* that of relating different structure of a channel protein (static) to its biological function (dynamic)?

Dr Smart responded: Yes the problem of relating structure to function for ion channels is particularly acute. Even for gramicidin, where, not withstanding recent controversy, we know the structure for the conducting form, there is the problem of understanding the closure event. This is known to involve dimer breakdown into a monomeric form. However, the exact monomeric conformation adopted is still unknown. For more complex behaviours this problem is more acute. It must be remembered that in single channel recordings we are watching a single molecule or molecular assembly in action. Very often there is a massive excess of "silent" molecules present. In these cases most other experimental techniques can be expected to yield information on the conformation of the closed states.

The high and low conductance states may correspond to a difference in oligomerization but this identification is put forward as a working hypothesis worthy of further study rather than a firm conjecture. It must be remembered that a major piece of evidence is the fit between the structure-based HOLE prediction of the effect of PEG on conductance and that further work is required to make identification more firm (the data reported here on gramicidin is the first part of this process). A coauthor of the paper has an alternative explanation that the change in conductane is principally due to an alteration in the ionization state of the channel (see ref. 50 of our paper). The switch between the states is rapid in the timescale of single channel recordings. However, closure events are only very rarely observed in the absence of divalent ions. To further complicate the matter single channel recordings suggest that there are two low conductance states which differ in their own selectivity (see ref. 50 and 51 of our paper).

Prof. Smith asked: How was the diffusion coefficient calculated? Was it from the time dependence of the mean-square displacements, and if so did this function exhibit the required linearity?

Finally, in which way is the timescale of the motion thus quantified relevant to conductance?

Dr Smart responded: The diffusion coefficients were calculated by Tieleman and Berendsen from a molecular dynamics simulation of OmpF. A full description of the method used is given in ref. 10 of our paper, but the diffusion coefficients D_z are derived from the mean square displacement of atoms using:

$$\lim_{t \to \infty} \langle \{z(t) - z(0)\}^2 \rangle = 6 D_z t$$

Each water molecule was assigned to belong to a slice of z coordinate space (1.2 Å thick) and its displacement measured over the next 5 ps. At the end of this time interval a reassignment was made. The diffusion coefficient was an average of all molecules within a slice. No information is given about the linearity you refer to.

The relevance of the timescale of the motion to conductance is an interesting question. Smith and Sansom[1] have shown that the diffusion coefficients of ions within channels are affected by factors of the same magnitude. My interest in using the data is in taking an empirical approach. A benefit of using the diffusion coefficient correction is that it results in the correct "boundary condition". As a channel gets sufficiently large to be regarded as macroscopic the diffusion coefficients of the water within it will tend to bulk values and therefore the correction will tend to 1 and the predictions will reduce to Ohm's law. The validity of this adaptation can only be proven by extending to all the systems analyzed with the original purely empirical correction and seeing whether it leads to improved predictions.

1 G. R. Smith and M. S. P. Sansom, *Biophys. J.*, 1998, **75**, 2767.

Dr Sansom responded: For the alamethicin simulations and for OmpF (ref. 1) the water diffusion coefficient was calculated from the mean square displacement over a 5 ps period. In an earlier study without a bilayer[2] we have looked at ion diffusion within pores and shown that diffusion coefficients, again on a *ca.* 5 ps timescale, are similarly reduced. However, clearly these timescales are short relative to that of ion permeation. The relevance of such motions to conductance will depend *inter alia* on the strength of direct ion–pore interactions.

1 D. P. Tieleman and H. J. C. Berendsen, *Biophys. J.*, 1998, **74**, 2786.
2 G. R. Smith and M. S. P. Sansom, *Biophys. J.*, 1998, **75**, 2767.

Prof. Klein commented: With regard to diffusion of water molecules through ion channels I mention that our MD study of LS2 suggested two types of water molecules. The majority diffuse more or less unhindered through the pore but at about one-third of the bulk liquid value. But a few waters are hydrogen bonded to the inner wall of the channel with residence times of hundreds of ps. In layer channels, with more pore water, these long-lived bound waters may be less important.

Dr Smart responded: Presumably the diffusion coefficient I have used here reflects an average over these two types of water molecule. I think that the presence of tightly bound water may on occasions be important. Given a tight enough binding the water molecule may, in effect, become part of the channel. This could have some effect on the overall conductance but be very important in size selectivity. A possible example is the nicotinic acetyl choline receptor channel which appears from electron microscopic results to have a pore much larger than one would expect from its size selectivity.

Dr Gilbert asked: Does HOLE take account of the possibility of both surface and bulk conductance phenomena within a channel such as that formed by staphylococcal α-toxin.[1] Would the surface/bulk conductance dichotomy be relevant in seeking to explain high and low (and intermediate) conduction states of such channels (especially considering the expected dynamic nature of their effective diameters)?

1 Y. E. Korchev, C. L. Bashford, G. M. Alder, P. Y. Apel, D. T. Edmonds, A. A. Lev, K. Nandi, A. V. Zima and C. A. Pasternak, *FASEB J.*, 1997, **11**, 600.

Dr Smart responded: As discussed in my earlier response to Dr Sansom α-toxin has a difference in charge selectivity between the high and low-conductance states. This has implications for interpreting the results of the PEG addition experiments in terms of a small pore size. As discussed in detail elsewhere (in ref. 5 of our paper) this interpretation is thrown into question by data of PEG addition to other channels. For example the difference in the PEG addition curves observed between the $C = 1$ and $C = 3$ states of alamethicin, which differ roughly ten fold in conductance, is smaller than that between the 'high" and 'low' conducting forms of α-toxin. It is almost universally agreed that the different conductance states for alamethicin reflect different oligomerization numbers of the peptide rather than a difference in the charge state of the peptide. The assertion, therefore, that the data for α-toxin is incompatible with a difference in the size of the pore beween the states can be seen to be weak. The HOLE calculations provide an explanation for the difference without the need to invoke surface effects. It is quite possible that both effects contribute in reality.

HOLE takes such effects into account in an average fashion when making predictions of absolute conductance as they presumably play a role in the training set used for the derivation of the correction factors. Given a much larger set of data to work with it may be possible to explicitly include a correction function which incorporates the number of ionizable charges on the channel. However, at present, I have no way of representing the differences of such an effect between two systems.

You state in passing that α-toxin is expected to have a dynamic effective diameter. Given the fact that the channel lumen is provided by a β-barrel with strong hydrogen bonding it is likely that the channel is one of the least dynamic in terms of pore dimensions.

Prof. Klein commented: Our molecular dynamics studies[1] of small peptide bundles suggest that the dynamical behaviour of the bundle itself may play a role in conduction. That is, it could be important to take an ensemble average over the modes of vibration of the bundle. Radial breathing and torsional motion may be coupled to the passage of water molecules. This issue is not usually discussed. In a recent paper involving a collaboration with the DeGrado group we draw attention to the possible role of dynamical fluctuations[2] at least for small bundles. Do you think this effect could have wider implications?

1 T. Husslein, P. B. Moore, Q. Zhong, M. L. Klein, D. M. Newns and P. C. Pattnaik, *Faraday Discuss.*, 1998, **111**, 201.
2 G. R. Diekmann, J. D. Lear, Q. Zhong, M. L. Klein, W. F. DeGrado and K. A. Sharp, *Biophys. J.*, 1999, **76**, 618.

Dr Smart responded: You make a good point. We know for instance that a caesium ion cannot fit through the gramicidin channel without a marked change in structure. However, the HOLE method does not try to account for the real behaviour of channels during the conductance of an ion. Rather an empirical approach is taken in which many factors are incorporated in an average way in the fitting process. A problem behind taking an empirical approach is that it does not lead to physical insights as to the actual processes involved. But it does have other advantages. In the area of prediction of binding affinities of ligands to receptors the "*ab initio*" approach whereby you start with a model, some interaction potential energy function and use chemical physics to understand the system has been shown to be of limited use. Rather Marshall and co-authors have shown that calculating relatively simple physicochemical properties for a set of complexes with known binding energies and applying a fitting procedure can result in high reliability methods for making predictions. There is not yet enough data to be able to take such a rigorous approach for channel conductance but HOLE is an attempt to start the process. To date results are good.

1 R. D. Head, M. L. Smythe, T. I. Oprea, C. L. Waller, S. M. Green and G. R. Marshall, *J. Am. Chem. Soc.*, 1996, **118**, 3959.

Prof. Roux commented: I agree with Prof. Klein's comment about the important structural fluctuations of channels formed by bundles of helices. But I would take this further by questioning the significance of the statistical fluctuations observed in MD in the absence of ions in the channel. The presence of ion in the channel may very much affect the structure of those flexible channels.

Dr Smart responded: As the questions in this discussion have revealed there are very many processes which affect the conductance of channels. The presence of an ion can be expected to have a marked effect which will vary as the ion moves through the channel. To understand and confidently predict this effect may eventually be possible. But my approach is to avoid it at present. It may be that solvated dynamics runs do not provide an improvement over a purely empirical approach because of the point you make.

Prof. Holzwarth asked: Can you predict dynamic changes in the micro- to millisecond time range caused by the mobility of membrane lipids next to channel forming peptides like gramicidin? We investigated the influence of peptides like gramicidin and an artificial peptide of 30 amino acids (2lys-gly-24leu-2lys-ala-amide) on the dynamics of the main phase transition of bilayer vesicles[1] and found very pronounced effects on the mobility and structure of the lipids near the peptides; I wonder how the mobility of the lipids might be reflected in the transport properties of the channel forming peptides.

1 A. Genz, T. Y. Tsong and J. F. Holzwarth, *Structure, Dynamics and Equilibrium Properties of Colloidal Systems*, *NATO ASI Series C*, ed. D. M. Bloor and E. Wyn-Jones, Kluwer, Dordrecht, 1990, vol. 324, pp. 493–515.

Dr Smart responded: It may be possible to use molecular dynamics techniques to predict the effects that peptides have on membrane lipids and *vice versa*. However, at present simulation practicalities limit the area of the lipid bilayer considered. I have no direct experience in the area.

Dr Burkhart commented: Crystallographic structures are a space and time average of the molecule and contain in their B-factors many of these static and dynamic features inherent in the molecular motion.

Dr Smart responded: Temperature factors provide some information as to the ensemble average dynamics of the molecule within the crystalline environment. How relevant this is to the lipid bound form of a channel is a debatable point. However, the high resolution crystal or NMR structure of a channel in its active conformation provides information which is unobtainable in any other way. The recent excitement over the structure for the KcsA potassium channel disproves the view that as ion conductance is a dynamic phenomenon that "static" structures are of limited importance.

Prof. Roux opened the discussion of Prof. Klein's and Dr Sansom's papers: Most of the helices forming bundle channels are amphipathic which could, presumably, be lying parallel to the membrane/bulk interface. The application of a transmembrane potential together with the presence of permeant ions could be the microscopic factor driving the formation of the bundle. This could suggest that the current MDs that you described correspond to a metastable state of the channel. Could you comment please?

Dr Sansom responded: The application of a transbilayer voltage is needed to induce helix bundle formation, both for alamethicin and for other peptides (*e.g.* the LS peptide[1]). However, once open these channels are metastable on a *ca.* 10 ms timescale (*i.e.* 10^6 times the simulation timescale) even at zero mV transbilayer potential, as shown by open-channel current–voltage curves (*e.g.* Woolley *et al.*[2] and Kienker *et al.*[3]). The effect of the presence of permeant ions is more difficult to comment on—certainly there is evidence for increased channel open times at elevated ionic strengths.[4]

1 J. D. Lear, Z. R. Wasserman and W. F. DeGrado, *Science*, 1988, **240**, 1177.
2 G. A. Woolley, P. C. Biggin, A. Schultz, L. Lien, D. C. J. Jaikaran, J. Breed, K. Crowhurst and M. S. P. Sansom, *Biophys. J.*, 1997, **73**, 770.
3 P. K. Kienker, W. F. DeGrado and J. D. Lear, *Proc. Natl. Acad. Sci. USA*, 1994, **91**, 4859.
4 W. Hanke, C. Methfessell, H. U. Wilmsen, E. Katz, G. Jung and G. Boheim, *Biochim. Biophys. Acta.*, 1983, **727**, 108.

Prof. Klein responded: The metastability referred to by Prof. Roux certainly applies to our MD calculations, and those of all other workers in the field.[1–3] The present situation has been likened to trying to ascertain how the human body functions by carrying out an autopsy on a cadaver.[4] Great strides were made in the early days of anatomy by studying non-functioning beings. Similarly, the study of an assembled bundle might be expected to yield clues as to the key elements of the functioning channel.

Naturally, we look forward to the day when larger systems with transmembrane potentials and ions will be amenable to study. For the present, we have a more modest aim and indeed focus on essentially pre-assembled bundles—with all of the many limitations this entails.

Prof. Roux is correct in pointing out that many helices that form channels are amphipathic in character and would thus likely prefer to be lying parallel to the membrane/bulk interface as isolated monomers. We have recently investigated the behavior of such a monomer—the Duff–Ashley M2 peptide and indeed find that on the nanosecond timescale, this 25-residue α-helical peptide is content to remain parallel to the interface albeit with some specific peptide–lipid anchoring interactions (see Fig. 1).

1 T. B. Woolf and B. Roux., *Proc. Nat. Acad. Sci. USA.*, 1994, **91**, 11631.
2 T. B. Woolf and B. Roux., *Proteins*, 1996, **24**, 92.
3 M. S. Sansom, *et al.*, *Biochem. Soc. Trans.* 1998, **26**, 438.
4 R. S. Eisenberg, 1998, personal communication.

Dr Bezrukov said: From your Fig. 2B I gather that you took great care over the ionization state of Glu18 sidechains and concluded that the net charge is rather small. What would be the effect of the higher charge on the channel structure? Also, what potential do you use to describe interactions between charges?

Dr Sansom responded: We have run simulations for $N = 6$ alamethicin bundles with either one Glu18 ionised[1] or with zero or six Glu18s ionised.[2] With six Glu18s ionised the bundle "falls

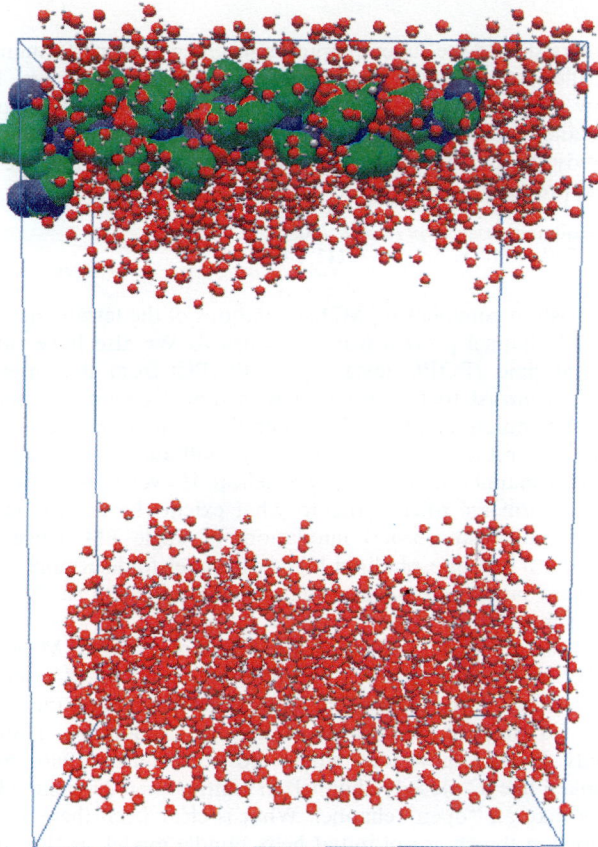

Fig. 1 The Duff–Ashley 25 residue peptide at the lipid/bulk water interface taken from an MD simulation initiated with the peptide lying parallel to the interface.

apart" during the simulation. With zero or one Glu18 ionised the bundle remains intact. This suggests that suppression of ionisation of sidechains may play a role in channel stability.

In studying the interactions between charges, in the pK_a calculations we used a screened interaction (by including a Debye length equivalent to 1 M KCl). But in the MD simulations a simple unscreened Coulombic interaction was used. Clearly this is an approximation, and may be a problem as it is known that *e.g.* melittin channels are stabilised by high ionic strength.

1 D. P. Tieleman, J. Breed, H. J. C. Berendsen and M. S. P. Sansom, *Faraday Discuss.*, 1998, **111**, 209.
2 D. P. Tieleman, H. J. C. Berenden and M. S. P. Sansom, *Biophys. J.*, 1999, **76**, in the press.

Prof. Neumann commented. There is no doubt about the importance of charged groups for the structure and for structural changes of macromolecules. Yet, for channel proteins the actual transport passage is usually hydrophobic without charged groups at some distance away from the channel part. See, for instance, Kukol and Neumann.[1]

1 A. Kukol and E. Neumann, *Eur. Biophys. J.*, 1998, **27**, 618.

Dr Bezrukov responded: Sometimes this is the other way around. See for example Forst *et al.*[1]

1 D. Forst, W. Welte, T. Wacker and K. Diederichs, *Nat. Struct. Biol.*, 1998, **5**, 37.

Prof. Klein responded: Current generation computers allow only *ca.* 5–10 nanosecond length trajectories, in most cases. However, there is one reported study by the Kollman group[1] spanning

1 μs. The next 3–5 years should see supercomputers reaching peak performance around 100 Teraflops, which should allow microsecond trajectories for membrane proteins. This will allow us in special cases to follow modest structural changes, caused for example by the passage of ions along channels. For less detailed models, the projected increase in CPU performance will enable coarse-grained models to study the assembly of model peptides into bundles and the response of the host membrane to external probes.[2]

1 Y. Duan and P. A. Kollman, *Science*, 1998, **282**, 740.
2 R. Lipowsky, *Progr. Colloid Polym. Sci.*, 1998, **111**, 34.

Dr Sansom said: I wish to comment on MD simulations of the tetrameric transmembrane (TM) helix bundle of the M2 channel protein from influenza A. We also have run such simulations,[1] albeit using a different lipid (POPC instead of diPhyPC) from that used by Klein and co-workers.[2] However, in contrast to the situation with model peptide channels (*e.g.* alamethicin) there is a problem with simulations of TM helix bundles from larger proteins, namely that of the exact extent of the helices. In the case of influenza M2, Duff and co-workers[3,4] have shown that a 25-residue peptide forms channels and is largely α-helical. However, it is not certain that all of the residues in the peptide form an α-helix, and to what extent the peptide mimics the intact M2 protein. We have used multi-nanosecond simulations of single TM helices of different lengths (from 18, 26 or 34 residues) in a phospholipid bilayer in order to determine the optimum length of the helix from M2. These simulations suggest that a region of length *ca.* 22 residues forms a helix stable throughout the simulation.[5]

The length of helix has a profound effect on the behavior of four TM helix bundle models in bilayer simulations. We have compared simulations with a bundle of 18-residue helices and with a bundle of 22-residue helices.[5] In both cases the helix bundles retained their left-handed supercoil structure, and gave a Cα rmsd of *ca.* 0.25 nm after a 4 ns simulation. However, the 18-residue bundle contained only 3 water molecules, which did not exchange with bulk water. Thus the 18-residue bundle looked like a "closed" channel. In contrast, the 22-residue bundle contained *ca.* 10 waters, and looked like an "open" channel. What is clear from these simulations is that one has to be a bit cautious in the choice of initial helix bundle model, as this may have a profound influence on the results of any subsequent bilayer simulation.

1 M. S. P. Sansom, D. P. T. Tieleman, L. R. Forrest and H. J. C. Berendsen, *Biochem. Soc. Trans.*, 1998, **26**, 438.
2 T. Husslein, P. B. Moore, Q. Zhong, D. M. Newns, P. C. Pattnaik and M. L. Klein, *Faraday Discuss.*, 1998, **111**, 201.
3 K. C. Duff and R. H. Ashley, *Virology*, 1992, **190**, 485.
4 K. C. Duff, S. M. Kelly, N. C. Price and J. P. Bradshaw, *FEBS Lett.*, 1992, **311**, 256.
5 L. R. Forrest, D. P. Tieleman and M. S. P. Sansom, *Biophys. J.*, 1999, **76**, in the press.
6 L. R. Forrest and M. S. P. Sansom, in preparation.

Prof. Okazaki said: It is interesting for me to see a decrease of membrane thickness and a decrease of alkyl chain order by the inclusion of the peptide. If you have reached some conclusion, please let me know about the mechanism. Are there any particular sites in the peptide which interact directly with the lipid molecule?

Prof. Klein responded: I am confident that our results are correct for the peptide concentration used in the MD simulation. Unfortunately, the ratio of peptide to lipid is only 1 : 8 in each leaflet of the bilayer. I am, therefore, concerned that the observed "thinning" of the membrane is related to channel–channel repulsions arising from interactions between the large helix dipoles. It would be useful to run more dilute samples, with say 128 and 256 lipids to quantify this effect more precisely.

Prof. Klein said: Did you find that the M2 (22-residue) peptide yielded a stable 4-helix bundle with a water pore or was it blocked at the His as suggested by various experiments and MD calculations[1] on a simpler system?

1 Q. Zhong, T. Husslein, D. Newns and M. L. Klein, *FEBS Lett.*, 1998, **434**, 265.

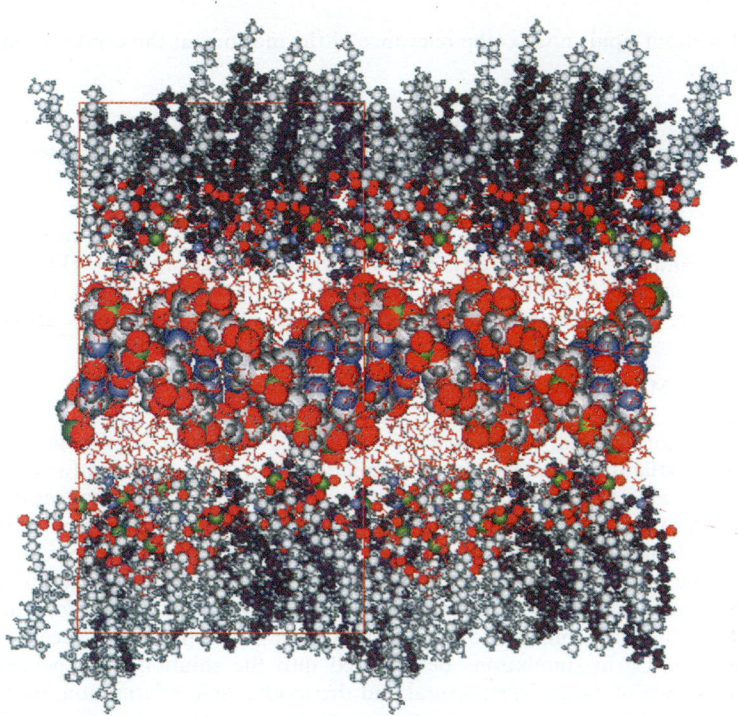

Fig. 2 A configuration taken from a multi-nanosecond MD simulation of the lipid–DNA complex. The two distinct lipids (DMTAP and DMPC) are drawn with light and dark shading.

Dr Sansom responded: The M2 (22-residue) helix bundle was "stable" in the sense that the left-handed supercoil was maintained, and the Cα rmsd at the end of 4 ns was *ca.* 0.25. Up to *ca.* 2 ns there was not a continuous water pore. Instead, there was a water-filled pore which was open to the surrounding environment at the N-terminal mouth but which was occluded towards the C-terminal mouth by the ring of His37 sidechains. However, after *ca.* 2 ns this pore opened up at the C-terminal as well. We suspect that this sort of behaviour might be rather sensitive (at least, on a 1 to 10 ns timescale) to the starting model used in the simulation, and so we are now exploring different starting structures.

Dr Amblard commented: Modelling the dynamics of macromolecules in solution or in membranes by coupling the elasticity of the macromolecule and viscous dissipation by the small surrounding molecules taken as a continuum could be a way of simulating molecular dynamics at much smaller frequency than the limit of MD classical simulation. On the other hand, Prof. Smith suggested earlier by MD simulation of bacteriorhodopsin, that this membrane protein could undergo dynamical transitions at well defined temperatures, revealed by the non-linearity of the crystal temperature factor with temperature. This suggests that bacteriorhodopsin does not behave as a set of harmonic oscillators, but in a much more complex and likely non-harmonic way.

To what extent could the complex elastic behavior of proteins be inferred from temperature-factor analysis, at least to provide a basis for modelling the dynamics at longer time-scales than with MD simulations?

Dr Sansom responded: I'm really not sure about using temperature-factor analysis for membrane proteins. My reservations are: (i) for a number of membrane protein structures the resolution is a bit low and so the B-factors may not be that reliable; (ii) most membrane proteins are crystallised without lipids and so the relevance of the motions in the crystal to motions in the bilayer is uncertain.

Prof. Bayerl asked: Does the MD simulation of DNA in DMPC/DMTAP give any indication of a spatial confinement of the cationic lipids motion due to the electrostatic interaction with the DNA?

Prof. Klein responded: Our MD simulations[1] have been run long enough to begin to quantify the nature of the lipid–DNA interactions. The zwitterionic headgroup of DMPC competes effectively with the cationic lipid DMTAP in "neutralizing" the DNA phosphates. Fig. 2 and 3 show the overall structure of the complex and the distribution of nitrogen atoms around the DNA phosphates.[1]

1   S. Bandyopadhyay, M. Tarek and M. L. Klein., in preparation.

Prof. Holzwarth commented: The molecular dynamics simulations provide very interesting information about structural changes on an atomic level for times from femto- to several nanoseconds, but are not able to reach longer times. If one inspects the energy changes connected with membrane processes it can be shown that most of the important changes in biological systems are occurring at much longer times.[1] (1) Is there any chance in the near future to reach 100 nanoseconds or better the microsecond time range with MD simulations? (2) How do you rate the chances for approaches which avoid the enormous computer power needed for full nanosecond simulations by starting with some reasonable assumptions Could this be an acceptable approach and how can Monte-Carlo simulations be included into the solution? My personal ideas are circulating in a triangle of dynamic, structural and thermodynamic information, trying to connect all three types of available results to construct a simpler basis for MD simulations.

1   J. F. Holzwarth in *The Enzyme Catalysis Process*, ed. A. Cooper, J. L. Houben and L. C. Chen, Plenum, London, 1989, pp. 383–412.

Dr Sansom responded: There is every chance in the near future of reaching 100 ns simulations, given the increasing performance of computers, and the development of "smarter" MD algorithms. Indeed, we have run one alamethicin helix bundle simulation for nearly 20 ns (ref. 1). In the non-membrane field, simulations of the folding of a small protein fragment (*ca.* 40 amino acids) have been run for *ca.* 1 μs (ref. 2) although this is still out of the feasible range for membrane simulations.

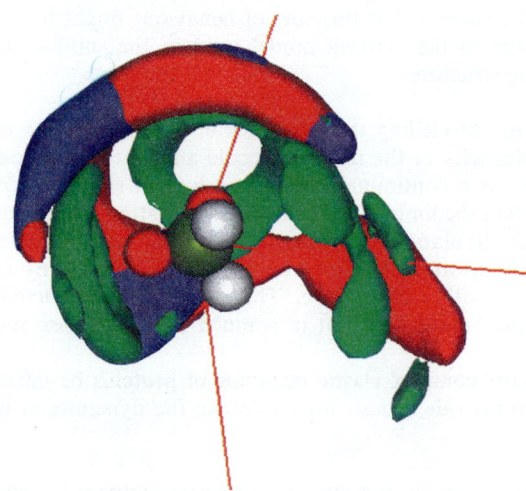

Fig. 3 Distribution of water molecules (green) and lipid N-atoms (red and blue) around a representative DNA phosphate group.

1   M. S. P. Sansom and D. P. Tieleman, unpublished data.
2   Y. Duan and P. A. Kollman, *Science*, 1998, **282**, 740.

Prof. Klein responded. As mentioned earlier, the Kollman group has reported on two microsecond long trajectories for a protein in water.[1] This was possible because of the availability of 256 CRAY T3E processors dedicated for a few months. The relentless progress in CPU performance should allow similar capability on a more routine basis in the 3–5 year time frame. The likely availability of these resources does not obviate the need to develop alternative approaches.

1   Y. Duan and P. A. Kollman, *Science*, 1998, **282**, 740.

Prof. Roux said: It is attractive to use a simplified description of the membrane environment in order to gain in computational efficiency and reach out to longer simulation times. However, it is important to keep in mind that one still doesn't know which details of the bilayer are going to be important in investigating the function of a membrane protein.

Dr Sansom responded: I fully agree with this. I think progress will be made by running more explicit bilayer simulations on a range of different membrane proteins, and then attempting a general analysis of which interactions are most important, at the same time as developing simplified descriptions as exemplified in the paper by Roux and co-workers.[1]

1   P. Lagüe, M. J. Zuckermann and B. Roux, *Faraday Discuss.*, 1998, **111**, 165.

Dr Amblard commented: When reading Dr Sansom's paper I was quite surprised by the fact that results of MD simulation are more often compared with results of other simulations than with experimental results. This is obvious in the 'Biological relevance' section of his paper. In the discussion of the last few papers dealing with MD simulation, not much was said about the connection with experimental data, about the kind of predictions generated by MD simulations at different time-scales, and about their experimental "testability" and the experimental models of interest. Could you and Prof. Klein comment on these points, and help non-specialists like me to grasp what the important questions and limitations are in this field, beyond the technicalities of MD simulations.

Dr Sansom responded: Maybe I was a bit too cautious in the 'Biological relevance' section of my paper, but I deliberately did not wish to over-interpret our results. I think some of the connections with experimental data for alamethicin have been described in Dr Smart's paper,[1] and so I won't duplicate them here. As I suggested in my paper, I think the motion of the pore-lining helices of alamethicin may have some relevance to gating of *e.g.* potassium channels such as KcsA. Indeed, recent spin-label studies of KcsA[2] provide experimental evidence of such helix motions, albeit on a much longer timescale.

1   O. S. Smart, G. M. P. Coates, M. S. P. Sansom, G. M. Alder and C. L. Bashford, *Faraday Discuss.*, 1998, **111**, 185.
2   E. Perozo, D. M. Cortes, L. G. Cuello, *Nat. Struct. Biol.*, 1998, **5**, 459.

Prof. Klein responded: You raise an excellent point. It is to be regretted if the results of a MD simulation are inaccessible to experimentalists. The predictive capability of the MD simulations is only useful to the extent that a range of experimental data can be accounted for. Alas, typically for membrane proteins, little is known beyond the basic structure. Even then, data is mostly related to the crystalline environments. The situation may change, as modern NMR methods achieve increasing success. Also, neutron and X-ray synchrotron experiments are likely to be important complements to modern NMR studies. I agree that it would be unfortunate if the focus shifted solely to technicalities of MD simulations.

Prof. Smith commented: Prof. Klein provides evidence here for a more fluid-like behaviour of the lipid in the presence of the protein and larger area per lipid, relative to the pure lipid. In preliminary results with Dan Mihailescu on gramicidin S binding to a DMPC bilayer we see the opposite effect in the lipid fluidity. Would you like to comment on whether findings similar to yours[1] have been observed elsewhere and on the physical origin of these effects?

1 T. Husslein, P. B. Moore, Q. Zhong, D. M. Newns, P. C. Pattnaik and M. L. Klein, *Faraday Discuss.*, 1998, **III**, 201.

Prof. Klein responded: I am sorry to say that I cannot comment on the gramicidin/DMPC system. Prof. Okazaki has been interested in this system for some time.

Dr Sansom responded: Tieleman and co-workers[1] have analysed lipid properties simulations of six-helix bundles of alamethicin. He saw a decrease in order parameters, corresponding to increased tilt of acyl chains of lipids close to the helix bundle. I suspect different results may be found for different systems (see response to next comment), and so we need to be cautious in making generalisations at this stage.

1 D. P. Tieleman, L. R. Forrest, M. S. P. Sansom and H. J. C. Berendsen, *Biochem.*, 1998, **37**, 17554.

Prof. Roux commented: The presence of protein in a membrane has been shown to increase the ordering of the acyl chains of the lipids. Rice and Oldfield showed that the deuterium quadrupolar splitting of specifically labelled DMPC was increased in the presence of gramicidin. Our results from molecular dynamics simulations are in good agreement with this observation.[2] However, it should be stressed that averaging from five independent trajectories was required to get convergence.

1 Rice and Oldfied, *Biochemistry*, 1979, **18**, 3272.
2 Woolf and B. Roux, *Prot. Struct. Funct. Gen.*, 1996, **24**, 92.

Prof. Klein responded: Prof. Roux brings out an important point: The issue of convergence of NMR order parameters obtained from MD simulations. It is also our observation that these are difficult quantities to obtain reliably. Unfortunately, my group has no specific data for the gramicidin/DMPC system.

Dr Sansom responded: Also, I think it is important to note that different results may be obtained for different proteins. Tieleman and co-workers have analysed lipid properties in simulations of several different systems. For single, transmembrane helices, the effects on surrounding lipids are relatively minor. For four helix (influenza M2) and six-helix (alamethicin) bundles, and for the porin OmpF (a large β barrel) a decrease in order parameters, corresponding to increased tilt of acyl chains, of nearby lipids was seen. This suggests that we need to gather data from a wider range of simulations before we can attempt any generalisations. However, I agree that to obtain reliable statistics either long simulations or multiple independent trajectories are needed.

1 D. P. Tieleman, L. R. Forrest, M. S. P. Sansom and H. J. C. Berendsen, *Biochem.*, 1998, **37**, 17554.

Prof. Okazaki commented: First, the calculated order parameter of gramicidin was absolutely dependent upon the initial configuration. No conformational changes were observed for the gramicidin molecule within our preliminary ns order MD calculation.

On the other hand, the calculated order parameter of the DPPC alkyl chain in the pure DPPC bilayer in the liquid crystal phase was larger than the experimental one. But the cumulative average of the order parameter did not converge within our several ns order simulation. The convergence is very slow.

The order parameter is, thus, very difficult to evaluate from the limited simulation time of the MD calculation.

Dr Sansom responded: I agree that longer simulations are needed to get proper estimates of order parameters. Also, one suspects, careful attention to the setup of the simulation (*i.e.* lipid–protein packing) is needed.

Dr Bezrukov commented: I want to highlight the importance of molecular dynamics simulations in ion channel studies. Unfortunately analytical methods for description of channel transport properties are far from satisfactory. For example, even a 'simple' question of how ionization of a single sidechain facing channel lumen influences channel conductance and selectivity is very hard to answer. Even in the case where all structural data is available (nothing to guess about here) I would not be surprised if the sign of the effect is difficult or just impossible to predict. Not only electrostatic *vs.* structural issues are involved here; the electrostatics itself at such short distances and high fields differs considerably from the continuous classic formulation. For this reason molecular dynamics simulations both of channel structure and its transport properties are much needed at present.

Dr Sansom responded: I completely agree that even "simple" questions are not resolved by simply looking at an X-ray structure. I feel one should be a bit more guarded about how quickly MD studies will progress. In principle, one should run a simulation which allows for dynamic protonation/deprotonation as an ion passes. I confess to not knowing how to do this at the moment, but I'm sure it will need long simulation times. Given the difference in timescale between current MD simulations (*ca.* 10 ns) and the permeation time of an ion (*ca.* 1 μs) I think we need to develop a hierarchy of theoretical descriptions of a channel in order to relate atomic resolution structures to physiological data from patch clamp experiments.

Prof. Holzwarth replied: Unfortunately the experimental techniques available are not specific enough to tackle the questions posed by Dr Bezrukov. On the other hand molecular dynamic simulations are not able to cover the time range beyond 10 ns. In summary I believe that the present experimental techniques are good in respect of the important time range from nanosecond to second but often lack molecular specificity; molecular simulations are fine for times shorter than a nanosecond but are not able to cover the really important "long time" phenomena. The future should improve this situation.

Dr Deber communicated: In the simulation of the four-helix bundles of M2, what were the features/motifs of interaction of the inter-helical faces/residues within the bundle (as depicted in Fig. 1 of your paper)? Are the interfaces predictable from primary sequence?

Prof. Klein communicated in response: To some extent, this issue has been discussed by my colleague, Prof. DeGrado and his collaborators,[1] in their analysis of mutagenesis data. Our MD study was designed to complement this experimental work and their analysis of the bundle structure. Although I am reluctant to speak for my colleagues on this issue, it is my impression that the inter-helical faces/residues do correspond to predictions based on the primary sequence of residues.

1 L. H. Pinto, *et al.*, 1997, *Natl. Acad. Sci. USA*, 1997, **94**, 1130.

Prof. Deber communicated: Is it possible to determine how helix–helix interactions vary (energetically, occurrences of residues at interfaces) as a function of alamethicin helical bundle size? A systematic approach here could enhance our understanding of the folding of polytopic membrane proteins.

Dr Sansom communicated in response: A very perceptive question! We are working on this, but haven't done such analysis as yet. One of the attractive features of these sorts of MD simulations is that they should allow analysis of helix–helix interactions in various "simple" bundles of trans-membrane helices.

Dr Bhakoo communicated: There are a vast array of isolated biomembrane systems and their interactions are being studied by numerous methods by a number of excellent research teams. One particular worry I had is the lack of correlation of data between groups and between techniques utilised. Perhaps this could be addressed by having closer collaboration between research groups.

Another key issue that requires addressing is relating the simple isolated membrane systems to real biological membranes. A number of times the detailed understanding of model systems will not lead to the understanding of biological membranes. I believe that there is an urgent need for a meeting between physical chemists, chemical physicists, biophysicists, biochemists, biologists, physiologists and microbiologists to discuss this issue.

Thus, a related question I have is as follows: How far do model membrane systems aid in providing understanding of real biological systems they are apparently mimicking?

Dr Sansom communicated in response: I guess this meeting is intended to address such worries. However, I think from a simulation perspective it is important that different groups work on similar systems so that one can get a feel for the extent to which the results converge. For example, going back to influenza M2, we have shown recently[1] that two independent modelling studies converge on a similar structure, which in turn is in agreement with experimental data.[2] This is encouraging.

In answer to your second question, clearly simple systems do not hold all of the answers. But, considering ion channels, gramicidin provided what turned out to be a good model for channel–ion interactions in the bacterial potassium channel.

1   L. R. Forrest, W. F. DeGrado, G. R. Dieckmann and M. S. P. Sansom, *Folding Design*, 1998, **3**, 443.
2   F. A. Kovacs and T. A. Cross, *Biophys. J.*, 1997, **73**, 2511.

Prof. Roux also communicated in reply to Dr Bhakoo: An important aspect of biophysics (experimental and theoretical) is the ability to use a reductionist approach in order to highlight the most important factors responsible for a phenomena. In doing so, it is useful to throw away as much detail as possible, as long as the essential elements are preserved. While I do not doubt that the complexity of biological membranes is much beyond our simple models, I think there is much to be gained by investigating simple model systems. Nonetheless, the intrinsic limitations of simple models should always be kept in mind.

In answer to your second question characterizing the magnitude of lipid-mediated protein–protein interactions is only a modest, though an important contribution to our understanding of the function of biological membranes.

Kinetics of the competitive response of receptors immobilised to ion-channels which have been incorporated into a tethered bilayer

Gillian E. Woodhouse, Lionel G. King, Lech Wieczorek and Bruce A. Cornell

Cooperative Research Centre for Molecular Engineering and Technology, Australian Membrane and Biotechnology Research Institute, 126 Greville Street, Chatswood, NSW 2067, Australia

Received 8th December 1998

A competitive ion channel switch (ICS) biosensor has been modelled yielding ligand mediated monomer–dimer reaction kinetics of gramicidin (gA) ion-channels within a tethered bilayer lipid membrane. Through employing gramicidin A, functionalised with the water-soluble hapten digoxigenin, it is possible to cross-link gramicidin to antibody fragments tethered at the membrane/aqueous interface. The change in ionic conductivity of the channel dimers may then be used to measure the binding kinetics of hapten–protein interactions at the membrane surface. The approach involves measuring the time dependence of the increase in impedance following the addition of a biotinylated antibody fragment (b-Fab′), which cross-links the functionalised gramicidin monomers in the outer layer of the lipid bilayer to tethered membrane spanning lipid. The subsequent addition of the small molecule digoxin, (M_r 781 Da), competes with and reverses this interaction.

The model provides a quantitative description of the response to both the cross-linking following the addition of the b-Fab′ and the competitive displacement of the hapten by a water-soluble small analyte. Good agreement is obtained with independent measures of the cross-linking reaction rates of the gramicidin monomer–dimer and the b-Fab : hapten complex. The rate and amplitude of the competitive response is dependent on concentration and provides a fast and sensitive detection technique.

Estimates are made of the concentration of gramicidin monomers in both the inner and outer monolayer leaflets of the membrane. This is used in the calculation of the gramicidin monomer/dimer equilibrium constant, K_{2D_3}. Other considerations include the membrane impedance limit set by the membrane leakage which is also a function of the concentration of the gA monomer concentration, and the two-dimensional kinetic association constant k_{2D_2}, of the hapten : b-Fab′ complex. The gA dimer concentration is dependent on both the concentration of gA-dig and of the tethered streptavidin : b-Fab′ complexes.

The model shows that the 2D dissociation constant $k_{2D_3}^{-1}$, must be at least 10 times faster than the 3D dissociation constant $k_{3D_2}^{-1}$ for digoxin to completely reverse the cross-linked hapten–receptor interaction at the membrane interface.

Introduction

Tethered bilayer lipid membranes (t-BLMs) offer improved robustness and simpler assembly pro-

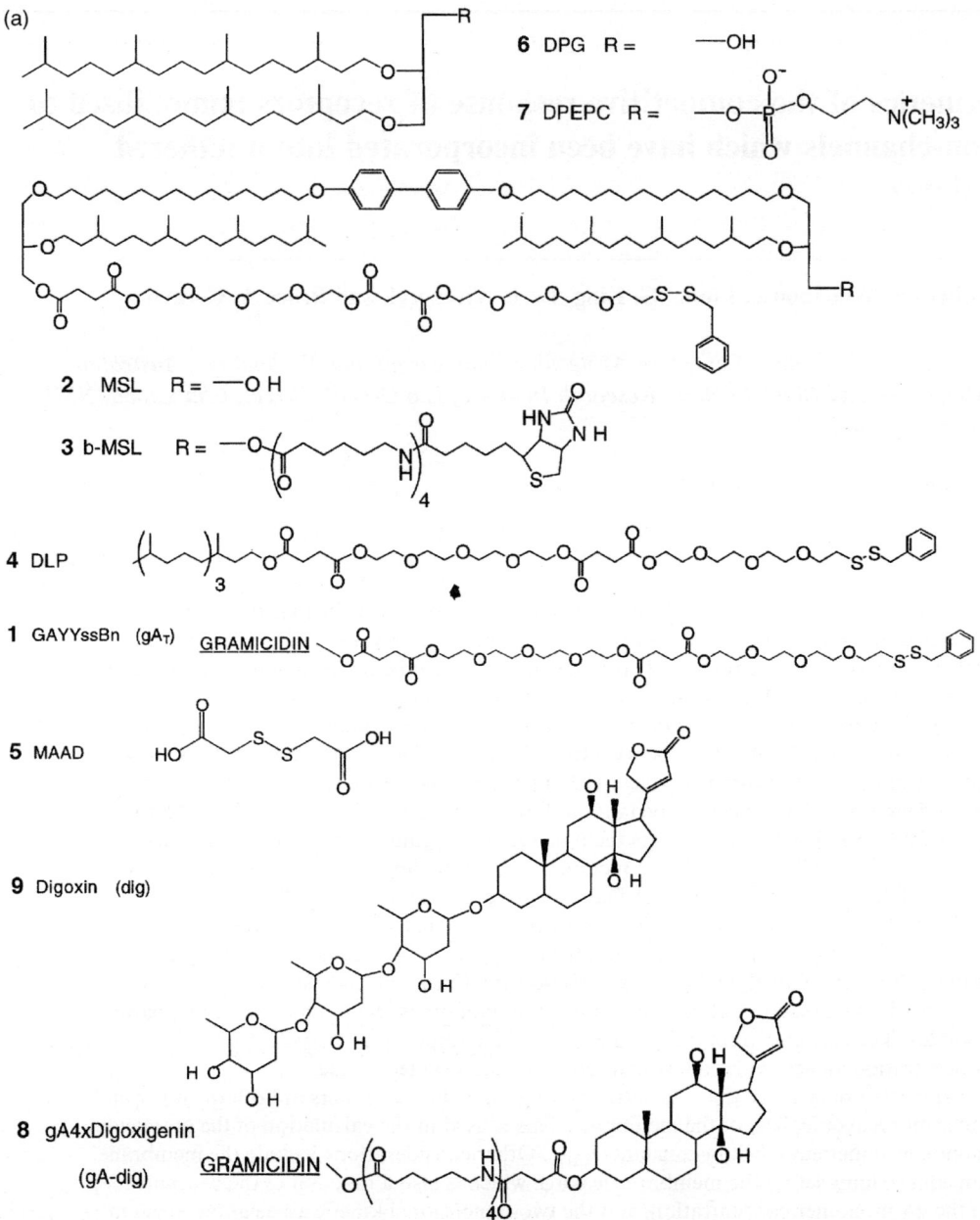

Fig. 1 (a) Chemical components for the competitive ICS biosensor. The immobilised components of the membrane comprise a mixture of tethered gramicidin A, gAYYSSBn (gA$_T$), double length reservoir half membrane spanning phytanyl lipids, (DLP), full membrane spanning lipids (MSL) and biotinylated MSL (b-MSL). The surface density of these tethered species is controlled by dilution with the small hydrophilic mercaptoacetic acid disulfide (MAAD). The outer mobile membrane leaflet consists of a 7 : 3 mole ratio mix of

cedures compared with solvent formed bilayer lipid membranes (BLMs), patch clamping techniques, Langmuir–Blodgett layers or supported bilayer lipid membranes (s-BLMs). They have many potential applications related to bioelectronics, biosensing, pharmaceutical screening, and physiological transduction studies. Tethered bilayers provide a novel tool for studying the molecular dynamics of channels incorporated into the membrane and of the binding kinetics of proteins

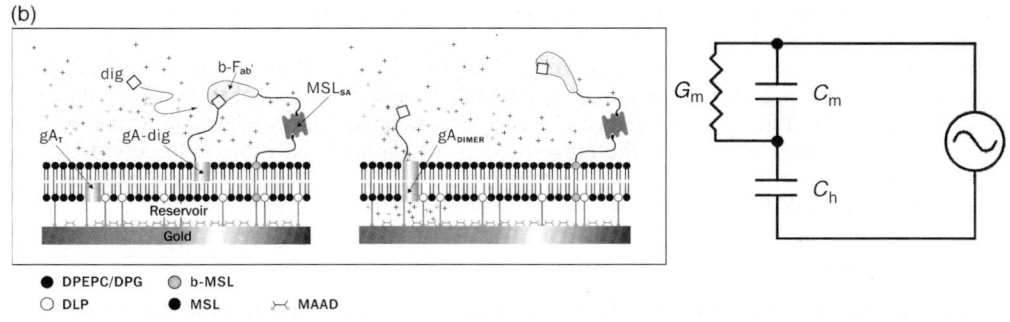

diphytanyl phosphatidylcholine (DPEPC): glycerodiphytanyl (DPG) which is mixed with a fraction of 1:10 000 mole ratio of digoxingenin derivatised gramicidin A (gA-dig). (b) Schematic of competitive ICS biosensor and its simple RC equivalent circuit. The equivalent circuit for the sensor may be approximated as an effective Helmholtz capacitance (C_h), in series with the capacitance of the membrane (C_m) and which is bypassed by the ionophore conduction (G_m).

at the membrane surface. The self-assembly mechanism of the tethered bilayer results in the formation of an ionic reservoir between the gold interface and the membrane. In the presence of conductive channels within the membrane, ions flow between the reservoir and the external solution, driven by an externally applied potential. Raguse et al.[1] have reported the use of valinomycin to characterise the ionic conductivity and capacity of the reservoir. As described by Cornell et al.,[2] the competitive biosensor consists of a tethered bilayer incorporating a reservoir, mobile and tethered ion channels and a family of molecular tethers.

Fig. 1(a), shows the inner layer species of the membrane, which comprises five components. Four of these components are amphiphilic, having hydrophobic, membrane forming end groups, which attach to the gold surface via a disulfide foot and an intervening hydrophilic portion constructed from two repeated units of tetraethyleneglycol succinic ester (TEGsuc). These are: tethered gramicidin A **1** (gA_T), membrane spanning lipid **2** (MSL), biotinylated membrane spanning lipid **3** (b-MSL), and a half membrane spanning lipid **4** (DLP). The fifth component is the small, hydrophilic disulfide molecule, dithiodiglycolic acid **5**, which acts as a spacer on the surface.

The outer lipid layer is composed of predominantly untethered lipids capable of translational diffusion within the 2D plane of the membrane. The outer layer lipids are: 3-di-O-phytanyl-sn-glycerol **6** (DPG) and 2,3-di-O-phytanyl-sn-phosphatidylcholine **7** (DPEPC). The b-MSL **3**, deposited with the first layer, spans the lipid bilayer and its terminal biotin group extends into the aqueous environment and, in the presence of streptavidin forms a complex on the membrane outer surface MSL_{SA}. Subsequent addition of biotinylated anti-digoxin antibody Fab' fragments (b-Fab') to MSL_{SA} affords the ternary b-Fab' : streptavidin : MSL complex (MSL_{SA} : b-Fab').

The outer layer membrane also includes digoxigenin derivatised gramicidin A (gA-dig) **8**, which can be cross-linked to MSL_{SA} : b-Fab', to form a further, quaternary complex (MSL_{SA} : b-Fab' : gA-dig). The gA-dig molecules in this quaternary complex are constrained to be out of register with the inner membrane gramicidins (gA_T), and thereby prevented from forming the ion-conducting gramicidin dimers (gA_{DIMER}). Addition of digoxin (M_r 781 Da) reverses this binding, releasing the gA-dig and allowing the channels to conduct.

Experimental

Materials

The structure of the membrane components have been described elsewhere: DLP,[2] MAAD,[2] DPG,[2] DPEPC,[2] MSL,[3] b-MSL,[3] gA-dig,[4] and gA_T.[4] All chemicals were purified by HPLC and their purity was shown to be better than 99% by MALDI mass spectroscopy, TLC and ^1H NMR. Analytical grade ethanol (Ajax) was used in preparing all the lipidic solutions. The components were stored for up to 6 months in solution at 4 °C. Digoxin (Sigma Chemical Co., St. Louis, USA) was dissolved in ethanol to 10 mg mL^{-1} and dilutions were made from this stock with phosphate buffered saline (50 mM PO^{4-}, pH 7.2) (PBS). Streptavidin solutions were prepared from a lyophilised powder (Boehringer Mannheim, Germany) and dissolved in PBS to prepare stock solution concentrations of 1 mg mL^{-1}. Anti-digoxin monoclonal antibody was obtained from Biodesign Ltd. (Maine, USA). (Fab')$_2$s were prepared by enzymic fragmentation which were further reduced to Fab's. Free cysteine exposed at the hinge region was biotinylated using iodoacetyl-LC-biotin (Pierce, Rockford, USA) to give a mono-biotinylated Fab' (b-Fab'). Purity was confirmed to be better than 98% (SDS-PAGE). The procedure has been described previously.[5]

Biosensor preparation

Clean room facilities were used for the preparation of the gold films and the deposition of the initial monolayer formation. All procedures were carried out at 21–22 °C unless otherwise specified. Glass microscope slides (Objekttrager, HD Scientific, Germany), were cleaned by soaking with detergent (Extran 300, BDH) for 2 h and rinsing thoroughly with milli-Q water. They were subsequently dried under a stream of dry nitrogen gas. A chromium adhesion film of 20 nm in thickness was evaporated at 0.1 nm s^{-1} onto each microscope slide, followed by a 100 nm film of gold at 2.5 nm s^{-1} using an Edwards High Vacuum Coating Unit (18SE4/S53) operating at 5 × 10^{-7} Torr.

Monolayers were chemically adsorbed onto the freshly prepared gold films by incubating the slides for 1 h in an ethanolic solution of 300 µM DLP, 150 µM MAAD, 0.043 µM gA$_T$, 4.5 µM MSL, 0.225 µM b-MSL. The slides were then rinsed thoroughly with ethanol and stored under ethanol at 4 °C until required.

Monolayer coated glass slides were removed from the storage solution and dried under a high pressure stream of dry, boil-off nitrogen gas. Slides were placed on a brass support (approximately 10 cm × 5 cm × 7 mm) and a second brass piece, containing eight holes spaced on a standard 96-well ELISA plate grid pattern (internal diameter 6 mm) into which machined Teflon tubes (3.6 mm internal diameter) had been inserted, was clamped onto the microscope slide and supporting brass block, to form sealed, discrete cells. The area of electrode exposed within the well had an internal area of 0.10 cm^2. A 10 µL volume of a 10 mM solution of DPG : DPEPC (3 : 7), containing 0.1 mM gA-dig, was dispensed into each cell sequentially, immediately followed by 100 µL phosphate buffered saline (pH 7.2, 50 mM). Excess lipid was washed out of the cell through exchanging the buffer volume at least 4 times, ensuring that the membrane interface was not exposed to air.

A 5 µL volume of a streptavidin solution at 0.05 mg mL^{-1} was injected into the cell to give a final cell concentration of 40 nM, and incubated on the membrane for 10 min. The excess material was removed by copiously washing with PBS. A 5 µL volume of a b-Fab' solution at 1 µM (PBS), was added to the cell, incubated for 10 min and rinsed by exchanging the volume with PBS.

Impedance spectra were obtained using a custom built 32-channel impedance spectrometer (Associative Measurements Pty. Ltd., Sydney, Australia) employing data analysis software commissioned for our laboratory (Aguilla Holdings Sydney, Australia). Frequencies were scanned between 1 Hz and 1 kHz over a period of 16 s. A bias of −300 mV was applied between the sensor electrode and a Pt counter-electrode while an ac ion excitation signal of 50 mV was applied between the measurement electrode and a Ag reference electrode. Up to 32 independent impedance trials could be conducted simultaneously.

After collecting a baseline of at least 10 spectra, test analyte was introduced to the cell by adding 100 µL into the 100 µL volume of PBS existing in the well and mixing thoroughly.

Modelling of competitive response

This paper describes the modelling of the kinetics of two reactions: the addition of b-Fab to a membrane containing mobile gA-dig and tethered MSL$_{SA}$ to form the complex MSL$_{SA}$: b-Fab' : gA-dig [eqn. (1)–(5) and (7)], and the disruption of this complex by solution digoxin to form MSL$_{SA}$: b-Fab' : dig + gA-dig [eqn. (6)]. Assuming adequate mass transport of analyte to the membrane, the behaviour of the system may be modelled by the following seven equilibrium reactions:

$$\text{gA}_\text{T} + \text{gA-dig} \xrightleftharpoons[k^{-1}_{2D_1}]{k_{2D_1}} \text{gA}_\text{DIMER} \quad \text{where } K_{2D_1} = k_{2D_1}/k^{-1}_{2D_1} \tag{1}$$

$$\text{gA-dig : b-Fab}' + \text{gA}_\text{T} \xrightleftharpoons[k^{-1}_{2d2}]{k_{2D_2}} \text{b-Fab}' : \text{gA}_\text{DIMER} \quad \text{where } K_{2D_2} = k_{2D_2}/k^{-1}_{2D_2} \tag{2}$$

$$\text{MSL}_\text{SA} : \text{b-Fab}' + \text{gA-dig} \xrightleftharpoons[k^{-1}_{2D_3}]{k_{2D_3}} \text{MSL}_\text{SA} : \text{b-Fab}' : \text{gA-dig} \quad \text{where } K_{2D_3} = k_{2D_3}/k^{-1}_{2D_3} \tag{3}$$

$$\text{b-Fab}' + \text{MSL}_\text{SA} \xrightleftharpoons[k^{-1}_{3D_1}]{k_{3D_1}} \text{MSL}_\text{SA} : \text{b-Fab}' \quad \text{where } K_{3D_1} = k_{3D_1}/k^{-1}_{3D_1} \tag{4}$$

$$\text{gA-dig} + \text{b-Fab}' \xrightleftharpoons[k^{-1}_{3D_2}]{k_{3D_2}} \text{gA-dig : b-Fab} \quad \text{where } K_{3D_2} = k_{3D_2}/k^{-1}_{3D_2} \tag{5}$$

$$\text{MSL}_\text{SA} : \text{b-Fab}' + \text{dig} \xrightleftharpoons[k^{-1}_{3D_3}]{k_{3D_3}} \text{MSL}_\text{SA} : \text{b-Fab}' : \text{dig} \quad \text{where } K_{3D_3} = k_{3D_3}/k^{-1}_{3D_3} \tag{6}$$

$$\text{gA}_\text{DIMER} + \text{b-Fab}' \xrightleftharpoons[k^{-1}_{3D_4}]{k_{3D_4}} \text{gA}_\text{DIMER} : \text{b-Fab} \quad \text{where } K_{3D_4} = k_{3D_4}/k^{-1}_{3D_4} \tag{7}$$

The constants k_{2D_1}, k_{2D_2}, k_{2D_3}, k_{3D_1}, k_{3D_2}, k_{3D_3}, k_{3D_4}, $k^{-1}_{2D_1}$, $k^{-1}_{2D_2}$, $k^{-1}_{2D_3}$, $k^{-1}_{3D_1}$, $k^{-1}_{3D_2}$, $k^{-1}_{3D_3}$, $k^{-1}_{3D_4}$, are the two- and three-dimensional association and dissociation reaction rates and K_{2D_1}, K_{2D_2}, K_{2D_3}, K_{3D_1}, K_{3D_2}, K_{3D_3}, K_{3D_4} are the equilibrium constants for reactions (1)–(7). The model assumes the binding of streptavidin to b-MSL to be irreversible.

Eqn. (3) describes the two dimensional reaction in which the complex $\text{MSL}_\text{SA} : \text{b-Fab}'$, cross-links to mobile gA-dig *via* two dimensional diffusion to form the ternary complex, $\text{MSL}_\text{SA} : \text{b-Fab}' : \text{gA-dig}$. In the presence of solution digoxin, mobile gA must compete for binding to the complex $\text{MSL}_\text{SA} : \text{b-Fab}'$ with the analyte in solution [eqn. (6)]. The competition between reactions (3) and (6) will determine the reaction kinetics and equilibrium of reaction (1).

The measurable result of this series of interactions is the concentration of gramicidin dimers, which is proportional to the admittance at minimum phase ($Y_{\phi\,\text{min}}$).

$$Y_{\phi\,\text{min}} \propto [\text{gA}_\text{DIMER}] \tag{8}$$

Electrical properties of the competitive ICS biosensor

The equivalent electrical circuit of the tethered membrane is given in Fig. 1(b) together with a schematic diagram of the competitive ICS biosensor. Through modelling the spectra (Fig. 2), values for membrane capacitance (C_m) and Helmholtz capacitance (C_h) have been derived and are in agreement with values reported by others.[1,6,7] The modulus of the admittance

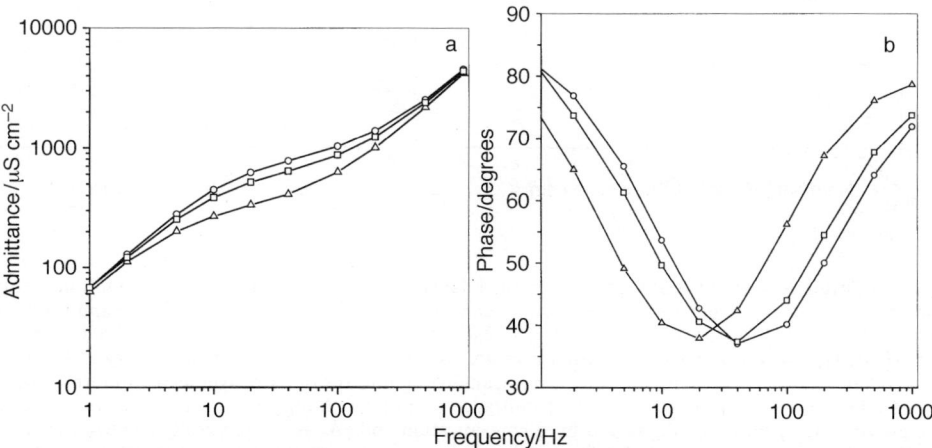

Fig. 2 (a) Impedance spectra and (b) phase *vs.* frequency plot for competitive ICS biosensor. Impedance and phase *vs.* frequency data of a tethered membrane with MSL_SA and gA-dig is shown (○). This shows the impedance and phase characteristics of the same membrane after 50 nM digoxin b-Fab' has been incubated for 10 min (△), and after excess b-Fab' has been removed and 0.38 ng digoxin has been added (□).

Table 1 Capacitance and conduction values of t-BLM before and after b-Fab′ : gA-dig cross-linking

	Before b-Fab′ addition	After b-Fab′ addition	After digoxin addition	Literature
Membrane capacitance (C_m)	0.52 μF cm^{-2}	0.50 μF cm^{-2}	0.52 μF cm^{-2}	0.68 μF cm^{-2} (ref. 6) 0.72 μF cm^{-2} (ref. 7) 0.52 μF cm^{-2} ^{a}
Helmholtz capacitance (C_h)	7.8 μF cm^{-2}	7.6 μF cm^{-2}	7.8 μF cm^{-2}	5 μF cm^{-2} [1]
Membrane conduction (G_m)	850 μS cm^{-2}	420 μS cm^{-2}	670 μS cm^{-2}	

^{a} Compared with $C_{GC} > 20$ μF cm^{-2} and $C_{SAM} \approx 2$ μF cm^{-2}, depending on electrolyte, where C_{GC} is the classical interfacial capacitance described by Guoy and Chapman[8] and C_{SAM} is the capacitance for a self-assembled monolayers described by Ulman.[9]

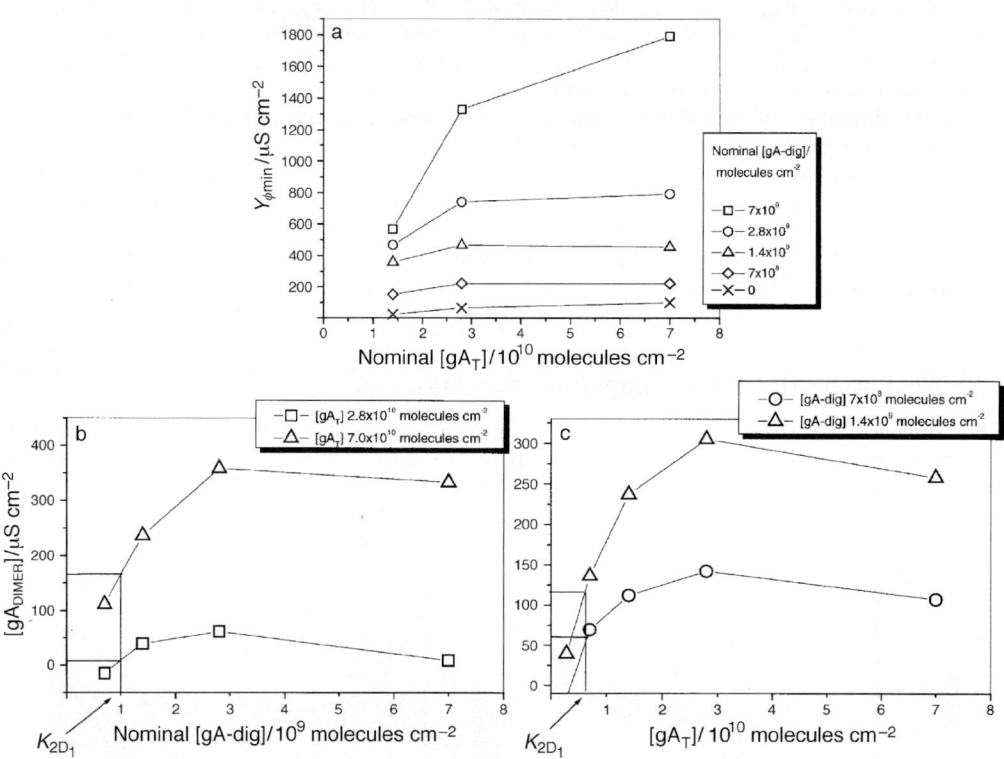

Fig. 3 (a) Titration of gA inner and gA outer in the bilayer. The data shows a titration of admittance at phase minimum, $Y_{\phi\,min}$ (μS), as a function of the nominal concentration of gA-dig and gA_T. The graph shows $Y_{\phi\,min}$ as a function of $[gA_T]$ for [gA-dig] = 7 × 10^9 (□), 2.8 × 10^9 (○), 1.4 × 10^9 (△), 7 × 10^8 (◇), 0 (×) molecules cm^{-2}. When [gA-dig] = 0 and the admittance of gA_T is increased in the bottom layer very little change in conduction occurs, demonstrating that gA_T is not transferring into the outer bilayer leaflet to form conducting $gA_T : gA_T$ dimers. The data suggest that the admittance given for [gA-dig] = 0 is due to conduction of monomeric gA. (b) and (c) When gA-dig is at a fixed concentration and gA_T is systematically increased in the inner layer (b), and conversely where $[gA_T]$ is fixed and [gA-dig] is systematically increased (c), the concentration of the bilayer increases to a point until a concentration which is equivalent to K_{2D_1}. The actual surface gA concentrations are not known. The nominal K_{2D_1} values are derived from $K_{2D_1} = 1/[gA]$ for eqn. (1), assuming that $Y_{\phi\,min} \propto [gA_{DIMER}]$ giving an apparent $K_{2D_1(gA-dig)} = 1 \times 10^9$ and apparent $K_{2D_1(gAT)} = 6 \times 10^9$ molecules cm^2. The ratio of $K_{2D_1(gA-dig)}/K_{2D_1(gAT)}$ allows the ratio of actual [gA] to be determined.

$|Y| = \omega C_h C_m/(C_h + C_m)$ is obtained over a range of frequencies (v), typically from 1 kHz to 1 Hz. A measure of the Helmholtz capacitance may be obtained from the modulus of the admittance, $|Y|$ at 1 Hz and the membrane capacitance from the modulus of the admittance $|Y|$ at 1 kHz. An approximate measure of the channel conductance may be obtained from the admittance, $|Y|$ taken where the slope $dY/d\omega$ is minimum. Given an electrode area of 0.10 cm^2, C_h and C_m were calculated and are shown in Table 1.

Upon the addition of the biotinylated anti-digoxin Fab′ there is little change in either the membrane or Helmholtz capacitance, by contrast, the conduction G_m decreases by a factor of 2. Similarly, upon subsequent addition of digoxin, the capacitance remains virtually unchanged whereas the conductance increases. These electrical changes correspond to those expected for a bilayer containing the reversible ionic switches as depicted schematically in Fig. 1(b).

gA dimerisation

A titration of the membrane admittance as a function of the [gA-dig] and [gA$_T$], Fig. 3(b) and (c), has been used to estimate the apparent monomer–dimer gA equilibrium constant, K_{2D_1}. The nominal concentration of gA molecules incorporated into each monolayer was initially assumed to be proportional to the ratio of the concentrations of gramicidin to those of the membrane forming lipids in the deposition solution. The total lipid surface concentration was taken as approximately 1.4×10^{14} cm^{-2}.[10] Monomers of gA$_T$ are seen to cause leakage due to conduction through the outer, sealed lipid monolayer [Fig. 3(a)]. This leakage is approximately 5–8% of the total conduction, for a bilayer with a nominal [gA-dig] of 1.4×10^9 molecules cm^{-2}. When one gramicidin population is fixed and the second titrated, the number of dimers reaches a saturating concentration, demonstrating that the amount of flip-flop is small. Furthermore, it is possible to estimate K_{2D_1} in terms of the apparent concentration of gA provided [gA] in the titrated layer is much greater than [gA$_{DIMER}$] when [gA] at the saturating admittance = $(K_{2D_1})^{-1}$. Fig. 3(b) and (c) show the graphical determination of a nominal K_{2D}. The leakage component in the outer layer was accounted for by subtracting the conduction values for nominal [gA-dig] from the data with various values of [gA$_T$]. Fig. 3(b) and (c) give estimates of nominal values for K_{2D_1} (gA-dig) of 1×10^9 molecules cm^{-2} and K_{2D_1} (gA$_T$) of 6×10^9 molecules cm^{-2}, respectively.

By taking the ratio of the apparent K_{2D_1}s in the two monolayers it is possible to cancel the uncertainties in actual concentration and obtain a ratio of the inner to outer layer gramicidin concentrations. This ratio may be used to constrain the modelling of the biosensor response.

Electrical response to analytes

$Y_{\phi\,min}$ was calculated using a spline fit of phase *vs.* frequency data. Fig. 4(a) shows the kinetic response at $Y_{\phi\,min}$ to the b-Fab′ addition. The time-course of the change in $Y_{\phi\,min}$ following the addition of the b-Fab′ and the digoxin were well approximated by the functions

$$Y_{\phi\,min} = Y_0(1 - e^{-t/\tau}) + Y_1 \quad \text{and} \quad Y_{\phi\,min} = Y_0 e^{-t/\tau} + Y_1$$

respectively. In addition, measures were obtained of the gating (%), given by $100 \times Y_0/(Y_0 + Y_1)$ and the normalised maximum slope (NMS), given by $(Y_0/\tau)/(Y_0 + Y_1)$, where Y_0 = the amplitude of the exponential, τ = time constant and Y_1 = the calculated admittance at infinity. The response elicited from b-Fab′ addition to the membrane arises from cross-linking the gA-dig molecules to the tethered MSL$_{SA}$ complex. The dependence of the admittance change on the presence of the MSL$_{SA}$ complex in the sensor confirms this mechanism. Omission of streptavidin, replacing the biotinylated MSL with a non-biotinylated MSL, or replacing the biotinylated Fab′ with a non-biotinylated Fab′, eliminates the response. Fig. 4(c) shows the time-course of the $Y_{\phi\,min}$ for a range of concentrations of the b-Fab′. Fig. 4(b) shows the subsequent competitive response elicited by digoxin addition to the bilayer. Digoxin (M_r 781 Da) causes an increase in the admittance as it competes with the b-Fab′ immobilised on the MSL$_{SA}$. In contrast, the introduction of biotin (M_r 244 Da), has very little effect on the membrane conduction. This reflects the high binding constant of MSL$_{SA}$: b-Fab′ (1×10^{15} M^{-1}),[11,12] compared with that for the gA-dig : b-Fab′ complex

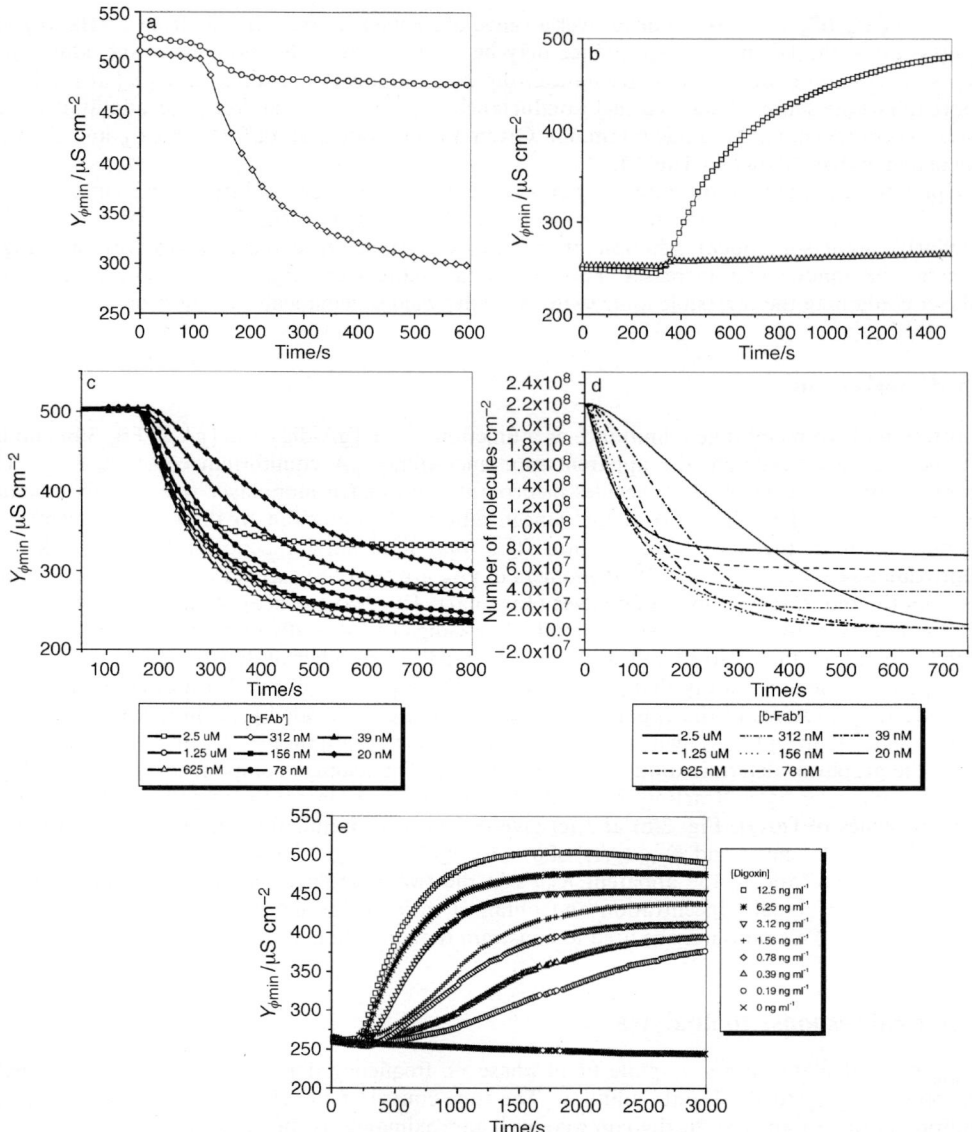

Fig. 4 (a) Comparison of $Y_{\phi\,\mathrm{min}}$ response to b-Fab' for membranes with and without MSL$_{SA}$ incorporated. The data shown are for the change in $Y_{\phi\,\mathrm{min}}$ upon addition of 50 nM digoxin b-Fab'. The data show b-Fab' responses for membranes with MSL$_{SA}$ (◇) and without MSL$_{SA}$ (○) incorporated. b-Fab' was added to the sensor at $t = 100$. The b-Fab' yielded a significant response with MSL$_{SA}$ and a negligible response for the membrane without MSL$_{SA}$. This shows the MSL$_{SA}$ is necessary for cross-linking. Similar controls have shown that the response is negligible if gA is used instead of gA-dig, if digoxin Fab' is used instead of digoxin b-Fab' and if thyroxin b-Fab' is used instead of digoxin b-Fab'. (b) Specificity of competitive response to digoxin. Data show the $Y_{\phi\,\mathrm{min}}$ response for 50 mM biotin (△) and 10 nM digoxin (□) demonstrating that the cross-linked complex gA-dig : b-Fab' : MSL$_{SA}$ is not disrupted with biotin but is with 7.8 ng mL^{-1} digoxin, reflecting the high binding constant of b-Fab : MSL$_{SA}$, the relative stability of gA-dig : b-Fab' : MSL$_{SA}$ in PBS and the fast apparent dissociation of gA-dig : b-Fab' : MSL$_{SA}$ to gA-dig and b-Fab' : MSL$_{SA}$. (c) Kinetics of response for titration of digoxin b-Fab' shown as $Y_{\phi\,\mathrm{min}}$ vs. time. The initial $Y_{\phi\,\mathrm{min}}$ for each concentration is taken as the mean of $Y_{\phi\,\mathrm{min}}$ at $t = 0$. Digoxin b-Fab' was introduced at $t = 150$ s into a phosphate buffered saline electrolyte,

Caption continued on next page.

50 mM PO^{4-} pH 7.2, 22 °C. For the low concentrations the first part of the response shows the k_{2D_1} and the reaction is limited by the b-Fab' concentration. At higher concentrations the initial rate is very fast and reaches a constant value when the response is limited by the equilibrium of the gA monomer–dimer reaction. A decrease in amplitude of response is seen when increasing b-Fab' concentration above 625 nM as the excess of b-Fab' results indirectly in gA-dig:b-Fab' and b-Fab':MSL_{SA} as well as gA-dig:b-Fab':MSL_{SA}. (d) Modelled kinetic responses for digoxin b-Fab'. The experimental data shown given in part (e) have been modelled and are shown here. The input parameters were: k_{2D_1} and $k_{2D_2} = 1 \times 10^{-11}$ molecules cm^{-2} s^{-1}; $k_{2D_1}^{-1}$ and $k_{2D_2}^{-1} = 100$ s^{-1}; $k_{2D_3} = 1 \times 10^9$ molecules cm^{-2} s^{-1}; $k_{2D_3}^{-1} = 1 \times 10^{-4}$ molecules cm^{-2} s^{-1}; $k_{3D_1} = 1 \times 10^5$ M^{-1} s^{-1}; $k_{3D_1}^{-1} = 1 \times 10^{-6}$ s^{-1}; k_{3D_2}, k_{3D_3} and $k_{3D_4} = 1 \times 10^4$ M^{-1} s^{-1}; $k_{3D_2}^{-1}$, $k_{3D_3}^{-1}$ and $k_{3D_4}^{-1} = 5 \times 10^{-4}$ s^{-1}. [MSL] = 8×10^8 molecules cm^{-2}, [gA_T] = 1×10^9 molecules cm^{-2} and [gA-dig] = 5×10^8 molecules cm^{-2}. (e) Kinetics of competitive response for titration of digoxin analyte shown as $Y_{\phi\,min}$ as a function of time. Digoxin b-Fab' introduced at $t = 200$ s into a phosphate buffered saline electrolyte, 50 mM PO^{4-} pH 7.2, 22 °C. The rate dependence of the competitive kinetic response, which is a function of the surface concentrations of digoxin and MSL_{SA}, is illustrated.

(2×10^8 M^{-1}).[13] Fig. 4(e) shows the concentration dependence of the time course for the competitive digoxin response. This response is the basis of a fast and sensitive competition assay for digoxin.

Modelling kinetic and equilibrium data

Fig. 5(a) and (b) show a superposition of the experimental and modelled kinetic and equilibrium responses. In this model it is assumed that membrane leakage is negligible and that the conduction is dominated by gramicidin dimers. Also, it is assumed that the solution volume above the membrane is completely mixed at all times and that mass transfer effects do not limit the kinetics. It is further assumed that the streptavidin–biotin interaction is essentially irreversible. A software package developed in this laboratory was used to numerically generate kinetic responses for the formation of the digoxin:b-Fab' complexes and the competitive response upon the addition of digoxin.

The values for the following parameters were derived from independent kinetic analysis of receptor analyte interactions.

$k_{3D_2} = 2 \times 10^4$ M^{-1} s^{-1} IAsys (affinity sensors) data[13]

$k_{3D_2}^{-1} = 5 \times 10^{-4}$ s^{-1} IAsys [13]

$k_{3D_1} = 1 \times 10^5$ M^{-1} s^{-1} surface plasmon resonance (SPR) data[12]

$k_{3D_1}^{-1} = <1 \times 10^{-5}$ s^{-1} SPR data[12]

Other input parameters were as follows:

$k_{2D_3} = 1 \times 10^8 - 1 \times 10^{10}$ molecules cm^{-2} s^{-1}

$K_{2D_3} = 7 \times 10^8 - 1 \times 10^9$ molecules cm^{-2} s^{-1}

$k_{2D_1} = 1 \times 10^{-10} - 1 \times 10^{-11}$ molecules cm^{-2} s^{-1}

$k_{2D_1}^{-1} = 0.033 - 0.1$ s^{-1}

The ratio [gA-dig]/[gA_T] was estimated as discussed previously. The adopted values of [gA_T], [gA-dig], [MSL_{SA}:b-Fab'] and K_{2D_2} were varied in the model until a best fit was obtained. Fits to the data, shown in Fig. 5(a) and (b), were obtained for specified values given in the figure caption. This model alone can not determine absolute concentrations of the bilayer components because the surface concentration of accessible binding sites can not definitively be correlated to the concentration of membrane components discussed previously. The model also does not provide an adequate description of the gating amplitude and other factors such as the membrane leakage level which can account for up to 10–20% conduction under certain conditions.

It is observed experimentally that the kinetic response (NMS) increases linearly with [b-Fab'] up to 5 and 30 nM for [gA-dig] 5×10^8 and 1×10^9 molecules cm^{-2}, respectively [Fig. 5(a)]. At this [b-Fab'] dependent portion of the titration curve, the binding of b-Fab' to MSL_{SA} is the rate determining step. Above these concentrations the rate of response is [b-Fab'] independent. The

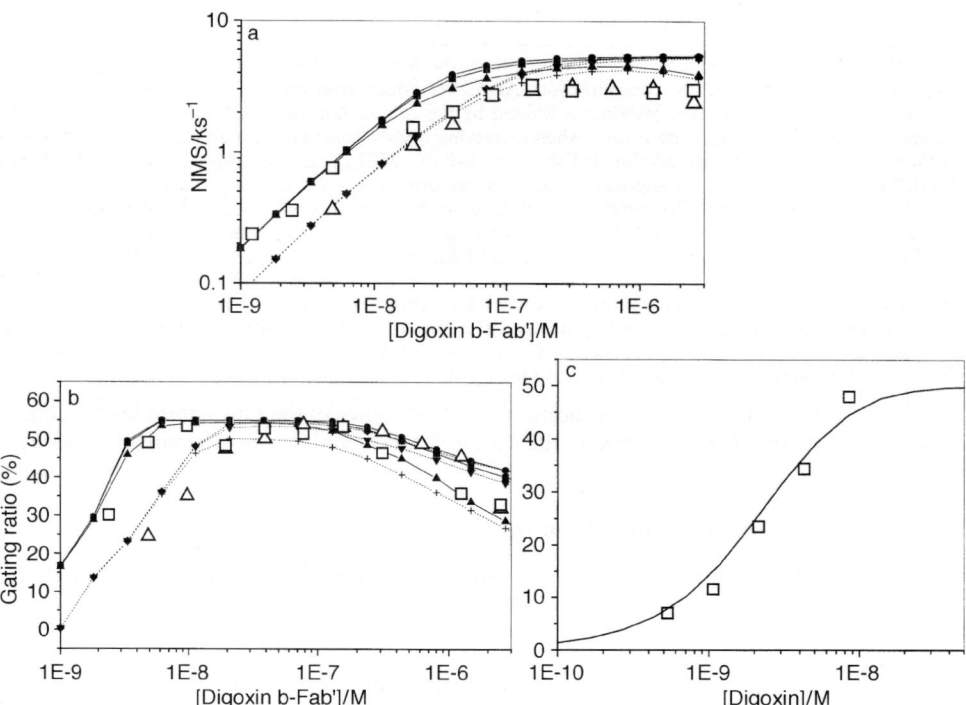

Fig. 5 (a) Digoxin b-Fab' response as normalised maximum slope (NMS). The graph shows the experimental and modelled titration curves for the ICS biosensors to digoxin b-Fab' as $(Y_0/\tau)/(Y_0 + Y_1)$ (NMS). The nominal concentrations for gA-dig are 1.4×10^9 (□) and 2.8×10^9 (△). The concentrations used in the model were 5×10^8 and 1×10^9 for gA-dig and 1×10^9 for gA_T and 8×10^8 for MSL. The values for k_{2D_3} were varied in the model as follows: for [gA-dig] at 1×10^9 molecules cm^{-2}, 5×10^{-8} (●), 5×10^{-9} (■), 5×10^{-10} (▲) and for [gA-dig] at 5×10^8 molecules cm^{-2}, 5×10^{-8} (◆), 5×10^{-9} (▼), 5×10^{-10} (+). Other inputs into the model were: k_{2D_1} and $k_{2D_2} = 1 \times 10^{-11}$ molecules cm^{-2} s^{-1}; $k_{2D_1}^{-1}$ and $k_{2D_2}^{-1} = 100$ s^{-1}; $k_{2D_3} = 5 \times 10^8$; 5×10^9 and 5×10^{10} molecules cm^{-2} s^{-1}; $k_{2D_3}^{-1} = 1 \times 10^{-3}$ molecules cm^{-2} s^{-1}; $k_{3D_1} = 1 \times 10^5$ M^{-1} s^{-1}; $k_{3D_1}^{-1} = 1 \times 10^{-6}$ s^{-1}; k_{3D_2}, k_{3D_3} and $k_{3D_4} = 2 \times 10^4$ M^{-1} s^{-1}; 0 $k_{3D_2}^{-1}$; $k_{3D_3}^{-1}$ and $k_{3D_4}^{-1} = 5 \times 10^{-4}$ s^{-1}. (b) Digoxin b-Fab' response as gating ratio. The graph shows the experimental and modelled titration curves for the ICS biosensors to digoxin b-Fab' as $100 \times Y_0/(Y_0 + Y_1)$ (gating ratio). The nominal concentrations for gA-dig are 1.4×10^9 (□) and 2.8×10^9 (△). The concentrations used in the model were 5×10^8 and 1×10^9 for gA-dig and 1×10^9 for gA_T and 8×10^8 for MSL. The values for k_{2D_3} were varied in the model as follows: for [gA-dig] at 1×10^9 molecules cm^{-2}, 5×10^{-8} (●), 5×10^{-9} (■), 5×10^{-10} (▲) and for [gA-dig] at 5×10^8 molecules cm^{-2}, 5×10^{-8} (◆), 5×10^{-9} (▼), 5×10^{-10} (+). k_{2D_1} and $k_{2D_2} = 1 \times 10^{-11}$ molecules cm^{-2} s^{-1}; $k_{2D_1}^{-1}$ and $k_{2D_2}^{-1} = 100$ s^{-1}; $k_{2D_3} = 5 \times 10^8$, 5×10^9 and 5×10^{10} molecules cm^{-2} s^{-1}; $k_{2D_3}^{-1} = 1 \times 10^{-3}$ molecules cm^{-2} s^{-1}; $k_{3D_1} = 1 \times 10^5$ M^{-1} s^{-1}; $k_{3D_1}^{-1} = 1 \times 10^{-6}$ s^{-1}; k_{3D_2}, k_{3D_3} and $k_{3D_4} = 2 \times 10^4$ M^{-1} s^{-1}; $k_{3D_2}^{-1}$, $k_{3D_3}^{-1}$ and $k_{3D_4}^{-1} = 5 \times 10^{-4}$ s^{-1}. (c) Digoxin analyte response as gating ratio. The graph shows the experimental and modelled titration curves for the ICS biosensors to digoxin as $100 \times Y_0/(Y_0 + Y_1)$ (gating ratio). Model inputs were: $k_{2D_2} = 1 \times 10^{-11}$ molecules cm^{-2} s^{-1}; $k_{2D_2}^{-1} = 100$ s^{-1}; $k_{2D_3} = 1 \times 10^{10}$ molecules cm^{-2} s^{-1}; $k_{2D_3}^{-1} = 1 \times 10^{-3}$ molecules cm^{-2} s^{-1}; $k_{3D_1} = 1 \times 10^6$ M^{-1} s^{-1}; $k_{3D_1}^{-1} = 1 \times 10^{-6}$ s^{-1}; $k_{3D_3} = 2 \times 10^4$ M^{-1} s^{-1}; $k_{3D_3}^{-1} = 5 \times 10^{-4}$ s^{-1}.

model shows that the transition from [b-Fab'] dependent to [b-Fab'] independent kinetics represents the transition from a three-dimensional to a two-dimensional rate determining step.

Eqn. (4), (5) and (7) describe the 3D reactions. Fig. 5(a) shows that the NMS is proportional to the $[gA_{DIMER}]$, i.e. the absolute rate is independent of $[gA_{DIMER}]$, so it follows that the rate determining step is the binding of b-Fab' to MSL_{SA} [eqn. (4)]. From Fig. 5(a), the gradient of NMS/[b-Fab'] (apparent on rate per dimer) gives 8×10^4 and 16×10^4 M^{-1} s^{-1} for 1×10^9 and 5×10^8 gA-dig molecules cm^{-2}, respectively. The ratio of MSL_{SA} to gA-dig can be derived from these values using the relationship $k_{apparent} = k_{3D_1}[MSL_{SA}]/[gA_{DIMER}]$. Using the independently measured value for $k_{3D_1} = 1 \times 10^5$ M^{-1} s^{-1}, the ratio of $[MSL_{SA}]/[gA_{DIMER}]$ has been derived and shown to be 0.8 and 1.6 for 1×10^9 and 5×10^8 gA-dig molecules cm^{-2}, respectively. These values compare with ratios of 25 and 50 for nominal [MSL$_{SA}$]/nominal [gA-dig]. While the low ratio may reflect actual ratios of deposited components (b-MSL/gA-dig), it may alternately be accounted for by lower than expected ratios of MSL$_{SA}$: b-Fab'/gA-dig or MSL$_{SA}$/gA-dig. In turn, this low ratio may reflect either functionality or accessibility of the species.

For the portion of the titration curve for which the response is 2D rate limited, either the cross-linking of gA-dig to MSL$_{SA}$: b-Fab' [eqn. (3)] or the dissociation of gA-dig or gA-dig : b-Fab' from gA$_{DIMER}$, [eqn. (1) and (2)] may be the rate determining step.

If the rate limiting step is the cross-linking of b-Fab' to gA-dig [eqn. (3)], fitting the model to the experimental data [Fig. 5(a)] shows the sensitivity of the response to the rate of surface analyte cross-linking.

Alternatively, if the rate-limiting step is the dimer dissociation, $k_{2D_1}^{-1}$ can be determined from the reciprocal of NMS in the 2D limited part of the curve [Fig. 5(a)]. By this analysis $k_{2D_1}^{-1}$ equates to 0.003 s^{-1}. This value is comparable to the independently measured gA dimer life-time reported by Anderson and co-workers.[14]

It is not possible to discriminate which of these 2D reactions are limiting from this data set.

The equilibrium response (gating ratio) also shows the 3D limiting region of the curve up to 5 and 30 nM for [gA-dig] of 5×10^8 and 1×10^9 molecules cm^{-2}, respectively [Fig. 5(b)], through the dependence of the response on surface or solution analyte concentration.

Two-dimensional rate constants, k_{2D_3} between 1×10^{-7} and 5×10^{-9} cm^2 mol^{-1} [Fig. 5(b)], do not impact the gating response significantly.

However, for rates of less than approximately 5×10^{-9} cm^2 molecules^{-1} s^{-1}, the response becomes limited by K_{2D_3}. When K_{2D_3} is limited, the gating ratio is reduced and there is a greater "hooking" effect. The data suggests that the kinetics is not strongly limited by K_{2D_3}.

Above approximately 200 nM b-Fab' the response decreases. This phenomena provides evidence for the binding of b-Fab' onto gA-dig and gA$_{DIMER}$. The binding of the b-Fab' onto the gA inhibits its ability to cross-link to the MSL$_{SA}$: b-Fab'. If the 2D limiting is surface analyte cross-linking, then the equilibrium response at this region will be dependent upon [MSL$_{SA}$], providing a test for the nature of the 2D limiting process.

The fitted data is sensitive to the number of accessible functional binding sites, presented on MSL$_{SA}$, in that a b-Fab' response is not observed when the concentration of b-MSL is decreased by a factor of 10. The model successfully predicts this and supports the low ratio of [MSL$_{SA}$]/[gA-dig] actual discussed previously.

The competitive kinetic response is rate and amplitude dependent which arises from the combined 2D and 3D kinetics. The rate is a function of the surface concentration of digoxigenin, [gA-dig] as free digoxin will bind directly to MSL$_{SA}$: b-Fab and compete for complexed b-Fab. When using the same parameters for modelling the competitive experimental gating response, there is again good agreement between the data and the model. The model is constrained to the values shown in Fig. 5(c). To obtain the final fit to the data the $k_{2D_2}^{-1}$ [eqn. (5)] must be equal to or less than 1×10^{-3} s^{-1}.

Conclusion

A biosensor based on a functional synthetic bilayer membrane has been constructed. The specific b-Fab' response has been characterised as a kinetic process occurring in both two- and three-dimensional planes, involving the capture of analyte onto the membrane surface and the cross-linking of the b-Fab. The modelling of the competitive ICS has demonstrated that the experimental data is in good agreement with our mechanistic model and has enabled the parameters within the model to be constrained.

By titrating gA in the inner and outer layer, the relationship between [gA$_T$], [gA-dig] and K_{2D_1} has been determined. Using this relationship in the model and using independent measures of 3D receptor binding constants, the actual concentrations of gA$_T$, gA-dig and MSL$_{SA}$ can be estimated.

Acknowledgements

We would like to thank Denise Thomas for her technical contributions and advice. We are grateful to Dr Frank de Hoog and Huu-Nuynh, CSIRO-Mathematical and Information Sciences, for helping to construct the mathematical model which was used for the analysis of the data reported herein. This work has been supported by the Cooperative Research Centre (CRC) program. The partner organisations within the CRC for Molceular Engineering and Technology are the Commonwealth Scientific and Industrial Research Organisation (CSIRO), the University of Sydney and the Australian Membrane and Biotechnology Research Institute (AMBRI). We thank all those within the CRC who have contributed to the invaluable discussions and who have supplied us with the quality materials that enabled the study to be undertaken.

References

1. B. Raguse, V. Braach-Maksvytis, B. A. Cornell, L. G. King, P. D. J. Osman, R. J. Pace and L. Wieczorek, *Langmuir*, 1998, **3**, 648.
2. B. A. Cornell, V. Braach-Maksvytis, L. G. King, P. D. J. Osman, B. Raguse, L. Wieczorek and R. J. Pace, *Nature (London)*, 1997, **387**, 580.
3. K. Raval, P. Culshaw, J. Prashar and B. Raguse, *Synthesis of functionalised membrane spanning lipid derivatives*, manuscript in preparation.
4. S. Kim, D. Bali and L. G. King, *Synthesis of functionalised gramicidin A derivatives*, manuscript in preparation.
5. G. W. Oddie, L. C. Gruen, G. A. Odgers, L. G. King and A. A. Kortt, *Anal. Biochem.*, 1997, **244**, 301.
6. A. L. Plant, M. Gueguetchkeri and W. Yap, *Biophys. J.*, 1994, **67**, 1126.
7. R. Benz, O. Frohlich, P. Lauger and M. Montal, *Biochim. Biophys. Acta*, 1975, **394**, 323.
8. P. Delahay, *Double Layer and Electrode Kinetics*, Wiley-Interscience, New York, 1965.
9. A. Ulman, *Introduction to Thin Organic Films: From Langmuir–Blodgett to Self-Assembly*, Academic Press, Boston, 1991.
10. A. Lewis and D. M. Engleman, *J. Mol. Biol.* 1983, **166**, 211.
11. N. M. Green, *Advances in Protein Chemistry*, Academic Press, New York, 1975, pp. 29–85.
12. S. Martin and G. Woodhouse, binding kinetics derived using surface plasmon resonance to determine binding rates of streptavidin binding to a monolayer of 2000 : 1 MSL : b-MSL molecules cm^{-2}, unpublished data.
13. G. Krishna and L. G. King, Binding kinetics of digoxin–b-Fab′ obtained through analysis of binding rates of digoxin b-Fab′ to NHS-xx-digoxigenin coupled to the carboxydextran matrix of an IAsys cuvette (Affinity Sensors), unpublished data.
14. D. B. Sawyer, S. Oiki and O. S. Andersen, *Biophys. J.*, 1990, 57, 100a.

Paper 8/09608B

Three-dimensional models of glutamate receptors

Michael J. Sutcliffe,*[a] Allister H. Smeeton,[a] Z. Galen Wo[b] and Robert E. Oswald[b]

[a] *Department of Chemistry, University of Leicester, Leicester, UK LE1 7RH*
[b] *Department of Pharmacology, College of Veterinary Medicine, Cornell University, Ithaca, New York 14853, USA*

Received 5th August 1998

Structural models of glutamate receptors have been produced as part of a multidisciplinary study of neuronal function—both ligand/receptor interactions and ion transport—at the atomic level. The models have concentrated on the agonist binding and transmembrane domains of ionotropic (*i.e.* ligand-gated) glutamate receptors (iGluRs), and have aided our understanding of the molecular determinants of (**1**) ligand binding and (**2**) channel activity. The model building process involved a combination of homology modelling, distance geometry, molecular mechanics, protein–ligand and protein–protein docking, electrostatic calculations and manual adjustment, in conjunction with restraints from site-directed mutagenesis, ligand binding and electrophysiological studies. The initial models were used to produce hypotheses which were tested experimentally; these models have been subsequently refined as part of an extremely effective multidisciplinary study using an iterative molecular modelling/experimental verification cycle in which restraints derived from experimental studies are used at all stages, and the findings from one round of modelling are used as restraints in the next. By studying a variety of agonists and antagonists, details have been built up of (**1**) those residues involved in ligand binding and (**2**) the role of agonist binding (*i.e.* agonist-induced conformational change) in channel gating. The models also aid our understanding of the conductance properties of the channels.

Introduction

Ion channels are key components in the activity of living cells. These channels are formed from membrane-bound proteins, and are commonly characterised in terms of their ionic selectivity and gating properties. Ion channels activated by the binding of a ligand—either internally (*e.g.* calcium ions, ATP, and cyclic nucleotides) or externally (*e.g.* the nicotinic acetylcholine receptor channel [nAChR], γ-aminobutyric acid [GABA] receptor channel, and the glutamate receptor channel)—are known as ligand-gated ion channels. Ligand-gated channels, activated by the action of neurotransmitters, are involved in fast synaptic transmission in the nervous system. These include glutamate receptors—the primary excitatory neurotransmitter receptors in the vertebrate brain—and play an important role in a wide variety of neuronal functions.[1]

Classification of glutamate receptors is based on their signal transduction mechanism—metabotropic glutamate receptors (mGluRs) are linked to GTP binding proteins and thus operate through second messengers,[2] whilst ionotropic glutamate receptors (iGluRs) function as ligand-gated cation channels. iGluRs are cation-selective channels, and are classified according to the

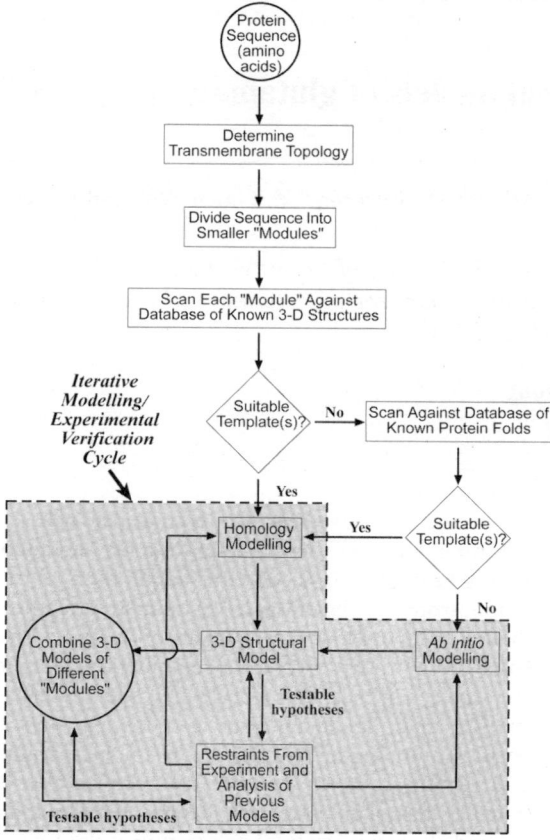

Fig. 1 Overview of model building procedure.

agonists by which they are selectively activated: (**1**) α-amino-3-hydroxy-5-methyl-4-isoxazolepropionate (AMPA) receptors (GluR1 to 4), (**2**) kainate receptors (GluR5 to 7 and KA1 to 2), and (**3**) N-methyl-D-aspartate (NMDA) receptors (NMDA-R1 to NMDA-R2D). In addition, kainate receptors of lower molecular mass (40–50 kDa, also known as kainate binding proteins) have been cloned from non-mammalian vertebrates [e.g. frog (fKBP),[3] chick (cKBP),[4] and goldfish (GFKARα and GFKARβ)[5]]; these exhibit considerable sequence homologies with the C-terminal half of the 100 kDa mammalian AMPA/kainate receptors.

We have produced structural models of the ligand-bound and ligand-free forms of the iGluRs GFKARα, GluR1 and GluR6, and the NMDARs NMDA-R1 and NMDA-R2C. These have aided our understanding of the molecular determinants of (**1**) ligand binding, and (**2**) channel activity. Our initial models of iGluRs[6] were used to produce experimentally testable hypotheses; these models have been subsequently refined[7,8] using an iterative molecular modelling/experimental verification cycle. There are essentially 4 stages in our modelling procedure (Fig. 1): (1) Determine the transmembrane topology of iGluRs. (2) Divide the amino acid sequence into smaller "modules" based on step 1. (3) Generate 3-D structural models of each individual "module". (4) Combine the 3-D models of the different "modules" to form a single structure.

Transmembrane topology

Knowledge of the transmembrane topology of a membrane-bound protein gives an indication of which regions of the sequence are likely to be close together in the three-dimensional structure—an important prerequisite in any modelling exercise. Thus, the first step in modelling iGluRs was to determine the transmembrane topology. Although originally thought to be structurally similar

to other ligand-gated ion channels,[9] experimental analysis of the position of native and engineered
N-glycosylation sites, protease sites, and epitopes (ref. 6 and references therein) demonstrated that
the transmembrane topology of iGluRs differs significantly from other ligand-gated ion channels.
Three membrane-spanning regions (denoted M1, M3 and M4) are present, the locations of which
were predicted using a consensus of the results from the programs MEMSAT,[10] TMAP,[11] and
PHD topology,[12] with the N-terminal region located extracellularly and the C-terminal region
located intracellularly. The region denoted M2, originally thought (based on hydropathy profiles)
to cross the membrane, does not span the membrane but, based on its role in ion conduction, is
most likely inserted into the ion conduction pathway in a manner similar to the P-segment of K^+
channels.[13]

Identification of smaller "modules"

The production of a structural model of an iGluR in a single step is currently unrealistic. Modelling is made much more tractable by division of the amino acid sequence into a number of smaller "modules", based on those regions of the sequence which are predicted to be close together in the three-dimensional structure. The iGluRs are composed of several evolutionary distinct modules:[14] (1) An extracellular portion that seems to have arisen from two different classes of bacterial protein—the N-terminal half is homologous to the leucine, isoleucine, valine-binding protein (LIVBP; absent from the 50 kDa kainate binding proteins), and the C-terminal half is homologous to the lysine, arginine, ornithine-binding protein (LAOBP).[15,16] The LAOBP-like domain in iGluRs consists of two distinct regions of amino acids, separated in sequence (but not in three-dimensional structure) by a large insert which includes two of the membrane spanning regions (M1 and M3). (2) A membrane-spanning, ion conduction pathway lining region that appears similar in topology to the pore lining region of K^+ channels. (3) A variable C-terminal regulatory domain of unknown origin which exhibits considerable diversity amongst subtypes.

Hitherto, we have concentrated our efforts on modelling two of the domains found in iGluRs—the LAOBP-like domain and the K^+ channel-like domain. We should stress that, as in any modelling study, these models represent a guess at the structure—the result is no substitute for a high-resolution experimentally derived three-dimensional structure, but nevertheless is extremely useful as a guide for subsequent experimental work, particularly when used as part of an iterative molecular modelling/experimental verification cycle.

Modelling the LAOBP-like domain

Identifying suitable 3-D templates

Once a structural module has been identified, a search is made for a similar protein(s) of known three-dimensional structure (determined either by X-ray crystallography or NMR spectroscopy). This is based on the observation that a good correlation exists between the level of similarity in the amino acid sequence and the level of similarity in the three-dimensional structure. The amino acid sequences of the LAOBP-like domain were searched against a database containing the amino acid sequences of proteins of known three-dimensional structure using the program BLAST[17] (this scans the LAOBP-like sequence against each sequence in the database in turn, returns a score for each pairwise comparison, ranks them and displays the most significant ones). However, no similarities of any significance were identified. An alternative approach was therefore adopted which is based on the premise that proteins with no obvious sequence similarity can still show remarkable similarities in their topologies (or "folds"), although fold similarities are not identified as reliably as sequence similarities. Both the TOPITS[18] and the UCLA-DOE Structure Prediction Server[19] returned histidine-binding protein (HBP) as the only significant hit (these scan the predicted secondary structure, solvent accessibility and amino acid environment of the LAOBP-like sequence against a database containing a representative set of known protein folds, rank the scores, and display the most significant ones). Thus, HBP was identified as a suitable template structure to use as the basis for modelling. The three-dimensional structure of HBP deposited in the Brookhaven Protein Data Bank[20] (PDB; an international archive of experimentally determined three-dimensional structures of protein molecules) is in the *holo* (ligand-bound) form (PDB accession number 1HSL[21]). Once a suitable fold is found, other proteins of known three-dimensional structure with the same fold are identified. Scanning this structure against DALI[22] identified the *holo* form of lysine, arginine, ornithine-binding protein (LAOBP; 1LST[23]) and the *holo* form of glutamine binding protein (QBP; 1WDN[24]) as additional template structures for modelling the *holo* form of the LAOBP-like domain. It also identified the *apo* (ligand-free) form of LAOBP (2LAO[23]) and the *apo* form of QBP (1GGG[25]) as suitable templates for modelling the *apo* form of the LAOBP-like domain of the iGluRs.

Sequence alignment

Once the structural templates are identified, the particular amino acids in the sequence of the iGluR which correspond to the amino acids in the template structures are identified. Multiple sequence alignment [using Clustal W[26] and CAMELEON (Oxford Molecular Ltd, Oxford, UK)] was used to align the amino acid sequences of 19 members of the iGluR superfamily with the amino acid sequences of LAOBP, HBP and QBP. The final consensus sequence alignment (Fig. 2) was produced manually within the program CAMELEON, subject to the constraints that (**1**) wherever possible, no insertion or deletion occurred within the crystallographically determined secondary structural elements of the templates; (**2**) the known N-glycosylation sites (ref. 6 and references therein) corresponded to surface positions [to be glycosylated, an amino acid (in this case, an asparagine residue) position must be accessible to the incoming sugar molecule]; (**3**) all those residues thought to be involved in ligand binding[27-29] were positioned in the binding site; (**4**) the disulfide bridge, which we showed by experiment is present,[30] could be formed; (**5**) some account was taken of the predicted secondary structure (determined using PHDsec[31]).

Mapping amino acid sequence onto 3-D templates

A set of three-dimensional models was then produced, based on this alignment, using the technique of homology modelling. The results of homology modelling are critically dependent upon (**1**) the choice of structural template(s) and (**2**) the sequence alignment used, emphasising the need for care in the preceding stages. Programs for homology modelling use one of two approaches—either a fragment-based stepwise approach (COMPOSER;[32,33] used for our early models[6]) or a single step approach (MODELLER;[34] used for our more recent models). Although both methods produce models of equally good quality, the latter is now our method of choice because it enables additional restraints—hydrogen bonds, charge–charge interactions, and hydrophobic interactions—to be added in the form of distance restraints *during* the modelling process, rather than in a *post hoc* fashion as with the fragment-based approach. The ligand binding site of the three-dimensional models was characterised using InsightII (an interactive molecular graphics program; MSI, San Diego, USA), GRID[35] (used to search for energetically favourable positions for functional groups in the ligand), DELPHI[36] (electrostatics calculations, used with "focussing" to enhance accuracy by using results from a low resolution calculation as boundary values for a higher resolution calculation) and GRASP[37] (for visualising the results from DELPHI). Once possible protein–ligand interactions are identified, these are then included in the homology modelling process in the form of distance restraints. For example, for a hydrogen bond the donor···acceptor distance (*e.g.* oxygen···nitrogen) is restrained to a range of 2.5–3.3 Å. Because of the less well defined nature of restraints between atoms thought to be involved in hydrophobic interactions, such restraints are initially entered relatively loosely, the resulting models analysed, and the hydrophobic restraints updated accordingly.

Ligand binding site

We have modelled the agonist glutamate into the binding site of the *holo* form of GFKARα, GluR1 and GluR6, NMDA-R1 and NMDA-R2C, and the agonists kainate and domoate into the binding site of the *holo* form of GFKARα. The modelling positioned the three "hot spots" known initially to affect the binding of agonist in both non-NMDA and NMDA receptors in the binding site (residues 27[411] VTTIKE 32[416], 69[461] DGRYG 73[465], and 260[662] GTIKDS 265[667] in GFKARα; here and below the corresponding residue number for GluR1 is given in

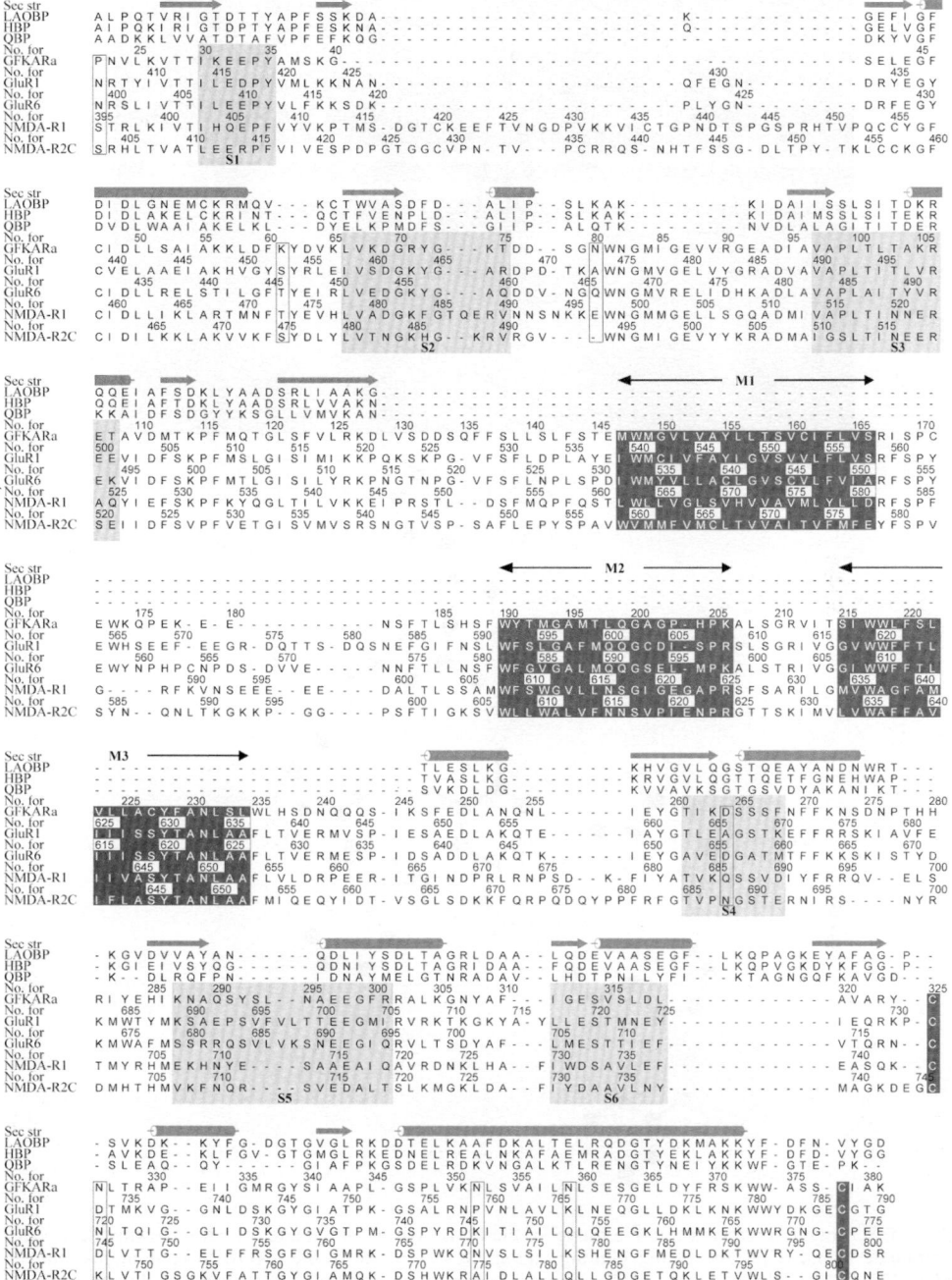

Fig. 2 Our current sequence alignment used for modelling the iGluRs. The region shown includes the LAOBP-like and the K⁺ channel-like domains of the iGluRs. The positions of M1, M2 and M3 are shown on a dark background; those regions involved in ligand binding (S1–S6) are shown on a light gray background; glycosylation sites are denoted by vertical boxes; and the two half-cystines are shown on a dark background. The consensus secondary structure observed in the crystal structures of the bacterial periplasmic amino acid binding proteins is also shown. (This figure was generated using ALSCRIPT[66].)

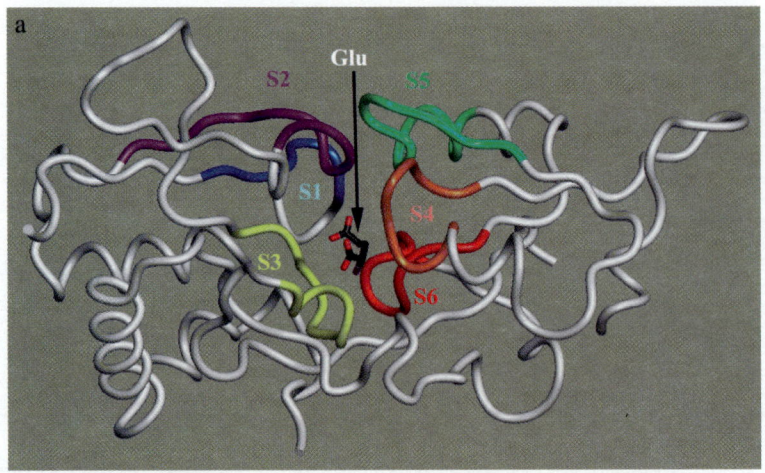
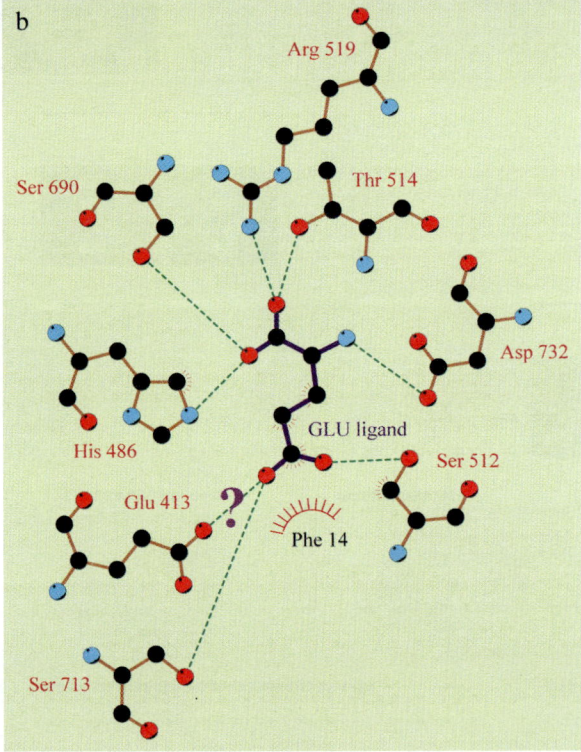

Fig. 3 (a) Schematic representation of a model of the NMDA-R2C–glutamate complex showing the positions of glutamate ligand (Glu) and ligand binding regions S1–S6 in one of the LAOBP-like domains (note that residues 425–454 have been omitted as these cannot be built with any confidence since they do not correspond to any amino acid residues in the template structures). (b) Schematic representation of the glutamate binding site in a model of the NMDA-R2C–glutamate complex showing one of the LAOBP-like domains. Hydrogen bonds and charge–charge interactions are shown as green dashed lines, hydrophobic interactions as red spoked arcs, carbon atoms as black balls, oxygen atoms as red balls and nitrogen atoms as blue balls. The purple "?" signifies the uncertainty in how the interaction between the two carboxyl groups is mediated. (This figure was generated using LIGPLOT.[67])

square brackets for reference—numbering begins at the start of the signal sequence), and gave insight into those amino acid residues in GFKARα likely to interact with the ligand. Our initial hypothesis, based on these models, was that E32[416], K71[463], Y72[464], R106[499] and R302[706] were involved in ligand binding (alongside other residues, to be investigated later). These suggestions formed the basis of experimental studies which partly confirmed protein–ligand interactions suggested by our initial models (E32[416], Y72[464] and R106[499]) and partly showed that some of the interactions had been modelled incorrectly (K71[463] and R302[706]). The orientation of the glutamate ligand was a concern at this stage—subsequent data[7] suggest that the glutamate ligand binds in the same orientation as the arginine, histidine and glutamine ligand in LAOBP, HBP and QBP, respectively (Fig. 3) rather than in the "upside down" orientation of our initial models[6] (see ref. 7 for a more detailed discussion). However, we are uncertain how the interaction between two negatively charged carboxyl groups (one from E32[416] [E413 in NMDA-R2C; note that this residue is a negatively charged glutamate in the majority of iGluRs, but a polar glutamine in NMDA-R1 and GFKARβ], the other from the ligand) could be mediated. We have discounted mediation by either a water molecule or a proton on the sidechain of E32[416] (*i.e.* E32[416] would have an unusually high pK_a) because an E at this position binds glutamate ligand approximately 5 times more strongly than does Q. Also, we have found recently that divalent cations do not affect the binding of kainate, suggesting that a divalent cation very likely does not mediate the interaction between the ligand carboxyl group and E32[416]. One remaining possibility is that the carboxyl group on the ligand becomes protonated.

Refinement using iterative modelling/experimental verification cycle

Subsequent to our initial modelling of iGluRs,[6] three additional "hot spots" for ligand binding have been identified, giving a total of six (Fig. 2 and 3; those that correspond to the "hot spots" included in the original modelling are denoted S1, S2 and S4). The current sequence alignment (Fig. 2) corresponds to the original sequence alignment[6] in 2 of these (S2 and S4), and differs by three residues in S1. Of the remaining three "hot spots", the current sequence alignment corresponds to the original in one (S3) and differs quite significantly in the remaining two (S5 and S6). Interestingly, all but one of these "hot spots" (S5) correspond to regions in the ligand binding site in the LAOBP-like crystal structures.[21,23,25] This further justifies the use of these crystal structures as structural templates. The absence of S5 in the bacterial periplasmic amino acid binding proteins can, based on our sequence alignment, be explained in terms of the insertion (2–6 residues) in the iGluRs lengthening this region of the structure, thereby allowing contact with the ligand. Positions of conserved amino acids in the alignment play an important role in this realignment, and all of the changes discussed leave the N-glycosylation sites accessible. A model of the LAOBP-like domain of the NMDA-R2C–glutamate complex, generated using our current sequence alignment (Fig. 2), is shown in Fig. 3.

Glycine binding *versus* glutamate binding

NMDA-R1 is thought to bind glycine and NMDA-R2 subunits are thought to bind glutamate.[38–40] One of these studies[39] suggests that this difference could arise from the presence of more bulky aromatic sidechains in the α-amino binding region of NMDA-R1. Our study, which is based on a different amino acid sequence alignment to this previous work, suggests a possible alternative explanation. Inspection of our models does not reveal any differences in the agonist binding site between these two types of subunit which would explain this likely difference in specificity. However, inspection of the sequence alignment (Fig. 2) reveals that there is a two residue insertion in a loop in S2 in NMDA-R1 (T486 to Q487) with respect to NMDA-R2 (there is also a three residue insertion in NMDA-R1—T486 to E488—with respect to the non-NMDA receptors). Although this does not manifest itself as a difference between our models of NMDA-R1 and NMDA-R2C, the two additional residues in NMDA-R1 (T486 and Q487) could result in a conformational change in this loop, thereby changing the position of the residue F484[464] (an important determinant in ligand binding) to form a constriction in the agonist binding site where an amino acid sidechain would be positioned, thereby preventing glutamate but not glycine from binding to NMDA-R1.

Modelling the K⁺ channel-like domain

Ab initio modelling

The amino acid sequences of the K⁺ channel-like domain of the iGluRs showed no significant sequence homology to any protein of known three-dimensional structure, nor did they show a significant propensity to adopt any of the known protein folds. The complete *ab initio* prediction of protein three-dimensional structures is not possible at present, and a general solution to the protein folding problem is not likely to be found in the near future. However, if some knowledge is available of (1) the secondary structural elements in the module being modelled, and (2) how these secondary structural elements pack together, the modelling exercise becomes far more tractable, despite remaining highly speculative. Using the similarity between M2 in the K⁺ channel-like domain and the P-segment (or H5 segment) of K⁺ channels,[14] we were able to produce initial models of the K⁺ channel-like domain using an *ab initio* approach[6]—distance geometry, followed by simulated annealing and energy minimisation (XPLOR;[41] built initially as an antiparallel β-barrel, with appropriate distance restraints for hydrogen bonds [NH $\cdots$ O 1.8–2.3 Å and N $\cdots$ O 2.5–3.3 Å] and dihedral restraints for β-strands [ϕ − 120° to −160° and ψ 115° to 155°], and restrained to be symmetrical), and subsequent manual refinement[6] (InsightII [MSI, San Diego, USA] and SCULPT[42]) to ensure that all residues were positioned consistent with experimental data. These original models of GFKARα, GluR1 and GluR6 were built with stoichiometries of 4, 5 and 6. The size of the pore in our early models lends support to either a pentameric stoichiometry or a hexameric stoichiometry (see ref. 6,7 for a more detailed discussion); our more recent models of NMDA receptors also suggest that a tetramer is too small (see below).

Satisfaction of experimentally derived restraints

An important consideration in building the models is the position of the crucial residue known to affect conductance in a number of iGluRs. This position in GluR2 and GluR6 (residue 198[600]; denoted the "Q/R/N site") can be changed from Q to R by RNA editing[43] and is known to be involved in the blockade of NMDA receptors by Mg^{2+} (N598 of NMDA-R1[44]). Electrostatic calculations (using DELPHI[36] and GRASP[37]) suggest that the Q form provides a good cation binding site, whereas the R form does not. This in turn suggests why the Q form is doubly rectifying (the D/E site is free to bind Ca^{2+}) whereas the R form is not (the predicted Ca^{2+} binding site is removed by formation of a salt bridge between the R and the D/E site; see ref. 6,7 for a more detailed discussion). Site-directed mutagenesis data[45–47] were also used for modelling M2.[6] In addition to M2, M1 and M3 are also likely to form part of the conductance pathway. M1 was modelled as an α-helix lining the ion conduction pathway[6]—consistent with the results of RNA editing of residues in M1 of GluR6.[48] M3 was also modelled as an α-helix lining the ion conduction pathway[6]—consistent with the suggested role of M3 in the channel blocking action of MK-801.[49]

NMDA receptor models

Our recent models of the K⁺ channel-like domain correspond to a refinement of our initial models,[6] into which we have incorporated the available experimental data for M2 from NMDA receptors.[46,50–53] Although the channels are known to exist as heteromers, they were modelled initially as homomers and then the homomers combined to form a series of heteromers—partly for simplicity of interpreting the models and partly because the number of each receptor type present in an active channel is unknown. The starting point for modelling the M2 region of both NMDA-R1 and NMDA-R2C was our recent model of the the P-segment (or H5 segment) of the inward rectifier K⁺ channel, Kir2.1[54]—although the crystal structure of a K⁺ channel has been determined recently.[13] The starting structure of M2 was refined automatically, using XPLOR[41] and distance restraints derived from the results of a block arising from applying methanethiosulfonate (MTS)-based thiol reagents to a NMDA-R1/NMDA-R2C channel following cysteine scanning mutagenesis[46]—if the block was observed at a particular position, the maximum diameter of the pore (measured using gamma atoms of the respective residue) was set at 20 Å; if the block did not occur, a minimum diameter of 15 Å was used. Channels were generated with

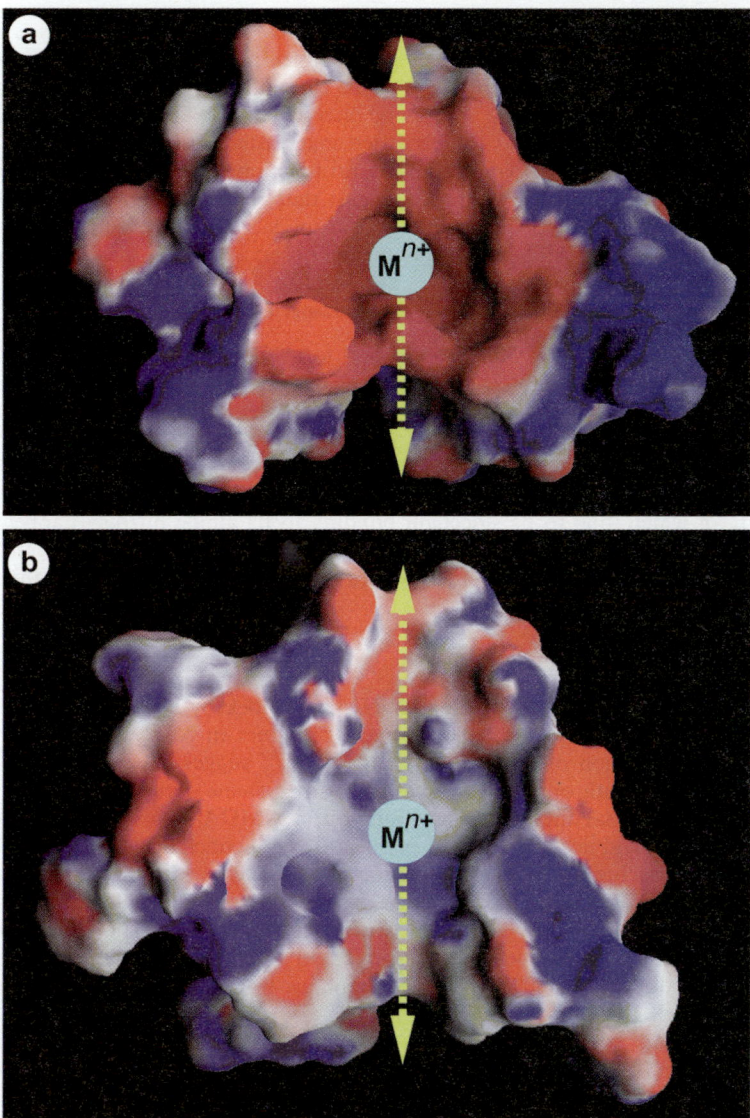

Fig. 4 Electrostatic potential on the surface of the narrowest part of the pore for (a) NMDA-R1 and (b) NMDA-R2C. Red corresponds to an electrostatic potential of $\leqslant -10\ k_B T/e$, white an electrostatic potential of $0\ k_B T/e$ and blue an electrostatic potential of $\geqslant +10\ k_B T/e$. In both cases, the view is from the side of the channel through the membrane with two of the four subunits cut away, leaving only those two "behind" the centre of the pore. Only the M2 region is shown, although calculations were performed with M1 and M3 also present. The yellow dashed line illustrates the location of the centre of the pore, and the light blue circle represents a cation.

both tetrameric and pentameric stoichiometries—residue accessibilities to the pore in the resulting models were consistent with experimental observations.

Selectivity filter

The cation selectivity filter (*i.e.* the pore lining component of M2) in these models consists of carbonyl oxygens from the polypeptide backbone (particularly G617[602], I618[603] and G619[604] in NMDA-R1), the sidechain amide oxygen of asparagine, and aromatic groups

(W608[592] and F609[593] in NMDA-R1). The presence of carbonyl oxygens is in agreement with the recent crystal structure of the KcsA K$^+$ channel,[13] in which the authors suggest that a carbonyl tunnel alone forms the atomic basis of the cation selectivity filter. Indeed, the predicted presence of carbonyl oxygens in the selectivity filter could explain why NMDA-R1 can form functional homomeric channels, and the predicted inaccessibility of the corresponding carbonyl oxygens could explain why NMDA-R2 cannot form functional homomeric channels (see below). Other experimental studies on K$^+$ channels suggest that an aromatic tyrosine residue lines the selectivity filter.[54,55] Thus, the selectivity filter in our models—involving both cation–oxygen and cation–π interactions—is consistent with experimental observations in other cation channels.

NMDA-R1, but not NMDA-R2, form functional homomeric channels

Our models also suggest why NMDA-R1 can form functional homomeric channels, but NMDA-R2 cannot.[56] Electrostatic calculations on our models of homomeric NMDA-R1 and NMDA-R2C channels (Fig. 4) show that the pore lining region of the NMDA-R1 channel, particularly in the region of the selectivity filter, is significantly more negative than NMDA-R2C. Analysis of the models reveals that this may result from the conserved proline in NMDA-R2 receptors (P619[604] in NMDA-R2C; this is a glycine in NMDA-R1), the restricted conformational freedom of which (due to its sidechain bonding to its amide nitrogen) appears to prevent exposure of carbonyl oxygens to the pore in this region. In NMDA-R1, however, the high level of conformational freedom of the two glycine residues (glycine has no sidechain) in this region could further contribute to this difference.

Mg^{2+} permeation

We have mentioned above the role of the Q/R/N-site in block by magnesium.[44] Additional sites also exist. Mutation of W607[592] to non-aromatic residues in NMDA-R2A and NMDA-R2B greatly increased permeation of extracellular magnesium,[52] suggesting that this residue lines the pore. This is consistent with our models, and also with cysteine scanning mutagenesis studies.[46] The same authors note that the equivalent mutation in NMDA-R1 (mutation of W608[592] to non-aromatic residues) did not affect Mg^{2+} permeation, and concluded that this residue does not line the pore. This conclusion is at odds with both the cysteine scanning mutagenesis studies[46] and our models. However, these models offer a possible alternative explanation. In NMDA-R1, the residue adjacent to W608[592] is F609[593] which is exposed to the pore, whereas in NMDA-R2A and NMDA-R2B this residue is a leucine. Thus, NMDA-R1 contains a second potential Mg^{2+} aromatic binding site (F609[593], which remains when W608[592] is mutated), whereas NMDA-R2A and NMDA-R2B do not. The models also present a possible explanation for the higher affinity for Mg^{2+} of the M1 region in NMDA-R2A and NMDA-R2B than either in NMDA-R2C or NMDA-R2D.[57] S561[551] in NMDA-R2A and NMDA-R2B is V561[551] in NMDA-R2C and NMDA-R2D. This residue is exposed to the pore and therefore could provide an additional Mg^{2+} binding site in NMDA-R2A and NMDA-R2B (due to its sidechain hydroxy group), but not in NMDA-R2C and NMDA-R2D (due to its hydrophobic nature).

Stoichiometry

Experimental determination of the stoichiometry remains somewhat ambiguous, suggesting either a tetramer[58–60] or a pentamer.[61] The models give insight into possible stoichiometry, but by no means provide a definitive answer. The narrowest diameter in the literature for NMDA channels is 5.5 Å[53]—this is consistent with our pentameric (~ 5.3 Å), but larger than our tetrameric channels (~ 4.3 Å), suggesting that perhaps the stoichiometry is pentameric.

Combining different modules

Once models had been produced for the different modules, the final stage (Fig. 1) was to combine them to create a model of the iGluR. This was achieved by ensuring that the model was consistent with available experimental data. A module comprising M1, M2 and M3 (*i.e.* the membrane-spanning region which lines the pore) was used as the starting point since this already contained

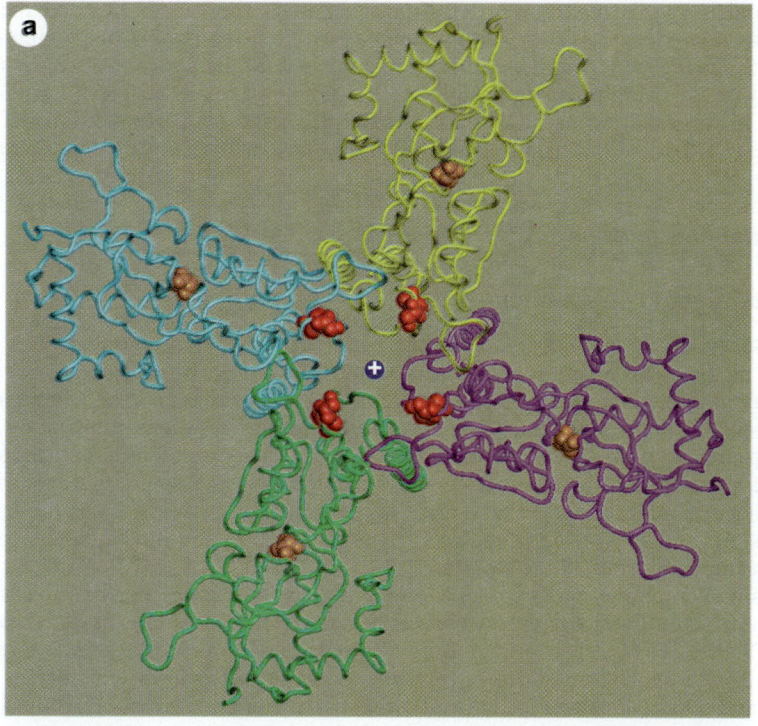

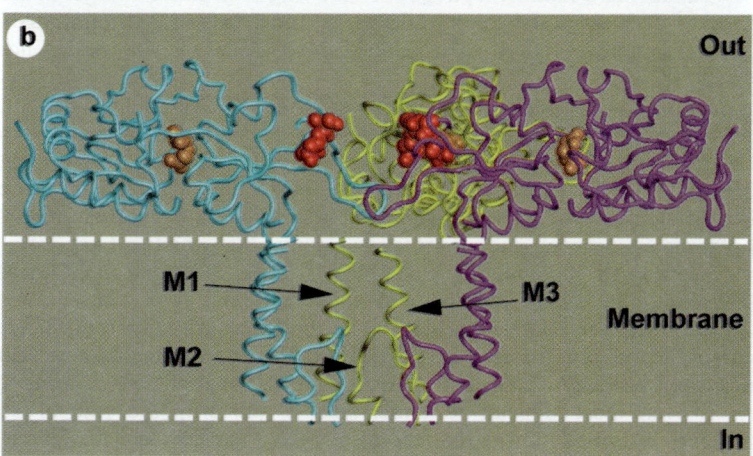

Fig. 5 Schematic representation of the tetrameric structure of the NMDA-R2C—glutamate complex viewed from (a) outside membrane and (b) along membrane [with the front (green) subunit removed for clarity]. Each subunit is in a different colour, glutamate ligand is in orange space filling representation and the conserved acidic amino acid and D668[652] and the adjacent conserved acidic/polar residue S667[651] are in red space filling representation.

the correct symmetry. The appropriate number of copies of the LAOBP-like domain (*e.g.* four for a tetramer) were added to this using interactive molecular graphics (InsightII). The model of the LAOBP-like domain was positioned empirically with respect to both the membrane and its symmetry related copies, so that (**1**) the consensus glycosylation sites were solvent accessible, (**2**) the agonist binding site was accessible, (**3**) the distance between the end of the N-terminal of the LAOBP-like domain and M1 was in a reasonable range (approximately 17 residues were missing from our models in this region), (4) the distance between the end of M3 and the start of the C-terminal section of the LAOBP-like domain was within a reasonable range (approximately 12 residues were missing from our models in this region), and (5) the domain was as close to the pore as possible without overlapping sterically with its symmetry related copies in both the *apo* and *holo* forms. In the resulting orientation, the long axis of the LAOBP-like domain was roughly parallel to the surface of the membrane. This positioning of the LAOBP-like domain, although not a unique solution (due to the limited experimental data), is not inconsistent with the currently available data and is, in fact, constrained to a large extent by the experimental results. An assembled tetrameric model of NMDA-R2C is shown in Fig. 5.

Binding and signal transduction

Once the models of the different modules had been combined, the molecular basis of the transduction of binding energy to ion channel opening could be investigated. The models were analysed, paying particular attention to those residues close to the ion conduction pathway in the "channel open" (ligand-bound) and "channel closed" (ligand-free) states of the model. If particular residues appeared to be involved in ion conduction, and they are conserved across the iGluRs, then these are possible candidates. It should be noted that the conformational change in the ligand binding domain assumed to take place in the transition between the ligand-bound and the ligand-free states is likely to be propagated to other parts of the model, and in particular the membrane-spanning portion of the structure.

The transition from the agonist-bound to the agonist-free states in our models results in a change in the orientation of a pair of conserved negatively charged amino acids. These correspond to residues S667[651] (which is a negatively charged residue in the non-NMDA receptors) and D668[652]. In the agonist-bound form, but not the agonist-free form, our models suggest that S667[651] and D668[652] are positioned near the opening, thus electrostatically attracting cations into the channel in a manner analogous to the ring of negative charges in nicotinic acetylcholine receptors.[62] This proposed role of S667[651] and D668[652] is supported by studies of NMDA-R1, which implicate the involvement of D669[652] in voltage-dependent spermine block.[63]

So far, the assumption has been made that the LAOBP-like domains exist in one of two states—either the ligand-free or ligand-bound states—an over-simplistic picture. In fact, there are at least two agonist-bound states—an initial channel open state and a subsequent desensitised (channel closed) state; the structural reasons for this remain unclear. It has been proposed[64] that the mechanism of agonist binding is akin to the "venus fly trap" mechanism observed in the bacterial periplasmic amino acid binding proteins—agonist binds first to Lobe 1 alone to give the "channel open" state, and subsequently binds also to Lobe 2 to give the "desensitised" state. However, in the case of the bacterial periplasmic amino acid binding proteins, in the initial ligand-bound "open cleft" form the relative orientations of the two lobes remain unchanged (see ref. 65 for a discussion of the different forms). Therefore, it is difficult to rationalise how ligand binding in such a manner could produce a sufficiently large conformational change to bring about channel opening. This is even more difficult to understand if our proposed role of S667[651] and D668[652] (residues in Lobe 2) in the channel open state is indeed correct (which it may not be), predicting a movement in Lobe 2 upon agonist binding. Our models suggest the following possibility, which is speculative but nevertheless not inconsistent with the experimental results.[64] Initial agonist binding causes the lobes to close together. This movement is impeded initially by the loop we denote S5 (Fig. 2 and 3), but which subsequently changes conformation to allow the two lobes to come closer together. This second change in the relative positions of Lobes 1 and 2 results in desensitisation, with the agonist now bound more tightly. Our structural models can neither prove nor disprove this notion, but merely suggest it as a potential mechanism that can be tested.

Conclusion

Structural models of iGluRs can be produced using molecular modelling techniques. This is achieved by dividing the amino acid sequence into a number of smaller "structural modules", producing models for these, and then combining the models of the "modules" to produce a model of the channel. The process is best achieved using an iterative modelling/experimental verification cycle. As in any modelling exercise, the results are no substitute for a high resolution three-dimensional structure. However, the resulting insights suggest directions for further study. The power of our approach is thus the iterative interplay between modelling and experiment.

Acknowledgements

M. J. S. is a Royal Society University Research Fellow. A. H. S. is funded by an EPSRC studentship. This work was supported in part by an National Science Foundation grant (IBN-9309480) to R. E. O.

References

1. D. T. Monaghan, R. J. Bridges and C. W. Cotman, *Annu. Rev. Pharmacol. Toxicol.*, 1989, **29**, 365.
2. M. Hollmann and S. Heinemann, *Annu. Rev. Neurosci.*, 1994, **17**, 31.
3. K. Wada, C. J. Dechesne, S. Shimasaki, R. G. King, K. Kusano, A. Buonanno, D. R. Hampson, C. Banner, R. J. Wenthold and Y. Nakatani, *Nature (London)*, 1989, **342**, 684.
4. P. Gregor, I. Mano, I. Maoz, M. McKeown and V. Teichberg, *Nature (London)*, 1989, **342**, 689.
5. Z. G. Wo and R. E. Oswald, *Proc. Natl. Acad. Sci. USA*, 1994, **91**, 7154.
6. M. J. Sutcliffe, Z. G. Wo and R. E. Oswald, *Biophys. J.*, 1996, **70**, 1575.
7. M. J. Sutcliffe, A. H. Smeeton, Z. G. Wo and R. E. Oswald, *Methods Enzymol.*, 1998, **293**, 589.
8. M. J. Sutcliffe, A. H. Smeeton, Z. G. Wo and R. E. Oswald, *Biochem. Soc. Trans.*, 1998, **26**, 450.
9. S. Nakanishi and M. Masu, *Annu. Rev. Biophys. Biomol. Struct.*, 1994, **23**, 319.
10. D. T. Jones, W. R. Taylor and J. M. Thornton, *Biochemistry*, 1994, **33**, 3038.
11. B. Person and P. Argos, *J. Mol. Biol.*, 1994, **237**, 182.
12. B. Rost, P. Fariselli and R. Casadio, *Protein Sci.*, 1996, **7**, 1704.
13. D. A. Doyle, J. M. Cabral, R. A. Pfuetzner, A. L. Kuo, J. M. Gulbis, S. L. Cohen, B. T. Chait and R. MacKinnon, *Science*, 1998, **280**, 69.
14. Z. G. Wo and R. E. Oswald, *Trends Neurosciences*, 1995, **18**, 161.
15. N. Nakanishi, N. A. Schneider and R. Axel, *Neuron*, 1990, **5**, 569.
16. P. J. O'Hara, P. O. Sheppard, H. Thøgersen, D. Venezia, B. A. Haldeman, V. McGrane, K. M. Houamed, C. Thomsen, T. L. Gilbert and E. R. Mulvihill, *Neuron*, 1993, **11**, 41.
17. S. F. Altschul, T. L. Madden, A. A. Schäffer, J. Zhang, Z. Zhang, W. Miller and D. J. Lipman, *Nucleic Acids Res.*, 1997, **25**, 3389.
18. B. Rost, in *TOPITS: Threading one-dimensional predictions into three-dimensional structures*, ed. C. Rawlings, D. Clark, R. Altman, L. Hunter, T. Lengauer, and S. Wodak, Menlo Park, CA, 1995.
19. D. Fischer and D. Eisenberg, *Protein Sci.*, 1996, **5**, 947.
20. E. E. Abola, F. C. Bernstein, S. H. Bryant, T. F. Koetzle and J. Weng, in *Protein Data Bank*, ed. F. H. Allen, G. Bergerhoff and R. Sievers, Data Commission of the International Union of Crystallography, Bonn/Cambridge/Chester, 1987.
21. N. H. Yao, S. Trakhanov and F. A. Quicho, *Biochemistry*, 1994, **33**, 4769.
22. L. Holm and C. Sander, *Proteins*, 1994, **19**, 165.
23. B. H. Oh, J. Pandit, C. H. Kang, K. Nikaido, S. Gokcen, G. F. L. Ames and S. H. Kim, *J. Biol. Chem.*, 1993, **268**, 11348.
24. Y. J. Sun, J. Rose, B. C. Wang and C. D. Hsiao, *J. Mol. Biol.*, 1998, **278**, 219.
25. C. D. Hsiao, Y. J. Sun, J. Rose and B. C. Wang, *J. Mol. Biol.*, 1996, **262**, 225.
26. J. D. Thompson, D. G. Higgins and T. J. Gibson, *Nucleic Acids Res.*, 1994, **22**, 4673.
27. A. Kuryatov, B. Laube, H. Betz and J. Kuhse, *Neuron*, 1994, **12**, 1291.
28. F. Li, N. Owens and T. A. Verdoorn, *Molecular Pharmacology*, 1995, **47**, 148.
29. S. Uchino, K. Sakimura, K. Nagahari and M. Mishina, *FEBS Lett.*, 1992, **308**, 253.
30. Z. G. Wo and R. E. Oswald, *Mol. Pharmacology*, 1996, **50**, 770.
31. B. Rost and C. Sander, *J. Mol. Biol.*, 1993, **232**, 584.
32. M. J. Sutcliffe, F. R. F. Hayes and T. L. Blundell, *Protein Eng.*, 1987, **1**, 385.
33. M. J. Sutcliffe, I. Haneef, D. Carney and T. L. Blundell, *Protein Eng.*, 1987, **1**, 377.
34. A. Sali and T. L. Blundell, *J. Mol. Biol.*, 1993, **234**, 779.
35. P. J. Goodford, *J. Med. Chem.*, 1985, **28**, 849.
36. A. Nicholls and B. Honig, *J. Comput. Chem.*, 1991, **12**, 435.
37. A. Nicholls, K. Sharp and B. Honig, *Proteins*, 1991, **11**, 281.
38. M. Honer, D. Benke, B. Laube, J. Kuhse, R. Heckendorn, H. Allgeier, C. Angst, H. Monyer, P. H. Seeburg, H. Betz and H. Mohler, *J. Biol. Chem.*, 1998, **273**, 11158.
39. B. Laube, H. Hirai, M. Sturgess, H. Betz and J. Kuhse, *Neuron*, 1997, **18**, 493.
40. H. Hirai, J. Kirsch, B. Laube, H. Betz and J. Kuhse, *Proc. Natl. Acad. Sci. USA*, 1996, **93**, 6031.
41. A. T. Brünger, *X-PLOR Manual*, Yale University, New Haven, 1997.

42 M. C. Surles, J. S. Richardson, D. C. Richardson and F. P. J. Brooks, *Protein Science*, 1994, **3**, 198.
43 B. Sommer, M. Köhler, R. Sprengel and P. H. Seeburg, *Cell*, 1991, **67**, 11.
44 N. Burnashev, R. Schoepfer, H. Monyer, J. P. Ruppersberg, W. Günther, P. H. Seeburg and B. Sakmann, *Science*, 1992, **257**, 1415.
45 R. Dingledine, R. I. Hume and S. F. Heinemann, *J. Neurosci.*, 1992, **12**, 4080.
46 T. Kuner, L. P. Wollmuth, A. Karlin, P. H. Seeburg and B. Sakmann, *Neuron*, 1996, **17**, 343.
47 N. Burnashev, A. Villarroel and B. Sakmann, *J. Physiol., London*, 1996, **496**, 165.
48 M. Köhler, N. Burnashev, B. Sakmann and P. H. Seeburg, *Neuron*, 1993, **10**, 491.
49 A. V. Ferrer-Montiel, W. Sun and M. Montal, *Proc. Natl. Acad. Sci. USA*, 1995, **92**, 8021.
50 J. Chao, N. Seiler, J. Renault, K. Kashiwagi, T. Masuko, K. Igarashi and K. Williams, *Mol. Pharmacol.*, 1997, **51**, 861.
51 K. Kashiwagi, A. J. Pahk, T. Masuko, K. Igarashi and K. Williams, *Mol. Pharmacol.*, 1997, **52**, 701.
52 K. Williams, A. J. Pahk, K. Kashiwagi, T. Masuko, N. D. Nguyen and K. Igarashi, *Mol. Pharmacol.*, 1998, **53**, 933.
53 L. P. Wollmuth, T. Kuner, P. H. Seeburg and B. Sakmann, *J. Physiol., London*, 1996, **491**, 779.
54 C. Dart, M. L. Leyland, P. J. Spencer, P. R. Stanfield and M. J. Sutcliffe, *J. Physiol., London*, 1998, **511**, 25.
55 Q. Lu and C. Miller, *Science*, 1995, **268**, 304.
56 C. J. McBain and M. L. Mayer, *Physiol. Rev.*, 1994, **74**, 723.
57 T. Kuner and R. Schoepfer, *J. Neurosci.*, 1996, **16**, 3549.
58 C. Rosenmund, Y. Stern-Bach and C. F. Stevens, *Science*, 1998, **280**, 1596.
59 I. Mano and V. I. Teichberg, *Neuroreport*, 1998, **9**, 327.
60 B. Laube, J. Kuhse and H. Betz, *J. Neurosci.*, 1998, **18**, 2954.
61 L. S. Premkumar and A. Auerbach, *J. Gen. Physiol.*, 1997, **110**, 485.
62 K. Imoto, C. Busch, B. Sakmann, M. Mishina, T. Konno, J. Nakai, H. Bujo, Y. Mori, K. Fukuda and S. Numa, *Nature (London)*, 1988, **335**, 645.
63 K. Kashiwagi, J. Fukuchi, J. Chao, K. Igarashi and K. Williams, *Mol. Pharmacol.*, 1996, **49**, 1131.
64 I. Mano, Y. Lamed and V. I. Teichberg, *J. Biol. Chem.*, 1996, **271**, 15299.
65 B. H. Oh, C. H. Kang, H. De Bondt, S. H. Kim, K. Nikaido, A. K. Joshi and G. F. Ames, *J. Biol. Chem.*, 1994, **269**, 4135.
66 G. J. Barton, *Protein Eng.*, 1993, **6**, 37.
67 A. C. Wallace, R. A. Laskowski and J. M. Thornton, *Protein Eng.*, 1995, **8**, 127.

Paper 8/06183A

Functional immobilization of biomembrane fragments on planar waveguides for the investigation of side-directed ligand binding by surface-confined fluorescence

Michael Pawlak,*†[a] Ernst Grell,[b] Eginhard Schick,[b] Dario Anselmetti[c] and Markus Ehrat[a]

[a] *Novartis Pharma AG, Preclinical Safety, DMPK, CH-4002 Basel, Switzerland.*
   E-mail: michael.pawlak@pharma.novartis.com
[b] *Max-Planck Institute of Biophysics, D-60596 Frankfurt, Germany*
[c] *Novartis Services AG, Scientific Services, Physics, CH-4002 Basel, Switzerland*

Received 26th August 1998

A method for the functional immobilization of Na,K-ATPase-rich membrane fragments on planar metal oxide waveguides has been developed. A novel optical technique based on the highly sensitive detection of surface-confined fluorescence in the evanescent field of the waveguide allowed us to investigate the interactions of the immobilized protein with cations and ligands. For specific binding studies, a FITC-Na,K-ATPase was used, which had been labelled covalently within the ATP-binding domain of the protein. Fluorophore labels of the surface-bound enzyme can be selectively excited in the evanescent field. A preserved functional activity of the immobilized enzyme was only found when a phospholipid monolayer was preassembled onto the hydrophobic chip surface to form a gentle, biocompatible interface. *In situ* atomic force microscopy (AFM) was used to examine and optimize the conditions for the lipid and membrane fragment assembly and the quality of the formed layers. The enzyme's functional activity was tested by selective K^+ cation binding, interaction with anti-fluorescein antibody 4-4-20, phosphorylation of the protein and binding of inhibitory ligand ouabain. The comparison with corresponding fluorescence intensity changes found in bulk solution provides information about the side-directed surface binding of the Na,K-ATPase membrane fragments. The affinity constants of K^+ ions to the Na,K-ATPase was the same for the immobilized and the non-immobilized enzyme, providing evidence for the highly native environment on the surface. The method for the functional immobilization of membrane fragments on waveguide surfaces will be the basis for future applications in pharmaceutical research where advanced methods for exploring the molecular mechanisms of membrane receptor targets and drug screening are required.

Introduction

Molecular interactions on biomembranes play a prominent role in the communication between cells and in signal transduction pathways.[1] Thereby, membrane receptors serve as the main

† Present address: Zeptogens AG, Benkenstr. 254, CH-4108 Witterswil, Switzerland.

targets,[2] able to recognize specific ligands selectively, which can trigger a cascade of functional cell responses such as the regulation of ion channel activity[3,4] or induction of secondary messengers, *e.g.* release of intracellular calcium upon G-protein coupled receptor activation.[5,6] Because of their regulatory mechanisms and their relevance in drug–target interactions, membrane receptors are, currently, the focus of detailed biophysical and biochemical investigations, to elucidate the relation between ligand binding and functional properties, or to resolve structure–activity relationships. In pharmaceutical research, membrane receptors are used as targets in the drug discovery process, where large numbers of new chemical entities are screened at an early phase.[7] For that purpose, a number of physical sensing techniques, preferentially based on fluorescence detection and designed for high throughput screening, have been developed recently.[8,9] Therefore, and due to the trend of integrating a larger and larger number of assays on a single detection plate,[8] leading to increased surface/volume ratios in an assay well, surface-sensitive fluorescence techniques[10–12] become increasingly important. However, detection of direct ligand–receptor interactions and of related protein functions is often limited, not only by the physical sensing technology, but also by the assay architecture, the non-specific binding, the loss of functionality upon immobilization of cells/receptors on surfaces and strong background signals from the bulk solution. Accordingly, the interfacing of biologically active receptor systems with such sensing surfaces, with preservation of the full functional protein activity and access to an optimum number of binding sites under controlled conditions, becomes more and more important, but still remains a major challenge.

Biosensor technology

Here, we present the advantageous combination of a highly sensitive technology for the detection of surface-confined fluorescence, excited in the evanescent field of planar waveguides,[13,14] and a new method for the functional immobilization of biomembrane fragments, which contain the receptor target, on such sensor surfaces under defined conditions. The high assay performance of evanescent wave sensors is achieved by: (i) the strong spatial discrimination by the evanescent field, between specific binding signals and bulk fluorescence causing background signals (penetration depth of the evanescent field is typically of the order of a wavelength); (ii) highly oriented and functional biomolecules (receptor targets) on the surface and (iii) the use of fluorescent labels for signal generation, independent of molecular weight. Assays demonstrating the real-time monitoring of *e.g.* immunorecognition at working concentrations as low as several fM have recently been published.[13]

Here, we report, for the first time, applications using an integral membrane protein which was functionally immobilized in its native membrane environment on waveguide sensor surfaces. Na,K-ATPase was chosen for a systematic investigation of the immobilization process as well as for probing molecular interactions on an immobilized protein.

Na,K-ATPase

The enzyme Na,K-ATPase is a protein that exists in the plasma membrane of all higher organisms. Its principal function as an alkali-metal ion dependent ATPase was discovered by Skou.[15] The free energy resulting from the hydrolysis of an intracellular ATP molecule is converted by this enzyme into the transport of 3 Na^+ ions out of and 2 K^+ ions into the cell. Both active cation transport processes occur against the existing alkali-metal ion concentration gradients of the plasma membrane. Furthermore, Na,K-ATPase exhibits electrogenic properties and thus contributes significantly to the regulation of the membrane resting potential. The protein itself is a heteromer and consists of the catalytic α-subunit, characterized by a molecular weight of *ca.* 100 kDa, and of the glycosylated β-subunit with a molecular weight of *ca.* 50 kDa. Besides its transporter function, Na,K-ATPase acts as the receptor for cardiac glycosides such as ouabain, which are bound to the extracellular side of the protein at very high affinity and lead to the inhibition of enzymatic activity.

Since the ionic composition of the medium differs between the cytoplasmic and the extracellular side and because the functional properties of the enzyme on both membane sides are different, the structure of the protein must also be asymmetric. Consequently, the binding properties of ligands and cations are expected to be side-directed. For example, the binding affinity of a ligand or of an alkali-metal ion to a site on one side of the membrane will depend on whether this ligand or cation is absent or present on the other side. Thus, if one wishes to understand the basic functional properties of the cation pump mechanism, such as that of Na,K-ATPase, the aspect of the sidedness is essential and calls for advanced investigation methods.

Na,K-ATPase was isolated as the major protein in nanoparticulate membrane fragments (discs of *ca.* 250 nm mean diameter) from specialized tissue such as kidney or salt glands. In contrast to solubilized systems, the protein is in its native membrane environment and retains its original biomembrane orientation (*cf.* Fig. 1). The isolated membrane discs, however, are surrounded by the same aqueous medium on both sides and are no longer capable of separating the two different aqueous cell compartments. Although such a preparation no longer allows the study of the aspect of sidedness, many relevant interactions, partial reactions and mechanistic aspects have been investigated. For example, a fluorescent marker, such as fluorescein-5-isothiocyanate (FITC) can be specifically bound to a single lysine residue, located within the ATP binding domain, for monitoring binding events by fluorescence quenching.[16-18] The fluorescence of the labelled Na,K-ATPase changes characteristically upon binding of different ligands and cations and was also used here for the analytical detection.

General working concept

Our concept is based on the formation of a biocompatible, sensitized waveguide chip surface and subsequent assembly of membrane fragments on this support under defined conditions. The combination with surface-confined fluorescence detection by planar waveguides allows the simultaneous investigation of side-directed ligand binding of and functional properties of, for example, Na,K-ATPase (Fig. 1). The initial goal of this work was the stable and functional immobilization of ATPase-rich membrane fragments using surface-assembly techniques. The on-line monitoring capability of the planar waveguide (PWG) sensor allowed us to carefully examine each step of surface preparation. Additional knowledge on the surface structure was obtained by *in situ* AFM, which delivered important 2D-resolved information about surface coverage and surface morphology of the sensitized chip. After successful immobilization of the membrane fragments, the functional activity of the immobilized protein was probed by monitoring the fluorescence changes upon specific, side-directed binding of cations and ligands using the FITC-labelled ATPase. The described approach offers great advantages because no detergent solubilization of the protein is

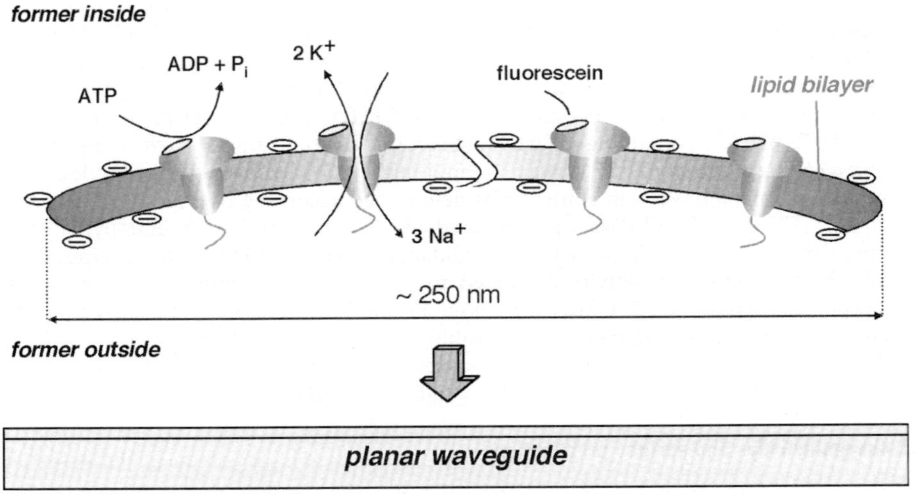

Fig. 1 Schematic illustration of a cross-section through a disc-shaped membrane fragment (diameter *ca.* 250 nm) containing FITC-labelled Na,K-ATPase, prior to adsorption onto a planar waveguide chip. The enzyme molecules consist of an α and β subunit. The whole protein is asymmetric with respect to function and protein moiety. The ATP binding side conserves a larger protein content and is located on the former, native inside of the biomembrane. Protein-to-lipid ratio of a membrane disc is typically *ca.* 1 : 250 (as determined for pig kidney membranes).

necessary, as is the case for other reconstitution procedures, because the assays can be optimized efficiently and because fast and reproducible measurements can be performed, due to the easy exchange of the aqueous media for the different assays. The orientation of the immobilized membrane fragments on the surface was finally concluded from the comparison of the fluorescence changes, measured with the PWG sensor, with those found with the same membrane sample in bulk solution.

Materials and experimental methods

Chemicals

Salts such as NaCl, choline chloride, imidazole and phosphates were purchased from Merck (Darmstadt, Germany) and Fluka (Buchs, Switzerland) and were of the highest purity obtainable (Ultrapure or Microselect). Organic solvents such as propan-2-ol were also from Merck and of UV-spectroscopy grade (Uvasol). 1-Palmitoyl-2-oleoyl-glycero-phosphocholine (POPC) and fluorescein-lipid (PE) were from Avanti Polar Lipids (USA). Mouse monoclonal anti-fluorescein antibody (4-4-20) was from Molecular Probes (Netherlands).

Lipid vesicle preparation

Lipid vesicles were prepared by extrusion (100 nm polycarbonate filters, Avestin Corp., USA) with 0.75 mM 1-palmitoyl-2-oleoyl-sn-glycero-3-phosphocholine (POPC, Avanti Polar Lipids), hydrated in 150 mM NaCl, 10 mM sodium phosphate buffer, 0.02% sodium azide at pH 7.5. The vesicle mean diameter was *ca.* 110 nm, as determined by dynamic light scattering (Zeta Plus, Brookhaven Instruments, USA).

Na,K-ATPase preparation

Na,K-ATPase-containing membrane discs were isolated from the dissected red outer medulla of pig kidney and from the rectal gland of dogfish (*Squalus acanthias*) in a two-step procedure. A microsomal fraction was prepared by tissue homogenization followed by differential-velocity centrifugations.[19] This fraction consisted essentially of closed membrane particles, where the majority of the ATP binding sites were not accessible from the external medium. The microsomal fraction was activated by a short sodium dodecyl sulfate (SDS) treatment which led to an opening and partial solubilization of the microsomal membranes. After this treatment, the Na,K-ATPase discs remained intact but the enzymatic activity of the preparations increased markedly. The discs were separated by density-gradient centrifugation using sucrose.[19] The pig kidney enzyme was kept in 25 mM imidazole in HCl, pH 7.5, containing 0.1 mM EDTA, and the dogfish enzyme in 30 mM histidine in HCl, pH 6.8, containing 25 wt.% glycerol at concentrations of *ca.* 2 mg ml^{-1}. The average Na,K-ATPase activity at 37 °C was 30 µmol mg^{-1} min^{-1}; it was reduced to less than 1% in the presence of ouabain. Protein content was determined according to a modified procedure of Lowry *et al.*[20] FITC-Na,K-ATPase was prepared as described in ref. 16, employing extended washing procedures[21] and was kept in 10 mM imidazole in HCl at pH 7.5 on ice. Upon labelling the sample the Na,K-ATPase activity decreased to <1 µmol mg^{-1} min^{-1} but the 2,4-dinitrophenylphosphatase activity at 37 °C was still about 19 µmol mg^{-1} min^{-1}. The degree of labelling was calculated from the absorbance in the visible range, assuming that the molar absorption coefficient of FITC remained unchanged at the same pH. The concentrations of the enzyme given here were calculated on the basis of a molecular weight of 150 kDa.

Planar waveguide chips and surface preparation

Planar waveguide chips (16 mm × 48 mm) consisted of a 150 nm thin, waveguiding surface layer of Ta_2O_5 (refractive index = 2.3), deposited on 0.5 mm thick glass substrates (Balzers AG, Liechtenstein). The self-assembling chemistry, chemical metal oxide modification and physicochemical analysis of the modified sensor surfaces is reported elsewhere.[22] Briefly, the metal oxide surfaces were routinely cleaned by ultrasonication in organic solvent, followed by a UV ozone treatment. The cleaned, hydrophilic surfaces were further modified to reach a stable hydrophobic character by self-assembling monolayers of mono-C_{16}-alkyl phosphates (Novartis Pharma AG) onto the freshly cleaned surfaces (3 day assembly at 1 mg ml^{-1} in propan-2-ol). Chips were thoroughly washed in propan-2-ol, sonicated, dried and stored under nitrogen. Contact angles of water on the hydrophobic surfaces were determined to be in the range of 100–110°.

AFM

Each step of our surface preparation was carefully examined by *in situ* AFM. For AFM experiments under defined buffer environment we used a commercial instrument (Nanoscope IIIa, Bioscope, Digital Instruments, USA) with Si_3N_4-cantilevers (Digital Instruments or Olympus) with nominal spring constants of 0.02–0.58 N m^{-1}. The AFM was operated in the tapping mode in order to reduce unwanted tip–sample interactions due to the imaging process. Scanning was typically done at an imaging line frequency of 1 Hz. The surfaces carrying the lipid layers and biomembrane fragments were always kept under buffer during mounting onto the AFM sample stage and were prepared under conditions comparable to those used in the sensor experiments. For the chip preparation, a cell made from silicon allowed careful rinsing of excess lipid and/or membrane material as well as a reproducible exchange of the buffer solution.

Fluorescence measurements with planar waveguide sensor

An in-house developed waveguide sensor instrument was used for the detection of surface-confined chip fluorescence measurements.[13] The schematic optical set-up is shown in Fig. 2. S-polarized (=transversal electric, TE-mode) excitation light at 488 nm from an Ar$^+$-ion laser (model 5490ASL, 0–30 mW, Ion Laser Technology, USA) was coupled into the waveguiding layer by means of a diffractive grating at a distinct angle of incidence. Since excitation light is efficiently guided in the waveguiding layer, excitation light intensities could be kept low, in the range of 10–100 µW. Part of the emitted, surface-confined fluorescence, excited by the propagating wave, was collected at the substrate side with a 1:1 imaging optics (numerical aperture = 0.19) and spectrally discriminated by means of two six-cavity interference filters transmitting in the range

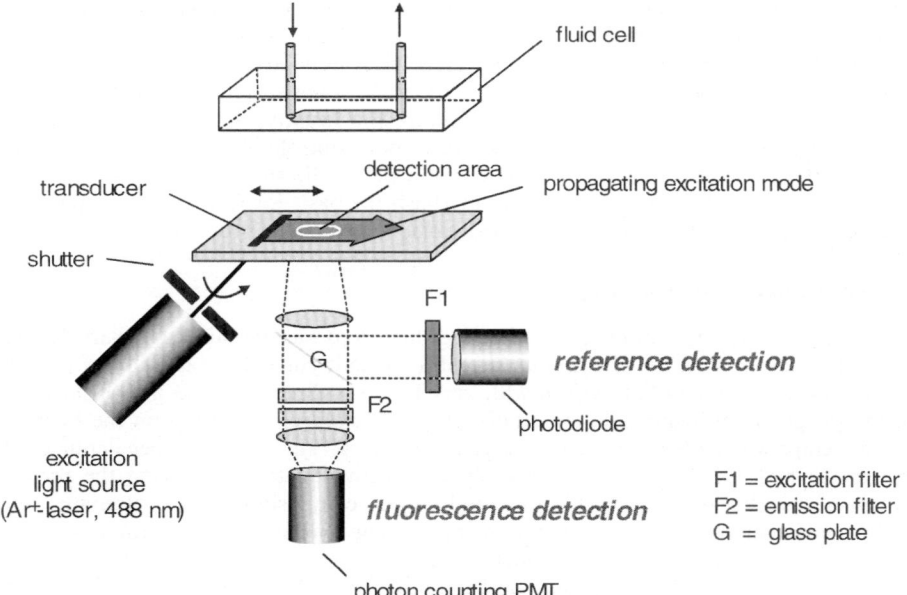

Fig. 2 Evanescent wave fluorescence measurement: surface-immobilized fluorophores are excited by laser light in the evanescent field of a planar Ta_2O_5 waveguide (propagating excitation mode). Fluorescence is detected to high sensitivity by a photon-counting photomultiplier (PMT) system. Fluorescence signals were referenced by measuring the excitation light within the detection area by means of a photo diode. Buffer and sample solutions were injected into the fluid cell with an automated fluid handling system (not shown).

515–545 nm (530DF30, Omega Inc., USA). Part of the surface-scattered light was used as a reference for the intensity of incoupled excitation light in the imaged area of interest. The field of detection was typically 5 mm × 1 mm, placed in the centre of the propagating beam by means of an aperture. A photon-counting PMT (H6240, Hamamatsu, Japan) in combination with a 225 MHz counter (model 53131A, Hewlett-Packard, USA) was used for the detection of fluorescence. Excitation and signal read-out was taken every 30 s with a sampling time of 1 s. Buffers and samples were applied to the sensor chip *via* a flow-through cell made from silicone elastomer (Sylgard 184, Dow Corning, USA). The cell volume was 6 mm^3 (15 mm × 2 mm × 0.2 mm). Fluid supply was performed automatically using a dispensing pump–six-way valve combination (Cavro, USA) for handling different working buffers and a sample injector (231XL, Gilson, USA). Injection volumes were in the range 50–100 mm^3 for stopped flow and up to 5 cm^3 for continuous flow applications. The fluorescence detection and sample handling was fully automated and controlled by a PC. The chip holder and optical unit were temperature controlled. All experiments were performed at 20 °C. The experimental buffer conditions were: buffer I (150 mM NaCl, 10 mM sodium phosphate buffer, 0.02% NaN$_3$, pH 7.5), buffer II (100 mM cholineCl, 10 mM imidazole, pH 7.5) and buffer III (buffer II + 3 mM MgCl$_2$). Errors given for results are determined from at least three repeated measurements.

Fluorescence measurements in bulk cuvette

Bulk fluorescence measurements were performed on a Spex Fluorolog 222 instrument, equipped with a thermostatted cuvette holder. Measurements were performed under conditions comparable to those in the waveguide sensor using identical protein preparations. The temperature was 20 °C in all cases.

Results

The main goal was to establish an experimental procedure for the functional immobilization of protein-containing membrane fragments in a defined and controlled manner. The concept of surface immobilization was based on a two-step preparation protocol shown schematically in Fig. 3: the first step (left) is the transfer of the bare, non-sensitized physical, transducer into a biocompatible interface, applying a self-assembly of a lipid monolayer to the hydrophobic metal oxide surface. The second step (right) comprises the stable physisorption of the membrane fragments onto the lipid monolayer interface, exploiting the negative surface charges and strong dipolar contributions[21] of the biomembrane particles in a controllable physico-chemical environment. It is important to control the salt content and pH of the surrounding media, as tools for optimizing this process. The intention was to establish the basis for future assay applications by the development of a well defined and optimized surface-preparation protocol.

Lipid monolayer formation on waveguides

The process of a defined lipid monolayer formation (first step) was optimized by characterizing the surface topography with AFM. The non-sensitized waveguide chips (bare as well as hydrophobic) showed a homogeneous and flat surface with very little roughness (0.2–0.3 nm rms), thus being well suited for potential high-resolution studies. Lipid monolayer formation on the hydrophobic metal oxide chips was performed *via* lipid vesicle spreading. The spreading behaviour of POPC vesicles on the hydrophobic metal oxide surfaces was investigated by varying the preparation techniques of vesicles (extrusion, dilution) and buffer composition (ionic strength, different inorganic/organic salts). The whole process of optimizing lipid monolayer formation on planar waveguides is reported elsewhere.[23] Under optimum spreading conditions, almost complete surface coverage with lipid monolayers could be realized (Fig. 4) using vesicles prepared by extrusion (diameter *ca.* 110 nm, as determined) at relatively high inorganic salt conditions (150 mM NaCl, 10 mM sodium phosphate buffer, 0.02% NaN$_3$, pH 7.5). Lipid (POPC) concentration was always 0.5 mg ml^{-1}. Up to 10% of the surface area was occupied by small monolayer defects, homogeneously distributed over the chip as single spots of 10–20 nm in diameter (Fig. 4). From these defects, the mean thickness of the lipid monolayer could be easily determined (1.8 ± 0.6 nm),

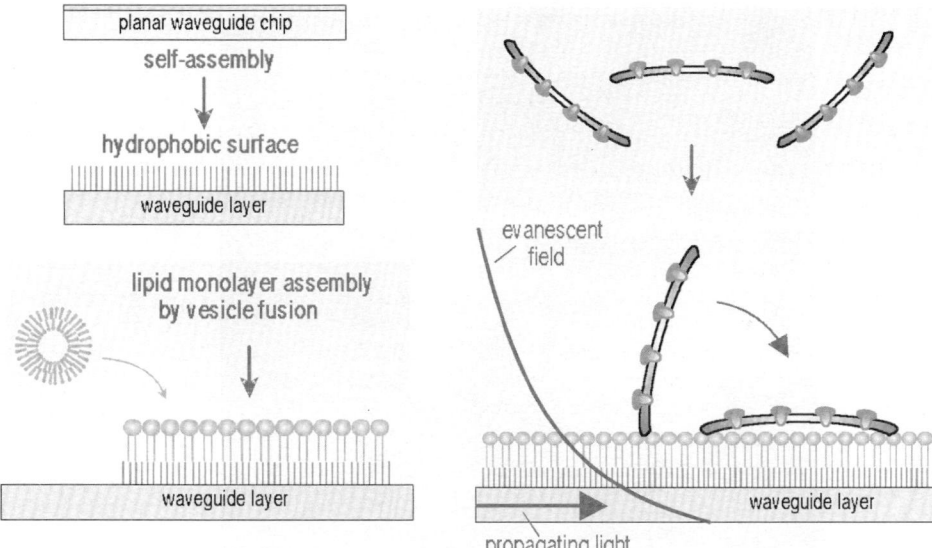

Fig. 3 Working concept for the functional immobilization of membrane fragments: first step (left), formation of a biocompatible interface by self-assembly of a lipid monolayer, after vesicle fusion, onto a hydrophobic waveguide surface; second step (right), controlled and stable physisorption of membrane fragments onto the preformed lipid interface on the waveguide. The penetration depth of the evanescent field of propagating excitation light is of comparable length to the size of the membrane discs. Upon surface-immobilization, the membrane discs can adopt two different orientations: former (cytoplasmic) inside surface up (as illustrated here) or former outside surface up. Drawn objects do not scale to real size.

corresponding well with the thickness of a single monolayer, as determined by other methods.[24] Since the diameter of the membrane discs for the second association process was found to be *ca.* 250 nm, 10–20 times larger than the size of the single defects (hydrophobic spots), the probability for contacts between biomembranes and such defects (leading to protein denaturation) in an otherwise closed lipid layer was considered to be negligible.

Membrane fragment immobilization

Having achieved a lipid-coated, biocompatible chip surface, surface-association of membrane discs, containing Na,K-ATPase, was carried out. Surface coverages and the morphology of the immobilized membrane fragments were investigated again by *in situ* tapping mode AFM. The kinetics of surface-association was monitored in real time with the fluorescence sensor using the FITC-labelled enzyme analogue. Typically, lipid monolayer formation and subsequent membrane fragment immobilization on a waveguide chip were performed in the PWG instrument and monitored in real time. The kinetics of these two processes is shown in Fig. 5 as fluorescence responses. After an initial phase of buffer equilibration (5 min at 0.5 ml min^{-1} flow), lipid vesicles (first step) or membrane discs (second step) were injected and incubated under stopped-flow conditions for 1 h. In a subsequent buffer wash (10 min at 0.5 ml min^{-1}, *i.e.* almost a thousand times exchange of the cell volume) excess material was efficiently removed.

For the lipid assembly, no fluorescence increase was observed, as expected. Interestingly however, the process of lipid spreading could, nevertheless, be followed by a clearly detectable increase of the reference signal (excitation light). In the case of membrane disc association, a strong and fast increase of fluorescence intensity was observed due to the association of the labelled membrane discs onto the surface. After 1 h, the signal slowly approached a stationary emission at *ca.* 10^6 cps. Only a minor portion of this signal disappeared on washing. This signal decrease was interpreted as the removal of loosely bound membranes in the vicinity of the surface. Final signal-to-noise ratios of the net fluorescence were 4000 ± 500. The established fluorescence signal, F_0, in buffer II was checked for its long-term stability, a prerequisite for subsequent assay

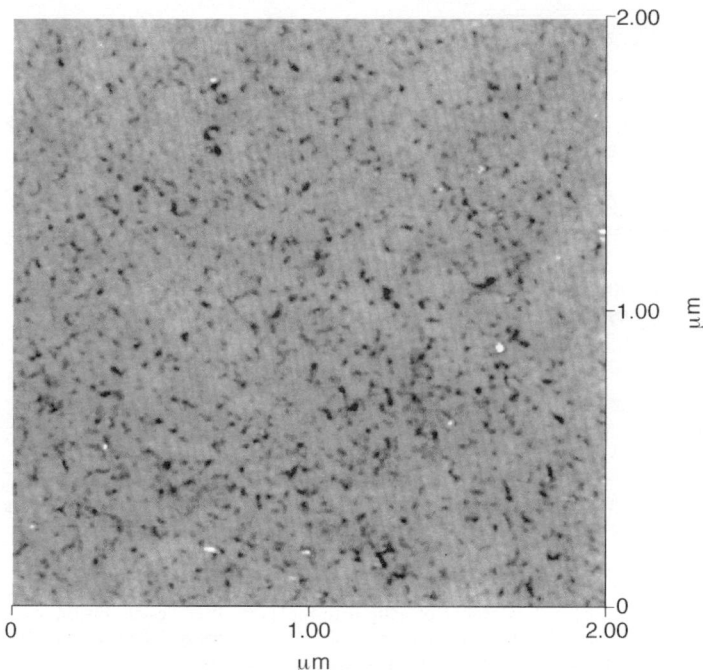

Fig. 4 Quality control of surface lipid assembly by *in situ* AFM: a lipid monolayer formed after spreading of POPC vesicles, prepared by extrusion (vesicle diameter 110 nm), to a hydrophobic Ta_2O_5 waveguide. The hydrophobic surface consisted of a self-assembled monolayer of mono-C_{16}-alkyl phosphate. Up to 10% of the total surface is covered by small-sized single defects. Preparation and imaging of the lipid monolayer were performed in 150 mM NaCl, 10 mM sodium phosphate buffer, 0.02% NaN_3 at pH 7.5. The gray scale of the image (black to white) corresponds to a height of 10 nm.

applications. Signals were stable as measured for up to 3 h in running buffer with signal readouts for 1 s every 1 min. Long-term drifts due to photo-bleaching were minimized using low illumination intensities (typically 40–100 μW before chip coupling). The concentration of membrane discs corresponded to a protein content of *ca.* 0.25 mg ml^{-1}. Assembly experiments have been carried out successfully with Na,K-ATPase prepared from both pig kidney and from dogfish rectal gland. However, in the following, the reported results refer only to the dogfish enzyme.

Membrane–surface association depends critically on the salt conditions of the surrounding buffer. In order to avoid an early loading of the cation binding sites of the enzyme, the use of Na^+ and K^+ cations in the assembly buffer was excluded. Therefore, surface immobilization was performed in 100 mM ChoCl, 10 mM imidazole, pH 7.5 (buffer II). High ionic strength strongly assisted the association of membrane particles to the surface and, in addition, provided a stable contact of the discs on the surface. Fig. 5 shows a representative experiment of lipid monolayer formation and membrane disc immobilization on the preformed lipid layer. Surface coverage and topology of such preparations were investigated under otherwise comparable conditions with tapping mode AFM in solution. Fig. 6 shows a representative AFM picture of membrane fragments immobilized flat on a lipid monolayer. The average coverages of such preparations ranged between 25% and 50%, depending on the scanning position on the chip and including chip to chip variation. In addition, coverages determined from fluorescence signals, with assembled membrane discs or with a phospholipid monolayer doped with fluorescein-labelled lipid (one label per 800 lipids), were in good agreement with the AFM results. This indicated the robustness of our preparation procedure.

Concerning the topology and stability of surface-associated membrane particles, membrane discs preferentially associated flat on the surface. The stability of the physisorption of the membranes was concluded from the fact that the disc-like structures could not be removed with the AFM tip from their positions during the process of scanning. At comparatively high forces the

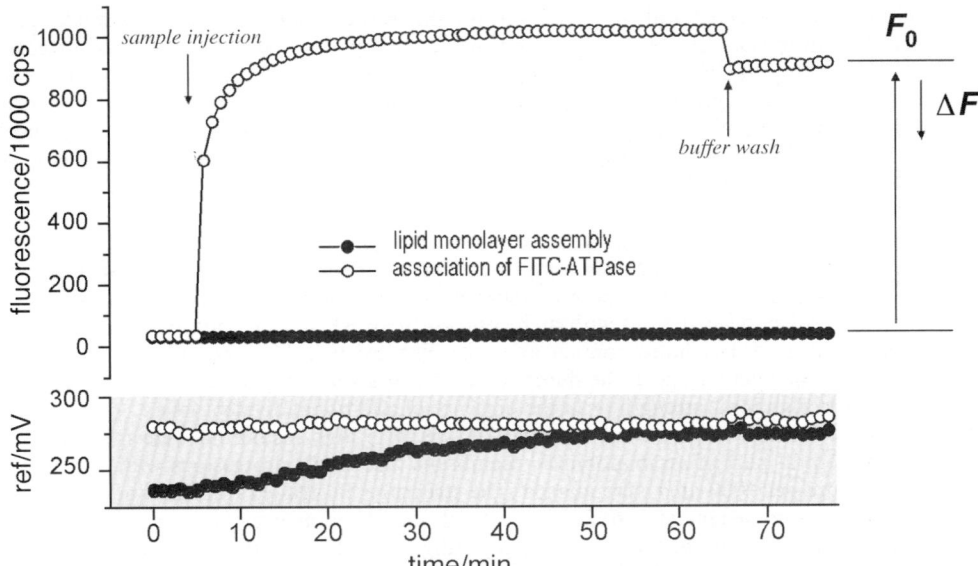

Fig. 5 Kinetics of lipid monolayer assembly (●) and subsequent surface-association of dispersed membrane fragments (○), containing FITC-Na,K-ATPase (protein concentration 0.22 mg ml^{-1}; prepared from dogfish rectal gland), on a waveguide chip. Measurements were performed in 10 mM imidazole/HCl, 100 mM choline chloride, pH 7.5 (at 20 °C). Fluorescence (in cps) and reference signals (in mV) were monitored after an initial 5 min of buffer exchange (under continuous flow), a 60 min sample incubation (under stopped flow) and final 10 min of buffer washing (under continuous flow). F_0 corresponds to the asymptotically reached emission intensity of surface-immobilized FITC-Na,K-ATPase, prior to the supply of samples containing interacting cations or ligands. Binding of the specific cations or ligands leads to the characteristic fluorescence changes ΔF.

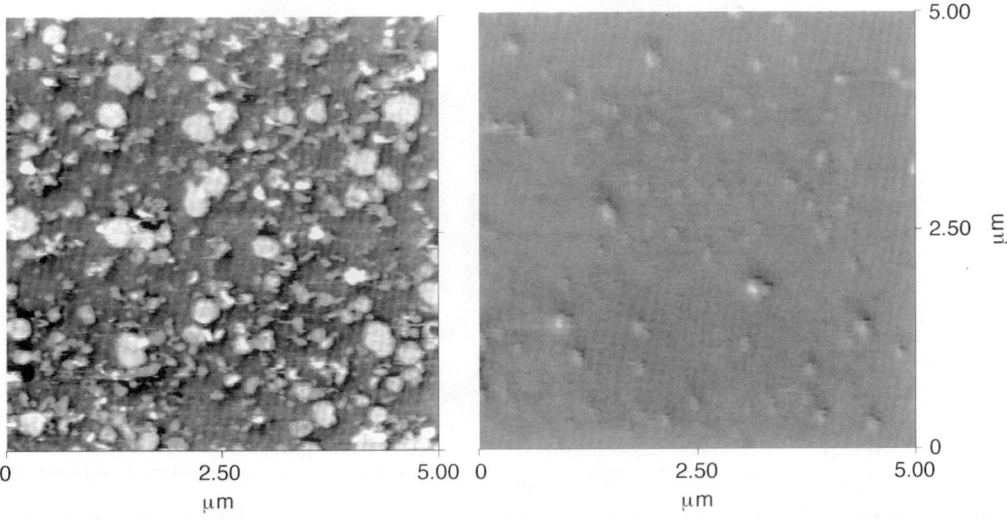

Fig. 6 AFM quality control of immobilized membrane fragments, containing FITC-labelled Na,K-ATPase, on sensitized waveguide surfaces. At high ionic strength (left) up to 50% surface coverage of assembled membrane discs was achieved; at low ionic strength (right) only few and loosely defined structures with lower contrast could be imaged. FITC-Na,K-ATPase was prepared from the dogfish rectal gland and used in concentrations of ca. 0.22 mg ml^{-1}. The gray scale of the image (black to white) corresponds to a height of 50 nm.

discs could even be dissected with the tip (data not shown). Further evidence for a stable membrane immobilization was the observation of a constant sensor fluorescence signal upon extensive washing with running buffer. The AFM experiments also indicated that the surface contact zone of a membrane disc was preferentially located close to the disc edge. The edges protruded an additional 4–5 nm into the solution. The central region of a membrane disc is assumed to be fairly flexible. This was concluded from the fact that the AFM resolution in the disc centre was lower than at the edge and that the height of the disc centre with respect to the substrate surface depended on the ionic composition of the buffer solution, for example by the addition of divalent cations. In the presence of 3 mM $MgCl_2$, the disc centre height was reduced, typically, from 9 to 7 nm (−25%). When membrane particles were assembled at low ionic strength (10 mM imidazole, pH 7.5), a much lower surface coverage was obtained (Fig. 6, right). Under these circumstances, it was difficult to resolve the structures with high contrast in an AFM scan, indicating that a weaker interaction, in terms of a reduced contact area per disc, existed. This was also evident by the observation of larger fluctuations of the fluorescence sensor signal and, additionally, of lower base signals F_0 than at high ionic strength or in the presence of additional divalent ions.

On individual membrane preparations, AFM images with high resolution were recorded (Fig. 7). Within the central part of an immobilized membrane disc single protrusions with a density of about one stucture per 1000 nm^2 were imaged. The structures protruded from the surface into the solution by an average height of 5–6 nm. The lateral size of the protrusions was in the range of 15–20 nm. Taking into account the size of the scanning tip (5–10 nm), the real size of the imaged structures must be *ca.* 10 nm. Such a size was interpreted as the membrane-external protein moiety of the ATPase. However, the lateral size of 10 nm is too large to represent an individual protein molecule. Therefore, we believe that the imaged structures correspond to aggregates of protein, equal or larger than a dimer. Considering aggregation, the final protein-to-lipid (P : L) ratio in the disc centre was determined to be in the range 1 : 100 to 1 : 1000. At the disc edge, P : L

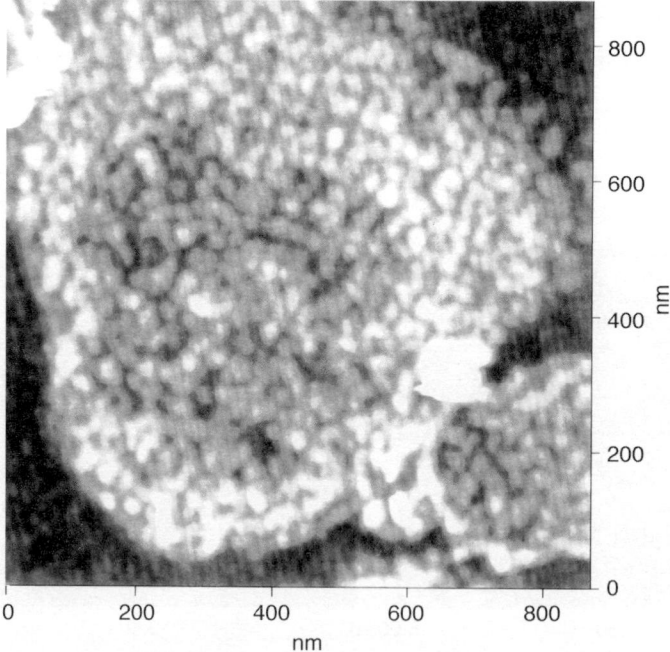

Fig. 7 High-resolution AFM picture of a membrane fragment, containing FITC-Na,K-ATPase (prepared from dogfish), immobilized on a sensitized waveguide chip. The central part of the disc shows individual protrusions with average lateral dimensions of 15–20 nm and an average height of 5–6 nm. Protrusions are interpreted as aggregated ATPase molecules (aggregation state > dimer). Protein : lipid ratios in the central part of the membrane disc were of the order of 1 : 100–1 : 1000, increasing towards and at the disc edge. The gray scale of the image (black to white) corresponds to a height of 20 nm.

ratios were still larger, leading to the assumption that, there, stronger aggregation or even first indications for protein crystallization occur.

Functional tests of immobilized Na,K-ATPase

Once the optimum conditions for a reproducible and stable immobilization of membrane fragments were identified, *i.e.* a stable and constant base fluorescence signal F_0 was generated on the chip, the functional activity of the Na,K-ATPase could be tested. This was done by using a set of complementary assays to probe the different binding sites of the protein. Each of these assays alters specifically, in a side-directed manner, the relative intrinsic fluorescence F/F_0 of the labelled enzyme. For a better overview, the characteristic fluorescence changes of the membrane discs in bulk solution upon variation of the buffer composition or additions of specific ligands are shown in Fig. 8. The specific binding of K^+ to the alkali-metal ion binding pocket was considered to be the most critical test for a preserved functional activity of the protein, since this binding is impaired with a characteristic change in protein conformation leading to a relatively large change of the fluorescein emission ($-\Delta F/F_0 \approx 30\%$).

Specific K^+-binding and binding isotherm. In all preparations, the specific K^+-effect was tested as a reference experiment. Small volumes of KCl in buffer II (maintaining the ionic strength constant) were applied under continuous flow. Upon K^+-contact, the fluorescence dropped instantly to lower, stable values, dependent on the KCl concentration (*cf.* Fig. 9). This fluorescence effect was fully reversible, yielding reproducible fluorescence levels after washing in the absence of KCl and in subsequent repeats. Titrations over six orders of concentration (0.1 µM to 100 mM KCl) were performed. From the specific fluorescence decreases ($-\Delta F/F_0$), a quantitative evaluation according to a binding isotherm was performed (Fig. 10). The data followed ideally according to a Langmuir binding kinetics, revealing a dissociation constant of $K_D = 180 \pm 20$ µM. Saturation of binding was achieved above 2 mM K^+, but only a $60 \pm 10\%$ fraction of the amplitude was reached when compared to the respective fluorescence changes in bulk solution (Fig. 8). On the other hand, the dissociation constant of the surface-confined measurements compared well to that found in bulk measurements. This indicates a preserved enzyme activity in the assembled state, but with a reduction of the number of K^+-binding sites per assembled protein molecule. This may imply that not all available sites are accessible by the buffer medium.

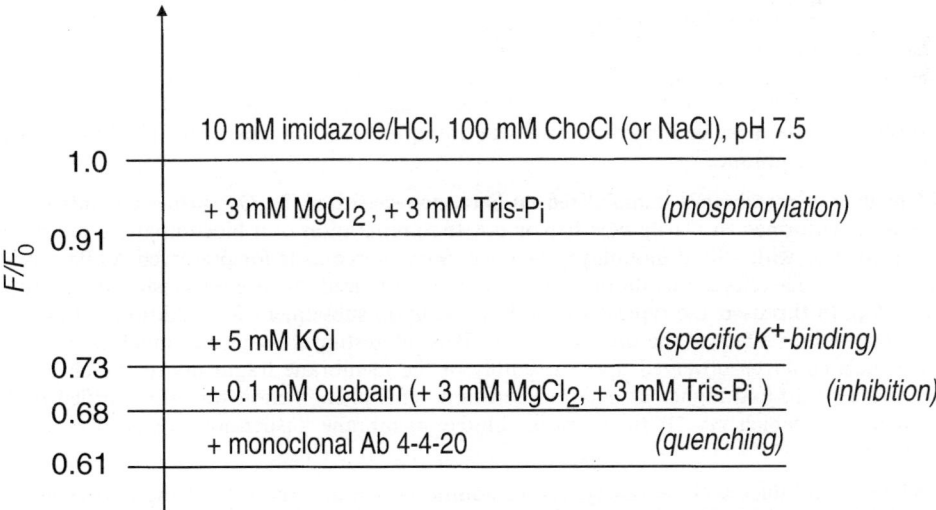

Fig. 8 Schematic illustration of the relative fluorescence intensity levels F/F_0 ($\lambda_{exc} = 488$ nm; $\lambda_{em} = 520$ nm) of FITC-Na,K-ATPase (prepared from dog fish rectal gland), as measured in bulk phase containing different cations or ligands (P_i = inorganic phosphate), relative to the base level F_0 in 10 mM imidazole in HCl, 100 mM choline chloride (ChoCl), pH 7.5 (buffer II) at 20 °C. The total ionic strength of all solutions was kept constant by adjustment with choline chloride (ChoCl).

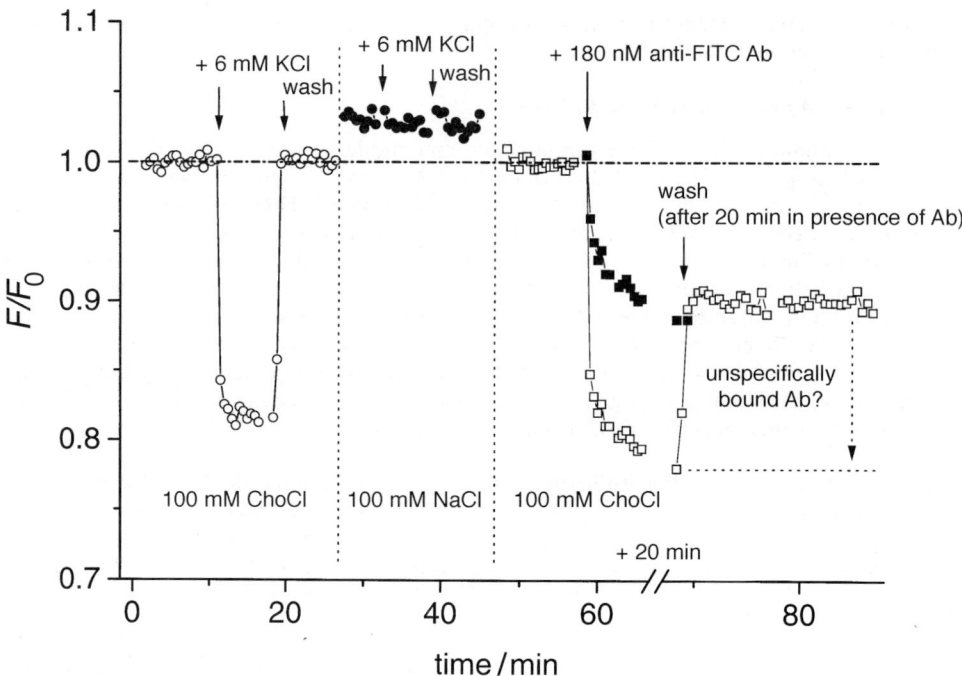

Fig. 9 Representative course of fluorescence signals in a typical assay sequence, measured after preparation of surface-immobilized, FITC-labelled Na,K-ATPase (dogfish) as shown in Fig. 5. (○) specific K^+-binding after injection of 6 mM KCl, (●) competitive inhibition of specific K^+-binding in the presence of high [Na^+], and signal decrease upon addition of 180 nM anti-fluorescein antibody 4-4-20 [(□) measured signal; (■) signal corrected for unspecific offset].

Specificity of K^+-binding. Further experiments were performed to check the specificity of K^+-binding. When the immobilized membranes were saturated with high concentrations of NaCl (100 mM), no further signal decrease was observed upon addition of KCl (cf. Fig. 9). Thus, specific K^+-binding is not observed in the presence of excess Na^+, which is consistent with bulk measurements. Furthermore, with excess KCl, >10 mM, an additional K^+-effect could be resolved, as visible from the observed deviations from constant saturation binding (cf. Fig. 10). This effect obviously suggests a weaker binding and was interpreted as an unspecific K^+-binding to the protein or the membrane.[18]

K^+-binding of membranes immobilized on bare, non-sensitized Ta_2O_5 surfaces. Control experiments were performed to clarify whether or not the preparation of a biocompatible interface, i.e. the chip covered with a lipid monolayer, is a necessary prerequisite for preserved ATPase binding characteristics. Therefore, membrane discs were immobilized on a non-sensitized, hydrophilic Ta_2O_5 chip. In this case, the typical signal decrease upon subsequent K^+-addition did not occur, except for a minor effect of the order of −2%. It is interesting to note that much larger fluorescence signals could be observed upon assembly of the membrane fragments to the chip surface, indicating a ca. 10-fold higher association constant. This phenomenon was also verified in AFM experiments, in which ca. 10 times more diluted membrane suspensions reached comparable surface coverages (25%–50%) as in the case of the lipid-coated chips.

Binding of anti-fluorescein antibody. As an additional feature, the side of the fluorescein label, placed within the ATP-binding domain of the enzyme, was probed by fluorescence quenching upon specific binding of monoclonal anti-fluorescein antibody 4-4-20 (Molecular Probes). This was done in a typical assay sequence (in working buffer II) after the initial testing for the specific K^+-binding as shown in Fig. 9. The antibody is able to quench the fluorescence of free fluorescein almost completely (>90%).[25] In the case of the FITC-labelled ATPase-analogue, fluorescence in

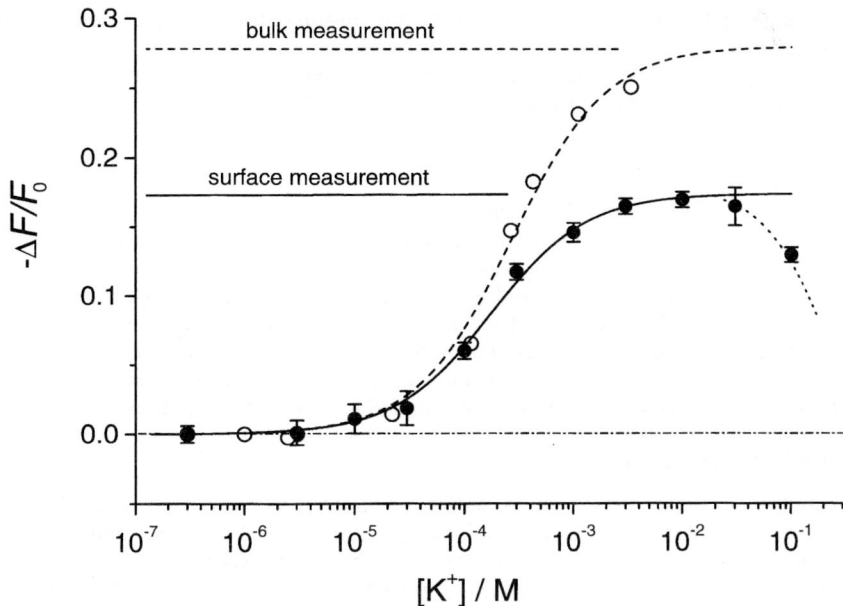

Fig. 10 Specific K$^+$-binding of FITC-Na,K-ATPase (prepared from dogfish) upon additions of KCl over five orders of concentration, maintaining the ionic strength constant. Experiments were performed with surface-immobilized enzyme (●) and with non-immobilized enzyme in bulk solution (○) in 10 mM imidazole in HCl, 100 mM choline chloride, pH 7.5 at 20 °C. The surface-immobilized enzyme reached only *ca.* 70% of the relative saturation amplitude of enzyme in solution. The concentration dependence in both cases was well described according to a Langmuir binding kinetics, leading to dissociation constants of $K_D = 180 \pm 20$ µM for the immobilized enzyme (———) and $K_D = 270 \pm 30$ µM for the enzyme in bulk (-----). The comparable K_D values indicate a preserved activity of the ATPase in the surface-immobilized state. A 70% relative saturation amplitude under a K_D comparable to solution suggests a 70% : 30% orientation of membrane fragments on the surface, with the specific K$^+$-binding site exposed to the solution.

bulk is only quenchable to a maximum extent of 30–40%, depending on the method of preparation. We found that the fluorescence in the case of the surface-immobilized protein is only one half of the quenching effect in solution (20% quench) upon incubation in saturating concentrations of the antibody (180 nM). An initial fast and a subsequent slower kinetic quench response could be resolved. The initial fast one was reversible upon subsequent washing with buffer, whereas the quench representative of the slower phase (within minutes) remained stable for minutes at $F/F_0 \approx 0.9$ (10% quench). The later effect was interpreted as the specific (direct) antibody binding to the fluorescein label (slow dissociation), whereas the fast response may represent an unspecific interaction to the fluorescein emission.

Phosphorylation and binding of inhibitory ligand (ouabain). To investigate the specific binding of an inhibitory ligand (ouabain) and phosphorylation, the enzyme activity was studied by another set of assays. Phosphorylation probes especially the ATP-binding site of the protein. The immobilized protein was first equilibrated in the presence of 3 mM MgCl$_2$ (buffer III). The base fluorescence signals F_0 of the sensor were higher in this buffer by at least 5–6%. This effect was explained by the induction of a closer contact of the assembled membrane discs to the chip surface in the presence of divalent cations (see AFM results). Phosphorylation after incubation of 3 mM inorganic phosphate led to a minor decrease in F_0 in the range 2–6%. Specific binding of inhibitory ligand was observed by measuring the decrease in the fluorescence signal upon continuous addition of 2.5 mM ouabain in the presence of 3 mM MgCl$_2$ and 3 mM inorganic phosphate. The kinetics of ligand binding was as slow as in bulk membrane dispersions (*ca.* 20 min to reach a stationary signal). The final signal decrease stabilized at *ca.* 40% of the effect measured in bulk solution.

Together with the K$^+$-binding studies (60% of the solution effect) and the antibody-induced quenching studies (40% of the solution effect), the results obtained with specific binding of ouabain (40% of the solution effect) are consistent with the assumption that the two orientations of membrane fragments upon immobilization on the waveguide (former inside up and former inside down) establish almost equally on the surface (cf. Fig. 3).

Discussion

We present a new method, showing that transmembrane proteins, embedded in natural biomembrane fragments, can be immobilized with well preserved activity, in a rationally defined and controlled manner, on a planar, metal oxide waveguide transducer. The optimization of the individual preparation steps for a defined and functional immobilization of membrane fragments was the main issue of this work. First examples are presented as to how the functional activity and the specific binding sites of immobilized Na,K-ATPase can be probed by surface-confined fluorescence. A new optical technique, based on the sensitive detection of fluorescence, excited in the evanescent field of thin planar waveguides, was applied for these investigations.

Evanescent field fluorescence detection

The detection of evanescent-field excited, surface-confined fluorescence has the inherent advantages of high sensitivity at large signal-to-noise (S : N) ratios with an almost complete suppression of background signals from the bulk environment. In the present case of immobilized fluorescein-labelled biomembrane particles, S : N ratios of up to 5000, at only partial surface coverages, were obtained. This implies that, under optimal sensing conditions, functional probing of membrane proteins in very small detection fields, approaching sizes similar to the scanning areas of the presented AFM pictures, may become possible. On the other hand, even in the case of macroscopic detection fields ($\geqslant 1$ mm^2), consumption of only low volumes of receptor samples or the use of very dilute samples, e.g. at low receptor yields, are required. Compared to bulk measurements, biological receptors immobilized on surfaces offer the great advantage that different buffer media and/or specific ligands or inhibitors can be sequentially applied to the same preparation and, subsequently, exchanged in a fast and easy manner. Thereby automation enables a high degree of reproducibility. A fast assay development and optimization,[26] and finally a high assay performance are the consequences, as demonstrated in the present case.

Functional surface immobilization

The surface immobilization of membrane proteins, under preservation of full protein activity and with presentation of a maximum number of accessible binding sites, is the key challenge for the investigation of protein functions with surface-sensitive detection schemes. We have pursued the concept of immobilizing intact membrane fragments where the proteins are embedded in their natural environment. There, a high likelihood consists that the protein function is maintained during the whole process of surface preparation. This is different to other methods, which employ the almost random immobilization of membrane receptors in purified and/or detergent-solubilized form on the surface[27] or, in a more defined way, the immobilization of solubilized receptors on functionalized surfaces via chemical affinity tags (His-Tag approach), positioned at distinct protein locations.[10] With the latter method, the ligand-binding activity was established as for the native receptor, but it has to be critically monitored for each preparation step, for changes in the protein environment (e.g. change of detergent) and for each reaction partner. The presented method is the preferred one to immobilize membrane receptors in their native environment under retention of the natural biomembrane orientation. However, in contrast to individual receptor molecules, the demands for the surface immobilization of nanoparticulate membranes in combination with surface-sensitive detection schemes are much higher. Especially, the stability of surface contacts seems to be very important, since the dimensions of the membrane fragments were of the same size as the penetration depth of the evanescent field (few hundred nm). Instabilities in the orientation of surface-associated membranes with respect to the evanescent field, especially when exposed to flowing buffer solution, would result in large fluctuations of the sensor signal (as observed in our first experiments or when applying too low ionic strength). We have achieved a stable and functional membrane fragment immobilization in a flat configuration.

Necessity for a biocompatible interface

The experiments clearly showed that the presence of a gentle, biocompatible interface is needed for a functional immobilization of Na,K-ATPase. The enzyme lost its specific cation binding activity completely when immobilized directly on the bare chip surfaces, although the protein was still embedded in its native membrane environment. The biocompatible interface was formed by covering the waveguide chip with a self-assembled phospholipid monolayer. Such a biomimetic layer protects the sensitive membrane proteins from potential denaturing contact with the bare, non-sensitized transducer surface. In addition, the chemical composition and the physico-chemical properties of such adlayers can be well controlled, since many natural and synthetic lipids and cofactors for their formation are available. The physico-chemical properties of the lipid surface and the diffusion properties of the monolayer can easily be modified and anchor groups for affinity binding can be introduced.[28] In addition, lipid layers are able to suppress the interaction of soluble substances with the surface thus reducing the non-specific binding.[24]

Stability of membrane fragment immobilization in the evanescent field

The conditions for a stable immobilization of the membrane fragments on the preformed biocompatible interface have been optimized by varying the physico-chemical properties of the surrounding buffer media. Buffer solutions at high ionic strength, as used in the course of this study, are known to overcome the long-range, repulsive electrostatic interactions (membranes as well as oxide surfaces are preferentially negatively charged) and to favour the strongly attractive van der Waals forces.[29] In the presence of the lipid monolayer, the membrane affinity to the coated surfaces was much lower compared to the non-sensitized metal oxide surfaces. However, a fully preserved protein activity established in the case of the preformed biocompatible surfaces was concluded from the comparable assay characteristics and binding constants of surface-assembled enzyme and 'free', non-immobilized enzyme in solution (see specific K^+-binding).

Aspect of sidedness

Besides the aspect of establishing a general procedure for the functional immobilization of membrane fragments, the other important interest in the investigation of a surface-immobilized Na,K-ATPase was the aspect of sidedness of the specific binding sites. To address this aspect, there are literature reports of electrophysiological measurements carried out with Na,K-ATPase, either with intact cells such as ventricular myocytes[30] or oocytes,[31] where only the extracellular medium can be changed extensively or the enzyme can be reconstituted in vesicular systems.[32–34] Such reconstitutions imply that the membrane-bound protein has to be solubilized with a suitable detergent with retention of its full enzymatic activity. These are obviously conditions in which the original orientation of the protein molecules in the disc membrane has been lost. The solubilized enzyme, which forms a mixed protein-detergent micelle, is then reincorporated in the lipid membrane of the vesicular systems. Besides the fact that it is difficult to find suitable detergents which do not alter irreversibly the structure and functional properties of the protein, the reconstituted enzyme can adopt two different orientations (inside-out and inside-in) in the vesicular lipid membrane. In addition, it can be adsorbed onto the lipid surface without being incorporated. In order to study side-directed properties of such systems it is necessary to introduce discriminations biochemically, for example by inactivating one state of orientation.

Two recently developed techniques provide information about directed properties of electrogenic as well as of electroneutral steps that are functionally coupled. These techniques are related to electrical measurements of Na,K-ATPase discs adsorbed onto or incorporated into black lipid membranes[35,36] and also employ the patch method on excised membrane fragments originating from cellular systems. Because the transport currents of transporter proteins are very small compared with those of ion channels, the development of the giant membrane patch method, which permits the inside-out and the outside-out membrane orientation, has led to interesting studies.[37,38]

In the case of a surface-immobilized ATPase, the sidedness of specific binding sites can be determined under conditions in which all membrane fragments are homogeneously oriented on the surface. Having achieved a functional immobilization of membrane fragments in a sensor configuration, under control of membrane fragment orientation on the sensor surface, the aspects of sidedness may be investigated in a comparatively simple and straightforward manner using the planar waveguide approach.

Acknowledgements

This article is dedicated to Prof. J. C. Skou, who discovered the Na,K-ATPase, on the occasion of his 80th birthday. The authors thank Mrs. A. Schacht for skillful enzyme preparations, Mr. A. Spielmann for experimental help, Dr. J. Fritz for part of the AFM studies, Mr. E. Lewitzki for numerous control measurements, and Dr. H. Ruf for a dynamic light scattering measurement. Many stimulating discussions with Dr. G. Duveneck, Dr. P. Oroszlan, Dr. G. Kraus, Dr. A. Abel, Dr. D. Neuschäfer, Dr. B. Klee, Dr. A. Cudd and Prof. Dr. H. Vogel are acknowledged. In addition, we are grateful to Prof. Dr. H. Vogel and Dr. A. Cudd for carefully reading the manuscript.

References

1. R. B. Gennis, in *Biomembranes*, ed. Ch.R. Cantor, Springer Verlag, New York, 1989.
2. D. Bray, *Annu. Rev. Biophys. Biomol. Struct.*, 1998, **27**, 59.
3. N. Unwin, *Cell*, 1993, **72**, 31.
4. J. P. Changeux, *Sci. Am.*, 1996, **6**, 499.
5. R. J. Lefkowitz, S. Cottechia, P. Samama and T. Costa, *Trends Pharmacol. Sci.*, 1993, **14**, 303.
6. A. G. Gilaman, *Angew. Chem. Int. Ed. Engl.*, 1995, **34**, 1406.
7. J. R. Broach and J. Thorner, *Nature (London)*, 1996, **384 (6604 Suppl.)**, 14.
8. J. J. Burbaum and N. H. Sigal, *Curr. Opin. Chem. Biol.*, 1997, **1**, 72.
9. P. Fürst and J. Heim, *BIOforum Int.*, 1998, **2**, 64.
10. E. L. Schmid, A. P. Tairi, R. Hovius and H. Vogel, *Anal. Chem.*, 1998, **70**, 1331.
11. J. Hodgson, *Biotechnology*, 1994, **12**, 31.
12. A. G. Frutos and R. M. Horn, *Anal. Chem.*, 1998, **70**, 449A.
13. G. L. Duveneck, M. Pawlak, D. Neuschäfer, W. Budach and M. Ehrat, *SPIE, Proceedings of Biomedical Systems and Technologies*, 1996, **2928**, 98.
14. G. L. Duveneck, M. Pawlak, D. Neuschäfer, E. Bär, W. Budach, U. Pieles and M. Ehrat, *Sens. Actuat. B*, 1997, **38–39**, 88.
15. J. C. Skou, *Biochim. Biophys. Acta*, 1957, **23**, 394.
16. S. J. D. Karlish, in *Na,K-ATPase Structure and Kinetics*, ed. J. C. Skou and J. G. Norby, Academic Press, New York, 1979, p. 115.
17. C. Hegyvary and P. L. Jørgensen, *J. Biol. Chem.*, 1981, **256**, 6296.
18. E. Grell, R. Warmuth and H. Ruf, *Acta Physiol. Scand.*, 1992, **146**, 213; 1993, **147**, 343.
19. P. L. Jørgensen, *Biochim. Biophys. Acta*, 1974, **356**, 36.
20. O. H. Lowry, N. J. Rosenbrough, A. L. Farr and R. J. Randall, *J. Biol. Chem.*, 1951, **193**, 265.
21. D. Porschke and E. Grell, *Biochim. Biophys. Acta*, 1995, **1231**, 181.
22. D. Brovelli, L. Ruiz, G. Kraus, G. Hähner, R. Hofer, A. Waldner, J. Schlösser, P. Oroszlan, M. Ehrat and N. D. Spencer, *Langmuir*, 1999, submitted.
23. M. Pawlak, E. Grell, D. Anselmetti and M. Ehrat, in preparation.
24. S. Terretaz, T. Stora, C. Duschl and H. Vogel, *Langmuir*, 1993, **9**, 1361.
25. E. Lewitzki, E. Schick and E. Grell, *J. Fluoresc.*, 1998, **8**, 113.
26. M. Pawlak, E. Schmid, R. Hovius, E. Grell, H. Vogel and M. Ehrat, in *IBC Conference Proceedings 'High Throughput Screening'*, IBC, Southborough, USA, 1998.
27. K. R. Rogers, J. J. Valdes and M. E. Eldefrawi, *Biosens. Bioelectron.*, 1991, **6**, 1.
28. C. Duschl, A. F. Sévin-Landais and H. Vogel, *Biophys. J.*, 1996, **70**, 1985.
29. D. J. Müller, M. Amrein and A. Engel, *J. Struct. Biol.*, 1997, **119**, 172.
30. D. C. Gadsby, M. Nakao, A. Bahinski, G. Nagel and M. Suenson, *Acta Physiol. Scand.*, 1992, **146**, 111.
31. W. Schwarz and L. A. Vasilets, *Cell Biol. Int.*, 1996, **20**, 67.
32. S. D. J. Karlish and U. Pick, *J. Physiol.*, 1981, **312**, 505.
33. A. Rephaeli, D. Richards and S. D. J. Karlish, *J. Biol. Chem.*, 1986, **261**, 6248.
34. F. Cornelius, in *The Sodium Pump: Structure, Mechanism, and Regulation*, ed. J. H. Kaplan and P. De Weer, Rockefeller University Press, New York, 1991, p. 267.
35. K. Fendler, E. Grell, M. Haubs and E. Bamberg, *EMBO J.*, 1985, **12**, 3079.
36. A. Eisenrauch, E. Grell and E. Bamberg, in ref. 34, p. 317.
37. D. W. Hilgemann, *Ann. N.Y. Acad. Sci.*, 1997, 260.
38. U. Eckstein-Ludwig, J. Rettinger, L. A. Vasilets and W. Schwarz, *Biochim. Biophys. Acta*, 1998, 1372, 289.

Paper 8/06704J

Analysis of membrane protein cluster densities and sizes *in situ* by image correlation spectroscopy

Nils O. Petersen, Claire Brown, Anna Kaminski, Jonathan Rocheleau, Mamta Srivastava and Paul W. Wiseman

Department of Chemistry, The University of Western Ontario, London ON N6A 5B7, Canada

Received 25th August 1998

Communication between cells invariably involves interactions of a signalling molecule with a receptor at the surface of the cell. Typically, the receptor is imbedded in the membrane and it is hypothesized that the binding of the signalling molecule causes a change in the state of aggregation of the receptor which, in turn, initiates a biochemical signal within the cell. Subsequently, many of the occupied receptors bind to membrane-associated structures, called coated pits, which invaginate and pinch off to form coated vesicles, thereby removing the receptors from the cell surface. The state of aggregation of membrane receptors is obviously in constant flux. Any useful approach to measuring the state of aggregation must, therefore, allow for dynamic measurements in living cells. It is possible to use fluorescently labelled signalling molecules or antibodies directed at the receptor of interest to visualize the receptor on the cell surface with a fluorescence microscope. By employing a laser confocal microscope, high resolution images can be produced in which the fluorescence intensity is quantitatively imaged as a function of position across the surface of the cell. Calculations of autocorrelation functions of these images provide direct and accurate measures of the density of fluorescent particles on the surface. Combined with the average intensity in the image, which reflects the total average number of molecules, it is possible to estimate the degree of aggregation of the receptor molecules. We refer to this analysis as image correlation spectroscopy (ICS). We show how ICS can be used to measure the density of several receptors on a variety of cells and how it can be used to measure the density of coated pits and the number of molecules per coated pit. We also show how the technique can be used to monitor fusion of virus particles to cell membranes. Further, we illustrate that, by calculating cross-correlation functions between pairs of images, we can extend the analysis to measurements of the distributions as a function of time, on the second timescale, as well as to measurements of the movement of the receptor aggregates on the surface. Finally, we illustrate that, by this approach, we can measure the extent of interaction between two different receptors as a function of time. This represents the most quantitative measurement of the extent of co-localization of receptors available and is independent of the spatial resolution of the confocal microscope. The theory of ICS and image cross-correlation spectroscopy (ICCS), focussing on the interpretation of the data in terms of the biological phenomenon being probed, is discussed.

Introduction

Structurally, the cell surface is a heterogeneous mixture of lipids and proteins in a constant state of flux. It contains a number of transmembrane proteins imbedded in a fluid lipid bilayer as well as a large number of membrane-associated proteins and lipid-anchored proteins. Many of these, in turn, bind to extracellular matrix proteins or intracellular networks of cytoplasmic proteins. All lipid components and lipid-anchored proteins exhibit a high degree of lateral mobility in the membrane. Transmembrane proteins, however, are subjected to several restrictions in their mobility. First, they move more slowly than anticipated from theory;[1] second, they are restricted by the cytoskeleton to move in sub-µm domains;[2] and third, they tend to aggregate with other proteins in specialized structures, such as the coated pits.[3] Recently, special organizations of lipids and proteins have been identified in cells (cavaolae)[4] and there is mounting evidence that glycolipids and lipid-anchored proteins can form detergent resistant membrane domains.[5-7]

Functionally, some lipids and many proteins in the membrane are involved as receptors to ligands that are in the medium, attached to the growth substrate or on neighbouring cells. In most cases, it is believed that the immediate consequence of binding of a ligand to a receptor is a redistribution of the proteins into aggregates (ranging from dimers to oligomers).[8] This association of receptors allows for biochemical interactions, such as trans-phosphorylation,[9] or physical interactions, such as enhanced binding.[10] Subsequently, the ligand–receptor complexes can be cleared from the surface by internalization through special structures such as the clathrin-coated pits[11] or the caveolin-associated invaginations.[12]

The complex structure and function of the cell surface components demands increasingly sophisticated tools to probe the specific intermolecular interactions that are coupled to specific functions. For example, it is argued that in order for a trans-membrane protein to be associated with a coated pit, it must have a specific amino acid sequence exposed in a particular fold that will allow it to bind to the adaptor protein in the coated pit.[13] The evidence for these and similar examples of specific intermolecular interactions is, for the most part, based on measurement of these protein–protein interactions *ex situ*, in extracted and purified preparations.[14,15] Clearly, it is desirable to be able to measure these interactions quantitatively *in situ* and in real time.

A number of techniques based on electron microscopy coupled with immunological labelling techniques have provided detailed pictures of molecular associations and structural organizations.[16] Prime examples are the clathrin-coated pits[17] and the caveolin-coated rosettes.[6] The high resolution and the specificity of antibody binding gives exquisite detail of organization. There are two drawbacks, however: it is very difficult to ensure and verify that all receptors are visualized and the nature of the specimen preparation precludes certain kinetic experiments. Immunofluorescence experiments provide the same specificity but at a much reduced resolution, even allowing for resolution enhancement techniques.[18] Laser scanning confocal microscopy introduces additional resolution in the third dimension and permits truly quantitative measurements of fluorescence intensity as a function of position.[19] This has permitted studies of the distribution and redistribution of receptor molecules at the sub-µm level in living cells with a time resolution of seconds.[20,21] More recently, techniques have been developed for studying the co-localization of two different receptors based on pixel-by-pixel mixing of colours representative of the fluorescence intensity: green for one receptor, red for the other receptor and yellow if both receptors provide intensity in the same pixels and by using point-by-point correlation analyses.[22,23]

We have introduced an autocorrelation function analysis of confocal images which allows us to estimate both the *density of receptor clusters* and the *number of receptors per cluster* irrespective of the actual physical dimensions of the clusters.[20,24] We have also introduced a cross-correlation function analysis of pairs of images which allows us to estimate the *kinetics of redistribution* of the receptor clusters (if the images are obtained at different times)[21] or the *fraction of one receptor co-localized with another* receptor (if the images are obtained for different probes).[25,26] The key advantage of these analyses is that they reduce the measurement from many observations within a single image to one number with a particular, understandable significance. This allows for statistically meaningful investigations of large populations of cells, giving rise to a more accurate measure of the relevant intermolecular interactions. Combination of the correlation analysis with more traditional image processing can also give quantitative information about the distribution of cluster sizes or about which sub-population of a receptor is interacting with other receptors.

Experimental

Cell preparations

Cells were grown in a T-25 flask in medium supplemented with fetal calf serum and antibiotics in a 5% CO_2 atmosphere. When needed for experiments, the cells were plated on glass coverslips at a low density. At the appropriate time in the procedures, the cells were fixed in 4% aqueous paraformaldehyde at pH 7 at room temperature for 30 min or by immersion for 5 min in cold methanol ($-20\,°C$) followed by 5 min in cold acetone ($-20\,°C$). At the appropriate time in the protocol, the cells were labelled with a primary antibody at saturating concentrations followed by a secondary antibody at saturating concentrations. The actual concentrations and times varied with the antibody and the cell type. Experiments performed with gangliosides were conducted on live cells contained in a temperature-controlled unit on the stage. Sendai Virus was labelled with the lipid dye, diQ (Molecular Probes) by simple mixing and separation on a size-exclusion column.

Microscopy

Laser scanning confocal microscope images were obtained on a Biorad 600 attached to a Nikon inverted microscope and equipped with an argon–krypton mixed gas laser. Each image was collected at Zoom 10 providing pixel resolutions of 0.03 µm and image dimensions of 15.5 µm in each direction. The images consisted of 515x512 pixels with 8 bit integer intensity data in each. The laser beam was focussed to *ca.* 0.35 µm at the e^{-2}-waist and was imaged through an expanded pinhole to give an effective depth of focus of *ca.* 1 µm. The fluorescein or nitrobenzoxadiazole dyes were excited at 488 nm while the Red-X and Di-Q dyes were excited at 568 nm. In dual-labelling experiments, images were collected separately at each wavelength to minimize cross-over fluorescence between the two channels. The images were collected in the photon-counting mode to ensure a linear intensity response. In all cases, the background level was set to deliver a non-zero count in each pixel and the gain was set to avoid saturating any of the pixels. This protocol ensures that no information is lost at either end of the intensity scale. As few as a single scan and as many as 25 scans were averaged, depending on the experiment, but for a given set of experiments, the same conditions were used for all samples. Images were collected periodically during an experiment to establish the background (the sample was not illuminated but the laser was allowed to reflect off all mirrors and lenses). Unlabelled cells were imaged to obtain images representing autofluorescence and, where possible, cells were labelled with the second antibody only to obtain images representing the sum of autofluorescence and non-specific labelling.

Image processing

All images were transferred to a MasPar 2 with 2048 parallel processors. The correlation calculations were performed using 2D FFT routines available on the MasPar. The resulting correlation functions were fit to a 2D, three-parameter Gaussian function using non-linear regression analysis (Marquandt Algorithms). The data output was processed using commercial software (QuatroPro, SigmaPlot, Photoshop, CorelDraw) as appropriate.

Theory

ICS

Let the intensity in each pixel in the image be $i(x, y)$. The spatial correlation function for the image is then given by eqn. (1)

$$g(\xi, \eta) = \frac{\langle(i(x, y) - \langle i(x, y)\rangle)(i(x + \xi, y + \eta) - \langle i(x, y)\rangle)\rangle}{\langle i(x, y)\rangle^2} \quad (1)$$

where the angular brackets indicate the average over all spatial coordinates.[27] It is known[28] that the limit of the correlation function as ξ and η vanish gives $g(0,0) = \mathrm{var}[\delta_n i(x, y)]$, the variance of the normalized intensity fluctuations,

$$\delta_n i(x, y) = \frac{i(x, y) - \langle i(x, y) \rangle}{\langle i(x, y) \rangle}$$

When the intensity is a true reflection of the concentration, this variance is also the variance in the concentration fluctuation, which equals[29] the inverse of the average occupation number, $\bar{N}_p$:

$$g(0, 0) = \text{var}(\delta_n i(x, y)) = \text{var}(\delta_n c(x, y)) = \frac{1}{\bar{N}_p} \quad (2)$$

The occupancy number represents the average number of fluorescent particles present in the observation volume defined in the confocal microscope by the convolution of the laser beam and the point spread function for the image pin hole. For molecules on thin, flat membranes, this corresponds to an area given by πw^2, where w is the e^{-2} beam waist. Thus the cluster density (CD) is defined as[27]

$$\text{CD} = \frac{1}{g(0, 0)\pi w^2} = \frac{\bar{N}_p}{\pi w^2} \quad (3)$$

Moreover, the intensity is proportional to the average number of fluorescent molecules so that $\langle i(x, y) \rangle = c\bar{N}_m$, where c is a constant reflecting optical parameters (molar absorption coefficients, quantum yields and the efficiency of collection of the confocal microscope).[30] Thus we can define a degree of aggregation (DA) as[31]

$$\text{DA} = \langle i(x, y) \rangle g(0, 0) = c \frac{\bar{N}_m}{\bar{N}_p} \quad (4)$$

The constant, c, can be determined experimentally[32] by assuming that the non-specific binding of fluorescently labelled secondary antibodies occurs as monomers i.e. $\text{DA}_{ns} = c$.

ICCS

Consider the intensity distribution for two images possibly collected at different times, $i(x, y, t)$ and $j(x, y, t + \tau)$. It is possible to calculate cross-correlation functions as[21]

$$g_{ij}(\xi, \eta, \tau) = \frac{\langle (i(x, y, t) - \langle i(x, y, t) \rangle)(j(x + \xi, y + \eta, t + \tau) - \langle j(x, y, t + \tau) \rangle) \rangle}{\langle i(x, y, t) \rangle \langle j(x, y, t + \tau) \rangle} \quad (5a)$$

$$g_{ji}(\xi, \eta, \tau) = \frac{\langle (j(x, y, t + \tau) - \langle j(x, y, t + \tau) \rangle)(i(x + \xi, y + \eta, t) - \langle i(x, y, t) \rangle) \rangle}{\langle j(x, y, t + \tau) \rangle \langle i(x, y, t) \rangle} \quad (5b)$$

where we expect $g_{ij}(\xi, \eta, \tau) = g_{ji}(\xi, \eta, \tau)$ in most cases.

If images are collected as a function of time, τ, from the same cell, observing the same chromophore throughout, then the cross-correlation functions will report the ensemble-averaged dynamics of the aggregation pattern on the surface. In the limit where the number of aggregates remains fixed, the zero-lag amplitudes of the cross-correlation functions are simple functions of time, i.e. $g_{ij}(0, 0, \tau) = g(0, 0, 0)f(\tau)$ where $f(\tau)$ depends on the dynamics as in normal fluorescence correlation spectroscopy [$f(\tau) = \exp(\tau/\tau_f)^2$ for 2D flow and $f(\tau) = (1 - \tau/\tau_d)^{-1}$ for 2D diffusion]. In these cases, $g(0, 0, 0)$ gives the cluster density as above.[21]

If images are collected at the same time from the same cell, observing two different chromophores, then $\tau = 0$ and the zero-lag amplitudes of the cross-correlation functions reflect the extent of co-localization of the two chromophores on the cell surface at a particular time t.[21,33] Specifically,[25]

$$g_{ij}(0, 0, t) = \frac{\bar{N}_{ij}}{(\bar{N}_i + \bar{N}_{ij})(\bar{N}_j + \bar{N}_{ij})} = \bar{N}_{ij} g_i(0, 0, t) g_j(0, 0, t) \quad (6)$$

where $\bar{N}_{ij}$ represents the average number of aggregates in which there are both chromophores present. Accordingly, we can define a cluster density of co-localized chromophores as

$$\text{CD}_{ij}(t) = \frac{g_{ij}(0, 0, t)}{g_i(0, 0, t)g_j(0, 0, t)\pi w^2} = \frac{\bar{N}_{ij}}{\pi w^2} \qquad (7)$$

Note that, in this case, the cluster density is proportional to the amplitude of the cross-correlation function. The cluster density of co-localized chromophores can now be calculated from the cross-correlation and the autocorrelation functions of a pair of images. By collecting pairs of images as a function of time, the evolution of co-localization can be followed.

At this time, we do not have a parameter corresponding to the degree of aggregation and, hence, the direct calculation of the number of each chromophore present in a co-localized aggregate is not available. However, we can calculate the fraction of clusters of one chromophore which participates in the co-localized aggregates. This we define as

$$F(i|j) = \frac{\text{CD}_{ij}}{\text{CD}_i} = \frac{g_{ij}(0, 0, t)}{g_j(0, 0, t)} \quad \text{or} \quad F(j|i) = \frac{\text{CD}_{ij}}{\text{CD}_j} = \frac{g_{ij}(0, 0, t)}{g_i(0, 0, t)} \qquad (8)$$

which we read as the fraction of clusters of type i that are co-localized with clusters of type j and vice versa.

Other considerations

Fitting. Confocal images of aggregates with dimensions smaller than that of the laser beam will consist of circular intensity spots with dimensions given by the size of the laser beam at the focal plane. If the confocal microscope is properly adjusted, the laser beam is in the TEM_{00} mode with a Gaussian transverse intensity profile.[34,35] In this case, the auto- and cross-correlation functions will have a 2D Gaussian decay centred at the origin and with a decay width given by the laser beam width, w. This can be measured independently.[30,34,35] Accordingly, we fit the calculated correlation functions as a three parameter function[27]

$$g(\xi, \eta) = g(0, 0)e^{-(\xi^2 + \eta^2)/w^2} + g_0 \qquad (9)$$

where the zero-lag amplitude, $g(0, 0)$, the beam width, w, and the offset, g_0 are the fitted parameters. The amplitude is the desired parameter. The width can be compared to the known width as a control for the quality of the fit. If the fit and known beam widths differ by more than 30%, the data normally should be rejected. In some cases, all the data show larger beam widths, suggesting that the clusters are comparable in dimension to the laser beam. Then the width from individual correlation functions is compared to the average width from all the relevant data. Individual data can be rejected if they fail the z-score test for outliers.[36] The offset in eqn. (9) is included to account for the empirical observation that the correlation function will not always decay to zero. This arises from the fact that the images have finite sizes and, at long distances, the information is correspondingly less accurate. For example, a length of 15.5 µm will only allow for *ca.* 50 characteristic fluctuations of a 0.35 µm laser beam across each dimension. The uncertainty in the offset is, therefore, of the order of $\sqrt{50/50} = 0.14$, corresponding closely to the uncertainty in the data from the best samples we can obtain. It has been shown that the optimum fit includes only the data in the correlation function to about three times the width.[37] This is implemented by using only data for which $(\xi^2 + \eta^2) < (3w)^2$, with $3w$ being *ca.* 32 pixel dimensions (*ca.* 1 µm).

The autocorrelation function will always fit to a maximum amplitude at the zero-lag coordinates. The cross-correlation function need not be maximum at the zero-lag coordinates since there may be systematic shifts between the images collected at two different excitation and emission wavelengths. Hence, the cross-correlation functions are fit to a maximum at $g(\xi_0, \eta_0)$ and the coordinates at the maximum are recorded. Systematic variations in these coordinates can be noted and accounted for. In fact, if there is a flow process, the changes in the coordinates of the maximum can be used to determine the flow velocity.[21] Random variations in the coordinates may result from small instabilities in the microscope and we accept data as being valid whenever these random variations are less than about one beam width (10 pixels) in either direction. If the cross-correlation is very small or absent, the fit will be to the largest correlation in the whole function. This may be at large lag values, since the information content is poorer because of the limited image dimension. We record the fraction of images for which the coordinates of the maximum of the cross-correlation function is greater than *ca.* 10 pixels and interpret this as the fraction of images which show no cross-correlation or cross-correlation below the random correlation level. The fraction of cross-correlation described above, [$F(i|j)$], is calculated only from those images for which the cross-correlation is maximum close to the zero-lag coordinates.

Spot densities. Many cells have between 100 000 and 2 000 000 receptors on their surface.[32,38–40] Typical surface areas are of the order of 5000–10 000 μm^2 suggesting densities of 10–400 receptors per μm^2. Our images (15.5 μm × 15.5 μm = 240 μm^2) would allow as many as 625 individual receptors to be resolved as distinct fluorescent spots [$\pi(0.35)^2$ μm^2 = 0.38 μm^2]. For most receptor systems the monomer density is, therefore, high enough for even high-resolution confocal images to show a very homogeneous distribution of fluorescence. In fact, if a monomer can be detected, it can be resolved in these images only if density is less than *ca.* 3 per μm^2. If the contrast and the signal-to-noise is good, it is possible to count the number of spots in the image and calculate a spot density, defined as SD = 'the number of spots per μm^2. This is difficult or impossible in cases where the signal is weak. The advantage of the correlation analysis is that it provides an approach to estimating the density, even in a noisy background.

Multiple populations. Whenever the spot density can be calculated, we expect agreement with the cluster density estimated from the correlation calculations. We have never seen SD > CD, but we routinely observe that CD > SD. We interpret this as an indication that there are at least two populations of receptor clusters: large clusters easily identifiable in the image and contributing to the spot density count and monomers or small clusters not discernible visually above the background, but contributing to the cluster density through their contribution to the correlation function.

In general, it is possible to account for contributions of multiple populations of receptors to the amplitude of the correlation function[30] since

$$g(0, 0) = \frac{1}{\bar{N}_m^2} \sum_i \bar{N}_i(\sigma_i^2 + \mu_i^2) \qquad (10)$$

where, for the *i*th population, μ_i is the mean number of monomers per aggregate (degree of aggregation), $\bar{N}_i$ is the mean number of such aggregates and σ_i is the standard deviation about the mean in the distribution of the number of monomers per aggregate. As before, $\bar{N}_m$ is the mean number of monomers. In the simplest case, we assume that there are two populations with vanishingly small standard deviations so that eqn. (10) simplifies to

$$g(0, 0) = \frac{1}{\bar{N}_m^2} (\bar{N}_1 \mu_1^2 + \bar{N}_2 \mu_2^2) \qquad (11)$$

Clearly, one measurement of $g(0,0)$ cannot provide sufficient information to calculate these five parameters, but there are often additional constraints depending on the model for aggregation in a particular system. Conservation of mass suggests that, to a reasonable approximation: $\bar{N}_m = \bar{N}_1 \mu_1 + \bar{N}_2 \mu_2$. Often, $\bar{N}_m$ is known from other sources, such as biochemical binding data for a ligand to a receptor. In a model where there are small and large aggregates, as suggested if CD > SD, then $\bar{N}_2$ can be estimated from the spot density, assuming that this is the large cluster population and the other population could be assumed to be monomers, so that $\mu_1 = 1$. With this particular model, we get

$$g(0, 0) = \frac{1}{\bar{N}_m^2} [\bar{N}_1 + (SD\pi w^2)\mu_2^2] \quad \text{and} \quad \bar{N}_M = \bar{N}_1 + (SD\pi w^2)\mu_2 \qquad (12)$$

which uniquely define all the parameters. The interpretation is very sensitive to the model chosen. Even so, the detailed calculations can provide some insight, especially when changes are observed as a consequence of treatment of the cells.

At this time, we have not extended the consideration of multiple populations to the cross-correlation calculations. This is important, but is likely to be complex.

Corrections. In general, several sources of fluorescence will contribute to the measured signal (m).[32,38] These include those of interest in the sample (s), those from non-specific fluorescence (ns), those from autofluorescence (a) and those from the background irradiation in the microscope (wn). The first three will exhibit fluctuations with the characteristic spatial dimension determined by the laser beam, whereas the last will have characteristics of white noise with zero amplitude in the correlation functions away from the cental pixel. If we assume that these are independent sources (a good assumption given that the origins of the signals are from different components of the sample) then

$$g_m(\xi, \eta)\langle i_m(x, y)\rangle^2 = g_s(\xi, \eta)\langle i_s(x, y)\rangle^2 + g_{ns}(\xi, \eta)\langle i_{ns}(x, y)\rangle^2$$
$$+ g_a(\xi, \eta)\langle i_a(x, y)\rangle^2 + g_{wn}(\xi, \eta)\langle i_{wn}(x, y)\rangle^2 \quad (13)$$

For true white noise, $g_{wn}(\xi, \eta) = 0$, except at the origin where it can be very large. Hence all fits exclude the data at the origin. Measurements on cells labelled with the secondary antibody only provide an estimate of the combined contributions from non-specific fluorescence and autofluorescence. These data can be used to make a simultaneous correction for the second and third terms in eqn. (13). If necessary, measurements on unlabelled cells provide an independent estimate of the contribution from autofluorescence, which can be used to extract the contribution for the non-specific fluorescence. This is useful for estimating the constant c, used in calculating the absolute degrees of aggregation as discussed above.

We perform the correction shown in eqn. (14) to get the proper estimate of the correlation function but, as implied, we apply the correction only to the zero-lag amplitude of the correlation function after it has been calculated from the fit of the raw correlation functions.

$$g_s(0, 0) = \frac{g_m(0, 0)\langle i_m\rangle^2 - [g_{ns}(0, 0)\langle i_{ns}\rangle^2 + g_a(0, 0)\langle i_a\rangle^2]}{(\langle i_m\rangle - [\langle i_{ns}\rangle + \langle i_a\rangle] - \langle i_{wn}\rangle)^2} \quad (14)$$

The best experiments require measurements of instrumental background, unlabelled cells, non-specifically labelled cells and specifically labelled cells under the same microscope conditions. To obtain statistically meaningful estimates, we routinely collect 40 images or more for each experimental condition. Frequently, the numbers are based on hundreds of images.

Results

Several applications of image correlation spectroscopy and image cross-correlation spectroscopy have been published.[20,21,25,27,31,38] Here we provide new data generated to illustrate the scope and limitations of these techniques for making measurements of cluster densities, degrees of aggregations and extent of co-localization of cell membrane associated proteins and lipids.

Cluster density of growth factor receptors

Platelet derived growth factor receptors. Table 1 shows the data obtained on a number of AG1523 cells fixed with 4% paraformaldehyde prior to the addition of monoclonal antibodies directed at the platelet-derived growth factor receptor (PDGF-R) and labelling the primary antibody with goat-anti-mouse IgG antibodies specific for either the Fc portion of the primary antibody or the Fab portion of the primary antibody. The former is labelled with fluorescein isothiocyanate, the latter with tetramethyl rhodamine isothiocyanate. The average intensity is much less in the fluorescein experiments since the chromophore bleaches rapidly and it is, therefore, only possible to collect and sum a few (here four) images on a given cell area. The degree of aggregation [eqn. (4)] is measured under conditions where every receptor is saturated by bound primary antibody and every primary antibody is saturated by bound secondary antibody. These values are corrected for non-specific fluorescence and autofluorescence. The degree of aggregation is also measured under non-specific labelling conditions and corrected for autofluorescence to give an estimate of the optical constant, c, for these illumination conditions. From these we can then estimate the mean number of fluorescent antibodies per cluster. This is calculated from eqn. (4) and (10), allowing for a Poisson distribution of cluster sizes about a single mean value (μ) with a variance equal to the mean ($\sigma^2 = \mu$) so that $\mu = [\langle i(x, y)\rangle g(0, 0)/c] - 1 = DA/DA_{ns} - 1$. It is clear that when the antibody is specific for the Fab portion of the primary antibody, μ is about twice the value obtained when the antibody is specific for the Fc portion of the primary antibody.

Table 1 PDGF-receptors on AG1523 cells

Antibody	FITC-IgG (Fc-specific)	TMRITC-IgG (Fab-specific)
$\langle i(x,y) \rangle$	0.2	4.4
DA	0.25	4.2
$DA_{ns} = c$	0.057	0.48
μ(antibody)	3.3	7.8
μ(receptor)	3.3	3.9
CD	2.2	2.3

Accounting for the fact that there are two Fab and one Fc segment per primary antibody we calculate 3.3 and 3.9 receptors per cluster. There is an excellent agreement between the two labelling approaches.

Binding data show that, on average, AG1523 cells express as many as 150 000 copies of the platelet derived growth factor receptor (PDGF-R) on the cell surface.[39] Given an average cell surface area[38] of *ca.* 11 000 μm^2, this corresponds to an upper limit of 14 receptors per μm^2. The ICS data yield as many as 2.3 clusters per μm^2 with 3.9 receptors per cluster corresponding to 9 receptors per μm^2. This is in reasonable agreement with the binding data and illustrates that the technique is able to detect most, if not all, of the receptors on the surface. Also, the small numbers of receptors per cluster clearly show that the confocal microscope combined with the correlation analysis enable us to image and characterize single receptors on the surface of adherent cells. Similar data can be obtained on living cells.[20]

Epidermal growth factor receptors. Table 2 shows the data for experiments on the mouse cell line A431 labelled with a primary antibody directed at the epidermal growth factor receptor (EGF-R) and a secondary antibody directed at the Fc segment of the primary antibody. The experiments were performed on cells which were incubated at different temperatures prior to fixing with paraformaldehyde. At 37 °C, the cluster density value is *ca.* 20, so we see *ca.* eight times more EGF-R clusters per unit area on A431 cells than for PDGF-R on AG1523 cells. There are, on average, 11 receptors per cluster, *ca.* 3 times as many for EGF − R on A431 compared with PDGF-R on AG1523. This corresponds to *ca.* 220 receptors per μm^2 for EGF-R compared to *ca.* 9 for PDGF-R

It is known that A431 cells overexpress EGF-R and as many as 2–3 million receptors are present per cell.[40] These cells are smaller than the AG1523 cells, with an average surface area of *ca.* 3000 μm^2, so we expect *ca.* 600–1000 receptors per μm^2. The ICS data confirm the excess expression. These measurements are made with saturating conditions, as illustrated in Fig. 1, yet only about one third of the receptors are detected. Nevertheless, the data show clearly that the overexpression of EGF-R on A431 cells leads to both a larger number of clusters and a larger number of receptors per cluster when compared to the normal expression of PDGF-R on AG1523 cells. The receptors to EGF are highly aggregated prior to exposure to the antibody and prior to exposure to the growth factor (these cells have been starved, *i.e.* deprived of any exposure to growth factors in the growth medium for 24 h prior to the experiment).

The average beam size measured by the autocorrelation functions is larger, by *ca.* 33%, than that expected from the laser beam (0.45 *cf.* 0.35 μm). This is consistent for all the samples at 37 and 22 °C and indicates that the dimension of the cluster is as large as, or slightly larger, than the laser beam. This suggests that these receptors may not be in molecular contact in the clusters but they could simply be confined to a domain in the membrane. Pre-existing clusters of receptors may be functionally important since they may enhance the rate of activation by the growth factor since

Table 2 EGF-receptors on A431 cells

	37 °C	22 °C	4 °C
CD	19 ± 4	13 ± 4	10 ± 2
μ(receptor)	11 ± 3	17 ± 3	33 ± 2
w	0.45 ± 0.03	0.46 ± 0.03	0.40 ± 0.01

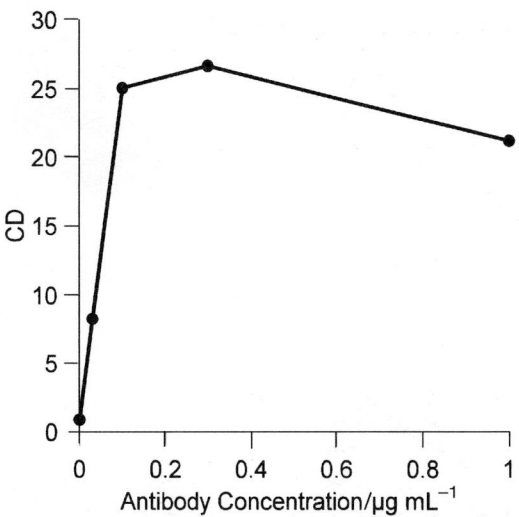

Fig. 1 CD of epidermal growth factor receptors on A431 cells measured as a function of concentration of the primary antibody in the incubation solution. Clearly, a maximum number of clusters is detected at concentrations greater than *ca.* 0.1 µg mL^{-1}. Further addition of the antibody does not reveal more clusters, but could reveal a larger number of receptors per antibody. In these experiments the DA reaches a maximum corresponding to *ca.* 11 receptors per aggregate at the same primary antibody concentrations.

the mean free path for collision would be significantly reduced. It appears to be a valid concept for EGF-R in A431 cells and, to a more limited degree, for PDGF-R in AG1523 cells. How general this observation is remains to be seen.

When the A431 cells are cooled to 4 °C for 30 min prior to fixation, the cluster density decreases by a factor of two and the mean number of receptors per cluster increases by a factor of three, indicating that cooling causes a further aggregation of the receptors. At the same time, the average dimension of the clusters also decreases significantly. While these data are still preliminary, they do suggest that the receptors associate more tightly at lower temperatures. A similar trend was observed previously with PDGF-R on AG1523 cells.[20]

Surface labelling. Note that the primary and secondary antibodies label only the top surfaces of the cells. This is illustrated in Fig. 2, which shows three confocal images at different heights through a cell. Some bright spots are clearly in focus on the top surface of the cell over the nucleus [Fig. 2(a)]. In contrast, there are no fluorescent spots in focus on the membrane under the nucleus [Fig. 2(b) and (c)] indicating that there is no labelling on that surface. This is shown more clearly in Fig. 2(d) which is a cross-section of the same cell that shows the fluorescence only on the top contour of the cell. This is a characteristic of all the experiments we have conducted with antibodies that are directed at the exterior surface of the membrane. Our estimates of cluster densities in those systems reflect this surface only and we have assumed that there is an equal expression of receptors on the bottom surface, but that the antibody cannot access them within the time frame of the labelling procedure.

In experiments where the antibody is directed at the cytoplasmic part of the membrane proteins, both the top and the bottom membrane of the cell is labelled. In the flat, peripheral regions of the cell, both membranes will be imaged simultaneously [they are separated by less than the focal depth of the confocal microscope (*ca.* 1 µm)]. This means that the cluster density reflects clusters on both surfaces and it is overestimated by a factor of two. We have made some measurements of cluster densities in regions under the nucleus, where only the bottom surface should contribute. These are complicated by the enhanced autofluorescence in the nuclear area but, qualitatively, the data are consistent with comparable cluster densities on each surface. Thus, for cytoplasmic proteins (see below), we divide the cluster density values by a factor of two to get a true density of clusters per µm^2 of membrane surface.

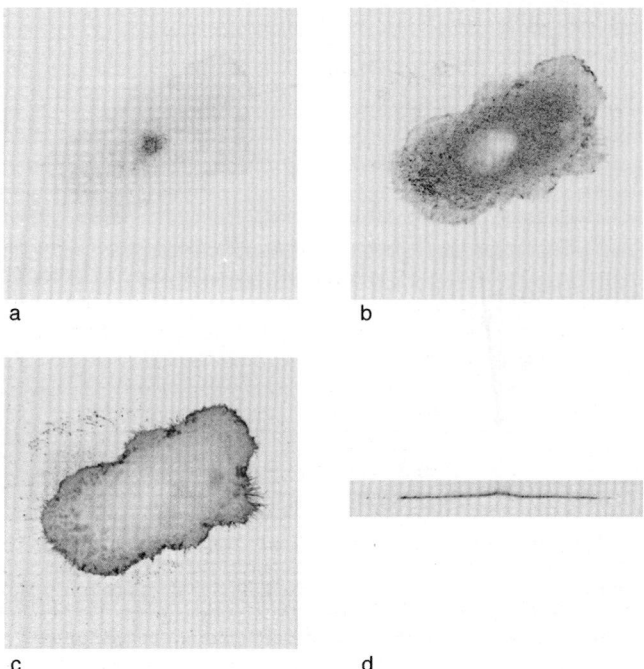

Fig. 2 Confocal images of A431 cells labelled to reveal EGF-R on the surface. (a) Section at the top of the cell, (b) section close to the top of surface of the cell at the periphery, (c) section at the bottom of the cell. The dark spots represent fluorescence (in a reversed image) and the light areas show lack of fluorescence labelling. The absence of fluorescence is evident under the nucleus in (b) and over a large region of the cell in (c) indicating that the antibody only labels receptors exposed on the top surface. (d) z-scan through the middle of the cell confirming that the labelling is confined to the top surface.

Distribution of coated pit proteins

Coated pits contain at least three protein components: membrane receptors targeted for internalization, adaptor proteins which bind to the receptors and clathrin which binds to adaptor proteins and form the structural framework of the coat in the pit and in coated vesicles.[41] The adaptor protein which is found at the plasma membrane is designated AP-2 and differs structurally and immunologically from AP-1, which is found exclusively at the Golgi and AP-3, which is the least understood member of the family.[42] Clathrin is found at the plasma membrane, in some cytoplasmic vesicles (before they get uncoated) and at the Golgi and as a free component in the cytoplasm.

Clathrin and adaptor proteins. Table 3 summarizes some of the data obtained for the distribution of clathrin and the adaptor protein AP-2 on monkey kidney fibroblast cells (CV-1 cells). Both clathrin and AP-2 are proteins associated with the cytoplasmic side of the membrane so they can only be labelled after the cells are made permeable, in this case through a methanol–acetone fixation at $-20\,°C$. The antibody will label both the top and the bottom membrane and the data in Table 3 have been adjusted by the factor of two as discussed above.

The cluster density is significantly larger than the spot density for both proteins. This suggests that there may be two populations of proteins, large aggregates that are clearly discernible as individual spots and small aggregates or monomers that blend with the background. Since the data have been corrected for the contributions from autofluorescence and non-specific labelling, we believe that the more disperse population is real. It could correspond to a monomer population in the thin layer of cytoplasm between the two membranes, or small aggregates associated with the membrane. To test these possibilities, cells were treated with saponin prior to fixation and labelling for clathrin or AP-2. Saponin is supposed to remove cholesterol from the membrane, rendering it permeable and permitting cytoplasmic components to escape the cell. Table 3 shows

Table 3 Coated pits on CV-1 cells

	Clathrin	Adaptor protein (AP-2)
CD	1.32 ± 0.08	1.4 ± 0.20
SD	0.55 ± 0.24	0.28 ± 0.04
CD after saponin	0.50 ± 0.10	1.5 ± 0.05

that, in the case of clathrin, the CD value is reduced to the same value as the SD following saponin treatment, confirming that the discrepancy between these values arises from a cytoplasmic component of the clathrin. In contrast, saponin does not affect the cluster density for the AP-2, suggesting that the small aggregate component of this protein is membrane associated.

The cluster density of the large aggregate population of clathrin is *ca.* 0.5, which compares favourably with the density of coated pits estimated by electron microscopy methods.[25,43] The ICS data, therefore, suggest that clathrin exists in two populations: coated pits and free cytoplasmic protein. Using eqn. (11) or (12), we estimate that *ca.* 70% of the clathrin is in the coated pits with *ca.* 120 clathrin triskelia per coated pit and 30% is in the cytoplasm as triskelia in these peripheral regions of the membrane.

The spot density for AP-2 is less than that of clathrin by close to a factor of two. We expect that these spots correspond to AP-2 associated with clathrin in coated pits. The smaller value for SD might then indicate that only half of the coated pits contain large amounts of AP-2. Alternatively, the spot density may be underestimated because a smaller number of AP-2 proteins per coated pit would give a poorer signal to background. It is proposed that the ratio of AP-2 to clathrin triskelia is one-third. The number of binding sites for antibodies on AP-2 may then be almost 10-fold less, making the spots less visible and harder to count. The cross correlation data discussed below support the latter alternative.

Model receptor proteins. In collaboration with Professors Henis (Tel Aviv University) and Roth (Southwestern Medical Centre), we have used the model receptor system obtained by transfecting the hemagglutination protein (HA) from the influenza virus into CV-1 cells.[44,45] The wild-type protein does not have the cytoplasmic signal sequence which allows for binding to the adaptor protein. It is, therefore, expected to be distributed homogeneously across the surface. The HA+8 mutant has been created by adding the eight amino acid sequence **YDYKSFYN**, which introduces the YXXΦ signal sequence.[45] This mutant will bind to the adaptor protein and is expected to be more aggregated as a result. Since the protein is transfected into the CV-1 cells, the cluster density may be expected to depend on the effectiveness of the expression. For comparable intensities, suggesting comparable surface expressions, the cluster density of the HA-wt is consistently six times or more than that of HA+8. The cluster density of the HA+8 is *ca.* 3.2 ± 0.7 for cells with good levels of transfection.[31] This value of the cluster density is significantly greater than the cluster density of either clathrin (1.5) or AP-2 (1.5) and the density of coated pits (0.5). This suggests that only a fraction of these receptors interact with coated pits at any given time.

Intermolecular interactions. Fig. 3 shows the auto- and cross-correlation functions from images of HA-wt and clathrin obtained from the same sections of a cell. It is evident that while the autocorrelation functions have large amplitudes [Fig. 3(a) for HA-wt and 3(b) for clathrin], there is no amplitude in the cross-correlation function [Fig. 3(c)], indicating a minimal co-localization. Fig. 4 shows the auto- and cross-correlation functions from images of HA+8 and clathrin. This time, the amplitudes of the autocorrelation functions [Fig. 4(a) for HA+8 and 4(b) for clathrin] and the cross-correlation function [Fig. 4(c)] are all significantly above background [Fig. 4(c)], indicating at least some co-localization. Table 4 summarizes the values of the cluster densities of co-localized aggregates (CD_{gr}), the fraction of receptors interacting with clathrin and AP-2 and the fraction of clathrin and AP-2 interacting with the receptor.

The amplitudes of the cross-correlation functions between the HA-wt and the clathrin or AP-2 were so small that they exceeded the random correlation function fluctuations in only *ca.* 20% of the measurements, and in those cases the amplitudes were very small. In fact, less than 2% of the receptors co-localize with either clathrin or AP-2, and *vice versa*.

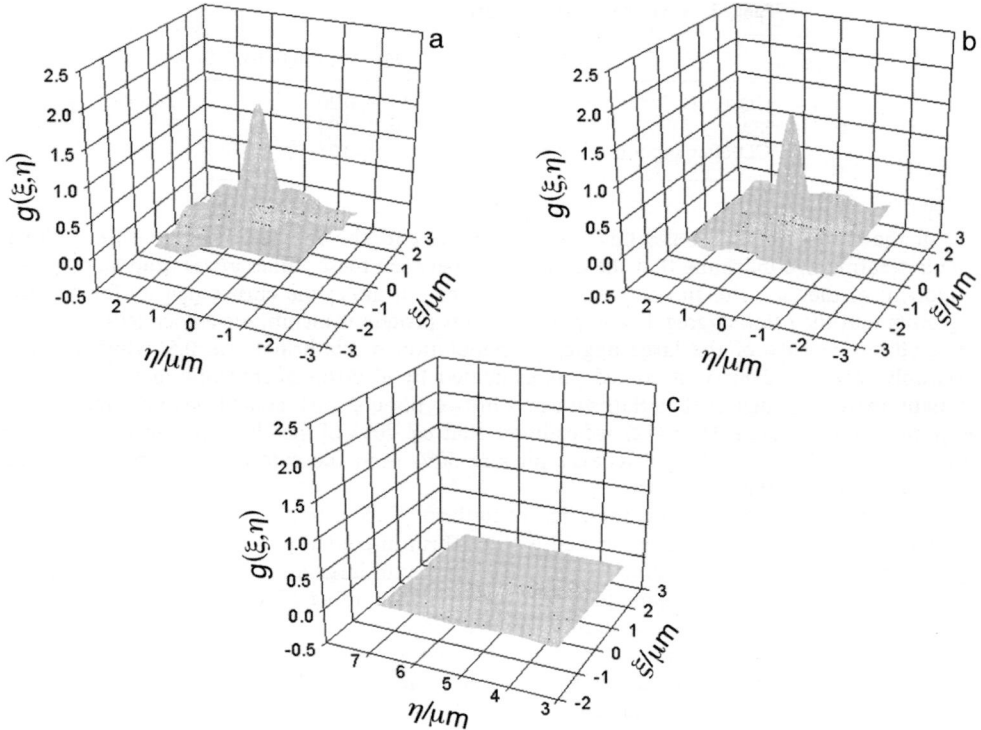

Fig. 3 Auto- and cross-correlation functions for images of HA-wt and clathrin. (a) Autocorrelation function for HA-wt. This is a small amplitude function since the HA-wt protein is relatively disperse. (b) Autocorrelation function for clathrin. This is a larger function since the protein is highly clustered. (c) Cross-correlation function for these two images. There is no detectable amplitude at zero-lag and the largest correlation occurs 5–6 μm from the origin.

The HA+8 system differs from the HA-wt since all the measurements yield cross-correlation functions with significant amplitudes at the zero-lag position. The density of clusters which contain both receptor and clathrin (CD_{gr} = 0.59) is the same as the density of clusters which contain both receptor and AP-2 (CD_{gr} = 0.57). These values correspond, in turn, very well to the density of coated pits determined above (Table 3) from the spot density and cluster density of the individual proteins. We therefore surmise that all the coated pits contain HA+8, clathrin and AP-2.

Table 4 shows that only *ca.* 25% of the HA+8 co-localizes with either clathrin or AP-2, suggesting that 75% exists outside the coated pits. This is consistent with previous diffusion data which suggest that *ca.* 25% of HA+8 is immobile whereas 75% moves freely and rapidly.[46] This

Table 4 Intermolecular interactions

Receptor	Protein	CD_{gr}	$F(H\|C)$ or $F(H\|A)$	$F(C\|H)$ or $F(A\|H)$	N
HA-wt	clathrin	NA[a]	<0.02	<0.02	40
HA-wt	AP-2	NA[a]	<0.02	<0.02	40
HA+8	clathrin	0.59 ± 0.06	0.25 ± 0.02	0.72 ± 0.06	175
HA+8	AP-2	0.57 ± 0.06	0.25 ± 0.03	0.69 ± 0.07	127

[a] NA = not applicable. Only *ca.* 20% of the cells imaged showed a cross-correlation at the zero-lag, indicating that there is very little co-localization. Of those that did have a cross-correlation function at the origin, the amplitudes were very small, leading to very small F-values. N represents those images that showed a cross-correlation.

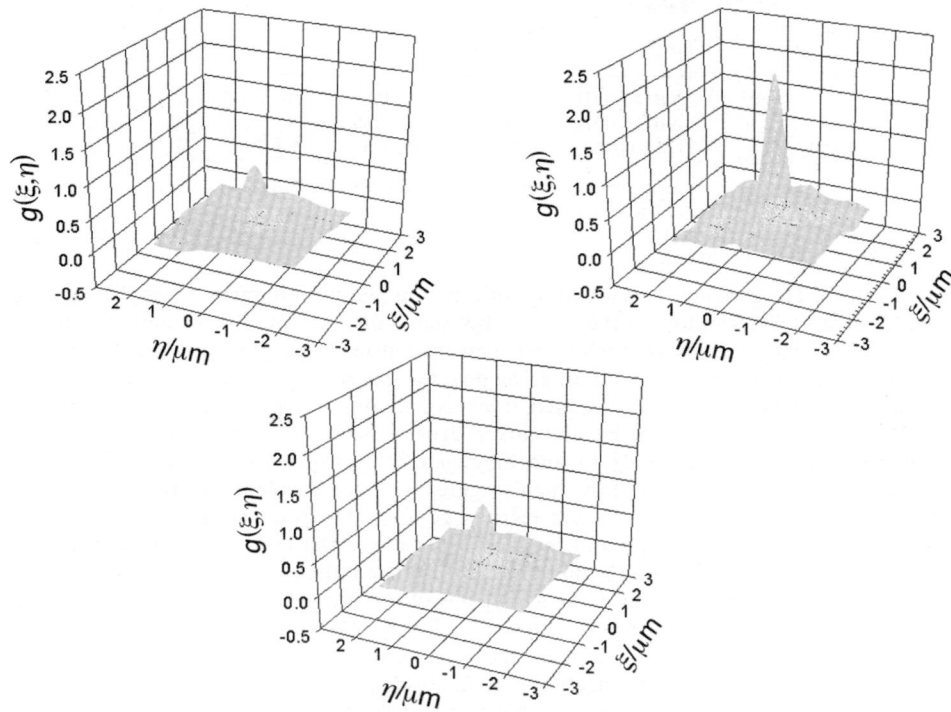

Fig. 4 Auto- and cross-correlation functions for images of HA+8 and clathrin. (a) Autocorrelation function for HA-wt. This is now a large amplitude function since the HA + 8 protein is partly clustered. (b) Autocorrelation function for clathrin. This is a larger function since the protein is highly clustered. (c) Cross-correlation function for these two images. There is a significant and measurable amplitude at zero-lag showing that there is at least some co-localization in these images.

is also consistent with the conclusion that only a fraction of the HA+8 interacts with the coated pits derived above from the cluster density data for the HA+8 (CD = 3.2). At this point we do not know the state of aggregation of the HA+8 outside the coated pits, since we do not know the level of surface expression. In fact, preliminary data suggest that the fraction associated with the coated pits varies with the level of surface expression in a manner predicted from a simple binding model, where monomers associate with a complex with many equivalent binding sites.

Approximately 72% of all the clathrin clusters co-localize with HA+8, suggesting that *ca.* 28% of these protein clusters exist outside the coated pits. This fits very well with the cluster density and spot density data from which we concluded that *ca.* 30% of the clathrin was in the cytoplasm as individual triskelia. Similarly, *ca.* 69% of all the AP-2 clusters co-localize with HA + 8, suggesting that *ca.* 31% of these proteins exist outside the coated pits. We have previously proposed that the AP-2 exists in coated pits and in smaller clusters which could be nucleation sites.[47] If this model is correct, then the cross-correlation data obtained here would be consistent with the AP-2 in coated pits interacting with HA+8, and the AP-2 outside the coated pits not binding to HA+8. This may, however, mean that the interpretation of the AP-2 outside the coated pits as nucleation sites is less certain.

Distribution of gangliosides

Gangliosides have been implicated in the formation of specialized domains in the membranes, such as caveloae and detergent-resistant membranes.[5-7] We have incorporated a fluorescent derivative of the ganglioside GD1a (NBD-GD1a) into CV-1 cells and studied the distribution on the surface of live cells before and after exposure to Sendai Virus at low temperature where no fusion

Table 5 Ganglioside distribution[a]

Probe	Treatment	SD	CD
NBD-GD1a	no virus	0.08	24
NBD-GD1a	virus	0.09	9
DiQ-virus	virus	0.3	0.6

[a] The absolute numbers depend on the concentration of ganglioside applied.

occurs (Table 5). The ganglioside is found in both the top and bottom membranes. To some extent it is in internalized vesicles close to the nucleus, but this is distant from the regions imaged in these cells. Clearly, the spot density is much lower than the cluster density, once again suggesting that the ganglioside exists in two populations, an aggregated one seen in the spots and a dispersed one in the background. When the virus is added, it labels only on the top surface. The spot density for the NBD-GD1a remains the same but the cluster density decreases by more than a factor of two. This suggests that the virus binds to and aggregates the disperse population of gangliosides without affecting the distribution of the highly aggregated ganglioside population or that it binds to the highly aggregated population and recruits gangliosides from the disperse population. In these experiments, the spot density and the cluster density of the virus are comparable, but both are highly variable and depend on the amount of virus added (these are not saturating conditions). There are more virus particles than large aggregates of the ganglioside, providing initial evidence that the large aggregates are not the only sites of binding for the virus.

Table 6 shows the co-localization data for the ganglioside and the virus. Less than half the cells measured show a significant co-localization of ganglioside and virus and, generally, the amplitude of the cross-correlation function is low. Nevertheless, in those cases where the amplitude can be measured, ca. 24% of the ganglioside is associated with virus and ca. 37% of the virus is associated with gangliosides. Visual inspection of the overlayed images shows that virtually none of the large ganglioside clusters overlaps with the virus (not shown). Thus the co-localization is confined to the disperse distribution of the ganglioside, in agreement with the change in cluster density shown in Table 5. For comparison, Table 6 also shows the co-localization of NBD-GD1a with the β-subunit of choleratoxin labelled with biotin. Choleratoxin is known to bind to the ganglioside GM1 and has been used as a marker for caveolae on the surface of cells.[49] It can be seen that over 80% of the cells measured show co-localization of NBD-GD1a with choleratoxin. The fraction of ganglioside co-localized with choleratoxin is still low (28%), but the fraction of choleratoxin clusters that contain NBD-GD1a is over 80%. This is consistent with the choleratoxin binding to the same domains where NBD-GD1a coexists with GM1. Inspection of the images also reveals a significant number of yellow spots in the overlapped images (not shown). For comparison, the lipid probe, NBD-PE, and choleratoxin show little cross-correlation, with no evidence for significant extent of co-localization.

In other work,[48] we have shown that the fusion of Sendai Virus to the cell membrane can be monitored by image correlation spectroscopy at the single cell level by following the dispersal of a dye initially in the virus membrane. This is an example of the reverse of an aggregation process.

Discussion

ICS is a simple image analysis tool which permits quantitative estimates of the density of fluorescent molecules on the surface of cells. When the images are collected with a confocal microscope

Table 6 Ganglioside co-localization

Ganglioside	Protein	Fraction fit	CD_{rg}	$F(G|P)$	$F(P|G)$
NBD-GD1a	virus	0.41	0.62	0.24	0.37
NBD-GD1a	cholera toxin	0.82	2.8	0.28	0.82
NBD-PE	cholera toxin	0.35	0.11	0.05	0.08

and when proper care is taken with control experiments, it is possible to detect receptors present in very low quantities. When the intensity information is used properly, it is also possible to estimate quite accurately the average number of fluorescent molecules per aggregate. This was demonstrated with the PDGF-Receptors on AG1523 cells which exist, on the average, in clusters of 3–4 proteins at a density of *ca.* 2–3 clusters per µm^2. It is possible in these systems to detect single receptors, however, so far none of the systems studied has consisted of monomers only.

The technique is precise enough that small, and possibly significant, changes in cluster densities can be determined when the cell system is perturbed. The study of EGF-Receptors on A431 cells shows that overexpression of the receptor in these cells leads to more and larger aggregates. The density and size of these aggregates is sensitive to temperature, and it is possible to monitor changes by factors of two and less, even when the aggregates are large. When the physical dimensions of the clusters increase, the correlation functions broaden. Deconvolution calculations can, in principle, be used to extract information *ca.* the actual dimensions of the clusters.

In many cases, the cluster density data calculated from ICS are inconsistent with the density of bright spots observed in the image. We argue that when the experiments are performed with the proper controls, this reflects two or more populations of the protein in question. The distribution of the proteins among these populations cannot be determined uniquely from these measurements. Nevertheless, system specific models can be introduced and tested. For example, clathrin should exist in at least two populations: coated pits and cytoplasmic monomers (triskelia). This model then provides enough information to give detailed estimates of the number of coated pits and the fraction of the clathrin they contain. A similar situation obtains with the adaptor protein, but here the data suggest that there are large and small aggregates present. The data are also compatible with a broad distribution of aggregates, but a single aggregate population is ruled out. The third bimodal system is represented by the HA+8 model receptor where some of the receptors are associated with the coated pits while others are free in the membrane.

Interactions between proteins can be studied effectively by using the cross-correlation approach. The amplitude of the cross-correlation function at zero lag is now proportional to the extent of co-localization. If the fraction of images, in which the amplitude of the cross-correlation function is significantly greater at zero lag than elsewhere, is close to one, then most cells exhibit a significant co-localization of the proteins. In these cases, it is possible to calculate the density of clusters in which the proteins interact and the fraction of the clusters of either protein which form the co-localized aggregates.

The density of clusters containing both HA+8 and clathrin is found to be the same as the density of clusters containing both HA+8 and AP-2. The fraction of HA+8 clusters associated with clathrin clusters is also the same as the fraction associated with AP-2 proteins. Finally, the fraction of clathrin clusters associated with HA+8 clusters is the same as the fraction of AP-2 associated with HA+8 clusters. The first two observations are consistent with the proposal that the HA+8-clathrin clusters are the same as the HA+8-AP-2 clusters, *i.e.* they are the coated pits which contain all three proteins. The third observation is probably a coincidence.

For technical reasons,† we have not yet succeeded in measuring the cross-correlation between clathrin and the adaptor protein. Given the data obtained above, we would anticipate a density of co-localized clusters between 0.55 and 0.60 and we expect each of the two proteins to be *ca.* 70% co-localized with the other.

The ganglioside derivative, NBD-GD1a, is found either dispersed or in large, bright clusters. The disperse population appears to be a better receptor for Sendai Virus, which binds and causes a change in the disperse population but not the aggregated one. The extent of co-localization between the ganglioside and the virus is low and there are very few of the virus particles which are co-localized with the large NBD-GD1a clusters. On the other hand, NBD-GD1a is co-localized to a much greater extent with the β-subunit of choleratoxin and there is significant co-localization in the bright spot. This suggests either that NBD-GD1a is a receptor for the toxin or that it is co-localized with GM1, which is known to be a receptor for the toxin. In either case, this co-localization is predominantly in the large aggregates. These may be either the caveolae or the

---

† This experiment requires simultaneous labelling intracellularly so the primary antibodies cannot be from the same source (mouse). While antibodies from other sources are available, we have not yet been able to establish the correct conditions. Hopefully, the problem will be solved soon.

putative detergent resistent membranes. Future cross-correlation experiments with caveolin and lipid anchored proteins may help distinguish these possibilities.

Conclusions

Careful measurements of fluorescence emission from cell surfaces using confocal microscopes can provide high quality images of the distribution of molecules in cells or on their surface. Calculation of the autocorrelation function of a single image yields quantitative estimates of the density of clusters and the number of molecules per cluster. Calculation of the cross-correlation function between a pair of images of different proteins yields an estimate of the fraction of co-localization and the density of mixed protein clusters. Measurements on a large number of cells give good statistical information and, hence, reliable conclusions about the protein distributions and the intermolecular interactions. Since the experiments can be performed on a rapid timescale (seconds to minutes), kinetics of redistribution of proteins and rates of intermolecular associations are accessible through these calculations. Complex systems, such as receptors interacting with coated pits or virus particles binding to and fusing with cells, can be investigated systematically to provide better insight into the mechanism of intermolecular associations.

ICS is an 'averaging' technique, in that all the fluorescence information in an image is reduced to a single number. This means that details are lost and specific information about the protein distributions is difficult to extract. However, the averaging provides great sensitivity to weak signals that may otherwise be overlooked. The ease of collecting images on modern confocal microscopes and the relatively simple calculations involved allow for a vast amount of data to be collected and processed in a short time. This means better experiments and better controls. Some argue that a single picture is worth a thousand words. ICS can reduce a thousand pictures to a single number. In many systems, this is important and worthwhile.

Acknowledgements

We greatly appreciate the many fruitful interactions with our collaborators: Karl-Eric Magnusson and Birgitta Rasmusson at Linkoping University in Sweden; Richard Epand at McMaster University in Canada; Tom Flanagan at the State University of New York at Buffalo in the United States; Yoav Henis at Tel Aviv University in Israel and Michael Roth at Southwestern Medical School in the United States.

References

1. D. Sheets, R. Simson and K. Jacobson, *Curr. Opin. Cell Biol.*, 1995, **7**, 707.
2. A. Kusumi and Y. Sako, *Curr. Opin. Cell Biol.*, 1996, **8**, 556.
3. S. L. Schmid, *Annu. Rev. Biochem.*, 1997, **66**, 511.
4. R. G. Parton and K. Simons, *Science*, 1995, **269**, 1398.
5. D. A. Brown and J. K. Bose, *Cell*, 1992, **68**, 533.
6. R. G. Parton, *J. Histochem. Cytochem.*, 1994, **42**, 155.
7. T. Harder and K. Simons, *Curr. Opin. Cell Biol.*, 1997, **9**, 534.
8. A. Sorkin and C. M. Waters, *Bioessays*, 1993, **15**, 375.
9. J. Schlessinger, *Trends Biochem. Sci.*, 1998, **13**, 443.
10. T. Kirchhausen, J. S. Bonifacino and H. Riezman, *Curr. Opin. Cell Biol.*, 1997, **9**, 488.
11. T. Kirchhausen, *Curr. Opin. Cell Biol.*, 1993, **3**, 182.
12. C. Lamaze and S. L. Schmid, *Curr. Opin. Cell Biol.*, 1995, **7**, 573.
13. I. V. Sandoval and O. Bakke, *Trends Cell Biol.*, 1994, **4**, 292.
14. W. Boll, H. Ohno, Z. Sangyang, I. Rapoport, L. C. Cantley, J. S. Bonifacino and T. Kirchhausen, *EMBO J.*, 1996, **15**, 5789.
15. H. Ohno, J. Stewart, M. Fournier, H. Bosshart, I. Rhee, S. Miyatake, T. Saito, A. Gallusser, T. Kirchhausen and J. S. Bonifacino, *Science*, 1995, **169**, 1872.
16. J. E. Heuser, *J. Cell Biol.*, 1980, **84**, 560.
17. J. E. Heuser and J. H. Keen, *J. Cell Biol.*, 1988, **107**, 877.
18. D. Gross and W. W. Webb, *Biophys. J.*, 1986, **49**, 901.
19. J. B. Pawley, *Handbook of Biological Confocal Microscopy*, Plenum Press, New York, 2nd edn., 1995.
20. P. W. Wiseman, P. Höddelius, N. O. Petersen and K. E. Magnusson, *FEBS*, 1997, **401**, 43.
21. M. Srivastava and N. O. Petersen, *Meth. Cell Sci.*, 1996, **18**, 47.

22 B. van Steensel, E. P. van Binnendijk, C. D. Hornsby, H. T. M. van der Voort, Z. S. Krozowski, E. R. de Kloet and R. van Driel, *J. Cell Sci.*, 1996, **109**, 787.
23 D. Demandolx and J. Davoust *J. Microsc.*, 1997, **185**, 21.
24 N. O. Petersen, *Can. J. Biochem. Cell Biol.*, 1984, **62**, 1158.
25 C. M. Brown, PhD Thesis, The University of Western Ontario, London, Canada, 1998.
26 C. M. Brown, *Biochim. Biophys. Acta*, 1998, submitted.
27 N. O. Petersen, P. L. Hoddelius, P. W. Wiseman, O. Seger and K. E. Magnusson, *Biophys. J.*, 1993, **65**, 1135.
28 B. J. Berne and R. Pecora, *Dynamics Light Scattering with Applications to Chemistry, Biology and Physics*, Wiley, New York, 1976, pp. 10–22.
29 N. Davidson, *Statistical Mechanics*, McGraw-Hill, 1962.
30 N. O. Petersen, *Biophys. J.*, 1986, **49**, 809.
31 E. Fire, C. M. Brown, R. G. Roth, Y. I. Henis and N. O. Petersen, *J. Biol. Chem.*, 1997, **272**, 29538.
32 P. R. St-Pierre and N. O. Petersen, *Biochemistry*, 1992, **31**, 2459.
33 Z. Foeldes-Papp, A. Schnetz and R. Riegler, *Biophys. J.*, 1998, **74**, A184.
34 N. O. Petersen, S. Felder and E. L. Elson, in *Handbook of Experimental Immunology*, ed. D. M. Weir, L. A. Herzenberg, C. C. Blackwell and L. A. Herzenberg, Blackwell Scientific, Edinburgh, 1985, ch. 24.
35 N. O. Petersen and E. L. Elson, in *Methods in Enzymology*, ed. C. H. W. Hirs and S. N. Timasheff, Academic Press, New York, 1985.
36 W. Mendenhall, *Introduction to Probability and Statistics*, 7th edn., 1987, pp. 64–65.
37 A. G. Benn and R. J. Kulperger, *Environmetrics*, 1996, **7**, 167.
38 P. W. Wiseman, PhD Thesis, The University of Western Ontario, London, Canada, 1995.
39 R. A. Seifert, C. E. Hart, P. E. Phillips, J. W. Forstrom, R. Ross, M. J. Murray and D. F. Bowen-Pope, *J. Biol. Chem.*, 1989, **264**, 8771.
40 T. Kawamoto, J. D. Sato, A. Le, J. Polikoff, G. H. Sato and J. Mendelsohn, *Proc. Natl. Acad. Sci. USA*, 1983, **80**, 1337.
41 M. S. Robinson, *Curr. Opin. Cell Biol.*, 1994, **6**, 538.
42 G. Odorizzi, C. R. Cowles and S. D. Emr, *Trends Cell Biol.*, 1998, **8**, 282.
43 R. G. W. Anderson, M. S. Brown and J. L. Goldstein, *Cell*, 1977, **10**, 351.
44 J. Lazarovits and M. G. Roth, *Cell*, 1988, **53**, 743.
45 D. E. Zwart, C. B. Brewer, J. Lazarovits, Y. I. Henis and M. G. Roth, *J. Biol. Chem.*, 1996, **271**, 907.
46 E. Fire, O. Gutman, M. G. Roth and Y. I. Henis, *J. Biol. Chem.*, 1995, **270**, 21075.
47 C. M. Brown, *J. Cell Sci.*, 1998, **111**, 271.
48 B. J. Rasmusson, T. D. Flanagan, S. J. Turco, R. M. Epand and N. O. Petersen, *Biochim. Biophys. Acta*, 1998, **14357**, 1.

Paper 8/06677I

Direct measurement of recognition forces between proteins and membrane receptors

Paul F. Luckham* and Kate Smith

Department of Chemical Engineering, Imperial College of Science, Technology and Medicine, Prince Consort Road, London UK SW7 2BY. E-mail: p.luckham01@ic.ac.uk

Received 9th September 1998

The interactions between the protein, cholera toxin B subunit attached to an atomic force microscope, AFM, cantilever, CTB and its receptor the ganglioside, GM_1 have been measured in a dilute electrolyte solution, pH 5.5. Although there is variation in the force separation data obtained, particularly on approach of the AFM tip to the GM_1 surface where usually, but not always an attraction is noted, an adhesion is always noted on separation of the surfaces. The strength of this adhesion varies from experiment to experiment, but appears to be quantised at a value of around 90 pN. Addition of cholera toxin to the aqueous electrolyte solution completely removes the attractive interaction and adhesion. This gives us confidence that in the earlier experiments, a specific interaction between the CTB and GM_1 was measured.

Introduction

Proteins are large, structurally diverse and complex molecules. Their interactions are governed by the subtle interplay of a variety of forces. Successful exploitation of the properties of proteins, important to fields as diverse as therapeutic drug development and washing powder technology, depends on a full knowledge of their behaviour. Many different methods have been employed in an attempt to measure the interactions between proteins directly. However, only one of these has come close to recording the interaction between two discrete molecules. This technique has ultimately evolved from the scanning tunnelling microscope (STM).[1]

Binnig *et al.* published their first STM data in 1982. The technique was purportedly capable of producing constant charge images, of electronically conducting sample surfaces, on an atomic scale. The work was ground-breaking, and earned them the Nobel Prize for Physics in 1986. However, Binnig *et al.* were not satisfied with the limitations of the STM imposed by its inability to image electrically insulating samples, such as biological materials. In fact, in the same year that they were awarded the Nobel Prize, Binnig *et al.* published the sequel to the STM; the atomic force microscope (AFM).[2] This improved version utilised a different tip–sample interaction from the STM, and produced constant tip–sample force images of the sample surface.

The change in the nature of the tip opened up two main avenues of research; a star burst of related force microscopes, all relying on a different type of tip–sample interaction, and adaptation of the function of the AFM to form a sensitive force sensor. This modification allowed the measurement of the interaction, on the nanometre and sub-nanoNewton scale, between any two materials that could be made to form a tip and a sample. Thus the modified AFM was born.

In this study we have used a modified AFM to measure the interaction between two molecules of biological importance. The molecules chosen were the 'cholera toxin B subunit' (CTB), which is responsible for infection of the human body by the disease, cholera, and 'ganglioside GM_1' (GM_1), which is known to be the natural receptor on human intestinal cells.

Cholera is a pandemic and epidemic disease that has afflicted the human population for over 2 millennia.[3] The disease can strike without warning and may kill within 6 h, although it requires a prior incubation period of between 12 and 28 h. The first symptom is abrupt, painless, and copious 'rice-water' diarrhoea; up to 20 l in 24 h.[4] This debilitating diarrhoea is later joined by vomiting, causing severe dehydration. Further symptoms are withered skin, falling blood pressure, and severe muscular cramps. The corpses of cholera victims have a characteristic blue–black, desiccated appearance.[5]

It is now known that *V. cholerae* adhere to and colonise the small intestine,[3,6] where they secrete an exotoxin, known as cholera toxin. The toxin consists of two polypeptide subunits, denoted cholera toxin A subunit (CTA), and cholera toxin B subunit (CTB). The B subunit binds to receptors on the epithelial cell membrane of the small intestine, particularly the jejunum and duodenum.

CTA does not bind to, and, on its own, is therefore not toxic to, intact cells. It is transferred across the cell membrane, where it triggers a biological cascade that ultimately results in cell death. Once inside the cell, however, the CTA can not infect further cells and is therefore lost when the cell dies. Thus cholera is self-limiting, continuing only as long as the *V. cholerae* bacteria remain in the gut, secreting replacement toxin.[5]

The accepted structure of the B subunit of cholera toxin (CTB), also known as 'choleragenoid', is pictured in Fig. 1.[7,8] CTB is a homopentameric protein, *i.e.* it consists of five identical monomers. It assumes the appearance of a doughnut-shaped ring, with a central pore formed by the long α-helices on each monomer. Viewed from the side, see Fig. 1, CTB has a smooth upper surface and a rough lower surface. The roughness of the lower face is the result of the extension of the long central helices beyond the monomer β-sheets. Synthetic peptides with the same structure as this rough region elicit production of protective antibodies *in vivo*. This ability implies that it is the rough surface that is responsible for binding target cells.

CTB stands between 3 and 4 nm high and has a quoted diameter ranging from 5.2 to 7.0 nm. The width of the central pore ranges from 1.1 nm at its smooth surface to 1.5 nm at the rougher side, and the length of the pore has been quoted as 3.0 nm.[7,9,10] The inner surface of the pore is highly hydrophilic, possessing no less than 25 positive and 15 negative charges. The pentavalent nature of CTB is achieved by the non-covalent interlinking of β-sheets on neighbouring monomers. As mentioned previously, it is CTB that is responsible for the binding of cholera toxin to membranes, targeting those of the mucosal cells of the small intestine. Alone, CTB is biologically inactive and will not induce symptoms associated with cholera. CTB, then, must act as a key to unlock the target cell for the toxin and effect entry of the active CTA. However, the exact mechanism by which CTA enters the cell is still a mystery.

The gangliosides are a structurally diverse group of lipids, found in the outer cell membrane of almost all vertebrate cells and tissues. They are a minor component in most cells,[11] except for

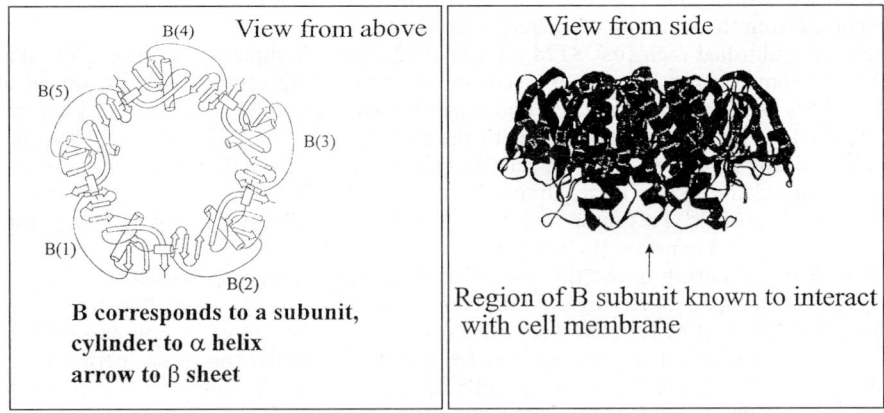

Fig. 1 Cholera toxin B subunit.

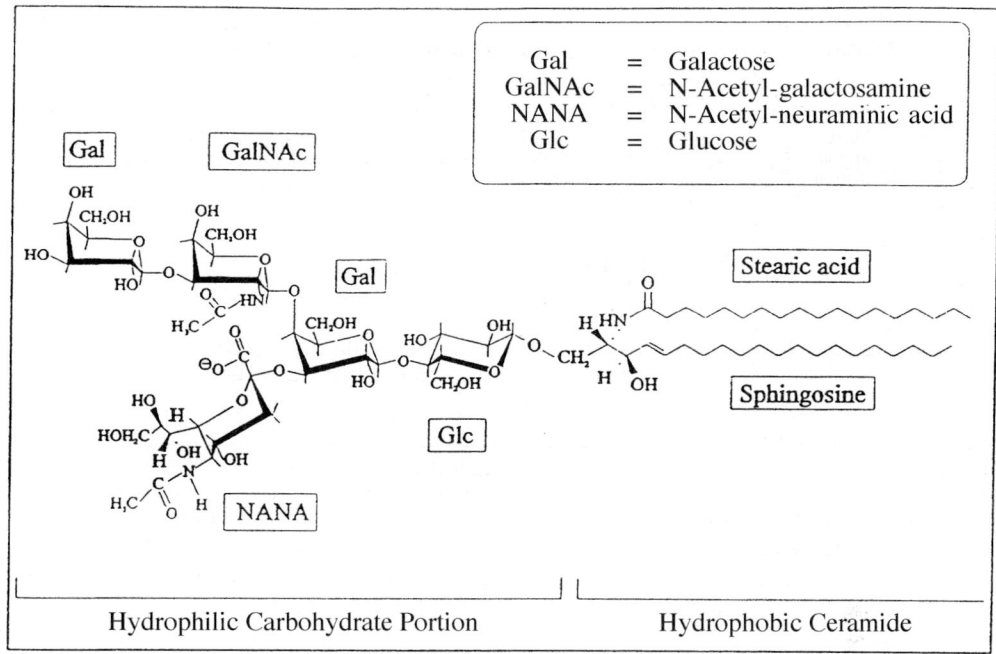

Fig. 2 Structure of ganglioside GM_1.

those found in nervous tissue where they are more common. Although they have been implicated in various cell recognition and signalling phenomena, their exact role in recognition remains a mystery. GM_1 is one of a family of sialic acid-containing glycosphingolipids. Its structure is depicted in Fig. 2. It is a ceramide-based glycolipid that possesses N-acetylneuraminic acid (NANA), in common with most gangliosides.[12] A ceramide is a sphingosine base that is amide-linked to a fatty acid, as indicated in Fig. 2. GM_1 is large in size relative to most other membrane lipids. It has a negatively charged head-group some 3.0 nm long, which extends out from the membrane surface by 2.5 nm.

Experimental

The GM_1, dipalmatoylphosphatidylcholine, DPPC, and the cholera B subunit were obtain from Sigma chemicals, all other chemicals were obtained from Aldrich Chemicals.

Having decided to study these two materials, the next task is to find a way in which they may be attached to surfaces in the correct orientation for a recognition interaction to occur. There are two criteria to be met; the surface–tip connection must be strong, and the molecule must be oriented correctly.

Careful inspection of the cholera toxin-B subunit's structure reveals that there is a preponderance of primary amine groups on the side that is not specific to GM_1. A method is therefore required that will link a silicon substrate specifically to an amine group.

There are many different ways by which proteins can be immobilised to a surface, one which meets the criteria was reviewed in 1994 by Williams and Blanch.[13] The method used, see Fig. 3, employs gluteraldehyde as a covalent cross-linker between the primary amine on a surface and a primary amine on the protein. This method was used to attach the CTB subunit to the silicon tip of the AFM cantilever used in these experiments.

Since the B-subunit molecules form a necessarily dense coverage on the cantilever tip, the sample coverage must be sparse in order to increase the likelihood of the detection of a discrete B-subunit–GM_1 interaction. A method that is perfect for the control of the spatial concentration of a molecule is Langmuir deposition.[14]

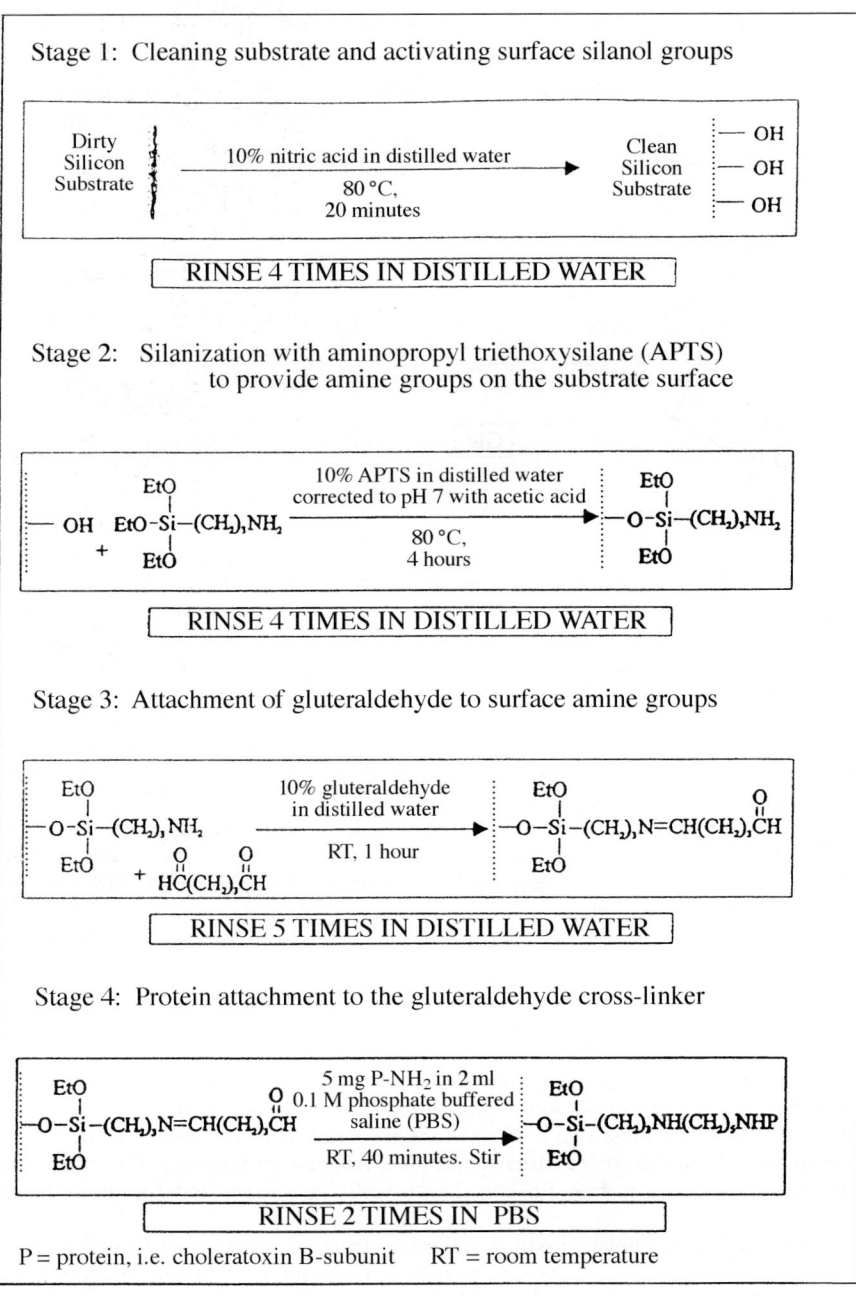

Fig. 3 Protocol adopted to attach the cholera toxin to a silicon atomic force microscope tip.

The GM_1 molecule has a hydrophilic head group and a hydrophobic tail. When embedded in a cell membrane *in vivo*, the tail group anchors the molecule in the lipid bilayer coating of the cell, and it is the head group that interacts with extracellular moieties, such as the B-subunit of cholera toxin.

The sample surface must therefore present the GM_1 head groups to the tip. This means that the sample substrate must be hydrophobic. Since glass is hydrophilic, some coating must be used to make the glass sample surface hydrophobic. Such a coating may be provided by silanisation.

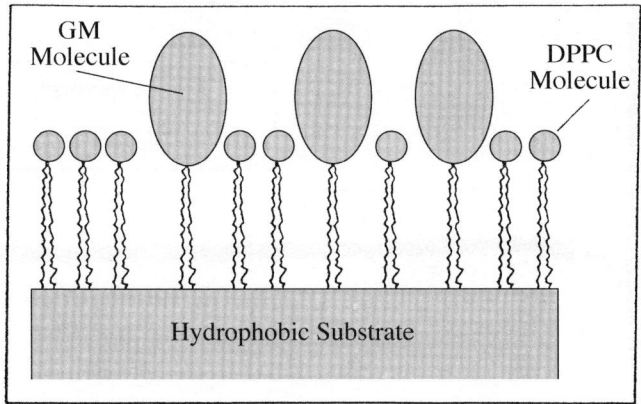

Fig. 4 Schematic representation of the way the mixed monolayer of GM_1 and DPPC molecules may appear on a hydrophobic glass surface.

As mentioned previously, the B-subunit-coated tip has a high density coverage of molecules. Thus the sample surface must have a low density coverage of GM_1 molecules. To ensure an even distribution of molecules on the substrate, however, the dipped monolayer must also be densely packed. This conflict may be overcome by diluting the GM_1 molecules in the monolayer with other, noninteractive, molecules. Such a molecule is dipalmitoylphosphatidylcholine, DPPC. Its hydrocarbon chains are very similar in length to those of GM_1, but its head group is shorter and narrower.

When a monolayer of a mixture of the two molecules is deposited onto a substrate such that the head groups of the molecules are uppermost, the GM_1 molecules tower above the DPPC molecules. This situation is represented schematically in Fig. 4.

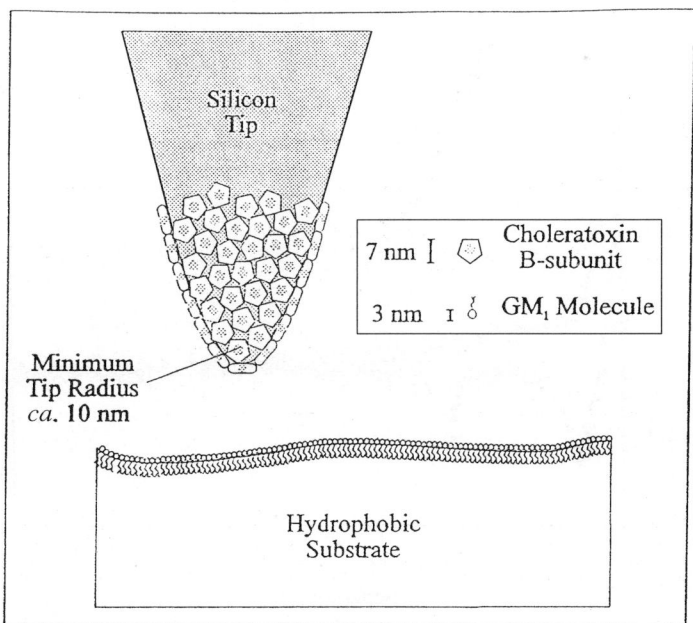

Fig. 5 Schematic representation of the way the modified AFM tip and the lipid monolayer may appear close to contact of the AFM tip to the lipid surface. This diagram is approximately to scale.

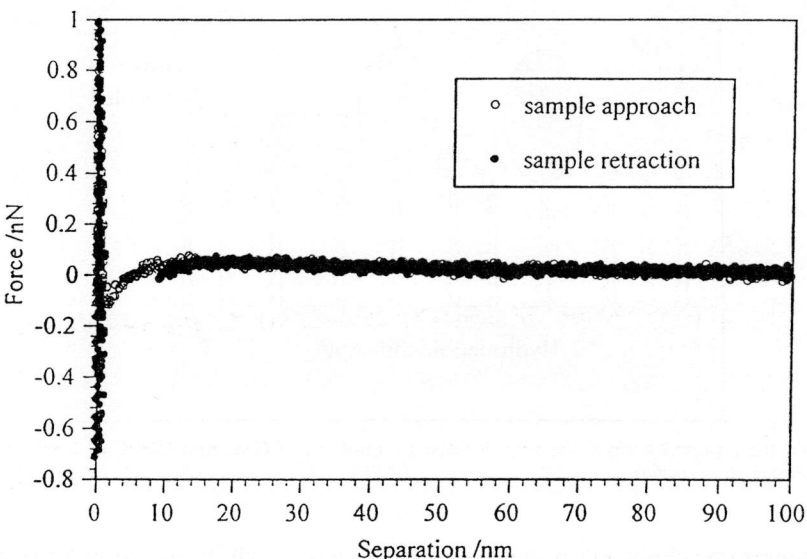

Fig. 6 Forces of interaction between an AFM tip bearing attached cholera toxin subunit B and lipid surface of GM_1 and DPPC (molar ratio 1:4) in water. The open symbols correspond to approach of the surfaces and closed symbols to separation.

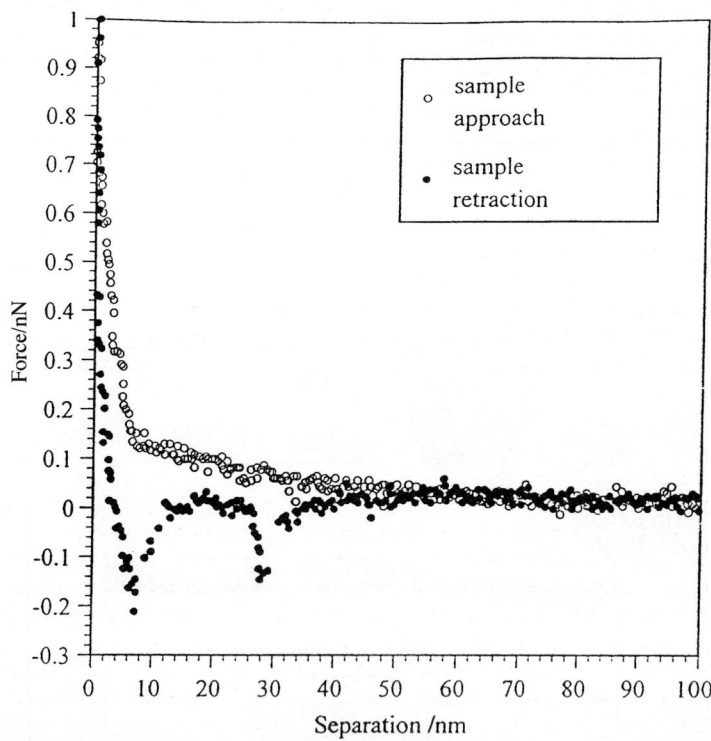

Fig. 7 Forces of interaction between an AFM tip bearing attached cholera toxin subunit B and lipid surface of GM_1 and DPPC (molar ratio 1:4) in water. The open symbols correspond to approach of the surfaces and closed symbols to separation.

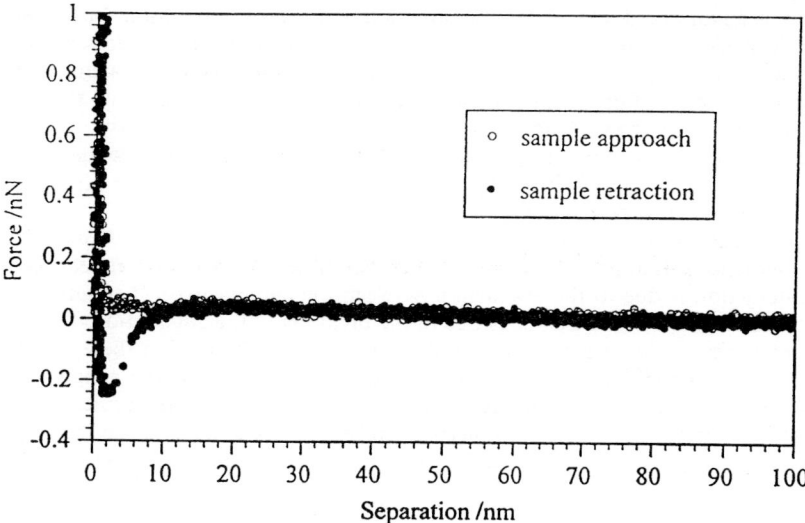

Fig. 8 Forces of interaction between an AFM tip bearing attached cholera toxin subunit B and lipid surface of GM_1 and DPPC (molar ratio 1 : 4) in water. The open symbols correspond to approach of the surfaces and closed symbols to separation.

The GM_1 molecules were therefore diluted to one fifth of the total number of molecules of GM_1 and DPPC in the spreading solution, a ratio of 1 : 4, GM_1 : DPPC. Isotherms of mixtures of the two species have been studied extensively, and have been found to exhibit features intermediate to those of pure GM_1 and pure DPPC. This indicates that the two molecules form a miscible monolayer.[15]

The solvent for the spreading solution was itself a mixture of 1 : 4 methanol : chloroform, as GM_1 and DPPC are not very soluble in chloroform alone. The substrate used was a glass microscope coverslip, small enough to be submerged fully into the trough. The substrate was dipped at a pressure of 30 N m^{-1}, corresponding to an area per molecule of 0.40 nm^2.[15] Once the substrate had been dipped, the surface of the trough was sucked clean, and the AFM fluid cell was introduced into the trough. The substrate and Teflon holder were then transferred, under water, to the fluid cell of the AFM. Thus, using these methods it was possible to orient both the CTB and the GM_1 in a manner similar to that which occurs *in vivo*, a schematic of the way these macromolecules are oriented, drawn to scale, is shown in Fig. 5.

The modified AFM used for these measurements was a purpose built force sensing piece of equipment similar to that described by Braithwaite *et al.*;[16] it is based around the principles of most commercial AFM equipment.

Results and discussion

The interaction between CTB molecules, attached to a silicon tip by gluteraldehyde cross-linkers, and GM_1 molecules, diluted by DPPC molecules and deposited onto a silanized glass substrate, was measured repeatedly at various sample sites, and at various sample speeds.

In each experiment over 100 compression and decompression profiles for the interactions between the modified AFM tip and the ganglioside GM_1 have been taken. The rate at which the profiles were obtained varied between 2 and 0.015 Hz. As will be seen when we examine the data, there is some difference between the compression/decompression force profiles, however there was no observable difference between the rates at which the compressions were obtained. Basically, three different forms of compression profiles could be obtained. In Fig. 6, for example we may see that there is no interaction between the modified tip and the GM_1 surface until the tip is some 10 nm from the ganglioside surface whereupon an attractive interaction is observed. On separation, an adhesion is noted, but more of this later. In Fig. 7, we note that there is a repulsion between the two surfaces commencing as a surface separation of around 30 nm but that at shorter separations an attraction is again noted. Once more an adhesion is noted on separation. In Fig. 8, there is no observable attractive interaction on approach of the surfaces until the surfaces come close to an intimate contact, whereupon a strong, hard wall, repulsion is observed and again an attraction is noted on separation. In Fig. 9, four compression/decompression profiles at the same site are plotted where we can see the variation that is observed in the data more clearly.

There are some features that are common to all the data sets shown here and that have been obtained through the course of these experiments. Firstly on separation an attraction is always noted, this adhesion may vary between 0.15 and 0.9 nN. Also a hard wall type repulsive interaction at short separations, less than 2 nm, is always observed. In roughly 3/4 of the data sets an attractive interaction on approach of the surfaces was observed. It would appear likely that this attractive interaction is due to the interaction between the cholera toxin B subunit and the GM_1 surface and that the adhesion observed is due to the breaking of this interaction.

Let us then consider the origin of the variation in the data observed in these experiments. The repulsion on approach of the cholera toxin coated AFM tip toward the ganglioside surface is likely to be steric in origin and be due to the cholera toxin B subunit molecules on the AFM tip being out of alignment with the GM_1 molecules on the sample surface. Since the CTB molecules are effectively tethered onto the AFM tip by molecular "ropes" they will possess a certain degree

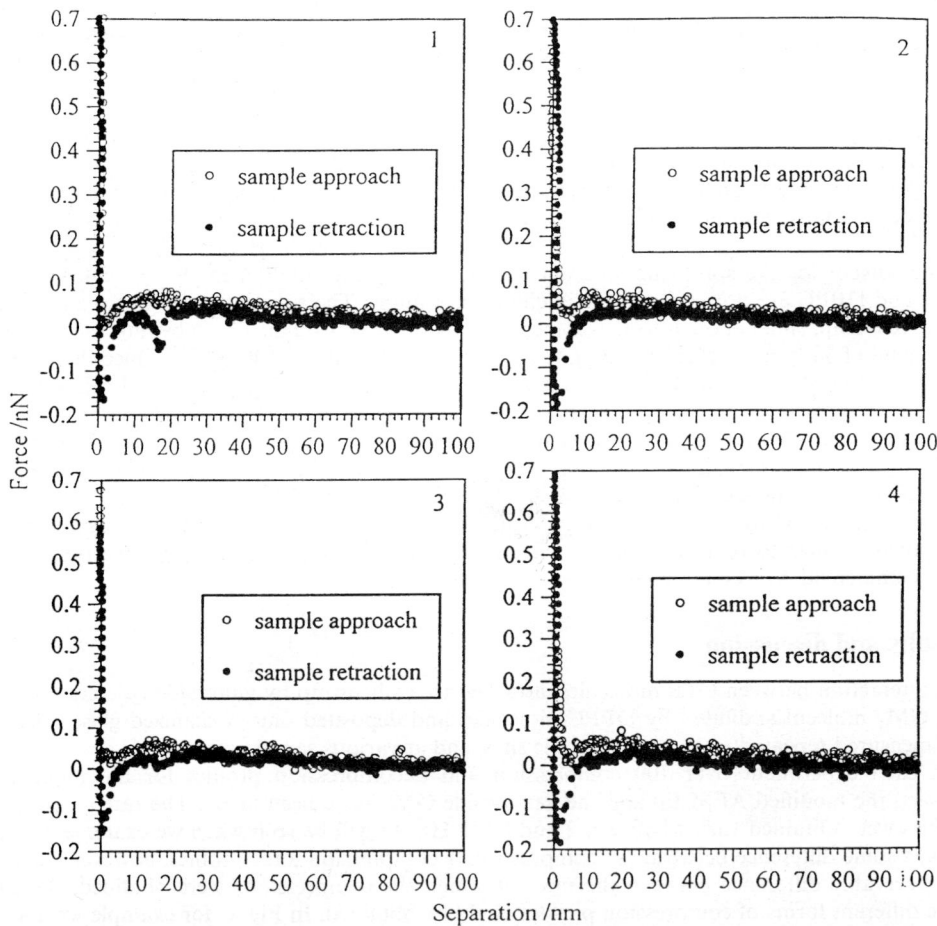

Fig. 9 Forces of interaction between an AFM tip bearing attached cholera toxin subunit B and lipid surface of GM_1 and DPPC (molar ratio 1:4) in water. The open symbols correspond to approach of the surfaces and closed symbols to separation. Here four repeat experiments at the same contact position on the surface are shown.

of rotational freedom. Clearly, the extent of this interaction will be different depending on the orientation of the molecule. Three mechanisms are proposed here for the AFM tip interaction with the GM1 surface and these are illustrated in Fig. 10.

Each interaction in Fig. 10 depends on the initial orientation of the CTB molecule. An initial side-on approach by the CTB molecule will result in a steric repulsion. If the CTB molecule is made to rotate by this repulsive force, then the CTB and GM_1 molecules will be favourably oriented, and an attraction may occur. Otherwise, the entire approach cycle will appear to be repulsive. Alternatively, if the CTB molecule happens to be oriented correctly on approach of the sample surface, then no initial repulsion will be experienced and only an attraction will be evident.

It is important to remember that the force–separation curves actually measured will not be as simple as those described in Fig. 10. In reality, the tip surface will be covered with CTB molecules,

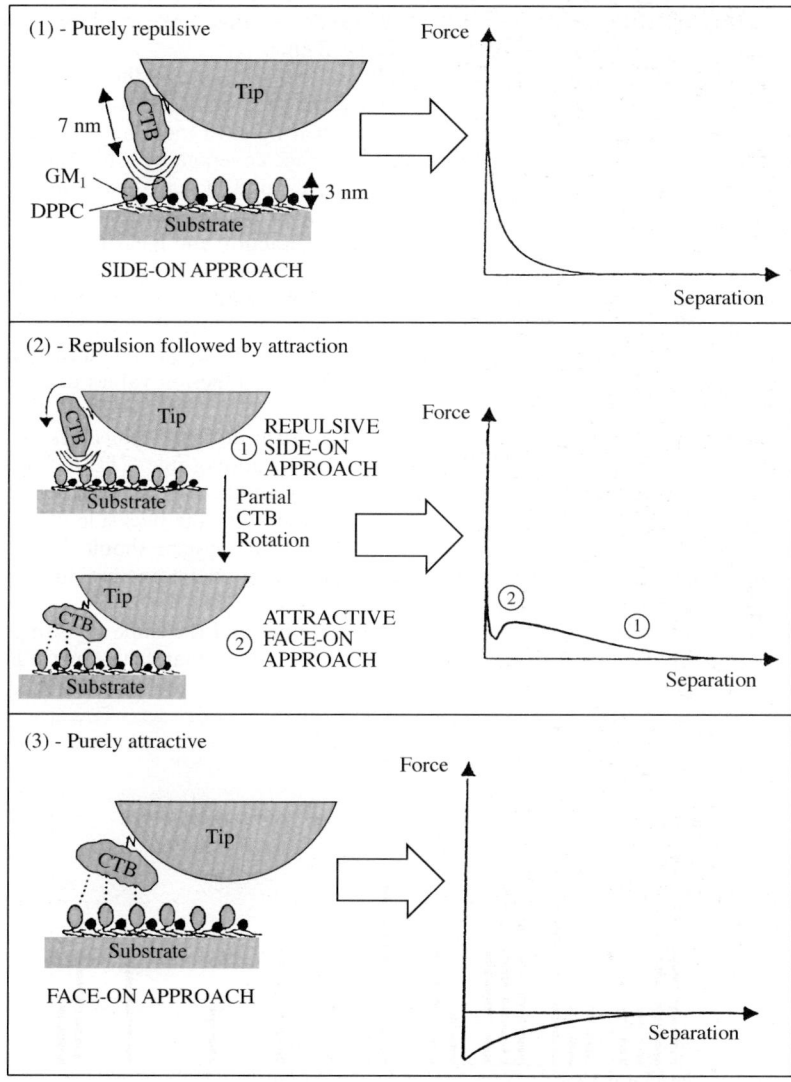

Fig. 10 Schematic representation of the different ways the CTB subunit and the GM_1 ganglioside can interact together with the predicted force curve for the approach of the surfaces. Note that in every case, following full compression an adhesion would be expected on separation.

and the resultant approach portion of the force–separation curve will most likely be a random convolution of more than one of the CTB–GM_1 interactions in Fig. 10.

Let us now look in more detail at the force profiles for separation of the AFM cantilever from the AFM tip. In most instances a clean separation, such as that observed in Fig. 6 is observed. However, on occasions, a more ripping like separation is noted, see for example Fig. 7 and 9 graph 1. It is likely that in these cases the cantilever has initially separated from some of the specifically interacting sites, but not all of them, and further separation of the AFM tip is required to fully separate the surfaces.

The likelihood of achieving the measurement of a single CTB–GM_1 interaction depends on the distribution of the GM_1 molecules on the sample surface. This may be calculated approximately from the isotherm of the DPPC–GM_1 mixture used. The dipping pressure used (25 mN m^{-1}) corresponds to an average area per molecule of 0.40 nm^2. Since the head groups of the GM_1 molecules tower above the DPPC molecules, effectively dominating the sample surface, the area occupied by the DPPC molecules may be added to the area available to the GM_1 head groups. The ratio of GM_1 to DPPC molecules is 4:1. Thus the average area available to each GM_1 molecule is five times the average area per molecule, *i.e.* 2 nm^2.

The approximate area occupied by a single CTB molecule is equal to the square of its radius, *i.e.* 49 nm^2. Assuming that each subunit of the CTB pentamer occupies a sixth of the total area, allowing space for the central barrel, the area commanded by each monomer in the CTB pentamer is around 8 nm^2. Thus the density of the sample surface coverage by GM_1 molecules is such that, not only is the tip–sample interaction likely to correspond at least to the interaction of an entire CTB pentamer, it is also very unlikely to correspond to the interaction of a single CTB molecule. Since it is not possible to view the tip–sample interaction and measure it simultaneously, there is therefore no direct way of determining the number of interacting CTB–GM_1 pairs. This problem may be overcome by repeating the measurement of the interaction many times. Collation of the interaction data should reveal a frequency distribution in which the interactions may occur in discrete groups. The spacing between these groups should be equal to the magnitude of the interaction of a single CTB–GM_1 pair. Fig. 11 reveals that the adhesion values occur in multiples of approximately 0.09 nN, as marked.

If the interaction measured is indeed that between CTB and GM_1 molecules, as postulated above, then it should be possible to block the interaction by adding free CTB molecules in solution to the sample chamber.[17,18] The free CTB molecules should cover the GM_1-covered surface completely, leaving the CTB molecules on the tip to interact with the backside of a CTB monolayer. Thus the initially attractive interaction of the CTB–GM_1 system should be replaced by a repulsive CTB–CTB interaction. Such a change in the measured interaction may be taken to imply that the initial CTB–GM_1 interaction has been blocked.

Free CTB molecules were injected into the sample chamber and the same tip–sample measurement was repeated. If the interaction measured above is indeed that between CTB and GM_1

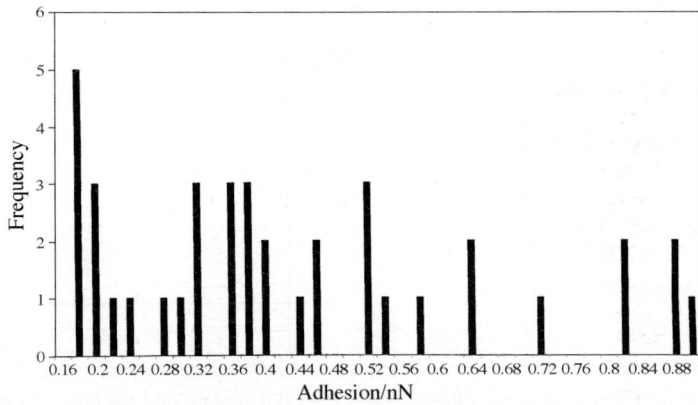

Fig. 11 Adhesion for the interaction between CTB and GM_1 plotted as function of the frequency of occurrence. Note that the adhesion seems to be roughly quantised in values of 0.09 nN.

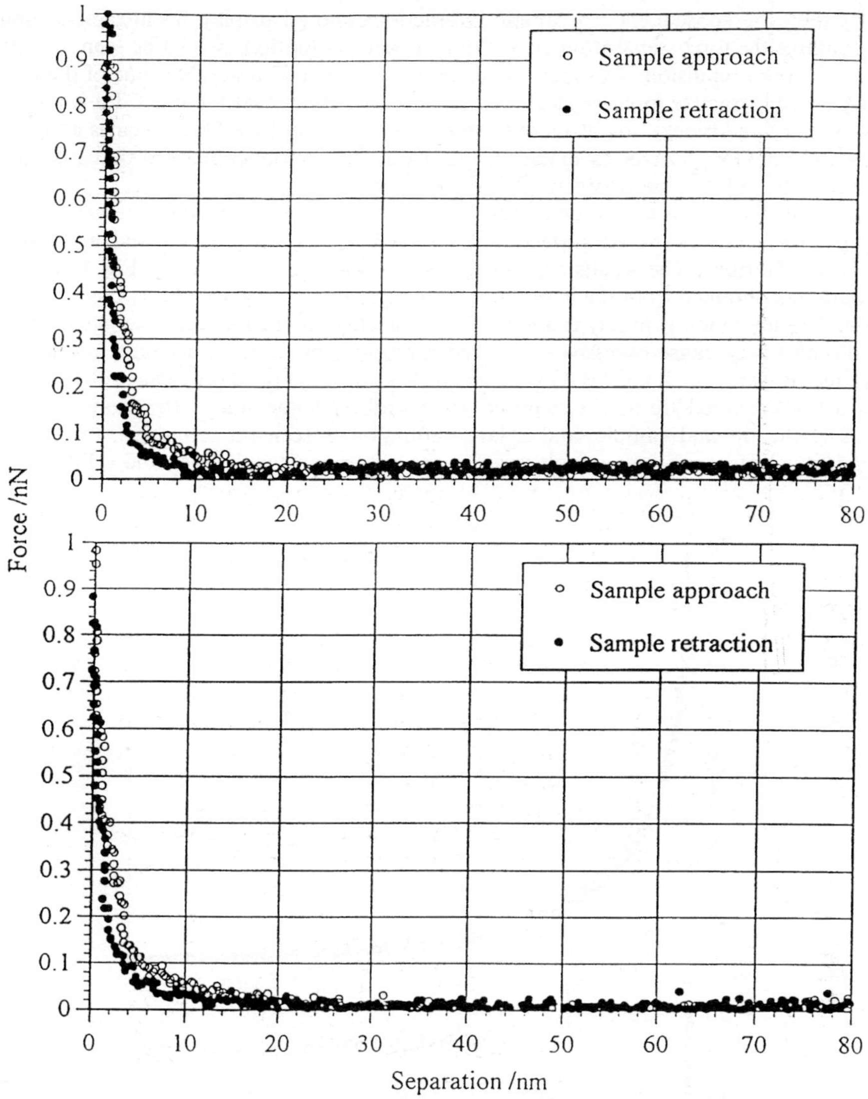

Fig. 12 Forces of interaction between an AFM tip bearing attached cholera toxin subunit B and a lipid surface of GM_1 and DPPC (molar ratio 1:4) in water. In this experiment a dilute solution of free CTB was added to the solution. Two force profiles are shown. The open symbols correspond to approach of the surfaces and closed symbols to separation.

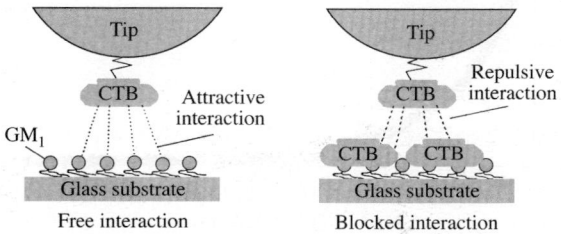

Fig. 13 Schematic representation of the way CTB in solution can block the interaction between the CTB coated AFM tip and the ganglioside GM_1.

molecules then the subsequent tip–sample interactions should display no attractive interaction. Fig. 12 contains the force–separation curves of just such a situation. As can be seen, the attraction is replaced by pure repulsion. This measurement was repeated 47 times. Not one of these measurements produced attractive force–separation curves. This implies that the free CTB is adsorbing on the GM_1 surface and blocks any attractive interaction between the CTB molecules attached to the AFM tip and the GM_1 surface as shown in Fig. 13. It is therefore contended that the earlier data correspond to the interaction between CTB and GM_1 molecules.

To further validate the above conclusion, as control experiments the interaction of a bare silicon tip with a GM_1-covered surface and between a CTB-covered tip and a silanized glass surface were performed. The results are shown in Fig. 14 and 15 respectively. Fig. 14 contains two typical data sets obtained from the interaction of a bare silicon tip with a GM_1-covered surface. In each case, the interaction is purely repulsive. Thus the effect of incomplete coverage of the tip with CTB molecules may cause excessive measured repulsion, but cannot account for the attractive interactions measured in the CTB–GM_1 system. Fig. 15 displays data typical of the interaction between a CTB-covered tip and a silanised glass surface. Some small attraction is observed on approach of the tip and sample, and a large adhesion is seen on retraction, of 320 pN. This adhesion is much stronger than any observed in the experiments with GM_1 and CTB.

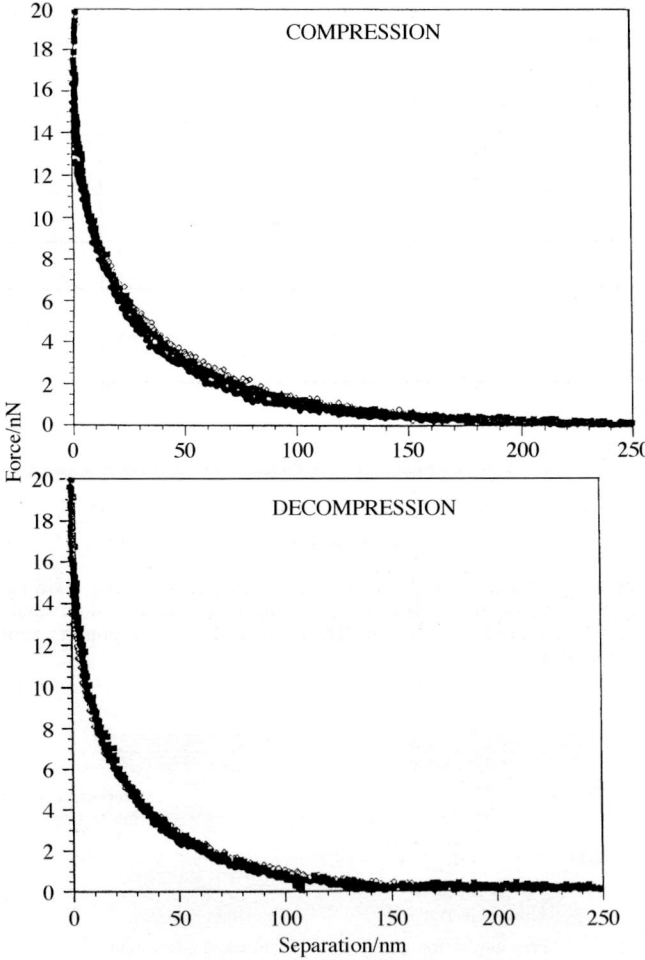

Fig. 14 Forces of interaction between a bare AFM tip and a lipid surface of GM_1 and DPPC (molar ratio 1:4) in water. The open symbols correspond to approach of the surfaces and closed symbols to separation.

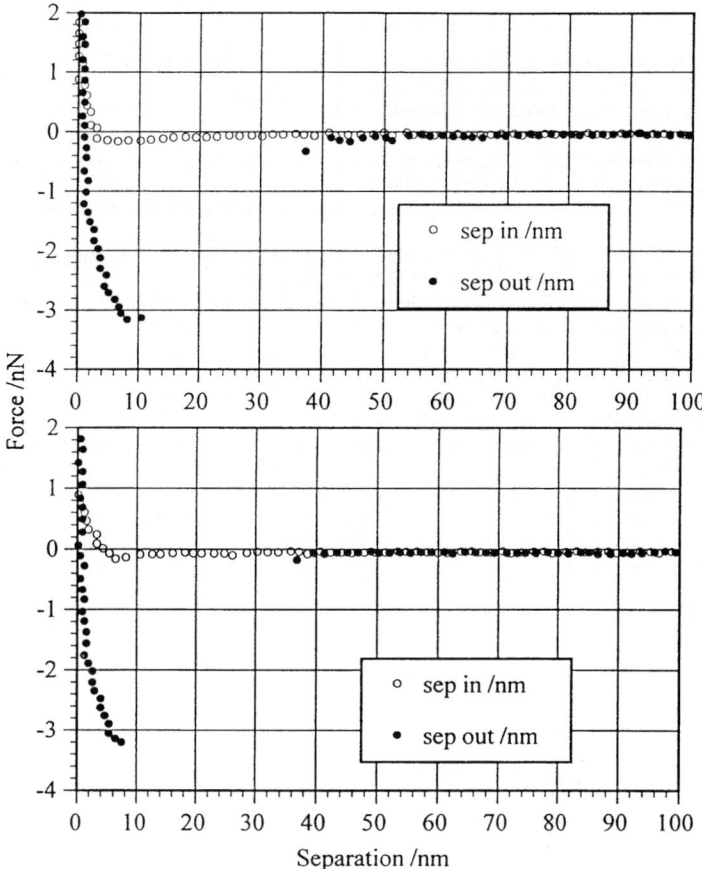

Fig. 15 Forces of interaction between an AFM tip bearing attached cholera toxin subunit B and a silanised glass in water. The open symbols correspond to approach of the surfaces and closed symbols to separation.

It is interesting to compare the strength of the interaction between GM1 and CTB of around 90 pN with intermolecular adhesions measured for other specific interactions. These are summarised in Table 1. It must also be mentioned though that it is possible that the value of 90 pN may correspond to the interaction between GM_1 and the whole pentamer of the cholera toxin. We cannot be unambiguous with the current data.

Table 1 Detachment forces for various intermolecular adhesions

Reference	Interaction	Detachment force/pN
19	Avidin–biotin	160 ± 20
	Avidin–iminobiotin	85 ± 15
	Streptavidin–biotin	257 ± 25
	Avidin–desthiobiotin	94 ± 10
	Streptavidin–iminobiotin	135 ± 15
20	Cell adhesion proteoglycans	40 ± 15
21	Adenine–thymine	54

References

1 Ch. Binnig G. Roher, C. Gerber and E. Weibel, *Phys. Rev. Lett.*, 1982, **49**, 57.
2 G. Binnig, C. F. Quate and C. Gerber, *Phys. Rev. Lett.*, 1986, **56**, 930.
3 P. Shears, *Ann. Trop. Med. Parasitol.*, 1994, **88**(2), 109.
4 J. Holmgren, *Nature (London)*, 1981, **292**, 413.
5 B. D. Spangler, *Microbiol. Rev.*, 1992, **56**(4), 622.
6 W. Curatolo, *Biochim. Biophys. Acta*, 1987, **906**, 137.
7 T. K. Sixma, S. E. Pronk, K. H. Kalk, E. S. Wartna, B. A. M. van Zanten, B. Witholt and W. G. J. Hol, *Nature (London)*, 1991, **351**, 371.
8 W. I. Lencer, C. Constable, S. Moe, P. A. Rufo, A. Wolf, M. G. Jobling, S. P. Ruston, J. L. Madara, R. K. Holmes and T. R. Hirst, *J. Biol. Chem.*, 1997, **272**(24), 15562.
9 J. Yang, L. K. Tamm, T. W. Tillack and Z. Shao, *J. Mol. Biol.*, 1993, **229**, 286.
10 J. Mou, J. Yang and Z. Shao, *J. Mol. Biol.*, 1995, **248**, 507.
11 P. H. Fishman, T. Pacuszka and P. A. Orlandi, *Adv. Lipid Res.*, 1993, **25**, 165.
12 W. Curatolo, *Biochim. Biophys. Acta*, 1987, **906**, 111.
13 R. A. Williams and H. W. Blanch, *Biosensors Bioelectronics*, 1994, **9**, 159.
14 J. Marra, *J. Colloid Interface Sci.*, 1985, **107**(2), 446.
15 P. F. Luckham, J. Wood, S. Froggatt and R. Swart, *J. Colloid Interface Sci.*, 1993, **156**, 164.
16 G. J. C. Braithwaite, P. F. Luckham and A. M. Howe, *Langmuir*, 1996, **12**, 4224.
17 V. T. Moy, E-L. Florin and E. Gaub, *Colloid Surf. A*, 1994, **93**, 343.
18 A. Chilkoti, T. Boland, B. D. Ratner and P. S. Stayton, *Biophys. J.*, 1995, **69**, 2125.
19 V. T. Moy, E-L. Florin and H. E. Gaub, *Science*, 1994, **266**, 257.
20 U. Dammer, O. Popescu, P. Wagner, D. Anselmetti, H-J. Güntherodt and G. N. Misevic, *Science*, 1995, **267**, 1173.
21 T. Boland and B. D. Ratner, *Proc. Natl. Acad. Sci. USA*, 1995, **92**, 5297.

Paper 8/07048B

Use of a laminar flow chamber to study the rate of bond formation and dissociation between surface-bound adhesion molecules: Effect of applied force and distance between surfaces

Anne Pierres, Anne-Marie Benoliel and Pierre Bongrand

Laboratoire d'Immunologie, INSERM U 387, Hôpital de Sainte-Marguerite, BP 29, 13274 Marseille Cedex 09, France. E-mail: bongrand@inserm.fr

Received 11th August 1998

It has recently been shown that much information on the behaviour of surface-bound adhesion molecules could be obtained by monitoring the motion of receptor-coated particles along ligand-derivatized surfaces in the presence of a hydrodynamic force of a few pN. This procedure is expected to allow direct monitoring of the formation and dissociation of individual bonds. We present experimental results on the interaction between streptavidin-coated spheres (1.4 μm diameter) and control or biotinylated mica surfaces in a laminar flow chamber. Moving spheres are found to display numerous arrests whose frequency is markedly increased (5–13-fold) in the presence of biotin groups. For a given shear rate, the binding frequency is strongly dependent on the sphere–surface separation. Indeed, this frequency displayed a 14-fold decrease when the velocity increased from 7 to 15 μm s^{-1} for a wall shear rate of 20 s^{-1}. Furthermore, the lifetime of observed arrests was of the order of several seconds, *i.e.* 5–50-fold higher than previously determined on models such as selectin–ligand, CD2–CD48 or cadherin–cadherin. Finally, this lifetime did not decrease when the wall shear rate was increased from *ca.* 10 to 40 s^{-1}.

Introduction

Cell adhesion is a process of prominent biological importance. It is usually mediated by the interaction of dedicated receptor and ligand molecules bound to interacting surfaces. It is, therefore, of obvious interest to develop suitable methods for predicting whether a cell-to-cell encounter will result in durable adhesion. The initial steps of cell attachment may be viewed as follows:[1] first, as a consequence of active or passive deformation and/or displacement, two biological surfaces are brought into binding distance. They are separated by a gap of fluctuating thickness. Second, when a first bond is formed, the distance between surfaces is likely to be decreased and maintained constant, as a consequence of the balance between bond properties and repulsive forces.[2] Third, the lifetime of the first bond subjected to a variety of disruptive forces is, therefore, a critical parameter. Indeed, if this bond is maintained until a second bond is formed, it is likely that the number of bonds will be progressively increased. Alternatively, if the first bond is ruptured before it is reinforced by additional interactions, membranes will become separated. In conclusion, the biologically relevant properties of a pair of membrane-bound ligand–receptor molecules are[3] the rate of bond formation as a function of the distance between the anchoring points of these molecules, and the rate of bond rupture as a function of the intensity of disruptive force.

While an impressive amount of information has been gathered on the rate of bond formation and dissociation between soluble reagents, little attention has been given to the properties of surface-bound molecules until Bell[4] elaborated a theoretical framework to relate these properties to the behaviour of soluble forms of these molecules. During the following years, Bell's ideas were subjected to direct experimental tests by several authors who studied the separation of surfaces bound by a few, or even a single, molecular bonds. Experimental methods included the use of hydrodynamic flow,[5] soft vesicle micromanipulation,[6] atomic force microscopy[7] or optical tweezers[8] (see ref. 9 for a review). These experiments led to results of immediate biological significance. Thus, it was shown that the remarkable capacity of flowing white blood cells to roll on the surface of blood vessels was mediated by a specific interaction between selectin molecules and their receptors,[10,11] and this interaction was characterized by a short lifetime[12,13] and high tensile strength.[13]

While the size and complexity of cell structure sets a limit to the accuracy of information that can be extracted from cellular models, it was recently found that much detailed information on the kinetics of bond dissociation[14,15] and formation[16,17] could be obtained by studying the motion of receptor-coated spheres along ligand-derivatized surfaces in a laminar flow chamber under very low hydrodynamic flow. The aim of the present paper is to illustrate the potential of this methodology by presenting preliminary experimental data on the interaction between streptavidin-coated spheres and biotinylated surfaces. The interest of this model is that it was extensively studied with powerful experimental methods such as the surface forces apparatus,[18] atomic force microscopy,[7,19] soft vesicle techniques,[20] as well as by theoretical approaches based on computer simulation.[21,22] The potential and limits of the flow chamber technology are also discussed on the basis of these and previous results.

Materials and methods

Molecules and surfaces

Streptavidin-coated beads of 2.8 µm diameter and 1.300 kg m^{-3} density (Dynabeads M280) were supplied by Dynal France (Compiègne). Freshly cleaved mica surfaces (Muskovite mica, Metafix, Montdidier, France) were sequentially incubated with 1mM $NiCl_2$,[23] then biotinyl-(gly)$_{12}$(His)$_6$-NH_2 (supplied by Neosystem, Strasbourg, France). The strong interaction between hexahistidine groups and nickel-treated mica surfaces was expected to allow the formation of a smooth surface of biotin groups. Using fluorescence determination,[14] the surface density of streptavidin groups on spheres was estimated at *ca.* 3460 molecules µm^{-2}. Further, mica surfaces treated with 10 µg ml^{-1} biotinylated peptide were found to bind 15 000 streptavidin molecules µm^{-2}.

Flow chamber

Our apparatus was described in a previous paper.[17] Briefly, the chamber was obtained by mechanical fixation of biotinylated mica surfaces against a drilled plexiglas block bearing a cavity of $0.1 \times 6 \times 20$ mm^3. The flow was generated by a syringe mounted on an electric syringe holder. The chamber was deposited on the stage of an inverted microscope (Olympus IX, with a long distance 40 × objective) bearing a CCD camera connected to a videotimer and a tape recorder for delayed analysis. In a typical experiment, the bead suspension (3×10^6 ml^{-1} in pH 7.2 phosphate buffer supplemented with 1 mg ml^{-1} bovine albumin) was driven through the chamber with a wall shear rate ranging between *ca.* 10 and 40 s^{-1}.

Particle tracking

The basis of our method has been described previously.[15] Briefly, an image processing system allowed real-time determination of the position of the centroid of the images of flowing beads with *ca.* 0.025 µm spatial accuracy, while the pixel size was 0.23 µm. Since the odd and even frames of each interlaced video image were analysed separately, the temporal resolution was twice the video rate: sequential positions were thus separated by a time interval of 20 ms. In a typical experiment, *ca.* 10 000 positions were stored by tracking 50–100 individual beads. Data were processed with dedicated software written in our laboratory.

Principle of data analysis

Since the details of our procedure have been described in previous papers,[14,17] we shall only outline the general strategy. We shall describe sequentially the study of bond formation and bond rupture.

Bond formation. The frequency of particle arrests was obtained by monitoring the motion of a series of particles and counting the binding events. The arrest frequency was simply equal to the number of counted arrests divided by the observation time. However, a number of steps were required to extract the on-rate of bond formation:

First, although particles were spheres of similar radius, their velocity displayed notable heterogeneity. This was ascribed to variations of sphere–surface distance, due to *e.g.* incomplete sedimentation. Indeed, basic results of fluid mechanics[24] show that the velocity of a free sphere moving in a laminar shear flow near a wall should vanish very slowly when the thickness of the gap between the sphere and the surface decreases. Thus, the average velocity of monitored spheres was determined on the preceding 160 ms interval after each step of 20 ms. The data recorded on a typical population of *ca.* 100 particles (yielding several tens of thousands of positions) were used to build the velocity frequency histograms of (i) all moving particles and (ii) all steps immediately followed by an arrest. The arrest frequency could thus be determined as a function of particle velocity by calculating the ratio of event numbers in both histograms for each velocity class.

Second, it would be tempting to use the known relationship between sphere velocity and distance to the surface[24] to obtain a relationship between particle arrest frequency and distance,[16] then to derive the rate of molecular association as a function of intermolecular distance d [*i.e.* $k_+(d)$] by determining the average density of binding sites on interacting surfaces.[16] However, the validity of this procedure is hampered by the occurrence of Brownian motion that is expected to result in marked variations of sphere–surface distance during a 20 ms interval. This motion is considerably complicated by (i) perturbation of the sphere motion by the effect of the wall on hydrodynamic drag, and (ii) the possibility of long-range electrodynamic (van der Waals) forces between spheres and surfaces. This difficulty was overcome as follows:[17]

(i) in each series of experiments, the *acceleration* of monitored particles was determined on periods of 2×160 ms, and the average acceleration was plotted *versus* velocity. Experimental curves were compared with theoretical curves obtained by computer simulation, which allowed accurate derivation of the wall shear rate G and the Hamaker constant A[25] for the interaction between spheres and the chamber floor.

(ii) The rate of bond formation was approximated as a step function:

$$k_+(d) = k_{+0} \quad (d \leq R) \qquad (1)$$
$$ 0 \quad (d > R)$$

where R is the interaction range. Binding curves were obtained by computer simulation, using the following formula:[17]

$$F(\delta) = 2\pi^2 k_{+0} a\sigma_1\sigma_2 R^3[2/3 - \delta/R + (\delta/R)^3/3] \qquad (2)$$

where $F(\delta)$ is the frequency of attachment between a sphere of radius a and a planar surface at distance δ, and σ_1 and σ_2 are the surface densities of binding molecules on the sphere and surface.

Bond rupture. The basic assumption[12] is that when receptor-coated beads are driven along ligand-derivatized surfaces with a force of the order of 1 pN, single ligand–receptor interactions should be able to generate detectable particle arrests. Further, in the absence of additional interaction, the probability $P(t)$ that a particle arrested at time zero remains bound at time t should be:[26]

$$P(t) = \exp(-k_- t) \qquad (3)$$

where k_- is the rate of bond dissociation (off-rate). Thus k_- was determined by measuring the duration of a sufficient number of arrests (typically 100 events) and plotting the number of particles remaining bound at time t after arrest *versus* t on a semi-logarithmic scale. The initial slope of the curve was used as an estimate of the off-rate. The experimental curve may not be a straight line if particle attachment is strengthened after arrest due to the formation of additional bonds or non-specific interactions.[12,26,27] Further, the dependence of k_- on applied force may be studied by varying the wall shear rate.[13–15] The validity of these basic assumptions will be discussed below.

Results

Flowing particles display numerous arrests of varying duration

First, we studied the motion of streptavidin-coated spheres along mica surfaces that had been treated with various amounts of biotinylated peptide. As exemplified in Fig. 1, the beads displayed a number of arrests of widely varying duration. Whereas most of these binding events lasted more than 1 s, arrests shorter than 0.1 s were occasionally detected (Fig. 2). In order to allow computer-assisted processing of binding events, a particle was defined as arrested at some point when it moved by less than 0.46 µm (*i.e.* two pixel units) during a time interval of 0.32 s. Visual examination of a number of trajectories showed that only a few very transient arrests were missed, and a minimal amount of false arrests were found.

Surface biotinylation markedly increases interaction with streptavidin-coated beads

In a first series of experiments, we compared the motion of spheres along surfaces treated with 100, 10, 1 or 0 µg peptide, in the presence of a wall shear rate of *ca.* 20 s^{-1}. In all cases, short or durable binding events were detected. However, several findings suggested that surface biotinylation substantially increased sphere–surface interaction:

(i) When streptavidin-coated spheres were driven along heavily biotinylated surfaces, only a very low proportion of these particles reached the microscope field, near the centre of the chamber.

(ii) As shown in Fig. 3, the average velocity of particles monitored on biotinylated surfaces was markedly higher than on the controls. Since particle velocity is known to be a strongly increasing function of distance from the surface,[24] this observation suggested that particles adhered to biotinylated surface as soon as they completed sedimentation, and observed particles were mostly incompletely sedimented spheres, in contrast to the controls.

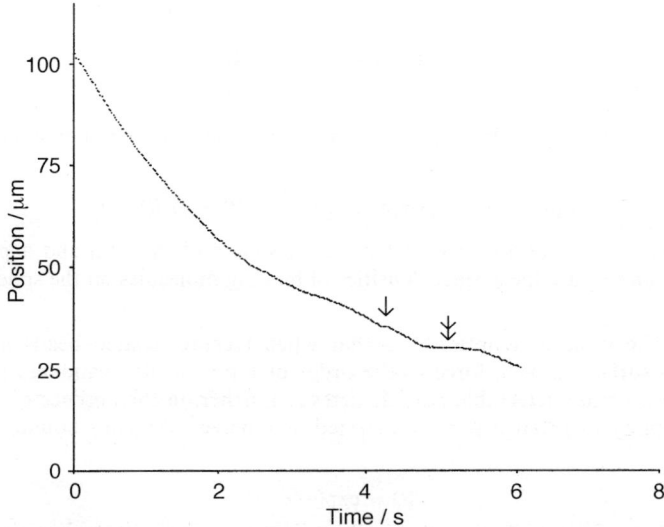

Fig. 1 A typical trajectory. The motion of a streptavidin-coated bead along a surface treated with 0.1 µg biotinylated peptide is shown. The wall shear rate is 17.2 s^{-1}. The position of the centre of gravity of the particle image was determined every 20 ms. A very short (arrow) and a longer (double arrow) arrest are shown.

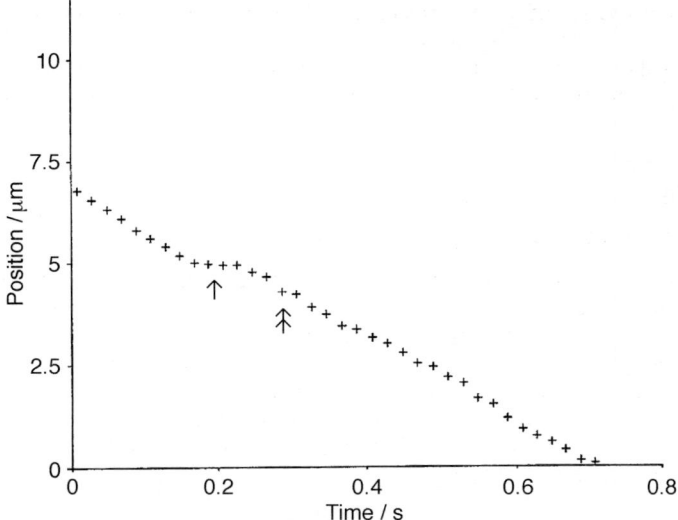

Fig. 2 Short arrests. An enlarged fraction of the trajectory shown in Fig. 1 is displayed. The short arrest shown in Fig. 1 is clearly visible (arrow). An artefactual discontinuity of the particle velocity is also shown (double arrow). Indeed, in the latter case, there is no displacement of the tangent to the trajectory.

(iii) Thus, in order to achieve a meaningful study of the relationship between surface biotinylation and binding efficiency, we calculated, in each series of experiments, the number of 20 ms intervals where the particle velocity was less than the product of the particle radius and the wall shear rate. The fraction of these intervals immediately followed by an arrest was then calculated: this fraction was, respectively, 0.0049, 0.0039, 0.0104 and 0.00077 when the mica surfaces were treated with 100, 10, 1 and 0 µg of biotinylated peptide. Thus, surface biotinylation increased the particle arrest by a factor ranging between 5 and 13.

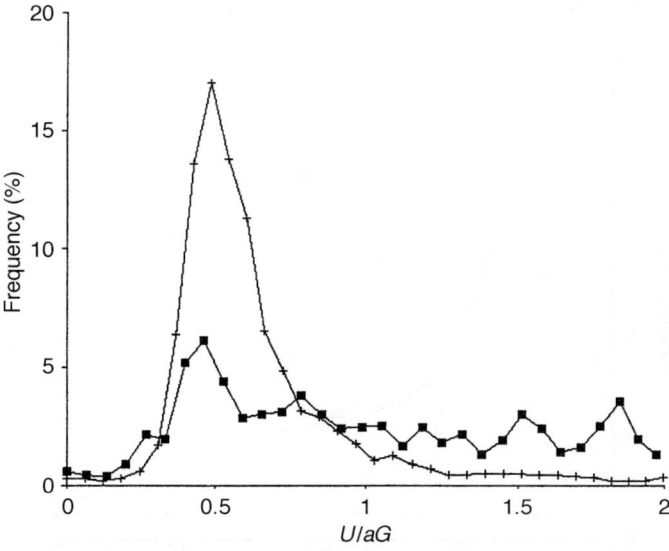

Fig. 3 Distribution of particle velocity. In two representative experiments, the motion of spheres on control (×) or biotinylated (■) surfaces was studied. The particle velocity U was determined every 20 ms, and the frequency distribution U/aG is shown (as % of total number of intervals). Clearly, particles flowing on the biotinylated surfaces display higher U/aG (i.e. higher separation from the surface) than the controls.

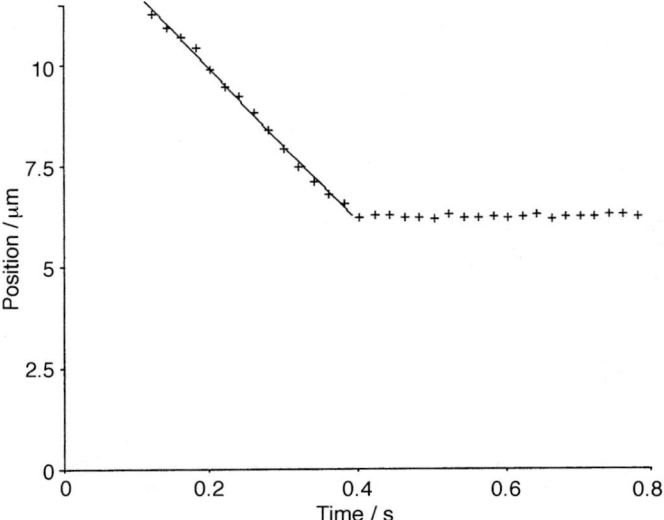

Fig. 4 Example of a clearcut arrest. A typical example of a clearcut arrest is shown. The particle velocity remains constant up to the binding event.

Arrest frequency is a strongly decreasing function of particle velocity

The particle velocity was determined for 160 ms periods preceding each arrest. As exemplified in Fig. 4, this was quite easy in many cases where the particle velocity was fairly constant immediately before arrest. However, in some cases (Fig. 5), the particle velocity progressively decreased before the observed stop, and it was felt that the definition of the velocity before arrest was somewhat arbitrary.

The results obtained in four series of experiments (wall shear rate 20 s^{-1}, 10 µg peptide per mica slide) were then pooled in order to obtain the experimental dependence of arrest frequency on particle velocity. As shown in Fig. 6, the binding frequency decreased by 93% when the ratio U/aG increased from 0.25 to 0.55. This result demonstrates that binding frequency is indeed

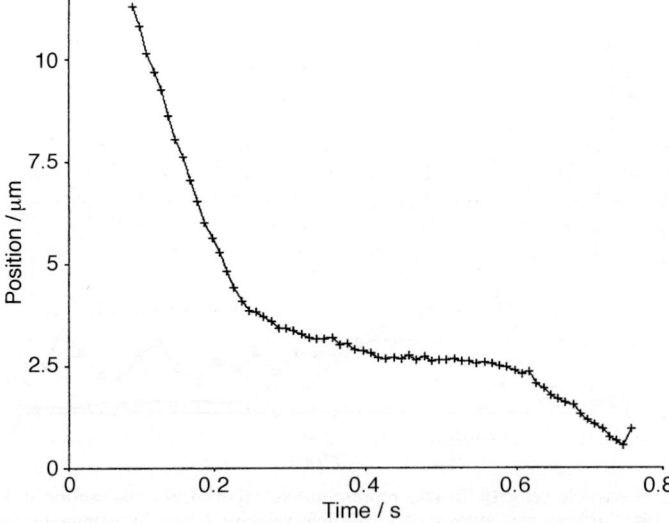

Fig. 5 Example of a progressive arrest. The motion of a particle binding to a mica surface is shown. The sphere displays progressive slowing, making it difficult to define unambiguously a "velocity before arrest".

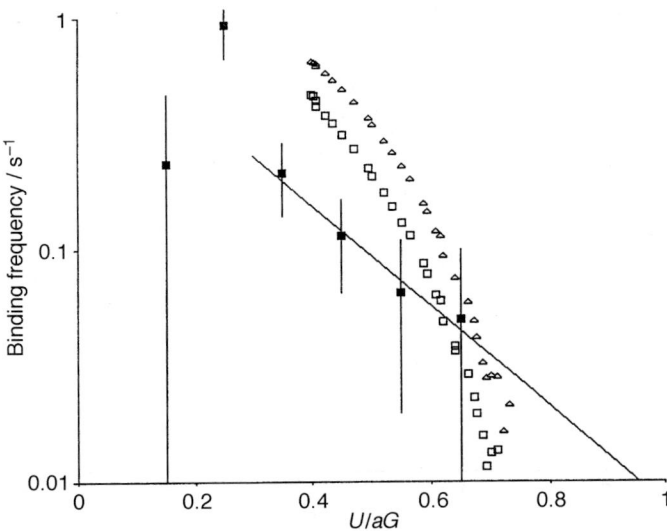

Fig. 6 Dependence of binding probability on particle velocity. In a typical series of experiments, the arrest of streptavidin-coated spheres on surfaces treated with 10 μg biotinylated peptide was studied (■). The histogram of particle velocity was built, and the arrest frequency was calculated for classes of increasing dimensionless velocity U/aG. The vertical bar length is a theoretical standard deviation calculated by assuming Poisson distribution. Theoretical curves corresponding to a bond range of 5 nm (□) and 40 nm (△) are also shown.

strongly dependent on sphere–surface distance. Further, due to the wide heterogeneity of particle velocities (Fig. 3), it does not seem warranted to define a mean binding frequency by retaining particles whose velocity falls below some arbitrary threshold.

Finally, we attempted to fit experimental data to theoretical results that were previously shown to account for the homophilic adhesion between cadherin-coated surfaces.[17] We used acceleration curves to estimate the Hamaker constant in ten separate series of experiments: no significant interaction between spheres and surfaces could be detected. As shown in Fig. 6, the slope of the theoretical curves (built with a zero Hamaker constant) was lower than the experimental one by a factor of two.

Binding efficiency is decreased when the shear rate is increased

The effect of the wall shear rate on the binding efficiency was studied. In three series of four separate experiments each, the interaction between spheres and surfaces treated with 10 μg biotinylated peptide was studied with a wall shear rate of *ca*. 10, 20 and 40 s^{-1}. The arrest frequency was determined for particles with an instantaneous velocity U lower than the product between the wall shear rate (G) and particle velocity (a): this frequency was 0.23, 0.15 and 0.10 s^{-1}, respectively. The binding frequency was thus a slowly decreasing function of particle velocity.

The lifetime of the interaction between streptavidin-coated beads and biotinylated surfaces is not markedly higher than that observed on controls

The lifetime of sphere–surface associations was studied by recording the duration of arrests and building detachment curves, as exemplified in Fig. 7. Assuming first-order detachment kinetics, the rate of particle detachment was derived from the fraction of particles remaining bound for less than 2 s.

In a preliminary series of experiments, the influence of surface biotinylation was studied. The rate of bond dissociation was, respectively, 0.20 (40 arrests), 0.16 (39 arrests), 0.28 (56 arrests) and 0.40 s^{-1} (44 arrests) when the mica surfaces were treated with 10, 1, 0.1 and 0 μg of biotinylated peptide.

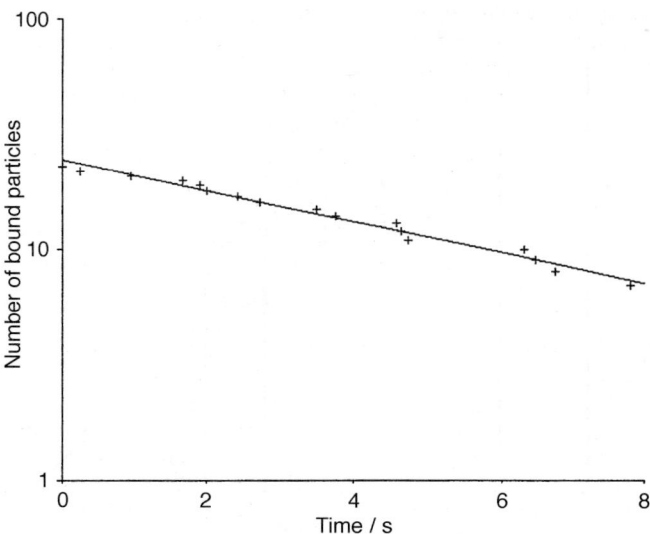

Fig. 7 In four series of experiments, streptavidin-coated beads were driven along surfaces treated with 1 μg biotinylated peptide. The wall shear rate was *ca.* 20 s^{-1}. A microscopical field was selected and the durations of observed binding events recorded. Results were used to plot the variations of the number of particles remaining bound at time t. The slope of the curve was defined as the detachment rate.

There is no obvious increase in the detachment rate when the shear rate is increased

The effect of the wall shear rate on the experimental detachment rate was also studied. When mica surfaces were treated with 10 μg biotinylated peptide and the wall shear rate was 10, 20 and 40 s^{-1}, the particle detachment rate was, respectively, 0.33, 0.22 and 0.09 s^{-1}.

Discussion

The aim of this work was to study the avidin–biotin interaction with the laminar flow chamber. It seemed of interest to compare the data concerning bond rupture with published reports based on other experimental techniques. Also, we wished to present new information on the rate of bond formation.

A first conclusion is that the flow chamber methodology may be more suitable to the study of weak and transient interactions such as CD2–CD48,[15,16] selectin–ligand[12,13] or cadherin–cadherin[17] association than the stronger avidin–biotin attachment. Indeed, in the present study, measured detachment rates were 5–50-fold lower than measured on the aforementioned systems and, presumably, many spheres displayed irreversible attachment to the surface as soon as contact occurred. Further, when binding site density was decreased, a substantial proportion of arrests were mediated by poorly defined non-specific interactions, which made it more difficult to relate the obtained results to the properties of avidin–biotin interaction. However, it seems reasonable to assume that interactions between single streptavidin and biotin groups were actually observed, since our methodology allows the monitoring of weak individual bonds between different adhesion receptors. Also, since interacting molecules are substantially shorter in the present work than in the aforementioned study, the occurrence of multiple binding events should be rarer than in those other cases.

Another possible difficulty with our system may stem from the relative shortness of the adhesion molecules. Indeed, in previous studies,[15,17] adhesion sites were connected to streptavidin groups through a fairly flexible molecular link of *ca.* 40 nm length. In contrast, in the present work, the length of biotinylated peptides was less than 5 nm, and no spacer was added to increase the distance between streptavidin groups and the particle surface.

Despite the aforementioned difficulties, biotinylation of the surface of the flow chamber markedly increased the efficiency of sphere to particle attachment, strongly suggesting than avidin–biotin interactions indeed occurred. However, a substantial fraction of observed attachments lasted only a few seconds, while the force exerted on the bonds did not exceed a few pN.[14] Several explanations may be suggested to account for this result.

(i) Since particle–surface separation results from the rupture of the weakest link between these structures, sphere detachment is perhaps due to the rupture of the interaction between biotinylated peptides and mica surfaces, or streptavidin–sphere association. However, this interpretation is not fully consistent with the observation that the particle detachment rate decreased when the hydrodynamic force was increased. Indeed, there is little experimental evidence that a disruptive force can increase the lifetime of a ligand receptor interaction although, in principle, this might happen.[28]

(ii) Another possibility would be that avidin–biotin association might involve intermediate binding steps, as previously reported in a study made on the antigen–antibody interaction with the flow chamber.[14] Indeed, since, as a rule of thumb, the relative velocity between the flow chamber and the sphere surfaces is about half the sphere velocity,[24] if the total length of interacting molecules is of the order of 5 nm, the time allowed for interaction is only 1 ms for a sphere of 10 µm s^{-1} velocity. Perhaps the paradoxical increase in bond lifetime observed when the wall shear rate was increased was due to the disappearance of putative intermediate binding steps that might become undetectable at higher shear rate. This possibility might be addressed by increasing the time resolution of our analysis, and/or adding a spacer to increase the length of the interacting structures.

A final point of concern is the discrepancy between the theoretical and experimental dependence of adhesion frequency on particle velocity, whereas our theoretical framework successfully accounted for the interaction between cadherin-coated surfaces.[17] Three points must be suggested to deal with this problem. First, it would be useful to check the smoothness of the peptide-coated mica surfaces, in order to exclude the possibility that we measured the frequency of surface defects rather than molecular associations. However, if this were the case, the binding frequency should increase when the shear rate was increased. Second, as illustrated by Fig. 5, there is sometimes a problem with the determination of particle velocity immediately before arrests. Perhaps our analysis might be improved by increasing the shear rate and temporal resolution. Third, the theoretical relationship between sphere velocity and distance to the surface[24] might no longer hold at low distance. This difficulty might be alleviated if we increase the length of receptors bound to the interacting surfaces.

In conclusion, the laminar flow chamber is a powerful means of studying the formation and dissociation of molecular bonds, with high temporal and spatial resolution. This seems ideally suited to the study of transient interactions. More work is needed to obtain new and accurate information on high affinity ligand–receptor association.

Acknowledgements

The expert technical assistance of Ms Dominique Touchard is gratefully acknowledged. This work was supported by a grant from the A.R.C.

References

1 A. Pierres, A. M. Benoliel and P. Bongrand, *Curr. Opin. Colloid Interface Sci.*, 1998, **3**, 525.
2 G. I. Bell, M. Dembo and P. Bongrand, *Biophys. J.*, 1984, **45**, 1045.
3 A. Pierres, A. M. Benoliel and P. Bongrand, *J. Immunol. Methods*, 1996, **196**, 105.
4 G. I. Bell, *Science*, 1978, **200**, 618.
5 S. P. Tha, J. Shuster and H. L. Goldsmith, *Biophys. J.*, 1986, **50**, 1117.
6 E. Evans, D. Berk and A. Leung, *Biophys. J.*, 1991, **59**, 838.
7 E. L. Florin, V. T. Moy and H. E. Gaub, *Science*, 1994, **264**, 415.
8 H. Miyata, R. Yasuda and K. Kinosita Jr., *Biochim. Biophys. Acta*, 1996, **1290**, 83.
9 A. Pierres, A. M. Benoliel and P. Bongrand, *Cell Adhesion Commun.*, 1998, **5**, 375.
10 M. B. Lawrence and T. A. Springer, *Cell*, 1991, **65**, 859.
11 U. H. Von Andrian, J. D. Chambers, L. M. McEvoy, R. F. Bargatze, K. E. Arfors and E. C. Butcher, *Proc. Natl. Acad. Sci. USA*, 1991, **88**, 7538.

12 G. Kaplanski, C. Farnarier, O. Tissot, A. Pierres, A. M. Benoliel, M. C. Alessi, S. Kaplanski and P. Bongrand, *Biophys. J.*, 1993, **64**, 1922.
13 R. Alon, D. A. Hammer and T. A. Springer, *Nature (London)*, 1995, **374**, 539.
14 A. Pierres, A. M. Benoliel and P. Bongrand, *J. Biol. Chem.*, 1995, **270**, 26586.
15 A. Pierres, A. M. Benoliel and P. Bongrand, *Proc. Natl Acad. Sci. USA*, 1996, **93**, 15114.
16 A. Pierres, A. M. Benoliel and P. Bongrand, *FEBS Lett.*, 1997, **403**, 239.
17 A. Pierres, H. Feracci, V. Delmas, A. M. Benoliel, J. P. Thiéry and P. Bongrand, *Proc. Natl. Acad. Sci. USA*, 1998, **95**, 9256.
18 C. A. Helm, W. Knoll and J. N. Israelachvili, *Proc. Natl Acad. USA*, 1991, **88**, 8169.
19 G. U. Lee, D. A. Kidwell and R. J. Colton, *Langmuir*, 1994, **10**, 354.
20 E. Evans, K. Ritchie and R. Merkel, *Biophys. J.*, 1995, **68**, 2580.
21 H. Grubmüller, B. Heymann and P. Tavan, *Science*, 1996, **271**, 997.
22 S. Izrailev, S. Stepaniants, M. Balsera, Y. Oono and K. Schulten, *Biophys. J.*, 1997, **72**, 1568.
23 C. R. Ill, V. M. Keivens, J. E. Hale, K. K. Nakamura, R. A. Jue, S. Cheng, E. D. Melcher, B. Drake and M. C. Smith, *Biophys. J.*, 1993, **64**, 919.
24 A. J. Goldman, R. G. Cox and H. Brenner, *Chem. Eng. Sci.*, 1967, **22**, 653.
25 J. N. Israelachvili, *Intermolecular and Surface Forces*, Academic Press, New York, 1991, p. 288.
26 A. Pierres, O. Tissot and P. Bongrand, in *Studying Cell Adhesion*, ed. P. Bongrand, P. Claesson and A. Curtis, Springer Verlag, Heidelberg, 1994, p. 157.
27 A. Pierres, A. M. Benoliel and P. Bongrand, in *Cell Mechanics and Cellular Engineering*, ed. V. C. Mow, F. Guilak, R. Tran-Son-Tay and R. M. Hochmuth, Springer Verlag, New York, 1994, p. 145.
28 M. Dembo, D. C. Torney, K. Saxman and D. Hammer, *Proc. R. Soc. London, Ser. B*, 1988, **234**, 55.

Paper 8/06339G

General Discussion

Dr Goñi opened the discussion of Mrs Woodhouse's paper: Could you tell us something about the stability of these tethered bilayers?

Ms Woodhouse responded: Our tethered bilayers are remarkably stable compared with black lipid membranes and supported bilayers. We have demonstrated that tethered bilayers comprising hydrated di-phytanyl derivatives are stable for at least 12 weeks at room temperature and longer when frozen. This has been characterised by measuring the capacitance at 1 Hz. Additionally, fully assembled biosensors have been dried using glycerol, trehalose or dextran and have been demonstrated to respond to analyte following reconstitution.

Dr Pawlak commented: Impedance measurements on isolating layers (here the lipid membrane) are very critically dependent on even a small number of layer defects, leading to unspecific leakage currents. The layer capacitance is hardly affected by these defects. How stable is your membrane preparation on the surface, with time, with potentials applied, and with biochemicals exposed to the membrane?
How long is the preparation time of the system until stable sensor signals can be measured?
How reproducibly can the system be prepared (chip-to-chip variation, impedance stability, amount of leakage current)?

Ms Woodhouse responded: The membrane stabilises very rapidly—less than one minute. The effect of potential on stability has not been quantified as such although we have performed multiple experiments on single membranes over hour time frames without apparent degradation of response. More than 100 biochemicals and chemicals, of many classes, have been exposed to the membrane: these range from blood and sera, to organic solvents and detergents. In many cases, the effect is concentration dependent. For example, the membrane preparation is stable in 100% v/v serum and blood, the effect of detergent depends on the protein/buffer matrix used.
The bilayer self-assembles very rapidly onto the monolayer. A stable signal can be measured immediately.
The variation in leakage current between batches of chips is about 15% coefficient of variation in terms of impedance at 1 Hz. However, reproducibility in terms of other parameters, such as conduction and analyte response is much better.

Dr Morigaki asked: Is the role of membrane-spanning anchor lipids to stabilize the supported bilayer system studied? How much effect, if any, do they have on stabilizing the membrane matrix?

Ms Woodhouse responded: The primary role of the membrane-spanning lipid is to provide a tether for attaching analytes to the membrane surface. The stability of the bilayer can be attributed to many parts of the biosensor design and it is difficult to elucidate which parts are more significant.

Dr Sansom asked: Does your sensor design allow you to distinguish between the double helix *vs.* helix dimer models of gramicidin channels/pores?

Ms Woodhouse responded: Absolutely: the data show the conduction measured for various sensor constructs is negligible when they are formed with gramicidin in either one layer or the other but not both. Additionally, the model fits with the notion that the gramicidin in each layer is in equilibrium between the monomeric and dimeric forms.

Prof. Bartlett asked: In the work presented in your paper you used a fixed bias of -300 mV. Have you looked to see if any of the rate constants derived from your analysis of the experimental data show a potential dependence?

In addition, from the text it is not clear to me exactly how the experimental impedance measurement was carried out. Is this a three electrode measurement using a potentiostat with the gold sensor electrode held at a bias of -300 mV with respect to the Ag reference electrode? Is this a real Ag/AgCl reference electrode or a Ag wire pseudo-reference electrode?

Ms Woodhouse responded: We have not investigated the potential dependence on rate. The potential applied improves the reservoir capacity and therefore the conduction and dynamic range of response.

To answer your second point, this is a three terminal measurement with 300 mV applied to the reference electrode which is a Ag wire pseudo-reference electrode. Measurements have been made with a real Ag/AgCl reference electrode, these have been indistinguishable from those employing a pseudo-reference electrode.

Dr Lakey asked: Does the data in Fig. 5c represent the Fab–digoxin affinity or are other factors determining this result?

By manipulation of the concentrations of certain components is it possible to alter the sensitivity of the biosensor to explore wider ranges of ligand concentration?

Ms Woodhouse responded: The data does represent the fast antibody off rates both in 3D and 2D as well as the surface concentration of MSL_{SA} : b-Fab′, gA-dig : b-Fab′ and gA_{DIMER} : b-Fab′.

Both the sensitivity and the dynamic range of the biosensor can be altered by the manipulation of the sensor components. High surface concentrations of MSL_{SA} compared with gA-dig results in an excess of surface bound b-Fab′ (MSL_{SA} : b-Fab′), compared with surface analyte, digoxigenin, so that the fraction of cross-linked b-Fab′ of the total is lower resulting in a lower sensitivity.

Dr Pawlak commented: You investigated the kinetics of the competitive channel response. Since a multiple set of processes is involved in your sensing mechanism, you have to know carefully the individual number of molecules which contribute to the signal.

What is the limit of quantitation (LOQ) of your sensor system in terms of the number of channels involved, assuming a 1 : 1 stochiometry of analyte-to-be-detected : channel?

Will 'single channel measurements' on supported lipid membranes be achievable in the future?

How easily and generally can your sensing principle be adapted to the detection of other analytes, *e.g.* antigens?

Is it possible to employ other channels, *e.g.* peptide bundles or larger receptor channels?

Ms Woodhouse responded: Theoretically we should be able to measure single channel electrical activity—practically we currently measure the total electrical response of an electrode containing 1 channel μm^{-2}. Through our model we can mathematically determine the number of channels involved in the gating response once we determine the actual surface concentration of MSL_{SA} : b-Fab employed.

Theoretically our technology could be scaled to achieve single channel measurements, we do not foresee any practical limitations. However, we are not pursuing this goal presently.

This sensing principle has been used for detecting many different analytes including thyroid stimulating hormone, ferritin, thyroxine and theophylline as described by Cornell *et al.*[1]

Raguse and co-workers has shown that valinomycin can be incorporated into the tBLM (referenced in the paper) and Yin has incorporated alamethicin and shown its selectivity to amiloride based inhibitors into the tethered bilayer. We are currently investigating methods for incorporating larger physiological channels.

1 B. Raguse, V. Braach-Maksvytis, B. A. Cornell, L. G. King, P. D. J. Osman, R. J. Pace and L. Wieczorek, *Langmuir*, 1998, **3**, 648.

Dr Atay asked: How do the binding constants measured in 2- and 3D compare? And have you looked at the energetics of the reaction at all?

Ms Woodhouse responded: The 2- and 3D kinetics are difficult to compare because the dimensions give different units of measurement. Also, as described by Hardt[1] the change in dimerisation is predictable and in agreement with our observation. We have not looked at the energetics of the reaction at all.

1   S. L. Hardt, *Biophys. Chem.*, 1979, **16**, 239.

Dr Atay communicated: I realise that the association constants mentioned in the paper were constants in 2- and 3D. Constants measured (or calculated) in 2- and 3D are rather difficult to compare due to their rather different units. We have, however, in a previous study shown a method of comparing these two. I suggest that the authors see the paper by Albery *et al.*[1] They might find the reference useful.

1   W. J. Albery, R. A. Choudery, N. Z. Atay and B. H. Robinson, *J. Chem. Soc., Faraday Trans. 1*, 1987, **83**, 2407.

Prof. Smith opened the discussion of Dr Sutcliffe's paper: The recently crystallographically determined GLU-receptor is, although related, not quite your target. Are there any important sequence differences between the two that might hamper your present validation of your model?

Dr Sutcliffe responded: The sequence identity is *ca.* 40% between the S1S2 domain of GluR2 (ref. 1, for which the crystal structure has been determined) and the goldfish kainate binding proteins I presented. Thus, we cannot be totally certain of the validity of our models based on this crystal structure. However, as a conservative estimate, we would suggest that well over 50% of our predicted protein–ligand interactions were identified correctly. Of the six regions involved in ligand binding (**S1** through **S6** in Fig. 2), the major difference with GluR2 comes in **S5**, where there are two additional residues in GluR2—thus we cannot validate the presence or absence of the proposed hydrogen bond involving **S5** and ligand.

1   N. Armstrong, Y. Sun, G. Q. Chen and E. Gouaux, *Nature (London)*, 1998, **395**, 913.

Dr Smart asked: What is the sequence homology between the recently published potassium channel structure (ref. 13 in your paper) and your "K^+ channel domain"? In addition, can I comment that your success in fold identification and homology modelling for a target domain with 20% homology to its nearest template is most impressive. Please could you state the overall root mean squared (rms) α-carbon deviation from the subsequently determined crystal structure of a closer homologue and what parts of the structure were best modelled.

Dr Sutcliffe responded: The KcsA crystal structure and the NMDA receptors have a sequence identity of *ca.* 20%—*i.e.* a similar level to that between the S1S2 domain and the bacterial amino acid binding proteins, which we have modelled to what you call an "impressive" degree of accuracy.

The crystal structure, which is of the S1S2 domain of GluR2 complexed with kainate, has not undergone full domain closure—therefore, a comparison of the rmsd is perhaps not that meaningful. The crystal structure does reveal mistakes in our sequence alignment, both in Lobe 1 and Lobe 2. Lobe 2, due to fewer misalignments, was "better modelled" than Lobe 1. With the exception of the region denoted **S1** in Fig. 2, these do not affect the ligand binding site to any great extent—resulting in a reasonable model of the binding site.

Dr Sansom said: Would you like to comment on the NMR studies of Opella and colleagues which suggest that the NMDA receptor pore-lining M2 segment adopts an α-helical conformation when in a membrane-mimetic environment, rather than the loop-like structure in your topological diagram?

Dr Sutcliffe responded: The glycosylation/epitope mapping/proteolytic data show that the S1S2 domain is indeed extracellular, thus M1, M2 and M3 must comprise of 2 membrane spanning, and one non-membrane spanning, components—site directed mutagenesis and constructs which delete M1, M2, or M3 suggest that M1 and M3 are transmembrane helices, and M2 a "re-entrant loop" similar to the P-segment in potassium channels. Thus, the different environments—isolated peptide *vs.* peptide in a protein environment—likely lead to the different conformation observed by Opella.

Prof. Holzwarth asked: Can your approach be described by using the known molecular composition of a functional polypeptide as bricks to form a house by making sensible assumptions about their arrangement to rooms and arranging the rooms to the house? After the first house was built by assumptions you test if the results from experiments can be verified; then you try to change your arrangement into the direction of better agreement until your circle results in an arrangement which could explain the experimental findings available so far.

Is there not the danger that you will go in the wrong direction, if the experimental findings are not accurate? In other words you strongly depend on the quality of experiments.

Dr Sutcliffe responded: This is a good analogy of the "iterative modelling/experimental verification cycle" we use, as shown in Fig. 1.

The iterative nature of our approach means that misleading experimental information, if not contradicted by other experimental data, would lead to us moving away from, rather than towards, the "correct" answer.

Prof. Smith commented: The probability of obtaining the correct tertiary fold in a homology model increases with the sequence identity between the aligned template and target structures. Can you give figures for this in your extra- and intra-membrane domains and maybe comment on their significance?

Dr Sutcliffe responded: The sequence identity for our LAOBP-based S1S2 domain models was *ca.* 20%. There is therefore some uncertainty in the identification of the fold. However, the recent crystal structure of the S1S2 domain of GluR2 has shown that use of the LAOBP-like fold was indeed correct. Similarly, our most recent models of the transmembrane domain are based on the crystal structure of the KcsA potassium channel, with which they share *ca.* 20% sequence identity. As with the S1S2 domain, we are not certain this is the correct fold but our models are consistent with all available experimental data—suggesting again that our models are roughly correct.

Prof. Evans asked: Although many things interfere with crystallization of proteins, could the inability to crystallize the structure surrounding the binding domain reflect 'flexibility' and if so, how would this affect your modeling concept?

Dr Sutcliffe responded: So far, although crystallisation trials have been run on similar constructs from glutamate receptors other than GluR2, only the S1S2 domain of GluR2 has been crystallised. As you say, many things affect crystallisation, and it is impossible to identify flexibility as the factor preventing crystallisation. At 300 K, all proteins undergo motion to some degree. It is likely that the ligand-free form of the receptors undergoes more motion that the ligand-bound form, as the latter is more "clamped together". Nevertheless, the presence of such motions does not change the interpretation of our models.

Prof. Roux asked: Based on your modelled structures of the ligand-bound and ligand-free form, how do you explain the gating mechanism?

Dr Sutcliffe responded: Ligand binding results in a conformational change of the S1S2 domain, and this likely propagates into the transmembrane domain. This conformational change causes the channel to gate. Our modelling suggests that channel opening positions conserved negatively charged residues in the extracellular entrance to the channel, thus electrostatically "opening" the channel. Our models also suggest that the region denoted **S5** might be involved in desensitisation of the channel. (For a more detailed discussion of this, see the section of our paper entitled "Binding and signal transduction".)

Prof. Laggner opened the discussion of Dr Pawlak's paper: In your paper you have addressed the problem of ensuring the correct sidedness of the membrane patches when they are deposited on the surface, but not really answered it. What are the actual right-side-out ratios reached in your present studies?

Dr Pawlak responded: In the described studies we have first concentrated on finding the optimum experimental conditions for yielding stable preparations of surface-immobilized membrane patches. Up to now we have achieved the situation of a 50% to 50% distribution on average of the two possible orientations of membrane patches. Single experiments however suggest that we have achieved up to 70% of the specific K^+ binding side exposed, assuming the binding of K^+ exclusively to one side of the membrane. We have to admit that the success of membrane orientation is, at the moment, much limited by the quality of the membrane patch preparation. The size distribution is broad and a varying portion of small particles, probably less defined associated with the surface, is always present. We expect a far more defined membrane orientation on the surface upon using membrane patches prepared with less variation in size. These studies will be undertaken.

Prof. Holzwarth asked: How good are your immobilised Na,K-ATPases containing fragments in comparison with the same fragments in solution?
Do you not introduce additional problems by attaching the fragments to a surface? In solution the fragments are accessible from both sides as in their natural surrounding, in your case you block one side.
What are the consequences of immobilisation for the activity of the pump?

Dr Pawlak responded: What do you mean 'good'? We found comparable affinity constants for specific binding of potassium to the Na,K-ATPase pump on the surface and in solution. Because K^+-binding is a very critical test for the protein conformation in its active form (a K^+-binding related structural change of the enzyme results in a large, up to 30% reversible change of fluorescein emission), we regard the surface-immobilized enzyme as being of comparable activity as in solution. However, you can only reach such a preserved activity by appropriate 'shielding' of the protein from the denaturating forces of the bare surface support. This we have achieved by introducing the biocompatible lipid layer.
In answer to your second question, to reach a controlled access only to one side of the membrane fragments is exactly our goal. Thus you can directly probe individual binding sites without a signal interference of binding sites on the other side of the membrane, which is not possible in solution.
To answer your third question, we do not know at the moment, since we haven't yet tested the pump activity. Experiments are planned in this direction.

Dr Keller asked: How do you orient the membrane fragments on the surface?

Dr Pawlak responded: The membrane fragments orient themselves on the surface by means of a self-assembly process in the presence of attractive forces between the strong dipole moments and/or charges of the membranes and the supported, planar lipid monolayer.

Prof. Evans said: Following on from Prof. Holzwarth's question, may I offer a brief comment on the sensitivity of evanescent-wave induced fluorescence techniques. Such techniques do indeed select strongly for the fluorescence of groups close to the interface (usually within 100 nm or so), but the practical problem is that there is always some scattered light from surface defects, and this light can excite fluorescence in the bulk solution. Such fluorescence can dominate, and needs to be carefully accounted for.

1   L. Fisher, in *Surface Analytical Techniques for Probing Biomaterial Processes*, ed. J. Davies, CRC Press, Boca Raton, 1996, pp. 43–66.

Dr Pawlak responded: You are right that such fluorescence 'can' (but must not, my comment) dominate, leading to an interfering bulk fluorescence background which will limit the sensitivity of a system. Thereby, the roughness of the chip surface plays a crucial role in determining the strength of this effect. In the case of our sensor, the material and surface properties of the waveguiding layer have been optimized over years and in many applications (see ref. 13, 14 and 26 in our paper). The roughness of our chip surfaces is very low, only 2–3 Å, as measured by AFM under high resolution (ref. 23). Therefore, the effect of bulk fluorescence induction by surface-scattered light is minimal and negligible for our chips. We could not observe such an effect in our applications. Especially, no shifts in fluorescence signals were detected when comparing signals in the presence and absence of fluorophores in chip rinsing solutions.

Prof. Bohne asked: Following the previous question could you comment on the requirement for the absolute quantum yield of the fluorophore and what intensity change you can detect.

Are there any limitations on the excitation wavelengths that can be used.

You commented that the technique can be time-resolved. Are there any limitations for exciting the sample with a pulsed excitation source?

Dr Pawlak responded: I do not see a limitation in the requirement of the absolute quantum yield of a used fluorophore, because we detect relative changes of an established base fluorescence signal, which is generated by the surface immobilization of a certain number of fluorophores on the surface. As you see from Fig. 5 in the paper we have a dynamic range of five to six orders of magnitude. If the quantum yield is too low, then you compensate by the number of fluorophores. We can detect a 1–2% change of established fluorescence. This is limited by signal noise.

Excitation in the far blue is limited by an increasing absorbance of the layer material.

To answer your final question, I don't see any limitations at the moment.

Dr Goñi commented: I am puzzled by the nature of your "discs". They appear to be intermediate structures between the intact vesicles and the lipid–detergent–protein mixed micelles. This is rather uncommon in membrane solubilisation. How have these discs been characterised in terms of size, composition and detergent concentration?

Also, a comment: Once the ATPase has been treated with SDS, however short, it will never go back to the native situation. Some SDS molecules will be irreversibly bound to the enzyme. This is something that should be taken into account when analysing the results.

Dr Pawlak responded: The discs have been characterized by electron microscopy and dynamic light scattering for the size and size distributions (mean diameter = 250 nm) and by lipid analysis for lipid composition. A considerable amount of cholesterol is present in the membrane. The SDS content of the preparation has been determined employing ^{14}C-labeled detergent, as reported by Jørgensen.[1] SDS is thought to be preferentially located at the edge of the membrane fragments in order to protect the hydrophobic section of the lipid bilayer. The membrane discs are stable in solution and cannot be regarded as intermediate structures between intact vesicles and detergent–protein mixed micelles. The short treatment of SDS in the preparation only breaks off unwanted cell membrane material around the typical, island-like ATPase accumulations as they appear in the natural membrane. It is not a process of protein solubilization.

In response to your comment, we are aware of this effect and will consider it in the data interpretation. We note that if the membrane discs are isolated upon a deoxycholate and saponin treatment, according to Skou and Esmann[2] instead of SDS, the same enzymatic properties are found. This indicates some independence of detergent treatment.

1 P. L. Jørgensen, *Biochim. Biophys. Acta*, 1974, **356**, 34.
2 J. C. Skou and M. Esmann, *Biochim. Biophys. Acta*, 1979, **567**, 436.

Dr Morigaki asked: The native membrane fragments seem to have some bending. What is the size distribution of them? Is it possible to access the curvature of them both in the suspended and the surface bound states? If the curvature (or undulation) remains at the surface bound state, it will affect the fluorescence quantification, since the evanescent light intensity decreases very sharply with the distance from the substrate.

What is generally the driving force for the membrane binding to the "bio-compatible" surface? How stable are the bound membrane fragments on the surface?

Dr Pawlak responded: The size distribution is broad. The particle diameters range from 50 to 600 nm. The bending in the surface-associated state must be minimal and hence will not affect the quantification very dramatically (the exponential decay length of the evanescent field is around 150 nm). In AFM images the central part of a disc seems to be held at a certain, but more or less constant, distance (10 nm) from the surface. Interestingly, this distance could be slightly reduced by addition of Mg^{2+} ions, which we interpret as the presence of a certain aqueous reservoir between surface and membrane. Indeed this distance reduction could be actually measured by an increase of fluorescence signal (see Fig. 9 in our paper).

Regarding the driving force for membrane surface association, please see the previous question and answer. We believe the contacts must be very strong, since the adsorbed discs could not be rinsed off in the sensor for hours and under strong hydrodynamic flow, nor be removed by the AFM tip applying increasing force. The contact regions seem to be preferentially located at the edge region of the discs.

Prof. Petersen asked: How will the topography affect the estimate of density in Fig. 7? Membranes on an angle will look in projection as if the density is greater (see Scheme 1).

Dr Pawlak responded: Height differences between edge and central regions are also small and in the range of several nm. We should also note that only the extramembranous protein parts are imaged. In summary, a strong influence of topography on the density estimate is not expected.

Prof. Neumann opened the discussion of Prof. Petersen's paper: Do you control the number of fluorophores per protein molecule? This should matter in quantifying the number of proteins from the fluorescence intensity.

Prof. Petersen responded: No. We estimate the number of fluorescent probes per protein molecule by standard methods (absorbance ratios). If the number of probes per protein is uniform, the effect is similar to having a larger quantum yield of fluorescence and there is no effect on the measurement of cluster densities. If there is a distribution of the number of probes per protein, this will add to the distribution of fluctuations, but it will not seriously affect the average cluster density estimate.

Prof. Barclay asked: Formaldehyde fixation can affect antigenic sites. Have you observed any loss of antigenic activity? I have seen reports where formaldehyde fixation might affect the distribution of cell surface antigens. As far as I am aware it is still not clear how formaldehyde fixation works. Have you been able to confirm your data on fresh cells?

Prof. Petersen responded: We have no direct evidence whether fixation affects our antigenic sites. We always perform titration curves so that we ensure that we work under saturation conditions while minimising non-specific fluorescence. Under these circumstances, the total number of

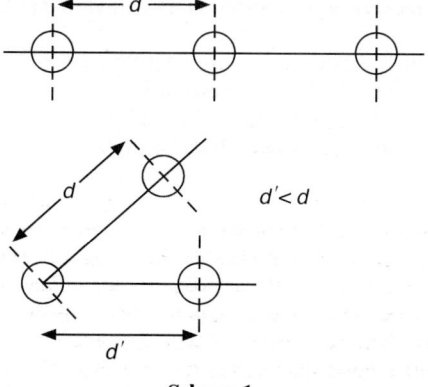

Scheme 1

receptors detected (for EGF- and PDGF-receptors) match nicely the numbers determined by standard biochemical techniques. This suggests that the effect is minor. We also know, in the case of PDGF-receptors, that the results are comparable for fixed and live cells.

I agree that formaldehyde fixation in principle is undesirable. We have observed the same results with formaldehyde fixation and methanol/acetone fixation in the case of adaptor protein distributions, so if there are artifacts, they are similar for the two procedures. We have seen similar results on fixed and live cells for PDGF-receptors, so to this point we have no evidence that fixation is a real concern.

Prof. Svetina asked: Is clustering of membrane proteins an effect of protein–protein interaction or possibly of their links to the cytoskeleton?

Prof. Petersen responded: Our measurements do not speak directly to the mechanism of formation of the clusters. I suspect that the answer depends on the system. For the association of HA + 8 to the adaptor protein, it is clearly an amino acid sequence that is allowing for the interaction *i.e.* it is a protein–protein interaction. In the other cases, (EGF- and PDGF-receptors) the clustering could arise from interactions with the coated pits, with cytoskeletal proteins or through the confinement into domains by the cytoskeletal 'fences'.

Dr Amblard commented: Clusters seen by your image correlation spectroscopy (ICS) methods are only resolved as such if their mean separation distance is larger than the optical resolution of the microscope. Because subwavelength resolution can be achieved by digital processing of images that contain enough signal, I am wondering how good the spatial resolution is for distinguishing clusters? Another question is related to this one, because photon statistics is involved, namely: how precisely can you resolve the average number of molecules inside the clusters.

Prof. Petersen responded: The spatial resolution of the ICS technique and in particular of the image cross-correlation spectroscopy (ICCS) technique is related to how well the peaks of two adjacent fluctuations are resolved. If they are not arising from the same spatial feature, the fluctuations will affect the *width* of the correlation functions. The resolution therefore is a question of how accurately the width can be determined. Our experience with a large volume of data is that we can determine the width, w, to within about 10%. Thus, in principle, the spatial resolution for the co-localisation is on the order of one-tenth of the laser width or about 35 nm.

Your second question is a difficult question to answer in general. We believe that the cell-to-cell variation is the largest source of uncertainty and this generally works out to about a 30% uncertainty. In systems where there is a single population of receptors (*e.g.* PDGF-receptors) we have determined by two approaches an average of four receptors per cluster with an uncertainty of about one—allowing us to distinguish dimers from tetramers from hexamers. However, there is clearly a distribution of sizes of clusters so that monomers co-exist with oligomers. This complicates a complete analysis. For highly aggregated systems (*e.g.* EGF-receptors) we have determined an average of 10 to 20 receptors per cluster depending on the temperature. We expect that these numbers are accurate to 2–4 receptors per cluster as indicated in Table 2 in the paper.

Prof. Holzwarth asked: Which mode of the laser did you use (TEM00) and what is the power density in the centre of the beam? I ask this question in order to judge if there is any danger of damage to your sample caused by the laser light. Secondly, can you gain any molecular information about the conformation of the sugar heads and their function for surface recognition?

Prof. Petersen responded: Indeed, the laser is used in the TEM00 mode to ensure a Gaussian transverse intensity profile. We use a 25 mW laser (total power at the source) which after attenuation and losses in the system provides a power at the sample in the microwatt region. There is always a photobleaching process with all of the possible side effects associated with generation of free radicals. Experience in other types of photomicroscopy suggests that exposure to microwatts of power for periods of seconds lead to minimal, if any, damage.

The simplest answer to your second question is no. It is possible that in particular instances the association of molecules is conformation dependent, but then the evidence would be very indirect.

Prof. Neumann asked: Can you identify rotational displacements of the chromophore residues on the macromolecules upon complex formation?

Prof. Petersen responded: It is possible to perform correlation analysis on fluorescence which is analyzed by polarization. This can yield rotational information. We have not attempted any of these experiments.

Dr Amblard said: You told us that the peak value of the spatial autocorrelation function, $g(0,0)$, gives directly the average number of molecules in the clusters. But the raw information given by your images are relative chromophore numbers on each pixel. To interpret such data, I understand that some independent information is necessary about the stoichiometry of the receptor–antibody–chromophore interactions. This sort of calibration leads to relative receptor numbers on each pixel. How do you then extract absolute receptor numbers for the cluster size without additional hypothesis? Don't you need for instance to assume that the background staining is due to receptor monomers?

Prof. Petersen responded: The peak value of the autocorrelation function is a measure of the number count *within the integrated area (volume) of the laser beam*. Thus, while there is an intensity count associated with each pixel in the image, this intensity reflects the intensity contributions from molecules with the entire area exposed by the laser beam when it is in that particular location. This is an absolute number without any assumptions needed. To extract the information about the number of monomers within each cluster we need to know more: either we need to know the exact quantum efficiency of the microscope, the chromophores and other optical parameters, or we need to determine the autocorrelation function for a known monomer distribution. We have so far done this by assuming that the non-specific staining by the antibody will reflect a monomeric distribution. This is a big assumption, but the best we can do at present. This issue is addressed in detail in a paper by Wiseman and Petersen to be published in the *Biophysical Journal* in February 1999.[1]

1 P. W. Wiseman and N. O. Petersen, *Biophys. J.*, 1999, **76**, 963.

Prof. Barclay asked: Does the addition of ligand such as PDGF affect the distribution of receptors and clusters?

As triggering through many receptors such as PDGF receptor is thought to involve bringing together receptor molecules and associated kinase domains, it is surprising that you find the receptors already clustered. This of course has functional implications. Would you like to comment?

Some of the cell times you describe have very large number of receptors. Have you looked at normal cells (untransformed)?

Prof. Petersen responded: We expect it to, but in our work on PDGF-receptors to date we have not seen any effect. This is the only system we have had a chance to look at so far.

In response to your second question, first it should be clear that our technique cannot establish whether the proteins within the clusters are in physical contact or are simply confined (by whatever mechanism) to a small domain. While we were surprised at the observation that receptors are already clustered, we can speculate that pre-existing, but inactive, complexes or domains of receptors, may control the activation process, particularly at low ligand concentrations. The probability of encounter of two occupied receptors may be low if they are free to move over the entire cell, but it can eventually occur. If the proteins are confined to domains, two proteins in a domain must be occupied at the same time. This will not likely occur at very low concentrations. Hence the pre-existing clusters can lead to a threshold below which activation will not occur. Some of the cell types you describe have very large numbers of receptors. Have you looked at normal cells (untransformed)?

In response to your third question, we have only looked at tissue culture cells, so these are all transformed cells. The A431 cells overexpress the EGF-receptor and therefore have an abnormally high number of receptors. The expression levels in the AG1523, CV-1 and other cells we have looked at are normal. We know we can detect as few as 100 000 receptors and believe that this can be extended to about 20 000 per cell but probably not fewer than that will our current technology.

Prof. Morantz opened the discussion of Prof. Luckham's paper: Have you seen any effects of varying the withdrawal rate? It there a possibility of re-grabbing between the surfaces during withdrawal?

Prof. Luckham responded: We have done some experiments at different approach rates, the results quoted in our paper correspond to approach rates between 2–0.02 Hz, *i.e.* over two orders of magnitude. We did not obviously observe any difference in the results over this somewhat limited range. We shall investigate this further.

Yes there is the possibility of re-grabbing the surfaces during withdrawal, this may explain the two minima in some of the withdrawal profiles, *e.g.* Fig. 9.1 in the paper.

Prof. Evans commented: In order to be confident that you've tested single molecular connections of any type, the frequency of "force events" has to be reduced (by dilution) to 1 out of 5–10 touches to a surface. Moreover, the force needed to break a multiple-bonded contact does not simply scale in proportion to number nor can the statistics of force events be analyzed to derive the strength of a single molecular bond. The reason is that it is not possible to know how the force is distributed amongst the multiple bonds. The best chance of accessing properties of single bonds is to use a very soft cantilever, dilute the surface concentration of reactive sites, touch the surface gently, and perform experiments over many orders of magnitude range in rate of force/time.

Prof. Luckham responded: I have no real comment to make other than to thank Prof. Evans for these comments and to point out that these comments largely relate to the detachment of the surfaces. In this paper we also report on the interactions as the cholera toxin approaches a GM_1 coated surface.

Dr Fisher asked: Why did you choose to work in pure water rather than to suppress double-layer interactions by adding salt? Secondly, it is possible to integrate the stepwise measurement of force during separation to obtain the work done. It would seem from the comments made during this conference that work may be a more useful measure that detachment force. Could you comment on this, both in general and in relation to your own measurements?

Prof. Luckham responded: No reason really, I hoped not to be asked that one! In retrospect it wasn't very clever, but initially we had problems with drying out and salt crystal formation on the cantilever and so Kate Smith the co-author of the paper, stuck to water. To answer your second question, conceptually this can be done rather straightforwardly. The problem being the value one should take for the distance over which the interaction is occurring on approach of the surfaces. As you can see from our data, sometimes we do not seem to get an attraction on approach, for the reasons outlined in our paper. However, if we say that the interaction is occurring over 1 nm, then the energy of interaction, or work done, E, will be given by

$$E = \text{force} \times \text{distance} = 20 \times 10^{-12} \times 10^{-9} = 2 \times 10^{-20} \text{ J}$$

or

$$\frac{2 \times 10^{-20}}{1.3 \times 10^{-23} \times 300} = 5kT$$

This is the order of the strength of a single hydrogen bond. Clearly there have been many gross assumptions, particularly in the separation that one should take, but it is still instructive. These comments apply to both our data and the work of others.

Dr Lee commented: The atomic force microscope promises to bring a molecular level of understanding to the physical properties of biological interfaces because: (1) the microscope is capable of producing physical and chemical images of a surface with nanometer scale resolution, (2) the microfabricated cantilever is a sensitive force transducer with ultrahigh temporal resolution. Progress in measuring single molecule interaction with the AFM has been slowed by the difficulty of mastering the many techniques necessary to make these measurements. That is, the measurement requires the mastery of surface chemical, microfabrication, optics and biological techniques. We have recently made two breakthroughs that should greatly accelerate these measurements. First, we have developed an assay that will determine the activity of a biological molecule as a function of load. The upper limit of force that can be applied to streptavidin and a DNA oligonucleotide was approximately 1.5 nN. Using this force as a threshold of the upper limit of force that should be applied between the probe and surface we have made up to more than 6000 force measurements with a single experiment. Second, we have developed a microfabricated probe array which can be used with probeless cantilevers to measure hundreds (if not thousands) of different interactions in a single experiment. This is a critical breakthrough because it allows us to make many measurements quickly with the appropriate controls. In conclusion, as the key technical issues have been resolved and many groups have perfected their surface chemistries, we feel there will be a rapid expansion in the number of studies made of intermolecular interactions with AFM and their quality.

Prof. Luckham responded: These tips would certainly make life considerably easier in performing these experiments, although we found that using the procedure outlined in the paper the success of each experiment was roughly 50%.

Prof. Holzwarth asked: How are the molecules attached to the tip (covalently or electrostatically) and how is the coverage verified?

Prof. Luckham responded: The protein is attached to the tip covalently as described in the paper. Basically it is through a glutaraldehyde linkage, a common means of attaching proteins to surfaces. As this is a common means of performing protein attachment we have not verified that there is any protein present apart from through the results presented. Any analysis of the tip is likely to destroy the tip, so we thought it best to just try it out. Roughly 50% of the tips show the results presented in the paper, the rest just show repulsive interactions. We have performed a whole series of control experiments, as shown in the paper, to support our thesis that we are studying the interaction between cholera toxin and GM_1.

Dr Amblard commented: To study the strength of intermolecular forces by AFM, the molecules must be surface-grafted in a very controlled way. To this end, silanisation techniques are very popular on glass surfaces. They are mostly carried out in liquid phase, and a distillation procedure is often required to avoid the condensation of silane compounds. In our hands, silanisation in gas phase, by vapor deposition gives much better results. Ultrasensitive dark-field microscopy reveals a very homogeneous coverage, which was systematically devoided of molecular aggregates very often seen after liquid phase silanisation.

Prof. Luckham responded: OK, but in AFM experiments one has to be careful not to damage the cantilever. I agree that the procedure may not actually graft the coupling agent directly to the glass surface, the silane may actually be polymerised by the presence of surface water on the silicon/silicon nitride surfaces. This may not be all bad though as this would give some flexibility to the protein that ultimately is grafted to the AFM cantilever.

Dr Fielden asked: Have there been measurements made in the macroscopic limit of radius on this type of system, *i.e.* using the surface force apparatus?

Prof. Luckham replied: No not really. We[1] have studied the interactions between GM_1 in DPPC bilayers. It must also be realised that it is much more difficult to modify mica surfaces than silicon surfaces so the immobilisation of the cholera toxin on mica would be very difficult.

1 P. F. Luckham and J. Wood, *J. Colloid Interface Sci.*, 1993, **156**, 173.

Prof. Evans opened the discussion of Prof. Bongrand's paper: How do you determine the 'hidden moment arm' or tether that transmits hydrodynamic force to the bond? What is the force history (over time) experienced by the bond up to rupture? Can you be sure that only single bonds are formed while a sphere is briefly captured by the surface? Once arrested, the sphere will be pushed against the surface as a reaction to the bond force, which along with tangential slip at the surface will greatly increase the likelihood of forming additional bonds.

Prof. Bongrand responded: The force F experienced by a single bond was calculated with standard mechanical reasoning (assuming an undeformed spherical bead with a bond under tension, and using known values of the force and torque exerted by hydrodynamic forces). As previously reported[1] the force is inversely proportional to the square root of the bond length. For a typical bond of 20 nm length, F (in pN) is about $0.5G$ (where G is the shear rate, in s^{-1}). Note that we reported an experimental way of estimating the effective length of receptors bound to the sphere surface.[2]

Under our experimental conditions, the bond lifetime is usually much higher than the passage time of the sphere surface along a binding site (a typical value of the relative velocity of the sphere surface is 5 µm s^{-1}: the length of a 20 nm bond is therefore spanned within 4 ms). Thus, it is probably warranted to assume that bond rupture occurs under constant load.

Clearly, it is very difficult to prove formally that arrests involve single molecular bonds. The following arguments may support this assumption: (a) when binding sites are diluted on the spheres, the arrest frequency is proportional to the first power of the site density; (b) the arrest duration is unaltered when binding sites are diluted. Now, suppose the receptor density is, say, 100 µm^{-2}; the distance between two neighbouring sites would be about 0.1 µm. Since a 0.1 µm displacement of a bead should be detected, the shift from a binding site to another should be detectable.

We altogether agree that once a sphere is arrested, it will be pushed against the surface, thus favouring the formation of additional bonds. In any case, we repeatedly found that once a cell or particle stopped, it displayed increased probability of stopping again.

1 A. Pierres, A. M. Benoliel and P. Bongrand, *J. Biol. Chem.*, 1985, **270**, 26586.
2 A. Pierres, A. M. Benoliel and P. Bongrand, *C. R. Acad. Sci. Ser. D*, 1995, **318**, 1191.

Prof. Neumann asked: The rate constant k_{on} for association steps or bond formations were given once in the units s^{-1}, another one in mol^{-1} s^{-1}, yet another one in µm^4 s^{-1} as compared to the unit l mol^{-1} s^{-1} for classical bimolecular association steps. Could you clarify the difference in the units?

Prof. Bongrand responded: If we consider two surfaces maintained parallel over a contact area A, with freely diffusing receptors and ligands of surface concentrations [R] and [L], the initial rate of bond formation (absolute number of bonds formed per unit of time) should read: k[R][L]A = number of bonds per second. Constant k should thus be expressed as m^2 s^{-1}, which is equivalent to the 3-dimensional form (expressed as m^3 s^{-1}). Now, if we lump the product kA into a single parameter, this will be expressed in m^4 s^{-1}.

However, we feel that this expression is not convenient when receptors or ligands are not freely diffusing (as occurs on the tip of an atomic force microscope or in our flow chamber). The intrinsic parameter is thus the frequency of bond formation between two adhesion molecules whose tails are maintained at distance d. This frequency should be expressed in s^{-1}.

Dr L. Fisher asked: Your method is clearly a very useful addition to the armoury of receptor force measurement techniques, but seems to be restricted to spherical hard particles because of the difficulty of calculating the force and force distribution in the case of a deformable particle such as a living cell. Does this mean that there is no value in laminar flow studies on living cells, or is there something still to be gained from such measurements?

Prof. Bongrand responded: The laminar flow chamber proved very suitable to study cell adhesion (*e.g.* ref. 12 by Kaplanski *et al.* as shown in our paper, or Pierres *et al.*,[1]) subject to some limitations: If our aim is to study the "natural lifetime" of bonds, there is no need to know the precise value of the applied force. Thus, we found it convenient to study the arrests of cells driven by very low shear rate—say less than 4 s^{-1}, along ligand-coated surfaces. This approach recently allowed us to analyze the effect of modulating antibodies on the lifetime of bonds involving integrins.[2] Alternatively, we may let cells adhere to a surface for a few minutes, then exert increasing forces to study detachment under microscopic control. In this case, cell deformability is certainly as important as bond density to determine detachment properties,[3] but the measured force is physiologically relevant, as shown *e.g.* in the recent paper by Palecek *et al.*[4] who demonstrated the relevance of cell-substratum binding strength to migration capacity. Second, the significance of binding frequencies measured on cellular systems is probably quite different from that measured on particles, and results seem to be dependent on the localization of receptors on the cell surface (*e.g.* tip of microvilli *vs.* cell body) as well as environment (*e.g.* thickness of cell coat). However, we feel that physiologically relevant information can be obtained with this technique.

1 A. Pierres, O. Tissot, B. Malissen and P. Bongrand, *J. Cell Biol.*, 1994, **125**, 945.
2 B. Masson-Gadais, A. Pierres, A. M. Benoliel, P. Bongrand and J. C. Lissitzky, *J. Cell Science*, submitted.
3 J. L. Mège, C. Capo, A. M. Benoliel and P. Bongrand, *Cell Biophys.*, 1986, **8**, 141.
4 S. P. Palecek, S. C. Loftus, M. H. Ginsberg, D. A. Lauffenburger and A. F. Horwitz, *Nature* (*London*), 1997, **385**, 537.

Prof. Holzwarth asked: Can you comment on the influence of the walls on your experiments? I have in mind the first two layers of molecules which are strongly attached to the wall and are not moved with the flow. Can these layers not influence the mobility of particles next to them?

Prof. Bongrand responded: This is a very important point. Indeed, the limitation of the flow chamber method is probably set by the quality of surfaces rather than measurement accuracy. There are two different problems: when we measure the bond lifetime and force dependence, it is important that rupture occurs at the ligand/receptor interface rather than on the wall. This may be a problem for strong interactions, since the only check for a given system is that higher forces and lifetimes are found when the ligand–receptor couple is replaced with a stronger one without changing the coupling procedure. In any case, as stated in the conclusion of our paper, the flow method seems to us better suited to weak interactions, as compared with atomic force microscopy. Secondly, when we are interested in bead-to-surface distance, it is important to know where the "hydrodynamic boundary" is located, since a low density of surface-bound molecules may be sufficient to drag a water layer near the bead or chamber surface. The main problem lies on the bead surface since mica is expected to be very smooth, and coated with a regular array of short oligopeptides. On the contrary, electron microscopy showed that the bead surface was fairly fuzzy (unpublished), and the supplier informed us that the binding sites were scattered in a region of about 6 nm depth. We are presently studying the effect of the length of receptor structures bound to the bead on adhesion efficiency. Indeed, in a previous paper (ref. 17, in our paper), we found that the range of interaction between beads and surfaces was about 10 nm, while binding sites were linked to the bead through a spacer comprising two immunoglobulin molecules of about 20 nm length each. Probably the difference between the effective and geometrical range is due to the location of the hydrodynamic surface.

Prof. Neumann asked: Is there an upper limit value for your k_{on} as there is a limit for diffusion controlled associations with $k_{on} = 10^9$ l mol^{-1} s^{-1}?

Prof. Bongrand replied: The upper limit of 10^9 l mol^{-1} s^{-1} that is reported for soluble molecules is set by diffusion (it is dependent on the viscosity of the solvent). This is not relevant to the binding frequency we estimate with the flow chamber.

Dr Fielden asked: Do you think it would be possible to make a complementary measurement with total internal reflection microscopy on this system?

Prof. Bongrand responded: It would certainly be very useful to combine total internal reflection microscopy and the flow chamber in order to study the initial steps of cell adhesion.

Concluding remarks and the challenge from the immune system

A. Neil Barclay

MRC Cellular Immunology Unit, Sir William Dunn School of Pathology, University of Oxford, South Parks Road, Oxford, UK OX1 3RE. E-mail: barclay@molbiol.ox.ac.uk

Received 25th January 1999

This meeting has covered a wide range of approaches to the study of interactions of biomembranes and one is struck by the progress in their analysis and the application of many new methods. The main theme of the meeting has been the structure of biomembranes and their components. Rather than pick highlights of these or attempt to summarise the findings in these Concluding remarks, I will instead summarise some of the interactions of the cells of the immune systems for which immunologists would like explanations at a molecular level. One of the features of the immune system is that it involves a variety of populations of cells that have complex migratory patterns and interactions that occur throughout life. The surfaces of these cells—the leukocytes—mediate interactions that are essential for the fine control of the immune system that ensures the rapid but controlled rejection of foreign materials such as viruses and bacteria. At the same time it must ensure that reactivity against self is prevented, otherwise autoimmune diseases such as rheumatoid arthritis and multiple sclerosis may result. Some of the features of the interactions involved are outlined in this short overview with more detailed analysis of the leukocyte cell surface given in ref. 1.

The role of the small lymphocyte

It is only around 40 years since the pioneering work of James Gowans first showed that the small lymphocyte was a key cell type in the immune response and that it migrated from blood to lymph through specialised "high walled" endothelial cells in the lymph nodes.[2,3] Fig. 1 shows a scanning electron microscope of a lymphocyte that has adhered to the endothelium prior to migrating between the endothelial cells and into the lymph node where it may encounter antigen and differentiate or leave by the lymphatics and back to the bloodstream.[4] Later it was established that each lymphocyte carried one receptor for antigen and the concept that lymphocytes patrol the body on the look out for pathogens was established. We are now beginning to understand the complexity of the interactions involved in terms of the proteins participating, even if not the biophysics.

What is at the surface of the lymphocyte?

Interactions at the surface of lymphocytes are central to mediating immune reactions. These interactions include the recognition of foreign antigens, interaction with specialised cells that "present antigens" such as macrophages and dendritic cells that are involved in the control of the immune response and interactions that control the migration of lymphocytes. Each of these events

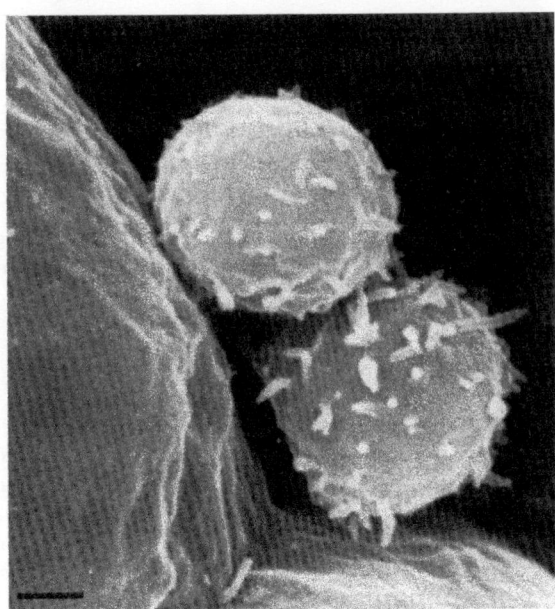

Fig. 1 Scanning electron micrograph showing adherent B lymphocytes binding specialised high walled endothelial cells of mesenteric lymph node (reproduced with kind permission from ref. 4).

is highly regulated and this is illustrated by the complex patterns of subpopulations of lymphocytes in secondary lymphoid organs such as lymph nodes and spleen resulting from their specific patterns of migration.[5] Lymphocytes are one type of white blood cell which are collectively known as leukocytes and include macrophages, neutrophils, mast cells and dendritic cells. However lymphocytes and their progeny are the only cell types which have specific receptors for foreign antigen although they share many other cell surface proteins with other leukocyte populations. The leukocyte cell surface has been studied extensively for two main reasons, first, because of its importance in medicine and secondly, because of the ready availability of populations of cells for biochemical analysis—compare the difficulty in obtaining isolated viable neurons! In a recent review we noted around 250 different proteins which were expressed on leukocyte surfaces but not widely on other cell types.[1] These are candidates for mediating leukocyte specific functions such as those described above. One would expect proteins required for general housekeeping, metabolism and ion transport might be shared with other cell types.

The functions of leukocyte surface proteins

The types of functions mediated by the extracellular parts of leukocyte surface proteins are illustrated in Table 1. The repertoire of proteins reflects the types of functions expected with many

Table 1 Frequency of functions associated with extracellular parts of leukocyte surface proteins

Activity	Frequency (%)
Unknown	49
Receptors for cell surfaces or extracellular matrix	25
Receptors for cytokines	13
Receptors for other soluble proteins (*e.g.*, complement receptors)	10
Enzymes	3
Others (*e.g.*, transporters)	1

Data from ref. 1.

Table 2 Binding constants for some leukocyte cell–cell interaction molecules

Interaction	T /°C	$K_d{}^a$/M	$K_{on}{}^b$/M^{-1} s^{-1}	$K_{off}{}^c$/s^{-1}	Ref.
CD2 with CD48 (rat)	37	$\sim 75 \times 10^{-6}$	$> 10^5$	> 10	7
CD2 with CD58 (human)	37	$\sim 15 \times 10^{-6}$	$> 4 \times 10^5$	> 4	19
CD48 with 2B4 (mouse)	37	16×10^{-6}	$> 8 \times 10^5$	3	20
CD48 with 2B4 (human)	37	8×10^{-6}	$> 10^5$	> 7	20
CD80 with CTLA4 (human)	37	$\sim 0.4 \times 10^{-6}$	$> 9 \times 10^5$	> 0.4	21
CD62L (rat) with glyCAM-1	37	100×10^{-6}	$> 10^5$	> 10	22
CD62P with PSLG-1 (human)	25	0.32×10^{-6}	4×10^6	1.4	23
LFA-1 with ICAM-1 (CD54) (human)	25	0.5×10^{-6}	2×10^5	0.1	24
Binding constants for some other macromolecular interactions					
Avidin to biotin	25	$\sim 10^{-15}$	$\sim 10^7$	$\sim 10^{-8}$	25
Interleukin-1 to its receptor	8	10^{-10}	10^6	$\sim 10^{-4}$	25
Antibody Fab to CD2 protein	37	5×10^{-8}	4×10^5	2×10^{-4}	25
Concanavalin A to trisaccharide	25	6×10^{-6}	—	—	25

^{a} K_d = Equilibrium constant. ^{b} K_{on} = Association rate constant. ^{c} K_{off} = Dissociation rate constant.

proteins involved in recognition events, *e.g.*, binding of antibodies, complement components and lymphokines. No clear functional data are available for about half the known proteins but many of these are expected to interact with other proteins in solution or at the surface of cells (see below).

The structures of leukocyte surface proteins

Leukocyte surface proteins are a very heterogeneous population of molecules. Their abundance varies from barely detectable levels (less than 1000) to 1 000 000 molecules per cell (CD90 on rodent thymocytes). The size of their extracellular regions varies from 12 amino acids (CD52) to 4400 (CD91). Most of the proteins are glycoproteins but the degree of glycosylation varies from 0% by weight to over 70%. Estimates of their size from electron microscopy indicated many small molecules in the range 10–15 nm but also proteins in the range 40–80 nm (*e.g.* CD43, CD45 and CD21). The amino acid sequences of the proteins contain regions that show similarity to other proteins and one can predict that many of these proteins are organised into arrays of domains. The most common domain type is the immunoglobulin-like (Ig-like) domain, which is known to be particularly suited to recognition events.[6] Ig-like domains are present in around one third of leukocyte surface proteins.[1]

Many interactions of leukocyte surface proteins are of low affinity

As the interactions between leukocytes and other cells are usually transitory in nature it is therefore not unexpected that the interactions are of low affinity. The first low affinity interaction between cell surface proteins to be characterised in detail was between CD2 and CD48 using new methods utilising the phenomenon of surface plasmon resonance and recombinant proteins corresponding to the extracellular regions of these proteins. This method allowed protein interactions to be followed in real time and was particularly suited to following weak interactions like that between CD2 and CD48 which has a K_d of around 75 µM at 37 °C with a particularly fast dissociation rate of greater than 6 s^{-1} (ref. 7). Thus the monomeric interaction has a half-life of a fraction of a second. Of course when the cells make contact the concentrations of the interactants are relatively high. There are extensive data to suggest that these weak interactions are of functional significance from using monoclonal antibodies to cell surface proteins in functional assays and methods involving genetic manipulation.[1] Interactions of this type and even weaker interactions could have major effects on the alignment of cells.[8]

One of the best-characterised leukocyte adhesion systems is the binding of neutrophils to endothelium (reviewed in ref. 9). This adhesion is triggered by changes in the endothelium which lead first to the weak attachment of the neutrophil which 'rolls' along. Some cells detach but others flatten and then the neutrophil can migrate between the endothelial cells into the tissue. One of

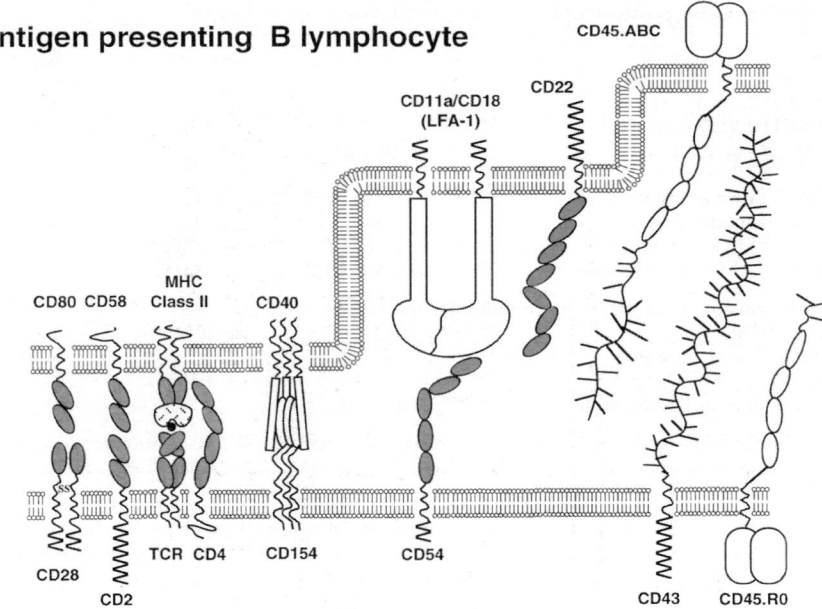

Fig. 2 Schematic view of some of the membrane proteins involved in interactions between a T lymphocyte and an antigen presenting B lymphocyte. The dimensions are based on the approximate sizes of the proteins determined by electron microscopy and X-ray crystallography. Ig-like domains are indicated by shaded ovals, domains in CD45 by clear ovals including three fibronectin type III domains; the CD40/CD154 dimensions are based on the structure of the TNF/TNFR. The peptide being presented to the TCR is indicated by a black spot between the MHC Class II and the TCR. N-linked carbohydrates are not indicated but regions with a high content of O-linked sugars are shown by solid bars in approximate accordance with site density. The carbohydrates are major features because of their bulk and this is discussed in more detail in ref. 1 and 18. Adapted from ref. 1.

the proteins involved in the rolling step is P-selectin (CD62P) and this is an example of a weak interaction (Table 2). Other proteins involved in later stages of adhesion include integrins. For example, LFA-1 (αLβ2) interacts with ICAM-1 (CD54); this has a somewhat higher affinity that the selectins (Table 2).

One of the consequences of the low affinity interactions is that it is difficult to identify novel interactions. For example, the half lives of several of the interactions in Table 2 are in the order of a fraction of a second so that binding studies with purified proteins will not withstand a washing step. It seems likely that many of the proteins for which functions are not known (Table 1) will fall into this category of mediating weak interactions between cells. However, it is possible to make the proteins multivalent by a variety of methods and identify and characterise weak interactions, *e.g.* ref. 10–12.

The distribution of leukocyte surface proteins

Fig. 1 shows that the lymphocytes are not simple spheres but have ruffles and microvilli. This adds another degree of complexity as local concentrations of proteins may vary in different regions of the cell surface. Thus immuno-electron microscopy has shown that CD62L was clustered at the tips of microvilli or membranes ruffles but was largely absent from the membrane of the cell body. In contrast the integrin αMβ2 was present mainly on the membrane of the cell body and seldom on the microvilli.[13] As discussed below this is consistent with the known roles of these proteins in that the CD62 is involved in the first interactions between the neutrophil in the blood stream and the endothelium that lead to rolling and tethering whilst the integrin is involved in later adhesion events once the cell has stopped rolling.

The size and shape of leukocyte surface proteins

The different sizes of proteins at the surface of cells have implications as to which ones are available to interact when cells come together. One of the most extensively studied cell interactions is that between T lymphocytes and those cells able to present antigen, such as dendritic cells and B lymphocytes. This interaction is crucial in determining the initiation of the immune response and hence is of major interest in both understanding the immune response itself and also in finding ways to manipulate it for medical benefit, *e.g.* in the treatment of autoimmune diseases and to facilitate organ transplantation without rejection. Some of the proteins involved are illustrated schematically in Fig. 2. It can be seen that several of the proteins involved are relatively small and the distance between opposing membranes when these interact is of the order of 15 nm.[1,14,15] These include the important interaction between the receptor for antigen on T cells (TCR) and the antigen peptide presented on major histocompatability antigens (MHC). When one considers that the large proteins such as CD45 and CD43 are not only 2–3 times this length but are also much more abundant than say CD4 or the TCR, it is evident that one cannot consider that specific interactions of the TCR without also considering what happens to the large abundant proteins. There is now clear evidence for the occurrence of redistribution of some of the membrane associated proteins[16,17] during cell contact.

Concluding remarks

The immune system utilises complex interactions at cell surfaces in a highly controlled manner. Many of the proteins involved are now identified and some of their interactions characterised in isolation. It is clearly more difficult to understand the events that occur when cells come into contact but in this meeting we have heard several systems that can be or have been applied to the particular challenges of the immune system.

Acknowledgements

I am grateful to support from the Medical Research Council and the EU Biotechnology programme grant for work on protein modules.

References

1 A. N. Barclay, M. H. Brown, S. K. A. Law, A. J. McKnight, M. G. Tomlinson and P. A. van der Merwe, *Leucocyte Antigens Factsbook 2nd edn.*, Academic Press, London, 1997.
2 J. L. Gowans, *Immunol. Today*, 1996, **17**, 288.
3 J. Gowans and E. Knight, *Proc. R. Soc. London Ser. B*, 1964, **159**, 257.
4 W. van Ewijk, N. H. Brons and J. Rozing, *Cell. Immunol.*, 1975, **19**, 245.
5 A. N. Barclay, *Immunology*, 1981, **42**, 593.
6 A. F. Williams and A. N. Barclay, *Annu. Rev. Immunol.*, 1988, **6**, 381.
7 P. A. van der Merwe, M. H. Brown, S. J. Davis and A. N. Barclay, *EMBO J.*, 1993, **12**, 965.
8 M. L. Dustin, D. E. Golan, D. M. Zhu, J. M. Miller, W. Meier, E. A. Davies and P. A. van der Merwe, *J. Biol. Chem.*, 1997, **272**, 30 889.
9 N. Hogg and C. Berlin, *Immunol. Today*, 1995, **16**, 327.
10 S. Preston, G. J. Wright, K. Starr and A. N. Barclay, *Eur. J. Immunol.*, 1997, **27**, 1911.
11 J. D. Altman, P. Moss, P. Goulder, D. H. Barouch, W. McHeyzer, J. I. Bell, A. J. McMichael and M. M. Davis, *Science*, 1996, **274**, 94.
12 I. Stamenkovic, D. Sgroi and A. Aruffo, *Cell*, 1992, **68**, 1003.
13 S. L. Erlandsen, S. R. Hasslen and R. D. Nelson, *J. Histochem. Cytochem.*, 1993, **41**, 327.
14 P. A. van der Merwe, P. N. McNamee, E. A. Davies, A. N. Barclay and S. J. Davis, *Curr. Biol.*, 1995, **5**, 74.
15 K. C. Garcia, M. Degano, R. L. Stanfield, A. Brunmark, M. R. Jackson, P. A. Peterson, L. Teyton and I. A. Wilson, *Science*, 1996, **274**, 209.
16 A. I. Sperling, J. R. Sedy, N. Manjunath, A. Kupfer, B. Ardman and J. K. Burkhardt, *J. Immunol.*, 1998, **161**, 6459.
17 C. R. Monks, B. A. Freiberg, H. Kupfer, N. Sciaky and A. Kupfer, *Nature (London)*, 1998, **395**, 82.

18　P. Rudd, M. Wormald, D. Harvey, M. Devasahayam, M. McAlister, M. Brown, S. Davis, A. Barclay and R. Dwek, *Glycobiology*, 1999, **9**, in the press.
19　P. A. van der Merwe, A. N. Barclay, D. W. Mason, E. A. Davies, B. P. Morgan, M. Tone, A. K. Krishnam, C. Ianelli and S. J. Davis, *Biochemistry*, 1994, **33**, 10 149.
20　M. H. Brown, K. Boles, P. Anton van der Merwe, V. Kumar, P. A. Mathew and A. N. Barclay, *J. Exp. Med.*, 1998, **188**, 2083.
21　P. A. van der Merwe, D. L. Bodian, S. Daenke, P. Linsley and S. J. Davis, *J. Exp. Med.*, 1997, **185**, 393.
22　M. W. Nicholson, A. N. Barclay, M. S. Singer, S. D. Rosen and P. A. van der Merwe, *J. Biol. Chem.*, 1998, **273**, 763.
23　P. Mehta, R. D. Cummings and R. P. McEver, *J. Biol. Chem.*, 1998, **273**, 32 506.
24　Y. Tominaga, Y. Kita, A. Satoh, S. Asai, K. Kato, K. Ishikawa, T. Horiuchi and T. Takashi, *J. Immunol.*, 1998, **161**, 4016.
25　P. A. van der Merwe and A. N. Barclay, *Trends Biochem. Sci.*, 1994, **19**, 354.

Paper 9/00659A

List of Posters

A molecular dynamics study of the free energy of water transport across the DMPC bilayer **T. Kawakami, W. Shinoda** and **S. Okazaki,** *Tokyo Institute of Technology, Japan* and **A. D. J. Haymet,** *University of Houston, USA*

Dilute detergent–lipid solutions: Towards an understanding of 2D crystallization of membrane proteins **J. B. Heymann** and **A. Engel,** *Biozentrum University of Basel, CH-4056 Switzerland*

Kinetics of lipid bilayer formation *via* adhesion and fusion **C. A. Keller, K. Glasmastar** and **B. Kasemo,** *Dept. of Applied Physics, Chalmers University of Technology* and *Goteborg University, Sweden*

Non-equilibrium perturbation response-kinetics of phospholipid bilayers in the biologically relevant L_α-phase **G. Pabst, M. Rappolt, H. Amenitsch** and **P. Laggner,** *Institute of Biophysics and X-ray Structure Research, Austrian Academy of Sciences, Austria* and **S. Bernstorff,** *ELETTRA, Synchrotrone Trieste, Italy*

Interaction of the pore-forming toxin pneumolysin with membranes **B. B. Bonev** and **A. Watts,** *Dept. of Biochemistry, University of Oxford, UK,* **R. J. C. Gilbert,** *Dept. of Biochemistry, University of Oxford, UK* and *Depts. of Biochemistry and Microbiology and Immunology, University of Leicester, UK* and **O. Byron,** *Division of Infection and Immunity, IBLS, University of Glasgow, UK*

Effect of electrolyte concentration gradient on curvature elasticity of bilayers and membrane electroporation **S. Kakorin** and **E. Neumann,** *Physical and Biophysical Chemistry, Faculty of Chemistry, University of Bielefeld, Germany*

A comparative study of maltoporin and sucrose porin **K. M. Ranatunga, C. Adcock, G. R. Smith** and **M. S. P. Sansom,** *Laboratory of Molecular Biophysics, University of Oxford, UK*

Structural studies on the transmembrane domain of phospholamban using rotational resonance magic angle sample spinning NMR spectroscopy **Z. Ahmed, C. Glaubitz, A. Watts** and **D. Middleton,** *Biomembrane Structure Unit, University of Oxford, UK*

Neutron and X-ray reflectivity studies at solid/liquid interface: The interaction of a peptide with model membranes **G. Fragneto, L. Perrino** and **E. Bellet-Amalric,** *Institut Laue-Langevin, Grenoble, France,* **F. Graner** and **P. Dubos,** *University of Grenoble I, St Martin d'Hores, France* and **L. Perrino, A. Braslau** and **J. Daillant,** *CEA (Saclay), France*

Molecular dynamics simulations of M2 protein from influenza A virus **L. R. Forrest** and **M. S. P. Sansom,** *Laboratory of Molecular Biophysics, University of Oxford, UK*

Structure and stability of lipid membranes on a biosensor surface **M. I. Fisher,** *CBSystems, Wiltshire, UK* and **T. Tjaernhage,** *Defence Research Establishment, Umea, Sweden*

Detection of complement activity using a robust tethered membrane **R. F. Castello, S. W. Evans** and **S. D. Evans,** *Universty of Leeds, UK* and **I. R. Peterson** and **J. Heptinstall,** *Nima Technology, Coventry, UK*

Cellulose I_α structure on the surface of native cellulose observed by AFN **A. A. Baker** and **M. J. Miles,** *University of Bristol, UK* and **W. Helbert** and **J. Sugiyama,** *University of Kyoto, Japan*

Measuring the adhesion between horse erythrocytes **F. R. Attenborough, K. Kendall** and **C. Stainton,** *Birchall Centre for Inorganic Chemistry and Materials Science, Dept. of Chemistry, Keele University, UK*

Kinetics of transport through microemulsion-based gel membranes **N. Z. Atay** and **A. Erkahraman,** *Dept. of Chemistry, Bogazici University, Istanbul, Turkey*

Protein crystallisation: A systematic method particularly for membrane proteins **I. Chohan, J. P. Derrick, R. C. Ford, A. F. Mustafa, W. L. J. Rowe** and **G. J. T. Tiddy,** *Dept. of Chemical Engineering, UMIST, UK*

Molecular dynamic simulation of KcsA, a potassium channel **I. H. Shrivastava, L. Forrest, M. S. P. Sansom** and **K. M. Ranatunga,** *Laboratory of Molecular Biophysics, University of Oxford, UK*

The effect of γ-radiation on the structure of mitochondrial membranes **M. G. Mironova, Y. I. Dreval** and **L. V. Sichevskaya,** *Dept. of Molecular and Applied Biophysics, Kharkov State University, Ukraine*

The modulation of enzyme activity by membrane torque tension: Insights into the mode of action of amphilic anti-cancer drugs **G. S. Attard, H. Beard, M. Dymond, W. S. Smith, J. W. Wan, A. Hunt** and **A. D. Postle,** *University of Southamption, UK,* **R. H. Templer,** *Imperial College, London, UK* and **S. Jackowski,** *St Jude's Children's Research Hospital, TN, USA*

Receptor binding and membranes translocation of nuclease colicins **C. Lemaître, A. J. Pommer, A. M. Hemmings, G. R. Moore, R. James** and **C. Kleanthous,** *Schools of Biological and Chemical Sciences, University of East Anglia, Norwich, UK*

Protein folding in biological membranes **A. R. Curran, P. J. Booth** and **R. H. Templer,** *Depts. of Biochemistry and Chemistry, Imperial College, London, UK*

Membrane ION transport by antibiotics and toxins **B. M. Burkhart, D. Ghosh, N. Li,** *Hauptman-Woodward Institute, New York, USA* and **V. Pletnev,** *Shemyakin Institute, Moscow, Russia*

A study of the headgroup dynamics of sphingomyelin using ^{31}P NMR and an analytically soluble model, **I. C. Malcolm, J. C. Ross** and **J. Higinbotham,** *Dept. of Applied Chemisty and Physical Sciences, Napier University, Edinburgh, UK* **F. G. Riddell,** *School of Chemistry, The Purdie Building, The University, St. Andrew's, UK*

Liposome adsorption on particulate surfaces **T. C. Bennett, C. Catuogno** and **M. N. Jones,** *School of Biological Sciences, University of Manchester, Manchester, UK,* **J. E. Creeth,** *Unilever Research, Port Sunlight Laboratory, Bebington, Wirral, Merseyside, UK*

Thermodynamics and kinetics of breakdown and formation of vesicles **S. Bucak, A. Fontana** and **B. Robinson,** *School of Chemical Sciences, University of East Anglia, Norwich, UK*

Nanomechanics of the polysaccharide pullulan **A. D. L. Humphris, T. J. McMaster** and **M. J. Miles,** *H. H. Wills Physics Laboratory, University of Bristol, Bristol, UK*

Predicting conductance properties for ion channel structures **O. S. Smart** and **G. M. P. Coates,** *School of Biochemistry, The University of Birmingham, Edgbaston, Birmingham, UK,* **M. S. P. Sansom,** *Laboratory of Molecular Biophysics, University of Oxford, Oxford, UK,* **C. L. Bashford,** *Division of Biochemistry, St. Georges Hospital Medical School, Cranmer Terrace, London, UK*

Effects of anaesthetics on the structure of a phospholipid bilayer MD investigation of halothane in the hydrated L_α phase of DPPC **M. Tarek, L. Koubi, D. Scharf** and **M. L. Klein** *University of Pennsylvania, USA*

Identification of specific carbohydrate–carbohydrate interactions in water by computer simulation **J. Westcott, S. Hanna** and **L. Fisher** *H. H. Wills Physics Laboratory, University of Bristol, Bristol, UK*

Tethering phospholipid bilayers to planar and porous substrates—progress towards biosensor development **S. Ogier,** *Dept. of Physics and Astronomy, University of Leeds, UK*

Probing the binding site of the nicotinic acetylcholine receptor using solid state NMR **P. T. F. Williamson, K. W. Miller** and **A. Watts,** *Biomembrane Structure Unit, Biochemistry Dept., University of Oxford, UK*

Shape transformations of axially loaded tubular phospholipid vesicles **S. Svetina** and **B. Žekš,** *Institute of Biophysics, Medical Faculty, University of Ljubljana, Slovenia* and *J. Stefan Institute, Ljubljana, Slovenia,* **B. Božič** and **V. Henrich,** *Institute of Biophysics, Medical Faculty, University of Ljubljana, Slovenia.*

Imaging blood cells under physiological conditions by atomic force microscopy **K. Nowakowski, P. Luckham** and **P. Winlore**, *Imperial College, London, UK*

Molecular dynamics simulation of gramicidin S in dimethylsulfoxide solution **D. Mihailescu,** *SBPM, DBCM, CEA Saclay, 91191, Gif-Sur-Yvette, France* and *Faculty of Biology, University of Bucharest, Romania* and **J. C. Smith,** *SBPM, DBCM, CEA Saclay, 911191, Gif-Sur-Yvette, France* and *Lehrstuhl für Biocomputing, IWR der Universitat Heidelburg, Germany*

A systematic study of a DMPC bilayer—preliminary results **G. La Penna,** *CNR, Genova, Italy,* **S. Letardi, V. Minicozzi, S. Morante** and **G. C. Rossi,** *Dept. of Physics, "Tor vergata", University of Rome, Italy* and **G. Salina,** *Infu Roma II, Italy*

Polyoxyethylene framework: A new approach to ionic transmembrane pores **E. Q. Morales, R. M. Rodríguez, M. Delgado, R. Pérez** and **J. D. Martin,** *Instituto de Investigaciones Químicas, C.S.I.C., Sevilla, Spain*

Modulation by cholesterol of the signal transduction chain *via* the µOPOID receptor **B. Lagane, J. F. Tocanne** and **L. Cézanne,** *IPBS, 118, Route de Narbonne, 31062 Toulouse, France*

List of Participants

Miss Z. Ahmed   *University of Oxford, UK*
Prof. M. Almgren   *Uppsala University, Sweden*
Prof. A. Alonso   *Universidad del Pais Vasco, Spain*
Dr F. Amblard   *Institut Curie, France*
Mr A. Asnacios   *Université Paris VII, France*
Dr Z. N. Atay   *Bogazici University, Turkey*
Dr G. Attard   *University of Southampton, UK*
Dr F. R. Attenborough   *Keele University, UK*
Dr A. A. Baker   *University of Bristol, UK*
Dr A. Bangham   *Cambridge, UK*
Prof. A. N. Barclay   *University of Oxford, UK*
Prof. P. N. Bartlett   *University of Southampton, UK*
Prof. T. M. Bayerl   *University of Wuerzburg, Germany*
Miss T. C. Bennett   *University of Manchester, UK*
Dr S. M. Bezrukov   *National Institutes of Health, USA*
Dr M. Bhakoo   *Unilever Research (Port Sunlight Laboratory), UK*
Prof. C. Bohne   *Max Planck Institut, Germany*
Prof. P. Bongrand   *INSERM Unité 387, France*
Dr C. M. Brown   *Institut Curie, France*
Miss S. Bucak   *University of East Anglia, UK*
Dr B. M. Burkhart   *Hauptman–Woodward Medical Research Institute Inc., USA*
Prof. A. Carrington   *University of Southampton, UK*
Dr S. Carrington   *University of Bristol, UK*
Dr C. Chen   *Institute of Food Research, UK*
Mr G. M. Coates   *University of Birmingham, UK*
Dr S. C. Crouzy   *CEA, France*
Dr M. Day   *Garland Publishing, UK*
Dr S. Dijkstra   *Imperial College, UK*
Dr W. L. Duax   *Hauptman–Woodward Medical Research Inst., Inc., USA*
Dr K. Edwards   *Uppsala University, Sweden*
Dr P. N. Edwards   *Zeneca Pharmaceuticals, UK*
Dr J. Essex   *University of Southampton, UK*
Prof. E. Evans   *University of British Columbia, Canada*
Dr M. L. Fielden   *University of Groningen, Netherlands*
Prof. J. L. Finney   *University College London, UK*
Mr D. Firmin   *Glaxo Wellcome R & D (Stevenage), UK*
Dr L. Fisher   *University of Bristol, UK*
Dr M. I. Fisher   *CBD Porton Down, UK*
Dr A. Fontana   *Universita Chieti, Italy*
Ms L. R. Forrest   *University of Oxford, UK*
Dr G. Fragneto   *Institut Max von Laue–Paul Langevin, France*
Mr P. L. Francis   *Glaxo Wellcome R & D (Stevenage), UK*
Mr T. P. Galbraith   *Birkbeck College, UK*
Mr F. Gathier   *Université Joseph Fourier, France*
Dr A. George   *Cerebrus Ltd, UK*
Mr R. J. Gilbert   *University of Leicester, UK*
Miss K. Glasmästar   *Chalmers University of Technology, Sweden*
Dr F. M. Goñi   *Universidad del Pais Vasco, Spain*
Mr S. Goodall   *University of Oxford, UK*
Prof. T. J. Haymet   *University of Houston, USA*
Dr B. Heymann   *University of Basle, Switzerland*
Prof. J. F. Holzwarth   *Free University Berlin, Germany*
Dr S. Horante   *Universita di Tor Uergata, Italy*
Mr A. D. Humphris   *University of Bristol, UK*
Dr R. S. Hutton   *Glaxo Welcome (R & D) Ware, UK*
Mr P. J. James   *University of Bristol, UK*

Mr C. R. Johans   *Helsinki University of Technology, Finland*
Dr M. N. Jones   *University of Manchester, UK*
Dr S. Kakorin   *University of Bielefeld, Germany*
Dr C. A. Keller   *Chalmers University of Technology, Sweden*
Prof. M. L. Klein   *University of Pennsylvania, USA*
Dr D. Klug   *Imperial College, UK*
Mr B. Lagane   *Université Paul Sabatier, France*
Prof. P. Laggner   *Austrian Academy of Sciences, Austria*
Dr J. H. Lakey   *University of Newcastle upon Tyne, UK*
Prof. A. G. Lee   *University of Southampton, UK*
Dr G. U. Lee   *Naval Research Laboratory, USA*
Dr C. Lemaitre   *University of East Anglia, UK*
Prof. J. T. Lewis   *University of Wales Bangor, UK*
Mr P. E. Liljeroth   *Helsinki University of Technology, Finland*
Dr H. Lu   *Imperial College, UK*
Prof. P. F. Luckham   *Imperial College, UK*
Dr J. E. Macdonald   *University of Wales (UWIC), UK*
Dr I. C. Malcolm   *Napier University, UK*
Mr A. R. Malloy   *University of Bristol, UK*
Dr T. McMaster   *University of Bristol, UK*
Dr W. Meijberg   *Imperial College, UK*
Dr D. Mihailescu   *CEA, France*
Dr E. Q. Morales   *CSIC–Consejo Sup. de Investigaciones Cientificas, Spain*
Prof. D. J. Morantz   *King's College, UK*
Dr K. Morigaki   *Max-Planck Institut für Polymerforschung, Germany*
Dr N. Myronova   *Kharkov State University, Ukraine*
Prof. D. Needham   *Duke University, USA*
Prof. E. Neumann   *University of Bielefeld, Germany*
Mr S. D. Ogier   *University of Leeds, UK*
Prof. S. Okazaki   *Tokyo Institute of Technology, Japan*
Dr C. Oldfield   *Napier University, UK*
Mr G. Pabst   *Austrian Academy of Sciences, Austria*
Dr M. Pawlak   *Novartis AG, Switzerland*
Dr J. Penfold   *Rutherford Appleton Laboratory, UK*
Prof. N. O. Petersen   *University of Western Ontario, Canada*
Prof. I. R. Peterson   *Coventry University, UK*
Mr S. P. Piotto   *ETH Zentrum, Switzerland*
Miss K. M. Ranatunga   *University of Oxford, UK*
Miss A. Renault   *Université Joseph Fourier, France*
Miss J. M. Rinuy   *Ecole Polytechnique Federale Lausanne, Switzerland*
Prof. B. H. Robinson   *University of East Anglia, UK*
Dr S. Roser   *University of Bath, UK*
Prof. B. Roux   *University of Montreal, Canada*
Mr W. L. Rowe   *UMIST, UK*
Dr G. S. Salina   *Universita di Tor Uergata, Italy*
Dr M. S. Sansom   *University of Oxford, UK*
Mr L. D. Schuler   *ETH Zentrum, Switzerland*
Prof. S. K. Scott   *University of Leeds, UK*
Mr J. Sharples   *University of Oxford, UK*
Dr I. H. Shrivastava   *University of Oxford, UK*
Prof. B. Simmons   *King's College, UK*
Dr O. S. Smart   *University of Birmingham, UK*
Prof. J. C. Smith   *University of Heidelberg, Germany*
Dr L. C. Smith   *Glaxo Welcome, Beckenham, Kent*
Dr M. J. Sutcliffe   *University of Leicester, UK*
Prof. S. Svetina   *University of Ljubljana, Slovenia*
Dr M. Tarek   *University of Pennsylvania, USA*
Dr R. H. Templer   *Imperial College, UK*

Prof. G. Tiddy *UMIST, UK*
Ms C. Trandum *Technical University of Denmark, Denmark*
Prof. G. E. Tranter *Glaxo Wellcome R & D (Stevenage), UK*
Prof. D. Walker *University of British Columbia, Canada*
Dr B. A. Wallace *Birkbeck College, UK*
Mr J Watts *University of Oxford, UK*
Dr J. T. Wescott *University of Bristol, UK*
Mr L. Whitehead *University of Southampton, UK*
Dr M. Wilkinson *SmithKline Beecham, UK*
Mr P. T. Williamson *University of Oxford, UK*
Ms G. Woodhouse *CRC Molecular Engineering & Technology, Australia*

Index of Contributors*

Aalouach, M., **95**
Alder, G. M., **185**
Almgren, M., 73, 77, 144
Alonso, A., **55**
Amblard, F., 69, 71, 241, 243, 338, 339, 341
Amenitsch, H., **31**
Anselmetti, D., **273**
Atay, Z. N., 332, 333
Barclay, A. N., 137, 156, 337, 339, **345**
Barger, W. R., **79**
Bartlett, P. N., 332
Basáñez, G., **55**
Bashford, C. L., **185**
Bayerl, T. M., **17**, 69, 70, 71, 75, 156, 242
Benoliel, A. M., **321**
Berendsen, H. J. C., **209**
Bezrukov, S. M., 70, 148, **173**, 230, 231, 232, 233, 238, 239, 244
Bhakoo, M., 245
Bohne, C., 73, 75, 149, 336
Boland, T., **79**
Bongrand, P., **321**, 342, 343
Breed, J., **209**
Brown, C., **289**
Bucak, S., 77, 148
Burkhart, B. M., 226, 228, 237
Castle, S. J., **41**
Chen, C., 139
Coates, G. M., **185**
Cornell, B., **247**
Curran, R. A., **41**
Deber, C. M., 157, 228, 245
Dijkstra, S., 72, 154
Duax, W. L., 225, 226
Dufrêne, Y. F., **79**
East, J. M., **127**
Edwards, P. N., 70, 74, 76, 142, 148, 149, 157
Ehret, M., **273**
Eichenbaum, G., **103**
Evans, E., **1**, 72, 138, 142, 147, 233, 334, 335, 340, 342
Fielden, M. L., 341, 343
Finney, J. L., 140
Fisher, L., 71, 138, 142, 231, 340, 342
Fontana, A., 148
Galbraith, T. P., **159**
Gilbert, R. J., 230, 236
Goñi, F. M., **55**, 76, 77, 78, 151, 156, 231, 331, 336
Grell, E., **273**
Haymet, T. J., 229
Hirn, R., **17**
Holzwarth, J. F., 69, 70, 71, 72, 75, 139, 140, 146, 153, 226, 229, 237, 238, 242, 245, 334, 335, 338, 341, 343
Husslein, T., **201**
Jones, M. N., 70, 146

Kakorin, S., **111**, 232
Kaminski, A., **289**
Keller, C. A., 335
King, L. G., **247**
Klein, M. L., 137, 140, 149, 155, **201**, 229, 236, 238, 239, 240, 241, 242, 243, 244, 245
Klug, D., **41**
Kriechbaum, M., **31**
Laggner, P., **31**, 71, 72, 74, 77, 152, 155, 227, 231, 334
Lagüe, P., **165**
Lakey, J. H., 332
Lee, A. G., 74, **127**, 153, 154, 155, 156, 157, 227, 232
Lee, G. U., **79**, 137, 138, 139, 340
Luckham, P. F. **307**, 340, 341
Mall, S., **127**
Mills, J., **103**
Moore, P. B., **201**
Morantz, D. J., 152, 340
Morigaki, K., 331, 336
Needham, D., **103**, 142, 144, 145, 146, 147, 148
Neumann, E., 69, **111**, 147, 149, 150, 151, 152, 153, 231, 239, 337, 339, 342, 343
Newns, D. M., **201**
Okazaki, S., 228, 240, 244
Oswald, R. E., **259**
Pabst, G., **31**
Parsegian, V. A., **173**
Pattnaik, P. C., **201**
Pawlak, M., 69, 139, **273**, 331, 332, 335, 336, 337
Petersen, N. O., 230, **289**, 337, 338, 339
Peterson, I. R., 72
Pierres, A., **321**
Rädler, O., **17**
Rand, R. P., **173**
Rapport, M., **31**
Robinson, B. H., 77, 145, 149
Rocheleau, J., **289**
Roux, B., 71, 73, 138, 153, **165**, 227, 228, 229, 230, 232, 237, 238, 243, 244, 246, 334
Ruiz-Argüello, M. B., **55**
Rumbles, G., **41**
Sackmann, E., **17**
Sansom, M. S., 69, 75, 140, **185**, **209**, 232, 233, 235, 238, 240, 241, 242, 243, 244, 245, 246, 331, 333
Schick, E., **273**
Schneider, J. W., **79**
Schuler, L. D., 141
Sharma, R. P., **127**
Simon, C., **95**
Smart, O. S., 141, **185**, 228, 232, 235, 236, 237, 238, 333
Smeeton, A. H., **259**
Smith, J. C., **95**, 140, 141, 142, 230, 235, 243, 333, 334
Smith, K., **307**
Srivastava, I. H., **289**

* The page numbers in **bold** type indicate papers submitted for discussions.

Sutcliffe, M. J., **259**, 333, 334
Svetina, S., 75, 76, 153, 338
Templer, R. H., **41**, 72, 73, 74, 75, 76, 137, 142, 233
Tiddy, G., 137, 141, 147, 152, 153, 156
Tieleman, D. P., **209**
Toensing, K., **111**
Vodyanoy, I., **173**

Wallace, B. A., **159**, 226, 227, 228
Wieczorek, L., **247**
Wiseman, P. W., **289**
Woodhouse, G., **247**, 331, 332, 333
Wo, Z. G., **259**
Zhong, Q., **201**
Zuckermann, M. J., **165**

General Discussions of the Faraday Society/Faraday Discussions of the Chemical Society

Date	Subject	Volume
1907	Osmotic Pressure	Trans. 3
1907	Hydrates in Solution	3
1910	The Constitution of Water	6
1911	High Temperature Work	7
1912	Magnetic Properties of Alloys	8
1913	Colloids and their Viscosity	9
1913	The Corrosion of Iron and Steel	9
1913	The Passivity of Metals	9
1914	Optical Rotary Power	10
1914	The Hardening of Metals	10
1915	The Transformation of Pure Iron	11
1916	Methods and Appliances for the Attainment of High Temperatures in a Laboratory	12
1916	Refractory Materials	12
1917	Training and Work of the Chemical Engineer	13
1917	Osmotic Pressure	13
1917	Pyrometers and Pyrometry	13
1918	The Setting of Cements and Plasters	14
1918	Electric Furnaces	14
1918	Co-ordination of Scientific Publication	14
1918	The Occlusion of Gases by Metals	14
1919	The Present Position of the Theory of Ionization	15
1919	The Examination of Materials by X-Rays	15
1920	The Microscope: Its Design, Construction and Applications	16
1920	Basic Slags: Their Production and Utilization in Agriculture	16
1920	Physics and Chemistry of Colloids	16
1920	Electrodeposition and Electroplating	16
1921	Capillarity	17
1921	The Failure of Metals under Internal and Prolonged Stress	17
1921	Physico-Chemical Problems Relating to the Soil	17
1921	Catalysis with special reference to Newer Theories of Chemical Action	17
1922	Some Properties of Powders with special reference to Grading by Elutriation	18
1922	The Generation and Utilization of Cold	18
1923	Alloys Resistant to Corrosion	19
1923	The Physical Chemistry of the Photographic Process	19
1923	The Electronic Theory of Valency	19
1923	Electrode Reactions and Equilibria	19
1923	Atmospheric Corrosion. First Report	19
1924	Investigation on Oppau Ammonium Sulphate-Nitrate	20
1924	Fluxes and Slags in Metal Melting and Working	20
1924	Physical and Physico-Chemical Problems relating to Textile Fibres	20
1924	The Physical Chemistry of Igneous Rock Formation	20
1924	Base Exchange in Soils	20
1925	The Physical Chemistry of Steel-Making Processes	21
1925	Photochemical Reactions of Liquids and Gases	21
1926	Explosive Reactions in Gaseous Media	22
1926	Physical Phenomena at Interfaces, with special reference to Molecular Orientation	22
1927	Atmospheric Corrosion, Second Report	23
1927	The Theory of Strong Electrolytes	23
1927	Cohesion and Related Problems	24
1928	Homogeneous Catalysis	24
1929	Crystal Structure and Chemical Constitution	25
1929	Atmospheric Corrosion of Metals, Third Report	25
1929	Molecular Spectra and Molecular Structure	26
1930	Colloid Science Applied to Biology	26
1931	Photochemical Processes	27
1932	The Adsorption of Gases by Solids	28
1932	The Colloid Aspect of Textile Materials	29
1933	Liquid Crystals and Anisotropic Melts	29
1933	Free Radicals	30

Date	Subject	Volume
1934	Dipole Moments	30
1934	Colloidal Electrolytes	31
1935	The Structure of Metallic Coatings, Films and Surfaces	31
1935	The Phenomena of Polymerization and Condensation	32
1936	Disperse Systems in Gases: Dust, Smoke and Fog	32
1936	Structure and Molecular Forces in (*a*) Pure Liquids, and (*b*) Solutions	33
1937	The Properties and Function of Membranes, Natural and Artificial	33
1937	Reaction Kinetics	34
1938	Chemical Reactions Involving Solids	34
1938	Luminescence	35
1939	Hydrocarbon Chemistry	35
1939	The Electrical Double Layer (owing to the outbreak of the war the meeting was abandoned, but the papers were printed in the *Transactions*)	35
1940	The Hydrogen Bond	36
1941	The Oil-Water Interface	37
1941	The Mechanism and Chemical Kinetics of Organic Reactions in Liquid Systems	37
1942	The Structure and Reactions of Rubber	38
1943	Modes of Drug Action	39
1944	Molecular Weight and Molecular Weight Distribution in High Polymers (Joint Meeting with the Plastics Group, Society of Chemical Industry)	40
1945	The Application of Infra-red Spectra to Chemical Problems	41
1945	Oxidation	42
1946	Dielectrics	42 A
1946	Swelling and Shrinking	42 B
1947	Electrode Processes	Disc. 1
1947	The Labile Molecule	2
1947	Surface Chemistry (Jointly with the Sociéitéi de Chimie Physique at Bordeaux Published by Butterworths Scientific Publications Ltd	
1947	Colloidal Electrolytes and Solutions	Trans. 43
1948	The Interaction of Water and Porous Materials	Disc. 3
1948	The Physical Chemistry of Process Metallurgy	4
1949	Crystal Growth	5*
1949	Lipo-proteins	6
1949	Chromatographic Analysis	7
1950	Heterogeneous Catalysis	8
1950	Physico-chemical Properties and Behaviour of Nuclear Acids	Trans. 46
1950	Spectroscopy and Molecular Structure and Optical Methods of Investigating Cell Structure	Disc. 9
1950	Electrical Double Layer	Trans. 47
1951	Hydrocarbons	Disc. 10
1951	The Size and Shape Factor in Colloidal Systems	11
1952	Radiation Chemistry	12
1952	The Physical Chemistry of Proteins	13
1952	The Reactivity of Free Radicals	14
1953	The Equilibrium Properties of Solutions on Non-electrolytes	15
1953	The Physical Chemistry of Dyeing and Tanning	16
1954	The Study of Fast Reactions	17
1954	Coagulation and Flocculation	18
1955	Microwave and Radio-frequency Spectroscopy	19
1955	Physical Chemistry of Enzymes	20
1956	Membrane Phenomena	21
1956	Physical Chemistry of Processes at High Pressures	22
1957	Molecular Mechanism of Rate Processes in Solids	23
1957	Interactions in Ionic Solutions	24
1958	Configurations and Interactions of Macromolecules and Liquid Crystals	25
1958	Ions of the Transition Elements	26
1959	Energy Transfer with special reference to Biological Systems	27
1959	Crystal Imperfections and the Chemical Reactivity of Solids	28
1960	Oxidation-Reduction Reactions in Ionizing Solvents	29
1960	The Physical Chemistry of Aerosols	30
1961	Radiation Effects in Inorganic Solids	31
1961	The Structure and Properties of Ionic Melts	32
1962	Inelastic Collisions of Atoms and Simple Molecules	33
1962	High Resolution Nuclear Magnetic Resonance	34
1963	The Structure of Electronically Excited Species in the Gas Phase	35
1963	Fundamental Processes in Radiation Chemistry	36
1964	Chemical Reactions in the Atmosphere	37
1964	Dislocations in Solids	38
1965	The Kinetics of Proton Transfer Processes	39
1965	Intermolecular Forces	40
1966	The Role of the Absorbed State in Heterogeneous Catalysis	41

Date	Subject	Volume
1966	Colloid Stability in Aqueous and Non-aqueous Media	42
1967	The Structure and Properties of Liquids	43
1967	Molecular Dynamics of the Chemical Reactions of Gases	44
1968	Electrode Reactions of Organic Compounds	45
1968	Homogeneous Catalysis with Special Reference to Hydrogenation and Oxidation	46
1969	Bonding in Metallo-organic Compounds	47
1969	Motions in Molecular Crystals	48
1970	Polymer Solutions	49
1970	The Vitreous State	50
1971	Electrical Conduction in Organic Solids	51
1971	Surface Chemistry of Oxides	52
1972	Reactions of Small Molecules in Excited States	53
1972	The Photoelectron Spectroscopy of Molecules	54
1973	Molecular Beam Scattering	55
1973	Intermediates in Electrochemical Reactions	56
1974	Gels and Gelling Processes	57
1974	Photo-effects in Adsorbed Species	58
1975	Physical Adsorption in Condensed Phases	59
1975	Electron Spectroscopy of Solids and Surfaces	60
1976	Precipitation	61
1977	Potential Energy Surfaces	62
1977	Radiation Effects in Liquids and Solids	63
1977	Ion–Ion and Ion–Solvent Interactions	64
1978	Colloid Stability	65
1978	Structures and Motion in Molecular Liquids	66
1979	Kinetics of State Selected Species	67
1979	Organization of Macromolecules in the Condensed Phase	68
1980	Phase Transitions in Molecular Solids	69*
1980	Photoelectrochemistry	70*
1981	High Resolution Spectroscopy	71*
1981	Selectivity in Heterogeneous Catalysis	72*
1982	Van der Waals Molecules	73*
1982	Electron and Proton Transfer	74*
1983	Intramolecular Kinetics	75*
1983	Concentrated Colloidal Dispersions	76*
1984	Interfacial Kinetics in Solution	77*
1984	Radicals in Condensed Phases	78*
1985	Polymer Liquid Crystals	79*
1985	Physical Interactions and Energy Exchange at the Gas–Solid Interface	80*
1986	Lipid Vesicles and Membranes	81*
1986	Dynamics of Molecular Photofragmentation	82*
1987	Brownian Motion	83*
1987	Dynamics of Elementary Gas-phase Reactions	84*
1988	Solvation	85*
1988	Spectroscopy at Low Temperatures	86*
1989	Catalysis by Well Characterised Materials	87*
1989	Charge Transfer in Polymeric Systems	88
1990	Structure of Surfaces and Interfaces as studied using Synchrotron Radiation	89*
1990	Colloidal Dispersions	90*
1991	Structure and Dynamics of Reactive Transition States	91*
1991	The Chemistry and Physics of Small Metallic Particles	92*
1992	Structure and Activity of Enzymes	93*
1992	The Liquid/Solid Interface at High Resolution	94*
1993	Crystal Growth	95*
1993	Dynamics at the Gas/Solid Interface	96*
1994	Structure and Dynamics of Van der Waals Complexes	97*
1994	Polymers at Surfaces and Interfaces	98*
1994	Vibrational Optical Activity: From Fundamentals to Biological Applications	99*
1995	Atmospheric Chemistry: Measurements, Mechanisms and Models	100*
1995	Gels	101*
1995	Unimolecular Reaction Dynamics	102*
1996	Hydration Processes in Biological and Macromolecular Systems	103*
1996	Complex Fluids at Interfaces	104*
1996	Catalysis and Surface Science at High Resolution	105*
1997	Solid State Chemistry: New Opportunities from Computer Simulations	106*
1997	Interactions of Acoustic Waves with Thin Films and Interfaces	107*
1997	Dynamics of Electronically Excited States in Gaseous, Cluster and Condensed Media	108*
1998	Chemistry and Physics of Molecules and Grains in Space	109*
1998	Chemical Reaction Theory	110*

* *Available for purchase, for current information on prices etc. please contact the Sales and Promotion Department, The Royal Society of Chemistry, Thomas Graham House, Science Park, Milton Road, Cambridge, UK CB4 0WF.*